IET CONTROL ENGINEERING SERIES 88

Distributed Control and Filtering for Industrial Systems

Other volumes in this series:

Volume 8 **A history of control engineering, 1800–1930** S. Bennett
Volume 18 **Applied control theory, 2nd edition** J.R. Leigh
Volume 20 **Design of modern control systems** D.J. Bell, P.A. Cook and N. Munro (Editors)
Volume 28 **Robots and automated manufacture** J. Billingsley (Editor)
Volume 33 **Temperature measurement and control** J.R. Leigh
Volume 34 **Singular perturbation methodology in control systems** D.S. Naidu
Volume 35 **Implementation of self-tuning controllers** K. Warwick (Editor)
Volume 37 **Industrial digital control systems, 2nd edition** K. Warwick and D. Rees (Editors)
Volume 39 **Continuous time controller design** R. Balasubramanian
Volume 40 **Deterministic control of uncertain systems** A.S.I. Zinober (Editor)
Volume 41 **Computer control of real-time processes** S. Bennett and G.S. Virk (Editors)
Volume 42 **Digital signal processing: principles, devices and applications** N.B. Jones and J.D.McK. Watson (Editors)
Volume 44 **Knowledge-based systems for industrial control** J. McGhee, M.J. Grimble and A. Mowforth (Editors)
Volume 47 **A history of control engineering, 1930–1956** S. Bennett
Volume 49 **Polynomial methods in optimal control and filtering** K.J. Hunt (Editor)
Volume 50 **Programming industrial control systems using IEC 1131-3** R.W. Lewis
Volume 51 **Advanced robotics and intelligent machines** J.O. Gray and D.G. Caldwell (Editors)
Volume 52 **Adaptive prediction and predictive control** P.P. Kanjilal
Volume 53 **Neural network applications in control** G.W. Irwin, K. Warwick and K.J. Hunt (Editors)
Volume 54 **Control engineering solutions: a practical approach** P. Albertos, R. Strietzel and N. Mort (Editors)
Volume 55 **Genetic algorithms in engineering systems** A.M.S. Zalzala and P.J. Fleming (Editors)
Volume 56 **Symbolic methods in control system analysis and design** N. Munro (Editor)
Volume 57 **Flight control systems** R.W. Pratt (Editor)
Volume 58 **Power-plant control and instrumentation: the control of boilers and HRSG systems** D. Lindsley
Volume 59 **Modelling control systems using IEC 61499** R. Lewis
Volume 60 **People in control: human factors in control room design** J. Noyes and M. Bransby (Editors)
Volume 61 **Nonlinear predictive control: theory and practice** B. Kouvaritakis and M. Cannon (Editors)
Volume 62 **Active sound and vibration control** M.O. Tokhi and S.M. Veres
Volume 63 **Stepping motors, 4th edition** P.P. Acarnley
Volume 64 **Control theory, 2nd edition** J.R. Leigh
Volume 65 **Modelling and parameter estimation of dynamic systems** J.R. Raol, G. Girija and J. Singh
Volume 66 **Variable structure systems: from principles to implementation** A. Sabanovic, L. Fridman and S. Spurgeon (Editors)
Volume 67 **Motion vision: design of compact motion sensing solution for autonomous systems** J. Kolodko and L. Vlacic
Volume 68 **Flexible robot manipulators: modelling, simulation and control** M.O. Tokhi and A.K.M. Azad (Editors)
Volume 69 **Advances in unmanned marine vehicles** G. Roberts and R. Sutton (Editors)
Volume 70 **Intelligent control systems using computational intelligence techniques** A. Ruano (Editor)
Volume 71 **Advances in cognitive systems** S. Nefti and J. Gray (Editors)
Volume 72 **Control theory: a guided tour, 3rd edition** James Ron Leigh
Volume 73 **Adaptive sampling with mobile WSN** K. Sreenath, M.F. Mysorewala, D.O. Popa and F.L. Lewis
Volume 74 **Eigenstructure control algorithms: applications to aircraft/rotorcraft handling qualities design** S. Srinathkumar
Volume 75 **Advanced control for constrained processes and systems** F. Garelli, R.J. Mantz and H. De Battista
Volume 76 **Developments in control theory towards glocal control** L. Qiu, J. Chen, T. Iwasaki and H. Fujioka (Editors)
Volume 77 **Further advances in unmanned marine vehicles** G.N. Roberts and R. Sutton (Editors)
Volume 78 **Frequency-domain control design for high-performance systems** John O'Brien

Distributed Control and Filtering for Industrial Systems

Magdi S. Mahmoud

The Institution of Engineering and Technology

Published by The Institution of Engineering and Technology, London, United Kingdom

The Institution of Engineering and Technology is registered as a Charity in England & Wales (no. 211014) and Scotland (no. SC038698).

First published 2013

The Institution of Engineering and Technology
Michael Faraday House
Six Hills Way, Stevenage
Herts, SG1 2AY, United Kingdom

www.theiet.org

British Library Cataloguing in Publication Data
A catalogue record for this product is available from the British Library

ISBN 978-1-84919-607-9 (hardback)
ISBN 978-1-84919-608-6 (PDF)

Typeset in India by MPS Limited
Printed in the UK by CPI Group (UK) Ltd, Croydon, CR0 4YY

In the Name of

GOD

the All-Compassionate, the All-Merciful.
"O Allah! Grant me benefit from what you teach me, teach me that which benefits me and enhance me in knowledge."

Dedicated
To My Loving Wife Salwa
To My Family: Medhat, Monda, Mohamed, Menna
To My Grandchildren: Malak, Mostafa, and Mohamed
MsM

Contents

Preface

In recent years technological advancements in the design and fabrication of integrated circuits have led to the development of cost effective, low power, thumb-size devices that can be used for sensing/actuating, communication, and computing. This trend is enabling a surge of new applications for which pervasive network architectures are being developed. A key feature of these systems is that they are decentralized and communication among different subsystems may be unreliable. From an engineering viewpoint, to ensure correct operation, the theoretical analysis requires a fundamental paradigm shift, as many of the typical assumptions of systems and control theory cease to hold.

Distributed control and filtering means the distribution of control and filtering capabilities throughout a system as opposed to *central control* in which control functions for a system are performed by a central computer (or by redundant, yet still centralized control computers). The rapid convergence of sensing, computing, and wireless communication technologies on cost effective, low power, miniature devices is enabling a host of new control applications. In recent years, we have witnessed wireless technologies replacing wired counterparts in all applications where it can be securely and reliably implemented. This book is about distributed control and filtering algorithms with particular emphasis on industrial applications.

An attractive approach for control of large-scale networked systems would be the decentralized approach that concentrates the design function at the subsystem level. This might, however, lead to severely deteriorated global performance. On the other hand, centralized optimization-based approach, for example, model predictive control, could be much better, but is often impractical due to communication constraints and an overwhelming number of decision variables. Distributed control/optimization, where the global control/optimization problem is decomposed into many smaller control/optimization subproblems that can be solved locally for each subsystem, is therefore appealing. In this distributed framework, the interaction between subsystems is taken into account, while the flexibility of the decentralized approach is still there.

Throughout this book, the following terminologies, conventions, and notations have been adopted. All of them are quite standard in the scientific media and only vary in form or character. Matrices, if their dimensions are not explicitly stated, are assumed to be compatible for algebraic operations. In symmetric block matrices or complex matrix expressions, we use the symbol $\bullet$ to represent a term that is induced by symmetry.

Acknowledgments

The subject matter of distributed control and filtering is attracting increasing attention. The various topics discussed in this book have constituted an integral part of my academic research investigations over the past several years. The idea of writing the book aroused and developed through an invitation from Dr Nigel Hollingworth at the IET. In writing this volume, I took the approach of referring within the text to papers and/or books that I believe taught me some concepts, ideas, and methods. I further complemented this by adding some remarks and notes within and at the end of each chapter to shed some light on other related results.

I am indebted to the colleagues who introduced me to the subject of distributed control and filtering and to the people who made the writing of this book possible. I owe a measure of gratitude to KFUPM management for their continuous encouragement and facilitating all sources of help. Particular appreciation goes to the deanship of scientific research (DSR) for providing a superb competitive environment of research activities through internal funding grants. It is a great pleasure to acknowledge the financial funding afforded by DSR through project no. **IN111033** and for providing overall support of research activities at KFUPM.

During the past five years, I had the privilege of teaching various graduate courses at KFUPM (Saudi Arabia). The course notes, updated and organized, were instrumental in generating different chapters of this book and the valuable comments and/or suggestions by graduate students were greatly helpful, particularly from those students who attended the courses **SE 509**, **SE 650**, **SE 652**, and **SE 658** offered at the Systems Engineering Department over the period 2007–2011. I deeply appreciate the effort of Rohmat Widodo, Mirza H. Baig, and Muhammad Sabih for their help in preparing portions of the manuscript.

Most of all, however, I would like to express my deepest gratitude to all the members of my family and especially my wife Salwa for her elegant style. Without their constant love, incredible amount of patience, and (mostly) enthusiastic support, this volume would not have been finished.

I would appreciate any comments, questions, criticisms, or corrections that readers may take the trouble of communicating to me at msmahmoud@kfupm.edu.sa or magdim@yahoo.com.

Dhahran, Saudi Arabia

Magdi S. Mahmoud
July 2012

Notations and symbols

$\|a\|$	the absolute value of scalar a
$\|A\|$	the induced Euclidean norm of matrix A
$\|A\|_p$	the induced ℓ_p norm of matrix A
$A(i,j), A_{ij}$	the ijth element of matrix A
A^{-1}	the inverse of matrix A
$A^{\dagger}$	the Moore-Penrose-inverse of matrix A
$det(A)$	the determinant of matrix A
e_j	the jth column of matrix I
I	an identity matrix of arbitrary order
$\lambda(A)$	an eigenvalue of matrix A
$\lambda_j(A)$	the jth eigenvalue of matrix A
$\lambda_m(A)$	the minimum eigenvalue of matrix A where $\lambda(A)$ are real
$\lambda_M(A)$	the maximum eigenvalue of matrix A where $\lambda(A)$ are real
$P > 0$	matrix P is real symmetric and positive-definite
$P \geq 0$	matrix P is real symmetric and positive semi-definite
$P < 0$	matrix P is real symmetric and negative-definite
$P \leq 0$	matrix P is real symmetric and negative semi-definite
$rank(A)$	the rank of matrix A
$\Re$	the set of real numbers
$\Re^n$	the set of all n-dimensional real vectors
$\Re^{n \times m}$	the set of $n \times m$-dimensional real matrices
$\rho(A)$	the spectral radius of matrix A
$trace(A)$	the trace of matrix A
x^t or A^t	the transpose of vector x or matrix A
$\|x\|$	the Euclidean norm of vector x
$\|x\|_p$	the ℓ_p norm of vector x

List of acronyms

ARE	algebraic Riccati equation
CMPC	centralized model predictive control
DC	distributed control
DecC	decentralized control
DF	distributed filtering
DFC	decentralized feedback control
DHC	decentralized H_∞ control
DIP	distributed information processing
DMPC	distributed model predictive control
DNS	decentralized nonlinear systems
DSMP	decentralized servomechanism problem
DTS	discrete-time systems
HC	hierarchical control
LBD	Lyapunov-based design
LKF	Lyapunov–Krasovskii functional
LMCR	liquid-metal cooled reactor
LMI	linear matrix inequality
LQC	linear quadratic control
MIMO	multi-input multi-output
MPC	model predictive control
OLC	overlapping control
OLD	overlapping decomposition
SISO	single-input single-output
SVD	singular value decomposition
TDS	time-delay system
TDUS	time-delay uncertain system

Magdi Sadek Mahmoud obtained his Ph.D. in systems engineering from Cairo University in 1974. He has been a professor of systems engineering since 1984 and is now a distinguished university professor and coordinator of the Distributed Control Group at KFUPM, Saudi Arabia. He has worked at different universities in Egypt, Kuwait, UAE, UK, USA, Singapore, and Australia, and has given invited lectures in Venezuela, Germany, China, UK, and USA. He has been actively engaged in teaching and research in the development of modern methodologies to distributed control, switched time-delay systems, fault-tolerant systems and information technology. He is the principal author of 30 books, inclusive book-chapters and the author/co-author of more than 490 peer-reviewed papers. He is the recipient of two national, one regional, and two university prizes for outstanding research in engineering, and is a fellow of the IEE, a senior member of the IEEE, the CEI (UK), and a registered consultant engineer of information engineering and systems (Egypt).

Chapter 1
Introduction

Rapid advances in information processing, communication, and sensing technologies have enabled more and more devices to be provided with embedded processors, networking capabilities, and sensors. For the field of estimation and control, it is now possible to consider an architecture in which many simple components communicate and cooperate to achieve a joint team goal. This distributed (or networked) architecture promises much in terms of performance, reliability, and simplicity of design; however, at the same time, it requires extending the traditional theories of control, communication, and computation and, in fact, looking at a unified picture of the three fields. A systematic theory of how to design distributed systems is currently lacking.

In this book, we will be interested in distributed estimation and control of cooperative networked systems. Such problems are important both because they show up in new engineering systems being envisaged today as well as because they provide an interesting setting where traditionally separate theories of control, communication, and computation have to be integrated. From the viewpoint of estimation and control that we will be undertaking in this book, such problems are, generally speaking, difficult because they challenge two standard assumptions:

1. There is no longer a central node that has access to all the information being generated. Agents have to make local decisions based on locally available information.
2. Information flow is no longer perfect. Any data that needs to be shared has to be transmitted over communication channels, which introduces effects like random delay, stochastic information loss, data corruption, and so on.

A viable approach would be a joint design of the information flow and the control laws to be followed by the agents. Within the constraints imposed by the communication channels, the design of the information flow provided an important degree of freedom and allowed us to benefit both in terms of stability as well as performance of the system.

Traditionally, control theory deals with dynamical processes in which the state is observed by a sensor that generates possibly noisy measurements. The measurements are processed by an estimator that generates an estimate of the state of the process. This estimate is fed to a controller that aims at minimizing some cost function. The controller generates a control input that is applied to the process through an actuator.

Over the last century or so, control theory has made spectacular advances both in understanding the underlying mathematical tools as well as in applying this model to many different problems and generating technological advances. However, the very success of this model means that the many assumptions inherent in the conventional block-diagram framework are often forgotten. As an example, one can spell out such assumptions as

- measurements are generated periodically by the sensor and transmitted to the estimator;
- the estimator and controller are able to process all the data that is generated;
- all the components run on the same clock;

and many others. At the heart of most of these features is the implicit assumption that almost unlimited amounts of information can be communicated and processed by the system components. In most traditional control systems, this assumption is largely held true. However, it is still somewhat surprising that more efforts have not been made to characterize and counter the effects of limited and imperfect communication sharing and processing, especially given the fact that, at its heart, control theory is an information science. The basic motif of feedback loop has as its underlying philosophy the use of information about the system to counter the effect of uncertainty in the environment.

1.1 Introduction

A wide class of control systems consists of interconnected controllers, sensors, and actuators that have been successfully implemented using point-to-point architecture where all components are directly wired to various controllers. The goal is to design flexible systems that can accomplish various tasks with minimum configuration cost, maximum reliability, and the desired performance. In feedback control there are two major operations to be accomplished: the transmission of signals (information flow) back and forth and the calculation of control actions (decision making). Various control architectures can be implemented to perform these operations in real plant: centralized, decentralized, and distributed control architectures. A centralized control system utilizes one centralized controller that collects data from all sensors, computes control decisions, and then dispatches actuation signals to actuators. It is well known that the centralized control can provide the best performance since it imposes the least constraints on the control structure under consideration. However, this system has a single point of failure [1] plus the computational and organizational complexity associated with centralized controllers often makes their implementation impractical, especially for plants with complex dynamics and/or spatially distributed in space such as the large-scale system (LSS).

These considerations have eventually motivated significant work on decentralized control (DecC). In DecC systems [2], multiple sensors and controllers are placed in a distributed structure, and the control decisions are made with or without the full set of response data. The controller nodes can be closely collocated with system actuators. The major difficulty in this structure is that there is no communication between controllers [3], therefore, the closed-loop performance of the plant may deteriorate, and in some cases stability may be lost. In this regard, significant research work [1,4–6] has explored in depth the benefits and limitations of decentralized controllers as well as possible ways of overcoming some of their limitations. The opposite approach in which there is a communication between controllers is called the distributed control system (DCS) and it stands nowadays as the technologically efficient infrastructure for many advanced control strategies in industry. DCSs are often implemented over local and proprietary networks. However, the great availability and ever-decreasing costs of the networked digital technology have been responsible for the replacement of the traditional point-to-point link for broadcast transmission. This motivates the implementation and the usage of multipurpose shared network to connect spatially distributed elements and results in flexible architectures, which help in reducing the installation and maintenance costs. Also, presents better characteristics in terms of modularity, scalability, and offers more design options [7–9]. By closing the feedback control loops over such communication network, this will form the networked control systems (NCSs) [10–13] in which the sensor, the actuator, and the controller are elements that share information by exchanging messages over the network.

Consequently, NCSs have been finding application in a broad range of areas such as mobile sensor networks [14], remote surgery [15], haptics collaboration over the Internet [16–18], automated highway systems (AHS), unmanned aerial vehicles, control of surveillance, and rescue robot teams to access hazardous environments and space exploration [19,20]. In such remote control applications, one major concern is the characterization of a sufficient amount of information transfer needed for a satisfactory performance. This information transfer can be between various components of an NCS. One necessity for satisfactory control performance is the ability for the controllers to track the plant states under communication constraints. Another challenge is the determination of the data rate required for the transmission of control signals, and the construction of dynamic encoding, decoding, and control policies meeting some criteria. One more important problem is the coordination among multiple sensors or multiple controllers/decision makers with the lowest information exchange possible. Moreover, the presence of a network in the control loop has the drawback of introducing time delays in the communications among field elements. Depending on the network configuration, these delays can exhibit random behavior and may even cause variation in the sampling periods, which in turn, results in a time-varying plant [13,21]. These delays deteriorate the performance and even the stability of the system as a whole [22–25].

Therefore, it is not a trivial task to design communication protocols or architectures for control systems since both communication delay and packet loss negatively impact estimation, closed-loop performance, and other parameters of controlled systems [26]. Currently, communications protocols and NCSs are designed separately. In particular, protocols are designed based on conservative heuristics, which, by and large, specify what the maximum time delay and maximum packet loss should be. The delays may not affect significantly an open loop control system but for the applications that require feedback data across the network, the traditional control strategies that can deal with constant time delays might not be suitable. Moreover, real-time implementation can introduce nonconstant time delay (jitter) in the control loop as the result of tasks preemption and priorities [27,28]. Each control task contains three parts; sampling, control algorithm, and actuation. The jitter can happen in sampling and/or actuation. The real-time scheduling can affect control system performance because of the jitter that is imposed by scheduling while the control system design can affect the real-time scheduling since the period of control task is determined during the control design stage. Hence, it is required to integrate the design of control system and real-time scheduling to eliminate the effect of jitter and achieve the desired requirements.

From these observations few questions arise [29]:

- How should we design estimators for networked systems that take into account simultaneous random delay and packet loss?
- How can we estimate their performance?
- When is the closed-loop system stable? How can we choose between a communication protocol with a large packet delay and a small packet loss and a protocol with a small packet delay and a large packet loss, in terms of best performance of a specific real-time application?
- Indeed, the situation becomes compounded when considering interconnected dynamical systems. How one can tackle it?

From this perspective, the objective of next section is to review a number of control architectures such as decentralized, distributed, networked, and hierarchical networked control and providing an overview for each type.

1.2 Control architectures

In what follows, we provide an overview of some of the known control architectures in order to identify the features and attributes of each. We start by completely DecC structures and then highlight the decentralized networked control and the quasi-decentralized structures. Next, we move to discuss their properties and the effects due to communication over a network. Later, we describe the main approaches for the design of DCSs and the distributed NCSs, where information is transmitted among local controllers via real-time network to achieve global stability and performance results. The relation between distributed control and game theory will be briefly illustrated. Then, we present NCSs and wireless networks as well as the main concerns about each architecture. After that, we introduce multiagents and hierarchical networked control structure where the actions of local decentralized controllers are coordinated by an algorithm operating at a higher level and communicating over a network.

1.2.1 Decentralized control

In this paradigm, the plant is decomposed into a number of simpler subsystems (typically based on functional and/or time-scale differences of the unit operations) with interconnections, and a number of local controllers are connected to each distributed subsystem with no signal transfer taking place between different local controllers [1]. DecC of multiunit plants can reduce complexity in the controller design and implementation, and can also provide flexibility in dealing with local controller failures. In a DecC system, since each controller communicates with sensors and actuators in its vicinity, the requirement on communication range can be significantly reduced, and communication latency decreases by reducing the number of sensors or actuators that each controller has to communicate with [3,30]. As the control decisions are computed and executed distributively by individual controllers, system redundancy and reliability can also be improved using DecC. On the other hand, since each controller may only have limited information and longer time delay using communication medium, the stability and optimality of DecC strategies need to be examined carefully.

In decentralized architectures we have the control (input u) and the controlled (output y) variables are grouped into disjoint sets. These sets are then coupled to produce nonoverlapping pairs for which local controllers are designed to operate in a completely independent fashion. The local controllers can be single-input single-output or multivariable (locally centralized) depending on the cardinality of the selected input and output groups. An example of a perfectly DecC structure is shown in Figure 1.1, which provides local posterior information to each controller with no information exchanged between local controllers. Knowledge of how the control actions of the local controller affect the overall system response is not known. Observe that the DecC structure implies that

$$\begin{bmatrix} u_1 \\ u_2 \\ \vdots \\ u_n \end{bmatrix} = \begin{bmatrix} K_1 & & & & \\ & \ddots & & & \\ & & K_j & & \\ & & & \ddots & \\ & & & & K_n \end{bmatrix} \begin{bmatrix} y_1 \\ y_2 \\ \vdots \\ y_n \end{bmatrix} \tag{1.1}$$

The system under control is assumed to be composed of n subsystems, with states, control, and output variables $x_i, u_i, y_i,\ i = 1, \ldots, n$, and the interaction between the subsystems is due to the mutual effect of the states. Once the decentralized

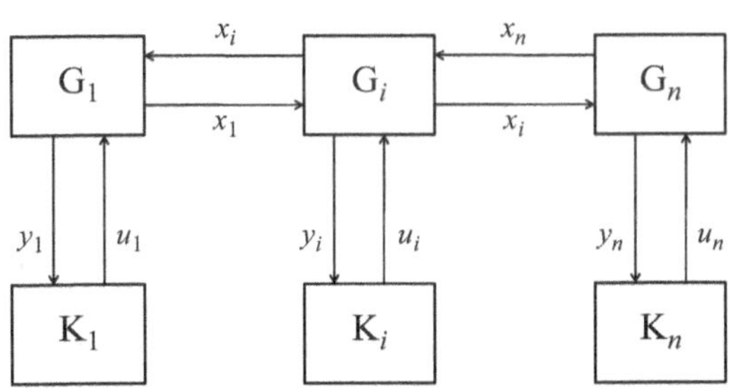

Figure 1.1 Perfectly decentralized control system

controller structure has been defined, the design of the local controllers $K_1, \ldots, K_n$, see Figure 1.1, becomes a routine task when the interactions among the inputs and the outputs of different pairs are weak. These interactions can either be direct (input coupling) or caused by the mutual effects of the internal states of the subsystems under control. On the contrary, it is well known that strong interactions can even prevent one from achieving stability and/or performance with DecC. In many practical situations, complete state measurements are not available at each individual subsystem for DecC; consequently, one has to consider decentralized feedback control based on measurements only or design decentralized observers to estimate the state of individual subsystems that can be used for estimated state feedback control. The general model for a decentralized linear observer that will track the plant states in the presence of bounded quadratic interconnections is shown in the following state equations:

$$\begin{aligned} \dot{\hat{x}}_i(t) &= A_i\hat{x}_i(t) + B_iu_i(t) + L_i[y(t) - \hat{y}_i(t)] \\ \hat{y}_i(t) &= C_i\hat{x}_i(t), \quad i = 1, \ldots, N \end{aligned} \tag{1.2}$$

where L_i is the observation gain matrix of ith subsystem and $\hat{x}_i(t)$ is the state observation of ith subsystem. Notice that the state observation structure of the global interconnected subsystems is completely decentralized since there is no information transfer among local observers. The observation error between the ith true state and the ith observer state is given by

$$e_i(t) = x_i(t) - \hat{x}_i(t), \quad i = 1, \ldots, N \tag{1.3}$$

The local control law of each subsystem is given by

$$u_i(t) = -K_i\hat{x}_i(t), \quad i = 1, \ldots, N \tag{1.4}$$

There have been numerous works in the literature in developing different decentralized techniques such as adaptive, robust, and non-linear DecC, see [31–35].

1.2.2 Interconnected systems

Interconnected systems are usually physically distributed over a wide area and the decentralized controllers are the most applicable control philosophy for LSS [2,36,37] and that the information exchange between subsystems of a LSS is not needed; thus, the individual subsystem controllers are simple and use only locally available information. Large-scale interconnected systems can be found in such diverse fields such as automated highway systems (AHS), autonomous vehicle systems (AVS), material handling systems (MHS), air traffic management systems (ATMS), manufacturing, power generation, and distribution [38]. Designing a centralized control for these systems may not be efficient due to the natural system's modularity, which may prevent a viable way on sharing information across the subsystems, and often it may be too costly when it is implemented. These limitations motivate the design of DecC schemes where the information exchange between interconnected subsystems of LSSs is not required. Thus, the control law implementation is more feasible and more economical. Similar to what we mentioned in section 1.2.1, in most of the decentralized state feedback control schemes the state variables are not available or costly to measure. Therefore, it is necessary to design a state observer able to reconstruct the individual states of each subsystem.

Two broad methods [38,39] can be used to design observer-based decentralized output-feedback controllers for LSSs:

1. Design local observer and controller for each subsystem independently and check the stability of the overall closed-loop system. In this method, the interconnection in each subsystem is regarded as an unknown input.
2. Design the observer and controller by posing the output-feedback stabilization problem as an optimization problem.

The aim of a control design is not only to stabilize the system but also to ensure satisfactory performances for this control on that system. For linear systems the quadratic cost is characterized by the linear quadratic (LQ) design, which offers an optimal solution. Whereas for non-linear and/or uncertain systems the guaranteed cost control, in the presence of admissible non-linearities and/or uncertainties, ensure that the closed-loop system is asymptotically stable and an upper bound of the quadratic cost is minimized. Hence, the results of these control designs are characterized in terms of parameterized algebraic Riccati equations that are difficult to solve.

The interconnected systems characterized by linear subsystems with added non-linear interconnections are modeled as follows:

$$\begin{aligned} \dot{x}_i(t) &= A_i x_i(t) + B_i u_i(t) + h_i(t, x(t)) \\ y_i(t) &= C_i x_i(t), \quad i = 1, \ldots, N \end{aligned} \tag{1.5}$$

where $x_i(t)$ is the state vector of ith subsystem, $u_i(t)$ is the control vector of ith subsystem, $y_i(t)$ is the output vector of ith subsystem, and $h_i(t)$ reflects the interconnection term illustrating the non-linearity of the subsystem. The matrices A_i, B_i, and C_i denote, respectively, the state matrices, the control matrix, and the output matrix of each subsystem with the following:

- The pair (A, B) is assumed to be controllable.
- The pair (A, C) is assumed to be observable.

The interconnected N subsystems in the global system can be represented by the following:

$$\begin{aligned} \dot{x}_g(t) &= A_g x_g(t) + B_g u_g(t) + h_g(t, x(t)) \\ y_g(t) &= C_g x_g(t) \end{aligned} \tag{1.6}$$

where $A_g = diag(A_i)$, $B_g = diag(B_i)$, $C_g = diag(C_i)$, for each subsystem i and

$$x_g = [x_1^T, \ldots, x_N^T]^T, \quad u_g = [u_1^T, \ldots, u_N^T]^T, \quad y_g = [y_1^T, \ldots, y_N^T]^T$$

are the states, inputs, and outputs vectors of the global system and $h_g = [h_1^T, \ldots, h_N^T]^T$ describes the non-linear interconnection functions. The non-linearity functions $h_i(t, x(t))$ should satisfy the following condition:

$$\|h_i(t, x(t))\| \leq \alpha_i \|x(t)\|, \quad \alpha_i > 0, \ldots, N \tag{1.7}$$

where $\alpha_i > 0, \ldots, N$ are the bounds of the interconnection parameters, to be maximized for reliable operation. In some cases of large-scale distributed plants, system (1.5) can be represented by

$$\begin{aligned} \dot{x}_i(t) &= A_i x_i(t) + B_i u_i(t) + \sum_{j=1, j\neq i}^{i-1} A_{ij} x_j(t) \\ y_i(t) &= C_i x_i(t), \quad i = 1, \ldots, N \end{aligned} \tag{1.8}$$

1.2.3 Quasidecentralized control

To tackle the problem where a DecC structure cannot provide the required stability and performance properties, and to avoid the complexity and lack of flexibility associated with traditional centralized control, a quasidecentralized control (QDC) (partially decentralized and not fully distributed) strategy with *minimum cross-communication* among the plant units offers a suitable compromise, see Figure 1.2 and it provides a way of ensuring partial knowledge of how the local controller is affecting the global system. The term QDC refers to a situation in which most signals used for control are collected and processed locally, although some signals (the total number of which is kept to a minimum) still need to be transferred between local units and controllers to adequately account for the interactions between the different units and minimize the propagation of disturbances and process upsets from one unit to another [36].

$$\begin{bmatrix} u_1 \\ u_2 \\ u_3 \\ u_4 \\ u_5 \end{bmatrix} = \begin{bmatrix} K_{11} & DK_{12} & 0 & 0 & 0 \\ DK_{21} & K_{22} & DK_{23} & 0 & 0 \\ 0 & DK_{32} & K_{33} & DK_{34} & 0 \\ 0 & 0 & DK_{43} & K_{44} & DK_{45} \\ 0 & 0 & 0 & DK_{54} & K_{55} \end{bmatrix} \begin{bmatrix} y_1 \\ y_2 \\ y_3 \\ y_4 \\ y_5 \end{bmatrix}$$

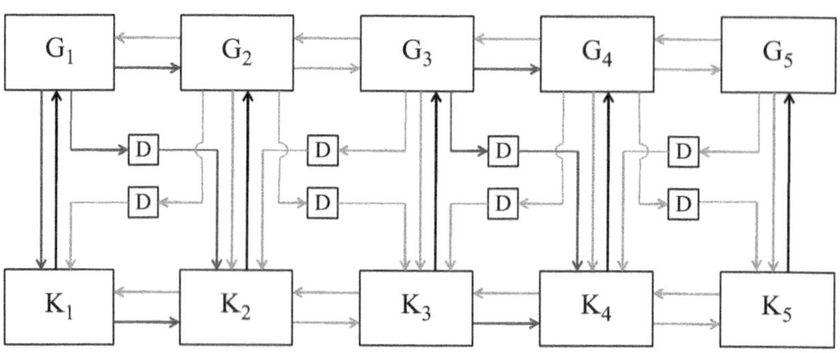

Figure 1.2 Quasidecentralized control system

One of the key problems that need to be addressed in the design of QDC systems [6,40] is how to coordinate the control and communication functions and how to account for possible limitations of the communication medium in the formulation and solution of the control problem especially if the communication is a shared medium like Ethernet. This is an important problem in view of the increased reliance in the process industries in recent years on sensor and control systems that are accessed over communication networks rather than hardwired.

The design of a QDC [41,42] strategy that enforces the desired closed-loop objectives with minimal cross-communication between the component subsystems is an appealing goal since it reduces reliance on the communication medium and helps save on communication costs. This is an important consideration particularly when the communication medium is a (potentially unreliable) wireless sensor network where conserving network resources is a key to prolonging the service life of the network.

1.2.4 Decentralized networked control

A system is said to be decentralized if there are multiple decision makers in the system (that is, controllers) and these decision makers have access to different and imperfect information with regard to the system they operate in, and they need to either cooperate or compete with each other. In such process control applications, a typical NCS has the structure of Figure 1.3.

For the purpose of control analysis and design, one major concern is the characterization of a sufficient amount of information transfer needed for a satisfactory

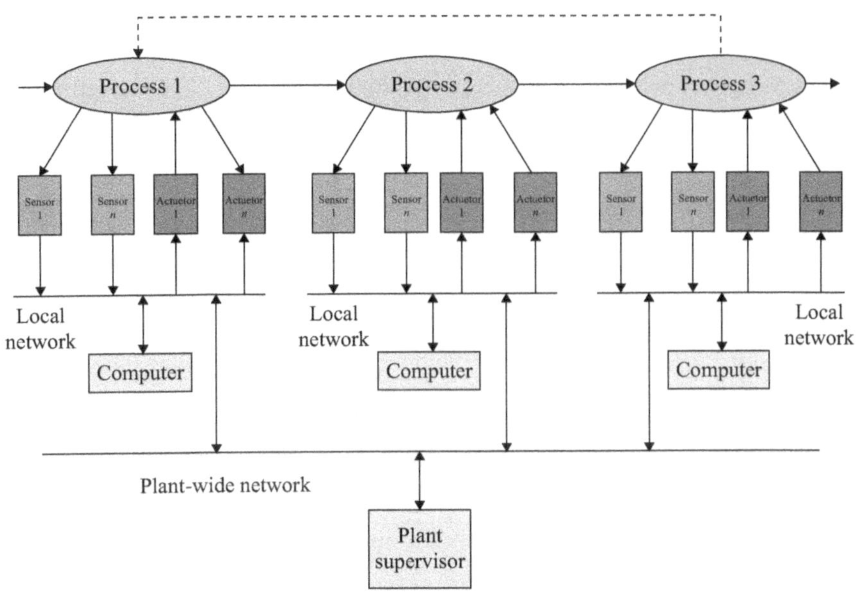

Figure 1.3 Process networked control system

performance. This information transfer can be between various components of an NCS and it will be called decentralized networked control system (DecNCS), see Figure 1.4. One necessity for satisfactory control performance is the ability for the controllers to track the plant states under communication constraints. One other challenge is the determination of the data rate required for the transmission of control signals, and the construction of dynamic encoding, decoding, and control policies meeting certain conditions. Another important problem is the coordination among multiple sensors or multiple controllers/decision makers with the lowest information exchange possible, which is the case in DecC system.

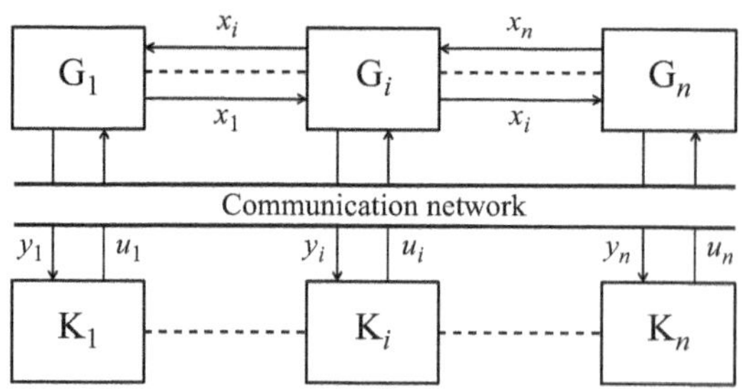

Figure 1.4 Decentralized networked control system

DecNCS combine the advantages of the centralized NCS and the DecC systems. Such a combination enables to cut unnecessary wiring, reduces the complexity and the overall system cost when designing and implementing control systems. DecNCS is an emerging research field for which developments are still on going to overcome several challenges raised by NCSs [43,45]. The control system stability in such system will require more attention to maintain the system stability [45–47] considering constraints like limited bandwidth or limited capacity channels [48], data rate constraints [49], and multicontrollers systems [50,51].

1.2.5 Distributed control

In distributed control (DC) structures, like the simple example shown in Figure 1.5, it is assumed that some information is transmitted among the local controllers so that each one of them has some knowledge on the behavior of the others. The information transmitted, for example, can consist of the future predicted control or state variables computed locally, so that any local controller can predict the interaction effects over the considered prediction horizon. A classification can be made depending on the topology of the communication network. Specifically, the following cases can be considered:

- Information is transmitted (and received) from any local controller to all the others (fully connected algorithms).
- Information is transmitted (and received) from any local controller to a given subset of the others (partially connected algorithms).

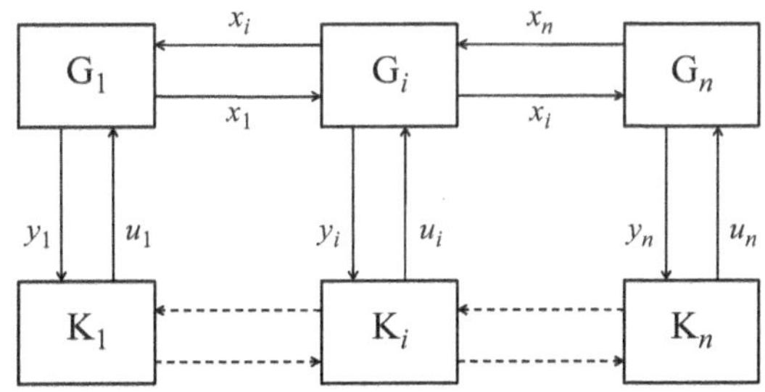

Figure 1.5 Distributed control system

DCS consisting of N subsystem discrete-time linear dynamic model of the subsystem i can be described as following:

$$x_i(k+1) = \sum_{j=1, j \neq i}^{N} A_{ij}x_j(k) + G_i w_i(k) \tag{1.9}$$

where x_i is the state of the subsystem i at instant k, w_i is the white noise process.

A partially connected information structure can be convenient in the case of large scale or complex systems made by a great number of loosely connected subsystems. In these cases, restricting the information exchange among directly interacting subsystems produces negligible performance deterioration. An interesting discussion on this point is discussed in [52], where reference is made to chemical processes composed by subsystems directly interacting only with their neighbors, possibly with additional recirculating flows. It is apparent that the amount of information available to the local controllers with iterative algorithms is higher, so that an overall iterative procedure can be set up to reach a global consensus on the actions to be taken within the sampling interval. To this regard, however, a further classification has to be considered:

- Distributed algorithms where each local controller minimizes a local performance index (independent algorithms).
- Distributed algorithms where each local controller minimizes a global cost function (cooperating algorithms).

1.2.6 Distributed networked control

When the control loops are closed over a real-time or lossy communication network, then this may introduce a new term called distributed networked control system (DNCS) [53–56] as in Figure 1.6. In a DNCS, a given subsystem uses its state and the states of its immediate neighbors to determine its control action [56] where the information is exchanged via the communication network also. Connecting the DCS components via a network can effectively reduce the complexity of the systems with nominal economical investments making the system scalable with efficient sharing of data [57]. Some parameters like induced delays, bit rate, packet size, packets drop, bandwidth, and sampling time will require more focus while designing the DNCS controllers. The best examples of such system include ad hoc wireless sensor networks

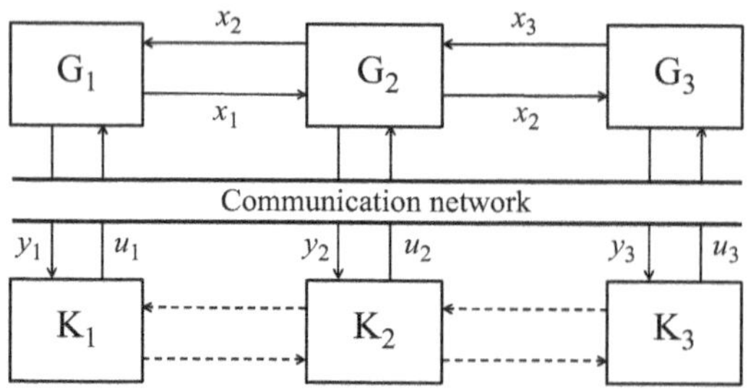

Figure 1.6 Distributed networked control system

and a network of mobile agents. However, the time complexity of the exact method can be exponential in the number of communication links.

The exchange of information among local controllers can be made according to different protocols:

- Information is transmitted (and received) by the local controllers only once within each sampling time (noniterative/aperiodic algorithms).
- Information can be transmitted (and received) by the local controllers many times within the sampling time (iterative/periodic algorithms).

Going back to system (1.9), we can express it considering that there is a communication between plants over a lossy communication network:

$$x_i(k+1) = \sum_{j=1, j\neq i}^{N} \alpha_{ij} A_{ij} x_j(k) + G_i w_i(k) \tag{1.10}$$

where α_{ij} is a random variable and it will set a value 1 at each time instant k a packet is received successfully by plant j from plant i otherwise it is zero.

1.2.7 Distributed control and game theory

The relationship between distributed control problems and team decision problems has recently gained renewed attention in the engineering literature. It has been shown that a collection of controllers with access to different sets of measurements can be designed using finite-dimensional convex optimization to act optimally as a team. Game theory has been used extensively as a quantitative framework for studying communication networks and DCSs among its other applications in engineering and economics. Game theoretic models provide a basis not only for analysis but also for design of network protocols and DecC schemes [58]. The noncooperative game theory has recently increasingly being used in engineering work on how to design games such that their outcome satisfies certain global objectives. Any game has three main components:

1. A set of players, $N = \{1, 2, 3, \ldots, n\}$.
2. Each player has a set of actions, $A_{ij} = \{a_{ij}\}$, and the joint action space $A = A_1 \times A_2 \times \cdots \times A_n$.

3. Each player orders the outcomes according to a utility or payoff function $u_i, u_i : A \rightarrow \Re, \forall i \in N$. All of them are working to achieve the global objective $G : A \rightarrow \Re$. When no player prefers a unilateral deviation to any of its other actions, the game is at a *Nash equilibrium*.

As discussed in [37] by means of game theory considerations, it is apparent that in iterative and independent algorithms each local controller tends to move toward a Nash equilibrium, while iterative and cooperating methods seek to achieve the Pareto optimal solution provided by an ideal centralized control structure. However, Nash equilibrium can even be unstable and far from the Pareto optimal solution, so that specific constraints have to be included in the control problem formulation to guarantee closed-loop stability. A stability constraint is included in the problem formulation, although stability can be verified only a posteriori with an analysis of the resulting closed-loop dynamics. Then, a min–max [59] approach aimed at minimizing local cost functions under the worst-case disturbance allows one to compute parameterized distributed control laws. A team of players is to optimize a worst-case scenario given limited information of nature's decision for each player.

1.2.8 Networked control systems

The research and developments on shared data networks have a long history started by principle data networks such as *Slotted* and *ARPANET* that were specially developed around 30–40 years ago [60,61]. Many industrial companies and institutes have shown interest in applying networks for remote industrial control purposes and factory automation because these networks enable remote data transfers and data exchanges among users, reduce the complexity in wiring connections, minimize the costs of media, and provide ease in maintenance. As a result of extensive research and development, several network protocols for industrial control have been released, for example, Controller Area Network (CAN), Profibus, Foundation Fieldbus, and Device-Net.

Most of these protocols are typically reliable and robust for real-time control purposes. Meanwhile, the technologies on general computer networks especially Ethernet have also progressed very rapidly. With the decreasing price, increasing speed, widespread usages, numerous software and applications, and well-established infrastructure, these networks become major competitors to the industrial networks for control applications. Thus, the control applications can utilize these networks to perform remote control at much farther distances than in the past without investing on the whole infrastructure. By having the feedback control system's loops closed through a real-time network, then it is called NCS, see Figure 1.7.

However, the insertion of the communication network in the feedback control loop makes the analysis and design of an NCS complex task. Conventional control theories with many ideal assumptions, such as synchronized control and non-delayed sensing and actuation, must be re-evaluated before they can be applied to NCSs. Specifically, the following issues need to be addressed. The first issue is the network-induced delay [62] (sensor-to-controller delay and controller-to-actuator delay) that occurs while exchanging data among devices connected to the shared

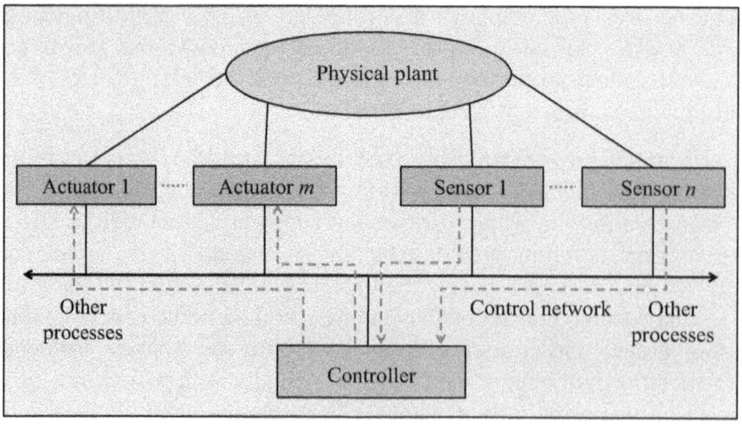

Figure 1.7 A typical NCS setup and information flows

medium. This delay, either constant (up to jitter) or time varying, can degrade the performance of control systems designed without considering the delay and can even destabilize the system [51]. Next, the network can be viewed as a web of unreliable transmission paths. Some packets not only suffer transmission delay but, even worse, can be lost during transmission. Thus, how such packet dropouts [63,64] affect the performance of an NCS is an issue that must be considered [65]. Another issue is that plant outputs may be transmitted using multiple network packets (so-called multiple-packet transmission), due to the bandwidth and packet size constraints of the network. Because of the arbitration of the network medium with other nodes on the network, chances are that all/part/none of the packets could arrive by the time of control calculation [8].

Depending on network protocols and scheduling methods, network-induced delays have different characteristics and can be constant, time varying, or stochastic [66]. There are essentially three kinds of delays that will affect the system as shown in Figure 1.8:

1. Communication delay between the sensor and controller, τ_{sc}.
2. Computational delay in the controller, τ_c.
3. Communication delay between the controller and actuator, τ_{ca}.

Due to the presence of network, delay concerns. For this purpose, there are various methodologies that have been formulated based on several types of network behaviors and configurations to control and maintain the stability of an NCS [60–62,67,68]. These methodologies are listed below:

- Augmented deterministic discrete-time model methodology
- Queuing methodology
- Optimal stochastic control methodology
- Perturbation methodology
- Sampling time scheduling methodology

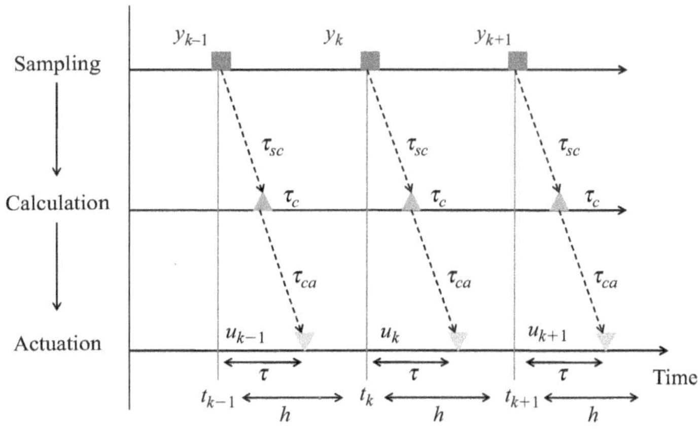

Figure 1.8 Network-induced delays

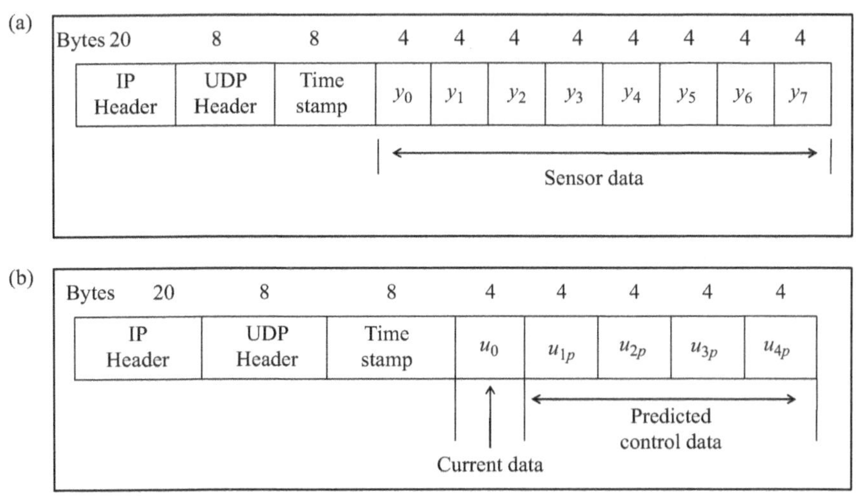

Figure 1.9 (a) Sensor data-packet; (b) control data-packet

- Fuzzy logic modulation methodology
- Event-based methodology
- End-user control adaptation methodology
- Robust control methodology
- Time-delay estimator methodology
- Predictive control methodology
- Markovian processes methodology
- Bandwidth management methodology

Another issue in feedback control systems, which is important, is that the sampled data should be transmitted within a sampling period in which the stability of control systems is guaranteed [69]. While a shorter sampling period is preferable in most control systems, for some purposes it can be lengthened up to a certain bound, maximum allowable delay bound (MADB), within which stability of the system is guaranteed in spite of the performance degradation [70]. The more information the controller can get the better control will be executed but this may degrade the network performance due to the high traffic generated with shorter sample period.

The NCS may have two main configuration [60,61] or structures:

1. **Direct/general structure:** The structure composed of a controller and several remote systems, each containing a physical plant, sensors, and actuators. The controller and the plants are physically located at different locations and are directly linked by a data network in order to perform remote closed-loop control as illustrated in Figure 1.10. In a practical implementation, multiple controllers can be implemented in a single hardware unit to manage multiple NCS loops in the direct structure. An example of this structure is DC motor speed control system.

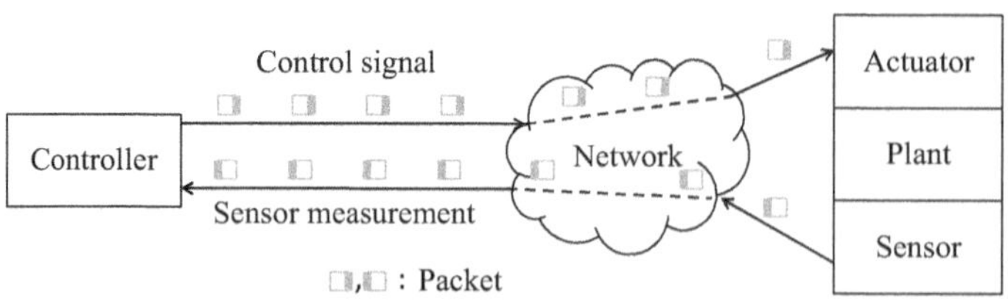

Figure 1.10 NCS in the direct structure

2. **Hierarchical/multilevels structure:** This structure may consist of a main controller and several remote closed-loop subsystems as shown in Figure 1.11. Each subsystem contains a set of sensors, a set of actuators, and a controller by itself. These system components are attached to the same control plant. In this case, a subsystem controller receives a set point from the central controller.

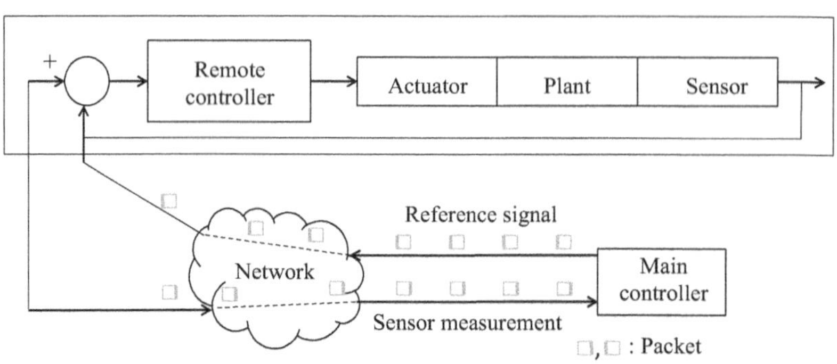

Figure 1.11 NCS in the hierarchical structure

The remote system then processes the reference signal to perform local closed-loop control and returns sensor measurement to the main controller for networked closed-loop control. Periodically, the main controller computes and sends the reference signal in a frame or packet via a network to the remote system. The networked control loop usually has a longer sampling period than the local control loop since the remote controller supposes to satisfy the reference signal before processing the newly arrived reference signal. This structure is widely used in several applications including mobile robots and tele-operation.

It should be noted that both structures have different advantages but the second structure is more modular, in which the control loop is simple to reconfigure and has better interaction because data is transmitted to components directly. A controller in the first structure can observe and process every measurement, whereas a (central) controller in the second structure may have to wait until the set point is satisfied to transfer the complete measurements, status signals, or alarm signals.

NCS have been attracting significant interest in the past few years and will continue to do so for the years to come. With the advent of cheap, small, and low-power processors with communication capabilities, it becomes possible to endow sensor and actuators with processing power and the ability to communicate with remove controllers through multipurpose networks. In view of this, we conjecture that in the near future NCSs will become the norm, replacing the current fixed-rate digital control systems that rely on dedicated connections between sensors, controllers, and actuators.

1.2.9 Wireless networked control system

Building a distributed or DecC system supported by a wireless network is a challenging task that requires a new design approach to both systems. Several problems, for instance, security, authentication, energy supply, signal path loss, transceiver operation mode, packet delay, and dropout etc., are explored in [71] for implementation of wireless networks in industrial applications. Wireless communication standards like IEEE 802.11 [72], IEEE 802.15.4/ZigBee, and IEEE 802.15.1/Bluetooth are used for wireless networked control system (WNCS). Most WNCS researches are based on mainly IEEE 802.11 standards and support data rates 1, 2, 11, 54 Mbps. IEEE 802.11 uses carrier sense multiple access with collision avoidance (CSMA/CA) as medium access control (MAC) protocol [73]. However, contention-based protocols, for example, CSMA/CA, are not appropriate for real-time communication as they require handshaking among the nodes [74] and do not guarantee bounded packet delay. For the ZigBee it is used for low distance, <10 m, and it has two types for high data rates with QoS and low data rates with low power consumption. Bluetooth offers low cost and low power requirement with a high degree of versatility. It has been used in some industrial applications such as sensor devices for monitoring, driver hands-free calling, etc.

Routing protocols determine how routes are established in wireless network and can be classified into proactive and reactive (on demand) protocols. A proactive protocol keeps up-to-date routing table by constantly requesting update information and sharing routing tables. The disadvantage of this strategy is that it produces huge traffic in the network [73–77]. Destination-sequenced distance vector (DSDV) routing is an example of proactive protocol for ad hoc networks. Reactive protocol attempts to establish a route when a node wishes to send a packet and there is no valid route in the route table. Routes are maintained until the destination becomes unreachable or the route is no longer required. The advantage is that less traffic is generated in the network. However, they have the disadvantages such as there is a delay in sending the packet and existing routes can become invalid without the node being made aware of it. Ad hoc on-demand distance vector (AODV) and dynamic source routing (DSR) are examples of reactive protocol.

End-to-end connection type communication over wireless network can be performed using either transmission control protocol (TCP) or user datagram protocol (UDP). TCP/IP is not suitable as it uses connection-oriented packet transfer. On the other hand, UDP offers low overheads as it does not maintain connections and discards obsolete or lost packets. Therefore, it is preferable for networked control applications. The other issue with wireless is the security where the wireless networks inherently suffer from security problems as signals are broadcast to all receivers. Two types of security issues can be identified: signal integrity and authentication. For signal integrity, the main concern comes from the interference from other radio transmitters. This problem can be crucial for IEEE 802.11 and Bluetooth technology as they both use the unlicensed ISM 2.4 GHz band. However, the spread spectrum techniques implemented by the standards can mitigate the interference in most cases. Moreover, as radio signals can be received by all nearby receivers, unauthorized users can exploit the resources of WNCS. The IEEE 802.11 standard offers a wireless local area network (WLAN) authentication mechanism called wired equivalent privacy (WEP) from the MAC layer. However, the security provided is not adequate. On the other hand, current Bluetooth technology specifies security in link layer and application developers have to choose the required security method. Again, bluetooth security is not strong enough to exchange sensitive data. For scheduling algorithms the following can be used: fixed priority (FP), rate monotonic (RM), and earliest deadline first (EDF).

1.2.10 Multiagents control systems

The co-design of control, computing, and communication for complex NCSs requires a new vision on complexity and new concepts and tools that will allow the designers to analyze and simulate how timing affects control performances and to determine the optimal structure of the hybrid distributed system with computing and communication constraints. New methods based on multiagent systems [78–80] could be used effectively for designing, modeling, simulating, and analyzing complex structures. Recently, the study on multiagent systems has been paid more attention due to its wide potential applications, such as platooning of vehicles in the urban transportation [81,82], the operation of the multiple robots [83], autonomous underwater vehicles [84,85], and the formation of aircrafts in military affairs [86,87]. Investigations for multiagent systems begin with studying the behavior of a large number of interacting agents with a common group objective.

1.2.11 Coordinated hierarchical control

An alternative to the distributed control schemes consists of two-level hierarchical control structures [88–90], like the one shown in Figure 1.11 for the example already considered in the previous sections. In this two-level structure, an algorithm at the higher level coordinates the actions of local controllers placed at a lower level. The basic idea is to describe the overall system under control as composed by a number of subsystems linked through some interconnecting variables, that is, the inputs of a given subsystem are the outputs or the states of another one. Then, for any subsystem an optimization problem is solved with minimizing a suitable local cost function under local state, input, and output constraints. If the computed local solutions satisfy the constraints imposed by the interconnecting variables, that is, if there is coherence among the values of the interconnecting variables computed by the local controllers, the procedure is concluded. Otherwise, an iterative 'price coordination' method is used: the coordinator sets the prices, which coincide with the Lagrange multipliers of the coherence constraints in the global optimization problem, by assuming as given the state, input, and output variables defined by the local regulators.

In turn, these optimal prices are sent to the low-level local optimizers that take them as given and recomputed the optimal trajectories of the state, input, and output variables over the considered prediction horizon. The iterations are stopped when the interconnecting variables satisfy the required coherence conditions. This conceptual iterative procedure must be specialized to guarantee its convergence as well as some properties of the resulting final solution. Finally, it must be noted that similar two-level structures are widely used in the intensive stream of research in computer science in artificial intelligence related to the so-called 'autonomous agents'. Basically, a number of agents must negotiate their actions through a 'negotiator' until a consensus on their actions is attained, see [91]. Figure 1.12 shows a communication network introduced to the coordinated hierarchical control system that brings all NCS issues to this type of control.

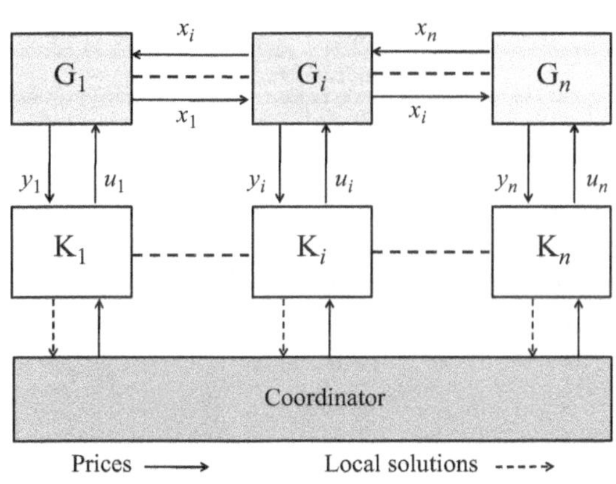

Figure 1.12 Hierarchical control for coordination of multi-input multi-output (MIMO) systems

1.2.12 Hierarchical control of multilayers systems

In hierarchical multilayer systems, the control action is performed by a number of controllers working at different time scales [36]. This can be useful at least in two cases: when the overall process under control is characterized by different dynamic behavior, that is, by slow and fast dynamics, or in plan to wide optimization when optimization and control algorithms working at a different rates compute both the optimal targets and the effective control actions to be applied. In industrial examples like a waste water treatment plant and a greenhouse control problem [92–94], the control can be performed at two different time scales. We can categorize the multilayer systems into the following:

- **Hierarchical control of multi-time-scale systems:** A controller acting at lower frequencies computes both the control actions (slow) of the manipulated variables that have a long-term effect on the plant, that is, the 'slow' control variables, and the references values of the 'fast' control variables, state variables, and output variables (u_{fast}^{ref}, x_{fast}^{ref}, y_{fast}^{ref}), respectively. A second controller takes these reference values as inputs and computes the 'fast' control variables u_{fast} solving a tracking problem at a higher rate. A conceptual scheme of this architecture for a two-layer structure is shown in Figure 1.13. Many industrial, economical, or sociological systems can be described by a hierarchical structure where the highest layer of the hierarchy corresponds to a dynamical system with slow dynamics. This system can be controlled by looking at its behavior over a long time scale, and its computed control inputs must be effectively provided by subsystems placed at lower layers of the hierarchy and characterized by faster dynamics. In turn, these subsystems must be controlled at a higher rate and can be placed at an intermediate layer of hierarchy. An example of a three-layer structure is shown in Figure 1.14. As a matter of fact, in these structures the regulator at a higher layer computes

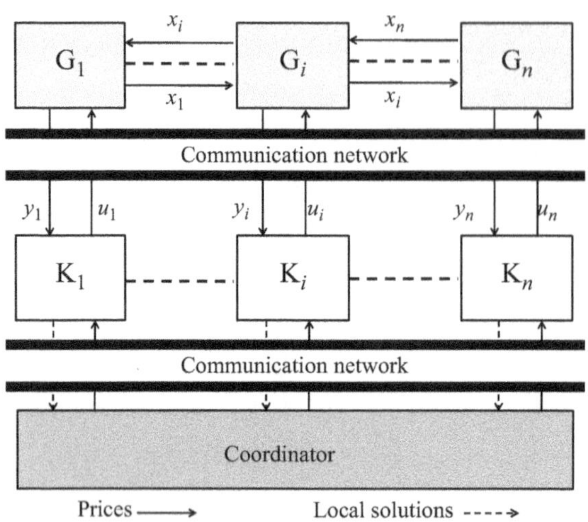

Figure 1.13 Hierarchical networked control for MIMO systems

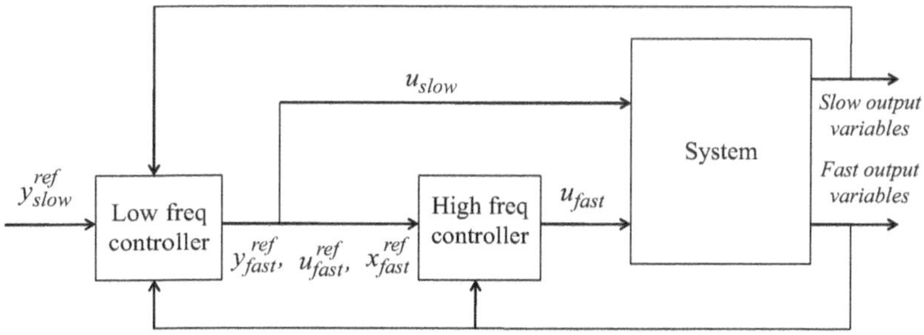

Figure 1.14 Control of system with slow and fast dynamics

its desired control inputs, which are the reference signals of the immediately lower layer. Moreover, the controllers of the subsystems at the lower layer must guarantee the solution of the corresponding tracking problems with an adequate level of accuracy, so that the mismatch between what is required by the higher level and what is provided by the lower one does not destroy some fundamental properties, such as stability and performance. From a control engineering point of view, this multilayer hierarchical structure corresponds to a classical cascade feedback control system as in Figure 1.15.

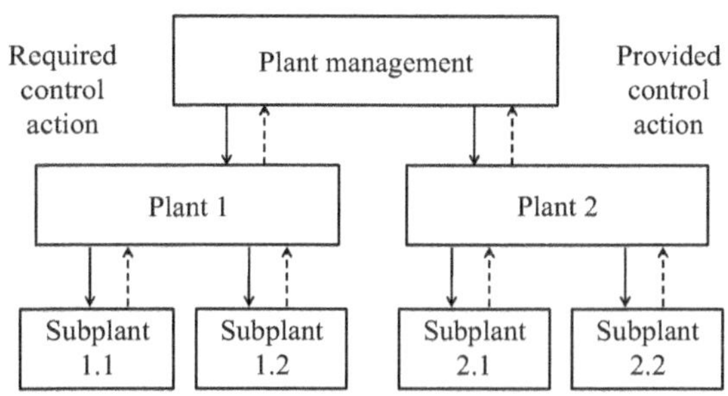

Figure 1.15 Hierarchical structure of a three-layer system

- **Hierarchical control for plant-wide optimization:** In the process industry it is common to design the overall control system according to the hierarchical structure as shown in Figure 1.16. At the higher layer, real-time optimization (RTO), which plays a fundamental role in process control, is performed to compute the optimal operating conditions with respect to a performance index representing an economic criterion. At this stage a detailed, although static, physical non-linear model of the system is used. At the lower layer a simpler linear dynamic model of the same system, often derived by means of identification experiments, is used to design a controller guaranteeing that the target values transmitted from the higher layer are attained. Also in this case, the lower level can transmit bottom-up information on constraints and performance. Moreover, the controller design

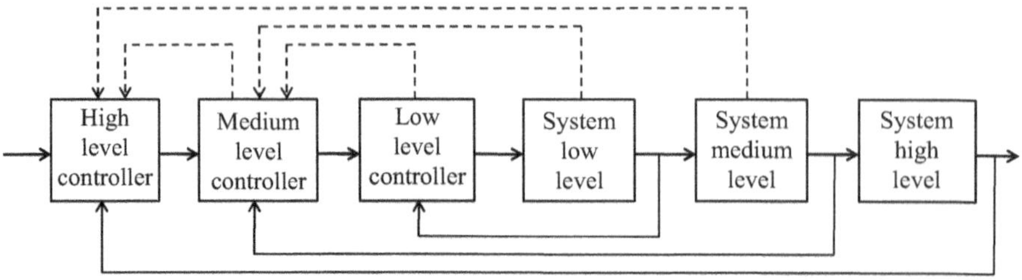

Figure 1.16 Three-layers cascade control structure

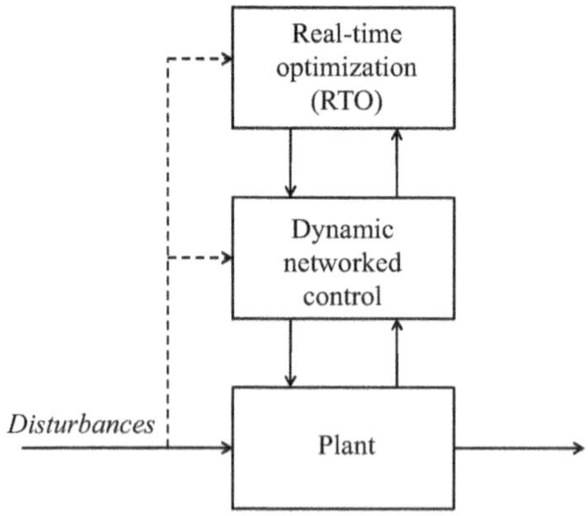

Figure 1.17 Hierarchical structure for plant-wide control and optimization

shall take care of constraints arises from closing the control loop over a shared communication network (Figure 1.17).

- **Hierarchical control system for dynamic resource management:** The hierarchical control system uses a set of utility functions to evaluate the performance of strings and missions in the system against current resource allocations [95]. The control system also uses the utility estimation function to estimate the desirability of various control actions with respect to the future performance and utility of the system. The control system chooses control actions that would result in a higher level of estimated utility [88]. If the system has enough unused system resources, the system could allocate resources to previously un-deployed missions or application strings to boost its overall utility and performance. Conversely, if resource contention were occurring due to an over deployment of missions (possibly due to resource failure among other possible causes), then the performance and utility of the deployed missions would drop. A drop in the measured utility indicates to the controllers that the allocation of resources should be adjusted in an attempt to relieve the resource contention and raise the measured utility. The control system uses a hierarchical control [10] philosophy that is fundamentally bottom-up. The low-level controllers are generally fast and responsive, while the high-level controllers have the ability to take more aggressive control actions.

Because higher level control actions are more invasive, the higher level controllers take more time to better estimate which of their control actions will maximize their local utility. Because local controllers in this design attempt to greedily maintain their local utility, the bottom-up control philosophy limits local, fast utility gains that are potentially detrimental to the overall system utility. String controllers perform fast low-level tuning of quality and throughput in order to maintain their local string utility. If a string controller is unable to maintain its local utility, its mission controller performs limited resource re-deployments to benefit local strings. If a mission controller is unable to maintain its local utility, the mission controller sends a request to the system controller to reinitialize system resources. The system controller has the ability to request that the Infrastructure Allocator perform a full re-initialization of system resources if the controllers are unable to take any action that would sufficiently raise the measured system utility. The controllers interact with each other through direct communications, but the controllers receive information about system resources or performance through resource status (Figure 1.18).

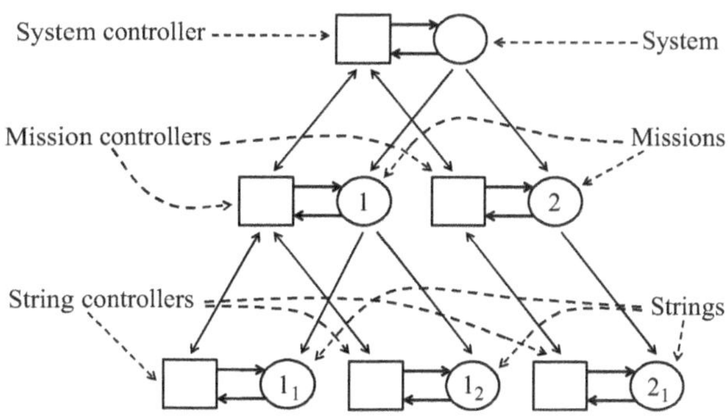

Figure 1.18 The control hierarchy

1.3 Related design problems and issues

After reviewing a number of architectures for control of interconnected and NCSs and considering a widespread industrial application of distributed, networked, and hierarchical solutions, many fundamental problems are still to be investigated. The following were selected.

1.3.1 New algorithms with guaranteed properties

Many theoretical contributions are required to develop efficient algorithms with guaranteed properties, such as stability and performance for DNCSs, where very few results are available for the DecC over network.

1.3.2 Selection of the control structure

Criteria must be developed for the selection of the proper control structure based on the relative improvements achievable by increasing the complexity [65]. For example, it is apparent that a distributed controller can stabilize systems that cannot be stabilized by a decentralized one because of the presence of fixed modes. However, if stability can be provided by both the schemes, it is still to be evaluated in term of performance whether it is worth considering a more complex structure, which requires more information to be exchanged among local control units. A second issue concerns the comparison of the performance provided by distributed controllers with respect to those achievable with the hierarchical and distributed networked control approaches.

1.3.3 Reconfigurable control structures and hybrid systems

With reference to the hierarchical structures we should explore the possibility to reconfigure the system, for example, by adding or removing actuators and sensors 'plug and play control', see [70,96,97]. This could be useful to consider time-varying performance requirements and to control systems described by a hybrid model. Finally, a flexible control configuration can better cope with the requirement of a high tolerance to faults.

1.3.4 Optimization algorithms

Many optimization algorithms have been developed to solve efficiently the minimization problems related to linear and non-linear centralized control. On the contrary, optimization methods for distributed and decentralized networked control are still lacking. This is an important and critical point where significant improvements are expected.

1.3.5 Distributed state estimation

This deals with the state estimation of a remote plant over a channel in which some of the packets are lost in a random fashion. Distributed control algorithms call for the availability of distributed state estimators guaranteeing the asymptotic convergence of the local state estimates. An approach called distributed and decentralized estimation and control (DDEC) was proposed in [98] and [99]. It was applied to some chemical industry showing results. However, further developments are required to include in the state estimation problem, the knowledge on state and noise constraints, as well as to link the convergence properties of the local estimates to local or global observability properties. Other works in [100] also shows some work on the states estimation with fault tolerance technique.

1.3.6 Band-limited channels

In communication systems, it is known that any communication network can only carry a finite amount of information per unit of time. This alleviates the issue of band-limited channels and a significant research effort has been devoted to the problem of determining the minimum bit rate that is needed to stabilize a linear system through feedback over a finite capacity channel [101–106]. Recently, some progress has also been made in solving the finite capacity stabilization problem for non-linear systems [107], and for linear systems with unknown parameters [108]. Performance limitations of feedback over finite capacity memory-less channels are addressed in [109], which obtains a general extension of Bode's integral inequality.

1.3.7 System partitioning

In the design of decentralized and distributed control the process under control must be partitioned a priori into subsystems properly defined to reduce the dynamic couplings and to facilitate the control design. In some cases partitioning is natural in view of the process layout, see for example [110] for power grids and chemical plants in [111] are considered.

1.3.8 Synchronization and communication protocols

Whenever the adopted control structure requires an exchange of information among local controllers, at the same or at different layers of a hierarchical structure, the achievable performance strongly depends on the adopted implementation, see [62] and communication protocols. Moreover, some fundamental problems related to these aspects must be considered, such as low transmission frequency or loss of information.

1.3.9 DecNCS and DNC limitations

Several limitations are due to the use of share communication network in terms of packet rate, sampling, network delay, packet dropout, bit rate, and quantization within decentralized systems. Further work is required to provide improved methods to lessen these limitations.

1.3.10 DecNCS and DNC stability

Another problem that need to be done is to examine the available results to determine the closed-loop stability of DecNCS/DNCSs in the presence of network sampling, delays, and packet dropouts [112] with the target being to develop improved robust stability methods using a variety of models including Markovian jump linear systems (MJLSs), linear time-varying systems, switched systems, non-linear systems with resets, asynchronous dynamical systems (ADSs), linear time-invariant systems with stochastic structured uncertainty, and linear systems with delayed inputs. Another problem is related to the stability of DNCs by taking into account a stochastic characterization for the inter-sampling times and consequently explores the major problem of performance and controller design.

1.4 Information flow

In recent years, a compelling need has arisen to understand the effects of information flow on estimation and control more fully. Technological advances in hardware and software over the past few decades have enabled cheap and small, yet powerful, communication and computation devices. This development, in turn, has made it possible to envisage systems in which multiple simple components cooperate to achieve a joint team goal. One can imagine three possible design architectures for such *multiagent systems* [497,498]:

1. Every component may be asked to share all its information with every other component in a reliable fashion, and all agents provided with ample computation power to process all this information to make (the same) globally optimal decisions. Even though multiple physical components may be present, this is still an example of a single agent system as far as the design is concerned. There is a rich theory already available to design such systems (modulo computational issues). We will refer to this architecture as a centralized system architecture. A simple example would be a group of aircrafts tasked with flying in a formation while either being allowed to communicate as much information as they desire with every other aircraft or communicating this information to a central ground unit that sends back the control inputs that they need to apply.
2. At the other end of the spectrum, all the components may be asked to depend only on the information that they possess to make their decisions. Thus, even though the components may have a joint team goal, they do not share information and instead try to coordinate their decisions by appealing to some correlated property of the environment that they are all observing. We will refer to this architecture as a completely decentralized architecture. Design of such systems reduces to design of multiple single agent systems. An example would be a group of aircrafts tasked with monitoring a region for threats where every aircraft generates its trajectory as if it is the only aircraft present.
3. Every component may be allowed to share information with some other components. Thus, the components try to coordinate their behavior by transmitting some of their information (and possibly plans) to some other components. We will refer to this approach as a cooperative, networked, or distributed system architecture. An example might be a group of aircrafts asked to achieve a common heading with every aircraft being allowed to observe the headings of the aircrafts only in its immediate vicinity (this task is often referred to as a consensus or an agreement task).

The distributed system architecture, were it to work, is very powerful since it allows the design of the individual units or components to be much simpler, while not compromising too much on the performance. Additional benefits include increased robustness to component loss, increased flexibility in that the components can be reconfigured for many different tasks, and so on. However, the design of such systems challenges the assumption of information of being unrestricted and reliable, that, as mentioned above, is at the heart of the traditional systems and control theory.

A typical design diagram for a networked system is shown in Figure 1.15. Such systems consist of the same components as a conventional system – dynamic processes, sensors, controllers, etc. – but there are many of them. Thus, there are many processes that need to be controlled. Their states are being sensed by many sensors. There is no one-to-one correspondence between processes and sensors. Many sensors might observe the same process and sensors may even switch from one process to another. The sensors may share information among themselves and also transmit it to the controllers. There might be many controllers with varying information and goals.

A simple example is an air traffic control situation where the controller in the control tower can observe all the aircrafts but cannot micromanage each one of them. On the other hand, the controller on-board the aircraft has limited information but has access to the actuators on-board the aircraft. All the components need to cooperate to achieve a team goal or to optimize a joint cost function and each entity can communicate with some other entities. The aspect of information as a central currency in the system is underlined by many new features that these systems present:

- **Information richness:** Because of so many sensors being present and entities transmitting information to each other, the system as a whole is *information rich*. At the same time, however, every component has access to a different information set (or has a different *world view*).
- **Information sharing:** Since they need to cooperate, the components have to reach some level of agreement and hence they need to communicate and share information. Because of the information-rich aspect, sharing of all the information is not feasible. If indeed all information could be shared among all the components, the situation would revert to a centralized architecture.
- **Presence of imperfect communication links:** Information sharing needs to be carried out over imperfect communication channels that have numerous limitations. As an example, such links may erase information stochastically, corrupt it, introduce random delays, and so on.
- **Limited information processing capability:** Components have limited information processing capabilities. Because of the huge amount of information being generated, components may need to decide which information to access and process at any time step.

These assumptions introduce information flow as both an enabler and a constraint in the system design. It is the sharing of information that allows components to cooperate and achieve better performance than that achieved by completely decentralized systems. However, the information flow is not perfect. There is too much information being generated so that all of it cannot be communicated or processed (information overload for the system); at the same time, components may not have access to the information they need to make globally optimal decisions because of constraints on the information flow (information paucity for the decision makers). These features are, by and large, not tackled in traditional systems and control theory and thus the design of such networked systems is quite different from conventional system design.

To fully exploit the benefits promised by such systems, there is, thus, a need to address the issues involved in networked system design and come up with a systematic theory for it. The design procedures at the moment tend to be ad hoc and do not allow us to make intelligent choices through an understanding of the inherent trade-offs and capabilities of the system. This chapter provides a new approach and new design principles to address some of the issues that are faced while designing networked systems from an estimation and control perspective. Besides the motive stated above, studying control and estimation problems in networked systems is useful for another reason. Since networked systems couple flow and processing of information with its use to estimate and control dynamic processes, a theory for such systems can be expected to unify concepts from information and communication theory, distributed and parallel computation as well as estimation and control theory. In other words, by focusing on common areas of traditionally separately studied fields of communication, computation, and control, networked systems can shed additional insight into all these fields.

As an instance, traditionally, information theory has grappled with the problem (as enunciated by Shannon [113]) of *reproducing at one point either exactly or approximately a message selected at another point.* While this field has had fascinating success in dealing with that question, it is also useful to recall a central assumption often made in it. Once again, to quote Shannon, 'Frequently the messages have meaning . . . , these semantic aspects of communication are irrelevant to the engineering problem'. This divorce of meaning of information from the reward obtained by transmitting it allows us to treat all information content alike. Furthermore, it allows long periods of use of a communication channel and render long delays irrelevant as long as the information is communicated. If we have to study the complexities that arise because of coupling between the flow of information and its use to estimate or control dynamical processes, we clearly cannot use results based on information theory in an off-the-shelf manner. As a simple instance, the same message may lose its value drastically if it is communicated to the controller with delay. This issue was eloquently raised by Witsenhausen [114] who used the term *information pattern* to denote the flow of information between various components (the terms *information flow* and *information pattern* are used interchangeably to denote this concept) and pointed out that the minimum cost achievable in a problem depends upon the prevailing information pattern. Thus, he suggested, information should be measured by its effect on optimal cost. This idea is a major extension of the traditional information theory, but is crucial if we want to fully grasp the interplay between information and dynamical systems. Networked control problems provide a natural source of problems in which such interplay happens.

Similarly, the fields of parallel processing and distributed computation have made amazing progress in dealing with problems such as speeding up computation by intelligently dividing it into operations that can be done concurrently by different processors or in calculation of functions of data held by many different processors. However, often these theories also make assumptions about all processors having access to a common memory unit or all processors being connected such that every processor can communicate with any other processor. Networked systems not only lack such features, they also pose problems of dynamics at the component level. Thus, components need to make decisions given whatever information they have without waiting for the coordination algorithm to finish. Traditional methods from parallel processing and distributed computation too need to be extended to fully understand the nexus between computation and control.

Thus, networked systems provide a playground where the traditional assumptions of communication, computation, and control are challenged. Any progress made in understanding the analysis and synthesis of such systems would then necessarily expand our knowledge about all these fields. In this book, we take a control- and estimation-oriented viewpoint and address some of the problems associated with the information flow.

1.5 Performance evaluation

The trend of modern industrial and commercial systems is to integrate computing, communication, and control into different levels of factory operations and information processes. For example, an integrated manufacturing system might include computer-numerically controlled machines, computer-aided design tools, supervisory controllers, and intelligent monitoring devices. The introduction of control network 'bus' architectures can improve the efficiency, flexibility, and reliability of these integrated applications, reducing installation, reconfiguration, and maintenance time and costs. The change of communication architecture from point-to-point to common-bus, however, introduces different forms of time delay uncertainty between sensors, actuators, and controllers. These time delays come from the time sharing of communication medium as well as additional functionality required for physical signal coding and communication processing. The characteristics of time delays could be constant, bounded, or even random, depending on the network protocols adopted and the chosen hardware. This type of time delay could potentially degrade a system's performance and possibly cause system instability.

Research in this class of systems, called NCSs, has focused on two areas: communication protocols and controller design. A proper message transmission protocol is necessary to guarantee the network quality of service (QoS), whereas advanced controller design is desirable to guarantee the control quality of performance (QoP). There are a wide variety of different commercially available control networks such as ControlNet, DeviceNet, Ethernet, Profibus, Sercos, WorldFIP, for the implementation of an NCS architecture. During the NCS design, a performance chart can be derived as shown in Figure 1.19. This performance chart provides a clear way to choose the proper sampling period for an NCS. Figure 1.19 is the comparison of control performance versus sampling period for continuous control, digital control, and networked control. The worst, unacceptable, acceptable, and best regions can be defined based on control system specifications such as overshoot, steady-state error, and/or phase margin. The performance axis in Figure 1.19 could be chosen to reflect a

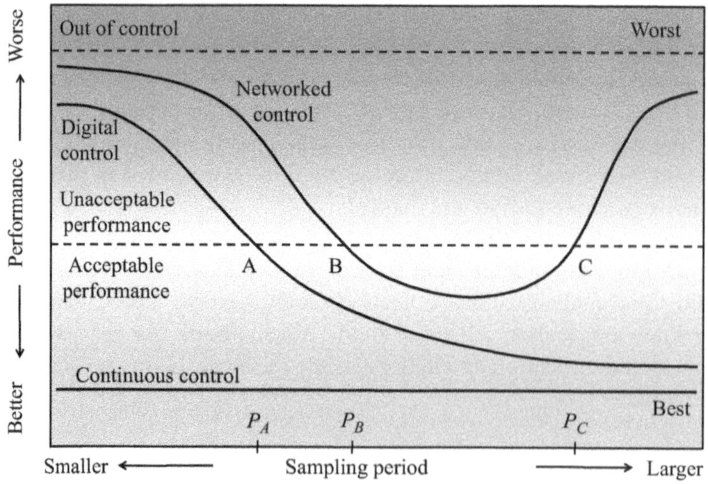

Figure 1.19 Performance evaluation

subset of these metrics. Since the performance of continuous control is not a function of sampling period, the performance index is constant for a fixed control law. For the digital control case, the performance only depends on the sampling period assuming no other uncertainties. The performance degradation point of sampling period in digital control could be estimated based on the relationship between control system bandwidth and sampling rate. For the networked control case, point can be determined by further investigating the characteristics and statistics of network-induced delays and device processing time delays. As the sampling period gets smaller, the network traffic load becomes heavier, the possibility of more contention.

1.6 Advantages of distributed control systems

Modern DCS have wide spectrum of practical applications where programmable logic controller (PLC) and SCADA play major role. While all control systems are distributed to a certain extent today and there is a definite merging of the concepts of DCS and despite the rapid growth in the use of PLCs and SCADA systems, some of the advantages of a DCS can still be said to be:

- **Integrity:** The expected process down-time caused by a conventional DCS is significantly less than with using a SCADA/PLC. One incident in a refinery can cost more than the difference in price between a DCS and SCADA/PLC. Reasons for this would include redundancy, fault tolerance, diagnostic alarming on I/O errors, system design, and others.
- **Engineering time:** A small SCADA/PLC system is easy to design and configure. As the system grows bigger, the effort involved to properly design and configure the system grows exponentially, and also the risks that things can go wrong.

To design and implement a single loop PID controller in a SCADA/PLC system is easy and quick. To design and implement the base layer control on a refinery using a SCADA/PLC system can be an absolute nightmare.

- **Abnormal situation management and intelligent alarm management:** This is a very important DCS issue that provides significant advantages over PLC and SCADA systems. Few DCSs do justice to the process; in terms of controlling for superior performance, most of them merely do the basics and leave the rest to the operators. Operators tend to operate within their comfort zone; they don't drive the process 'like Schumacher drives his Ferrari'. If more than one adverse condition developed at the same time and the system is too basic to act protectively, the operator would probably not be able to react adequately and risk a major deviation. Not only is the process control functionality normally underdeveloped but online process and control system performance evaluation is rarely seen and alarm management is often badly done. Operators consequently have little feedback on their own performance and exceptional adverse conditions are often not handled as well as they should be. This workshop gives suggestions on dealing with these issues. Why are DCSs generally so under utilized? Often because the vendor minimizes the applications software development costs to be sure of winning the job, or because he does not know enough about the process or if it is a green-field situation, enough could not be known at commissioning time but no allowance was made to add the missing functionality during the ramp-up phase. Often the client does not have the technical skills in-house to realize the desired functionality is missing or to adequately specify the desired functionality.

1.7 Outline of the book

During the past several years, considerable research has been directed to the intersect of three main fields: communication, computation, and control. One fundamental field that emerges from this effort is NCSs where all the needed functions are distributed among units and resources are shared through communication network. An integral part of these functions in distributed control and estimation algorithms that arose from different views. This book is written about the distributed control and estimation algorithms and the wide spectrum of its applications.

1.7.1 Methodology

Throughout the book, our methodology in each chapter/section is composed of five-steps:

1. **Mathematical modeling:** in which we discuss the main ingredients of the state-space model under consideration.
2. **Definitions and/or assumptions:** here we state the definitions and/or constraints on the model variables to pave the way for subsequent analysis.

3. **Analysis and examples:** this signifies the core of the respective sections and subsections that contains some solved examples for illustration.
4. **Results:** are provided most of the time in the form of theorems, lemmas, and corollaries.
5. **Remarks:** are given to shed some light of the relevance of the developed results vis-a-vis published work.

In the sequel, theorems (lemmas and corollaries) are keyed to chapters and stated in *italic* font with **bold titles**, for example, **Theorem 3.4** means Theorem 4 in Chapter 3, and so on. For convenience, we have grouped the references in one major bibliography cited toward the end of the book. Relevant notes and research issues are offered at the end of each chapter for the purpose of stimulating the reader. We hope that this way of articulating the information will attract the attention of a wide spectrum of readership.

This book aims at providing a rigorous framework for studying analysis, stability, and control problems of LSS while addressing the dominating sources of difficulties due to: dimensionality, information structure constraints, parametric uncertainty, and time delays. The primary objective is threefold: to review past methods and results from a contemporary perspective, to examine presents trends and approaches, and to provide future possibilities, focusing on robust, reliable distributed design methods.

In brief, the main features of the book are:

- It provides an overall assessment of distributed control and filtering algorithms over the past several years.
- It addresses several issues that arise at the intersect of communication, computation, and control.
- It presents key concepts with their proofs followed by efficient computational methods.
- It gives questions and end-of-chapter problems.
- It treats representative industrial applications.

1.7.2 Chapter organization

Interconnected systems have been investigated for a long time in the control literature and have attracted increasingly more attention for more than three decades. The literature grew progressively and quite number of fundamental concepts and powerful tools have been developed from various disciplines. With the rapid technological progress, many fundamental problems are brought up that call for further exploration. Among the core issues is that decentralized versus distributed in representation, analysis, design, and implementation. In particular, there still lacks a unified framework that can cope with the core issues in a systematic way. This motivated us to write the current book. The book presents theoretical explorations on several fundamental problems for interconnected systems.

The book is primarily intended for researchers and engineers in the systems, control, and communication community. It can also serve as complementary reading for elective courses for distributed control and estimation at the post-graduate level. The material of the book is divided into nine chapters:

Chapter 1 is an introductory chapter in which the different control architectures are examined in detail that are formulated to treat problems of interconnected systems. Following this is a discussion about the related design issues, information flow, performance, and finally the advantages of the DCSs.

Chapter 2 is devoted to recent developments in distributed model predictive control for interconnected networked systems. Several approaches are presented including dissipativity-based approach, stability-constrained approach, cooperative game approach, and feasible cooperation approach. Applications are subsumed of power plant control, quadruple-tank process control, and automatic generation control.

Chapter 3 introduces methods for designing distributed linear quadratic controllers for various classes of systems. First the class of identical decoupled dynamical systems is examined, then the class of semistable discrete-time systems, and finally the class of non-negative systems. Application to power load frequency control is provided.

Chapter 4 deals with design-distributed dynamic output-feedback controllers and it is divided into two major sections: one is on generalized framework for observer-based control and another is on approximate distributed feedback control. An application to chemical reactors is presented.

Chapter 5 establishes a detailed characterization of distributed consensus control methods. It investigates the dynamic consensus of multiagent systems, consensus control of time-delay systems, and robust consensus of multiagent systems.

Chapter 6 provides results on distributed estimation techniques and this include minimum variance estimation design, asynchronous multirate multismart sensors, and non-linear estimation.

Chapter 7 presents some recent results on distributed Kalman filtering methods. These include self-tuning fusion with and without feedback and filtering with intermittent communications. Both are under distributed Kalman environment.

Chapter 8 outlines the experimental setups and simulation of pilot-scale industrial processes.

Chapter 9 is an appendix containing some relevant mathematical lemmas and basic algebraic inequalities that are used throughout the book.

Several illustrative numerical examples are provided within the individual chapters.

1.8 Notes

This book covers a wide spectrum of design tools for distributed control and filtering in dynamical interconnected systems. For the purpose of unifying the results, we provide theoretical results of LSSs that are developed for linear, stationary, continuous-time deterministic systems. Afterward, whenever deemed convenient, we establish the discrete-time counterpart. Some other extensions to classes of non-linear systems are also presented.

In most of the recent developments, we start the model description and analysis for LSSs with time delays within the subsystems and along the coupling links. We feel that this set-up provides a general framework from which we can extract several important LSS descriptions of interest. For example, suppressing the delayed-coupling terms yields a class of LSS with time-delay subsystems having instantaneous interactions. On the other hand, deleting the delay terms within the subsystems results in a class of LSS where the subsystems have delayed communications due to finite processing and/or use of networks.

The analytical tools developed in this volume are supplemented with rigorous proofs of closed-loop stability properties and simulation studies. The material contained in this book not only organized to focus on the new developments in the control methodologies for continuous-time and discrete-time LSS systems, but it also integrates the impact of the delay factor, complexity, and uncertainties on important issues like delay-dependent stability and control design. For the purpose of clarity, it is intended to split the book into self-contained chapters with each chapter being equipped with illustrative examples, problems, and questions. The book will be supplemented by an extended bibliography, appropriate appendices, and indexes. At the end of each chapter, a summary of the results with the relevant literature is given and pertinent discussions of future advances and trends are outlined. Some selected problems have been organized for the reader to solve. The bibliographical notes cannot give a complete review of the literature but indicate the most influential papers/books in which the results presented here have been described and provide interested readers with adequate entry points to the still-growing literature in this field. The symbols and conventions used throughout the book are summarized in the List of Symbols.

It is planned while organizing the material that this book would be appropriate for use both as graduate-level textbook in applied mathematics as well as different engineering disciplines (electrical, mechanical, civil, chemical, systems), a good volume for independent study or a suitable reference for graduate students, practicing engineers, interested readers, and researchers from wide spectrum of engineering disciplines, science, and mathematics.

Chapter 2
Distributed model predictive control

This chapter presents different approaches to distributed model predictive control (DMPC) strategy for interconnected networked systems. In one approach, the dissipativity concept is employed as the vehicle for system analysis and design. Each node of the network is represented by a linear state space model designated as a subsystem herein. These subsystems are regulated by disparate multivariable controllers using an model predictive control (MPC) algorithm. In another approach, the focus is on information pattern, and a suboptimal linear quadratic Gaussian (LQG) design is developed. A third approach is based on MPC with stability constraint in which cooperative game theory is the core. In all of these approaches, detailed analytical treatments are provided to reach the desired results.

2.1 Overview

In model predictive control (MPC), also called receding horizon control, the control input is obtained by solving a discrete-time optimal control problem over a given horizon, producing an optimal open-loop control input sequence. The first control in that sequence is applied. At the next sampling instant, a new optimal control problem is formulated and solved based on the new measurements. The theory of MPC is well developed; nearly all aspects, such as stability, non-linearity, and robustness, have been discussed in the literature [115–118]. MPC is very popular in the process control industry because the actual control objectives and operating constraints can be represented explicitly in the optimization problem that is solved at each control instant. Many successful MPC applications have been reported in the last two decades [116,118]. Typically, MPC is implemented in a centralized fashion. The complete system is modeled, and all the control inputs are computed in one optimization problem. In large-scale applications, such as power systems, water distribution systems, traffic systems, manufacturing systems, and economic systems, it is useful (sometimes necessary) to have distributed or decentralized control schemes, where local control inputs are computed using local measurements and reduced-order models of the local dynamics [119,120]. The goal of the research described here is to realize the attractive features of MPC (meaningful objective functions and constraints) in a decentralized and distributed architectures. Previous work on DMPC is reported in [121–128]. In some applications, multiple low-level controllers are simply implemented using MPC, just as one might use proportional-integral-derivative (PID) controllers to close local feedback loops [127]. For water distribution systems, full-scale centralized MPC computations have been decomposed for decentralized computation, using standard coordination techniques (e.g., augmented Lagrangian) or estimation and prediction schemes to obtain solutions at each control instant [125,126,128]. Decentralized team strategies for linear-quadratic problems were considered in [122] and [123]. This method is suitable for linear time-invariant (LTI) systems. If the optimal control problem is an LQG problem, an analytical solution can be found. Some researchers [123] suggest using a neural network to approximate stationary controllers for non-linear systems, but this approach is not suitable for large systems.

2.2 Dissipativity-based approach

Research on complex networked systems has been increasingly attractive in various fields of science and engineering. Due to the finite speed of transmissions, traffic congestions or spreading, a time delay, called coupling delay herein, is often associated with signals, material flows, or influences traveling through a network [129,130]. The networked system notion, in a broad scene, is extended over many application domains such as, but not limited to, networked robots [131], infrastructure systems [132], power networks [133], and chemical processes [134]. A networked system can be modeled as a large set of interconnected nodes, in which a node is a fundamental unit with specific contents [130], linking to each other in an arbitrary connection topology. When a delay occurs in the interconnections between nodes or subsystems, the overall performance and system stability are seriously affected. This chapter is concerned with a distributed predictive control strategy for networked systems that suffer from an arbitrary coupling delay.

This section presents new dissipativity-based stability conditions in the time domain and the constructive method of receding-horizon stability constraints for DMPCs for networked systems. The novelty of the proposed approach is attributed for by the disposal of controller closed-form models in the development of stability conditions. The concept of dissipativity has been used in robust stability analysis [135], and control synthesis with state feedback and output feedback methods (see [136]). It has not, however, been effectively applied to the online-update controllers for networked systems having control and state constraints, as of our knowledge. In the present approach, the evolutions of control inputs, measurement outputs, and their phase and gain properties prescribed in the quadratic dissipativity of controller trajectories are used for the control manipulation purposes rather than for robust stability performance. Preliminary results have been presented in [137–139].

The DMPC strategy advocated in this chapter tackles the coupling delay problem by acquiring the property of delay-robust dissipativity for all periods of time delays that are bounded by a maximum delay time. A single random coupling delay is considered in this development, that is, the coupling delays between network nodes are assumed identical. The storage function of the Lyapunov–Krasovskii functional form is employed in both delay-dependent and delay-independent dissipativity criteria. The interconnection stability condition for delay-robust dissipative systems is subsequently derived in association with the dissipativity of controller trajectories. The MPC stability constraint that is developed from this dissipativity property is updated online based on the current state vectors. By employing the state predictions up until the end of the predictive horizon, and rolling after each time step, the present stability constraint provides a promising non-conservative solution to the coupling delay problem.

Time delay systems have emerged as an important research strand in the control literature (see [2,140]), and a plethora of research papers lately. Along with delay-independent and delay-dependent solutions employing various types of the Lyapunov–Krasovskii functional form, passivity-based approaches have been adopted for different time delay applications (see [141] and references therein). Recent works on delayed interconnected systems are presented in [2] and [3]. While laying firm foundations for the field, the solutions to distributed control problems in these approaches usually require artificial or assumed constraints on the coupling (interactive) vectors or on the objective functions. These constraints may, however, create conservativeness to dynamically coupled subsystems. In process control applications where the couplings are material or energy flows, such constraints will add unnecessary limitations to, or even worse, diminish the overall control performance. In the present approach, the stability of the networked system is achieved without having any artificial constraints on the coupling vectors, yet only decoupled controls are used. Furthermore, we provide an efficient solution with guaranteed feasibility to the distributed control problem of delayed interconnected systems that have both input and state constraints. Based on the input and output properties, the proposed approach can be directly extended to the problem of multiple state and control input delays.

In a distributed control architecture, the quadratic optimal control synthesis has been identified with major difficulties [2]. The obstacles around the constructive methods of $\mathcal{H}_2$ or $\mathcal{H}_\infty$ distributed optimal control lie in the non-convexity of optimization problems. Different results have been presented in the past, such as the computations for uplifting the $\mathcal{H}_\infty$ performance based on the finite horizon optimization in [142], or the sufficient condition upon which the distributed $\mathcal{H}_2$ control problem is convex [143]. An input-to-state small-gain theorem for nonlinear networked systems has been presented in [144]. Among various solutions to the distributed control problem, MPC has been identified as a potential candidate (see [138,145–148] and the references therein). MPC has been widespreadly used in the industrial process control because of its natural ability to systematically address the issues of input and state constraints, windups in control actions, as well as system sensitivities [117,149]. Initially developed for slow response and time delay linear systems, MPC can find many applications in process control nowadays [150]. MPC has also been extended to networked systems in various decentralized and distributed architectures, as shown in a survey [36].

2.2.1 Control system models

Consider a networked system Σ consisting of n_s subsystems, each denoted as $\mathcal{S}_i$, $i = 1, \ldots, h$ and has discrete-time state space model of the form

$$\mathcal{S}_i : \begin{cases} x_i(k+1) &= A_{ii}x_i(k) + B_iu_i(k) + g_i(x_j(k)), \quad i \neq j \\ y_i(k) &= C_ix_i(k) \\ g_i(x_j(k)) &= \sum_{i \neq j}^{n_s} A_{ij}x_j(k) \end{cases} \tag{2.1}$$

where $x_i \in \mathbb{X}_i \subset \mathbb{R}^{n_i}$, $u_i \in \mathbb{U}_i \subset \mathbb{R}^{m_i}$ and $y_i \in \mathbb{R}^{q_i}$ are, respectively, the state, control input, and measurement output vectors, with $\mathbb{X}_i$ is a closed and bounded subset on $\mathbb{R}^{n_i}$, $\mathbb{U}_i$ is a bounded subset on $\mathbb{R}^{m_i}$.

The input $v_\xi(k)$ of subsystem $\mathcal{S}_\xi$ and the output $w_\zeta(k)$ of subsystem $\mathcal{S}_\zeta, \xi \neq \zeta$ are connected to each other in and arbitrary topology. Hence, an arbitrary delay between subsystem nodes of the networked system is considered. The large-scale system Σ is then represented by a state space model of the block diagonal system $\mathcal{S}$ formed by h subsystems $\mathcal{S}_i$ and the interconnection process H as follows:

$$\Sigma : \begin{cases} x(k+1) &= Ax(k) + Bu(k) + g(x(k)) \\ y(k) &= Cx(k) \end{cases} \tag{2.2}$$

where $A = \text{diag}[A_{ii}]_1^{n_s}$, $B = \text{diag}[B_i]_1^{n_s}$, $C = \text{diag}[C_i]_1^{n_s}$, $E = \text{diag}[E_i]_1^{n_s}$, $F = \text{diag}[F_i]_1^{n_s}$, $L = \text{diag}[L_i]_1^{n_s}$, $x = [x_1^T \cdots x_h^T]^T$, $u = [u_1^T \cdots u_h^T]^T$, $y = [y_1^T \cdots y_h^T]^T$, $v = [v_1^T \cdots v_h^T]^T$, $w = [w_1^T \cdots w_h^T]^T$, $d = [d_1^T \cdots d_h^T]^T$, τ is the number of delay time steps, $\tau \in \{1, \ldots, \tau_{\max}\}$, and H is the global coupling/interconnection matrix.

From a centralized point of view, the model Σ can be written as

$$\Sigma : \begin{cases} x(k+1) &= Ax(k) + EHF\, x(k-\tau) + Bu(k) + +g_i(x_j(k)) \\ y(k) &= Cx(k) \end{cases} \tag{2.3}$$

In a distributed control scheme, the models of stand-alone subsystems are requisite. When all the interaction inputs and outputs vanish (i.e., $v_i = 0$ and $w_i = 0$), a stand-alone subsystem $\mathcal{S}_i$, denoted as $\mathcal{S}_i^s$, has the state space model of the form

$$\mathcal{S}_i^S : \begin{cases} x_i(k+1) &= A_ix_i(k) + B_iu_i(k) + +g_i(x_j(k)) \\ y_i(k) &= C_ix_i(k) \end{cases} \tag{2.4}$$

Assumptions:

1. All stand-alone subsystems $\mathcal{S}_i^s$ (2.1), and large-scale systems Σ (2.3) without state delays, are controllable and observable.
2. The updating time instants and all invocations are synchronized between all subsystems and their controllers.

The problem of designing h disparate model predictive controller for h subsystems $\mathcal{S}_i$ in a distributed configuration that are subject to an arbitrary coupling delay is the main concern of this section. The inter-subsystem communication between local controllers is not required for exchanging online data by the optimization solvers. The stability constraint approach is used for guaranteeing the stability of the MPC systems.

2.2.2 Interconnection stability

This section presents the main results of this chapter on establishing the new interconnection stability conditions in the real-time domain for large-scale networked systems subject to an arbitrary coupling delay. The dissipativity of open-loop subsystems is first defined in Section 2.2.3. In what follows, a subsystem $\mathcal{S}_i$ (2.1) is said to be delay-robust dissipative if it is quadratically dissipative with respect to the following quadratic supply rate:

$$\xi^{si}(y_{si}, u_{si}) = y_{si}^T Q_{si} y_{si} + 2y_{si}^T S_s i u_s i + u_{si}^T R_s i u_s i \tag{2.5}$$

where

$$u_{si}^T(k) = [u_i^T(k) u_i^T(k) d_i^T(k)], \quad y_{si}^T(k) = [w_i^T(k-\tau) y_i^T(k)]$$

$$Q_{si} = \begin{bmatrix} Q_{\tau i} & 0 \\ 0 & Q_i \end{bmatrix}, \qquad R_{si} = \begin{bmatrix} R_{\tau i} & 0 \\ 0 & \begin{bmatrix} R_i & 0 \\ 0 & R_i^d \end{bmatrix} \end{bmatrix},$$

$$S_{si} = \begin{bmatrix} S_{\tau i} & 0 \\ 0 & [S_i \quad 0] \end{bmatrix}$$

A subsystem $\mathcal{S}_i$ is said to be quadratically dissipative with respect to the quadratic supply rate $\xi^i(y_{si}, u_{si})$ if there exists a nonnegative C^1 function, addressed as storage function, $V_i(x_i(k))$ such that for all u_{si} and all $k \in \mathbb{Z}^+$, the following dissipation inequality is satisfied [151]:

$$V_i(x_i(k+1)) - V_i(x_i(k)) \leq \xi^i(y_{si}, u_{si}) \tag{2.6}$$

It is noted that in the above supply rates, the passivity terms describing the phase property between y_i and d_i are not considered. The dissipativity conditions for linear systems are well established in the control literature [135]. The storage function of a Lyapunov–Krasovskii functional form of the following will be considered in this chapter:

$$\begin{aligned} V_{si}(k) &= x_i^T(k) P_i x_i(k) + \sum_{j=k-\tau}^{k-1} x_i^T(j) P_{\tau i} x_i(j), \\ P_i &= P_i^T > 0, P_{\tau i} = P_{\tau i}^T > 0 \end{aligned} \tag{2.7}$$

Based on these definitions, we derive the delay-independent and delay-dependent dissipativity conditions for open-loop subsystems stated in the next theorems. It should be emphasized that a linear matrix inequality (LMI)-based condition is called delay-independent if the delay variable τ does not appear in the derived LMIs. Likewise, an LMI-based condition is called delay-dependent if the delay variable τ appears in the derived LMIs. The following theorem provides a delay-independent dissipativity condition for $\mathcal{S}_i$.

Theorem 2.1: *Consider a subsystem S_i (2.1) and the storage function of the Lyapunov–Krasovskii functional form (2.7). Subsystem S_i is delay-independent delay-robust dissipative with respect to the supply rate $\xi^{si}(y_{si}, u_{si})$ (2.5), if the following LMI*

$$\begin{bmatrix} \mathcal{P}_i & 0 & A_i^T P_i E_i & A_i^T P_i B_i - C_i^T S_i & A_i^T P_i L_i \\ * & -P_{\tau i} - F_i^T Q_{\tau i} F_i & -F_i^T S_{\tau i} & 0 & 0 \\ * & * & E_i^T P_i E_i - R_{\tau i} & E_i^T P_i B_i & E_i^T P_i L_i \\ * & * & * & B_i^T P_i B_i - R_i & B_i^T P_i L_i \\ * & * & * & * & L_i^T P_i L_i - R_i^d \end{bmatrix} \leq 0,$$

$$\mathcal{P}_i = A_i^T P_i A_i - P_i + P_{\tau i} - C_i^T Q_i C_i \tag{2.8}$$

is feasible in $P_i, P_{\tau i}, Q_i, S_i, R_i, R_i^d, Q_{\tau i}, S_{\tau i}, R_{\tau i}$.

Proof. By expanding the dissipation inequality of the following form:

$$V_{si}(k+1) - V_{si}(k) \leq \xi^{si}(y_{si}(k), u_{si}(k))$$

using the state space model (2.1) and the storage function of the Lyapunov–Krasovskii functional form (2.7), the resultant delay-robust dissipativity criterion cast in LMI will be obtained after some matrix manipulations, as detailed in the sequel. Substituting (2.7) to the above dissipation inequality gives

$$\begin{aligned} & x_i(k+1)^T P_i x_i(k+1) - x_i(k)^T P_i x_i(k) \\ & \quad - x_i(k-\tau)^T P_{\tau i} x_i(k-\tau) + x_i(k)^T P_{\tau i} x_i(k) \\ & \leq y_i^T(k) Q_i y_i(k) + u_i^T(k) R_i u_i(k) + d_i^T(k) R_i^d d_i(k) \\ & \quad + 2 y_i^T(k) S_i u_i(k) + w_i^T(k-\tau) Q_{\tau i} w_i(k-\tau) \\ & \quad + 2 w_i^T(k-\tau) S_{\tau i} u_i(k) + u_i^T(k) R_{\tau i} u_i(k) \end{aligned}$$

By dropping out all time index k, for sake of convenience, and substituting (2.1) to the above, we have

$$\begin{aligned} & (x_i^T A_i^T + u_i^T B_i^T + u_i^T E_i^T + d_i^T L_i^T) P_i (A_i x_i + B_i u_i + E_i u_i + L_i d_i) \\ & \quad - x_i^T (P_i - P_{\tau i}) x_i - x_i^T(-\tau) P_i x_i(-\tau) \\ & \leq x_i^T C_i^T Q_i C_i x_i + u_i^T R_i u_i^d + d_i^T R_i^d d_i \\ & \quad + 2 y_i^T S_i u_i + x_i^T(-\tau) F_i^T Q_{\tau i} F_i x_i(-\tau) \\ & \quad + 2 x_i^T(-\tau) F_i^T S_{\tau i} u_i + u_i^T R_{\tau i} u_i \\ & \leq 0 \end{aligned}$$

This is equivalent to

$$
\begin{aligned}
& x_i^T(A_i^T P_i A_i - P_i + P_{\tau i} - C_i^T Q_i C_i)x_i + u_i(B_i^T P_i B_i^T)u_i \\
& \quad + v_i^T(E_i^T P_i E_i - R_{\tau i})u_i + d_i^T(L_i^T P_i L_i - R_i^d)d_i \\
& \quad + 2x_i^T(A_i^T P_i B_i - C_i^T S_i)u_i + 2x_i^T A_i^T P_i E_i u_i + 2x_i^T A_i^T P_i L_i d_i \\
& \quad + 2u_i^T B_i^T P_i E_i u_i + 2u_i^T B_i^T P_i L_i d_i + 2u_i^T E_i^T P_i L_i d_i \\
& \quad + x_i^T(-\tau)(-P_{\tau i} - F_i^T Q_{\tau i} F_i)x_i(-\tau) - 2x_i^T(-\tau)F_i^T s_i u_i \\
& \leq 0
\end{aligned} \tag{2.9}
$$

This can be further expressed in compact form as

$$
B_i^t \Xi_i \beta_i < 0,
$$

$$
\Xi_i = \begin{bmatrix} \mathcal{P}_i & 0 & A_i^T P_i E_i & A_i^T P_i B_i - C_i^T S_i & A_i^T P_i L_i \\ \bullet & -P_{\tau i} - F_i^T Q_{\tau i} F_i & -F_i^T S_{\tau i} & 0 & 0 \\ \bullet & \bullet & E_i^T P_i E_i - R_{\tau i} & E_i^T P_i B_i & E_i^T P_i L_i \\ \bullet & \bullet & \bullet & B_i^T P_i B_i - R_i & B_i^T P_i L_i \\ \bullet & \bullet & \bullet & \bullet & L_i^T P_i L_i - R_i^d \end{bmatrix}
$$

$$
\beta_i = \begin{bmatrix} x_i & x_i(-\tau) & u_i & u_i & d_i \end{bmatrix}^t
$$

which is fulfilled by (2.8). □

The following result provides a robust delay-dependent dissipativity condition for $\mathcal{S}_i$. It will be seen from the proof that the delay-dependent condition is an extension of the delay-independent condition in Theorem 2.1.

Theorem 2.2: *Consider a subsystem $\mathcal{S}_i$ (2.1) and the storage function of the Lyapunov–Krasovskii functional form (2.7). Subsystem $\mathcal{S}_i$ is delay-dependent delay-robust dissipative with respect to the supply rate $\xi^{si}(y_{si}, u_{si})$ (2.5), if the following LMI is feasible in $P_i, P_{\tau i}, Q_i, S_i, R_i, R_i^d, Q_{\tau i}, S_{\tau i}, R_{\tau i}$ for $\tau = 1, \ldots, \tau_{\max}$:*

$$
\begin{aligned}
\Pi_i &= \begin{bmatrix} \Pi_{1i} & \Pi_{2i} \\ \bullet & \Pi_{3i} \end{bmatrix} < 0 \\
\Pi_{1i} &= \begin{bmatrix} \mathcal{P}_{\tau i} & A_i^{\tau^T} \mathcal{P}_i B_i & A_i^{\tau+1^T} P_i E_i - F_i^T S_{\tau i} \\ \bullet & B_i^T P_i B_i & B_i^T A_i^T P_i E_i \\ \bullet & \bullet & E_i^T P_i E_i - R_{\tau i} \end{bmatrix} \\
\Pi_{2i} &= \begin{bmatrix} A_i^{\tau^T}(A_i^T P_i B_i - C_i^T S_i) & A_i^{\tau+1^T} P_i L_i \\ B_i^T(A_i^T P_i B_i - C_i^T S_i) & B_i^T A_i^T P_i L_i \\ E_i^T P_i B_i & E_i^T P_i L_i \end{bmatrix} \\
\Pi_{3i} &= \begin{bmatrix} B_i^T P_i B_i - R_i & B_i^T P_i L_i \\ \bullet & L_i^T P_i L_i - R_i^d \end{bmatrix}
\end{aligned} \tag{2.10}
$$

where

$$\begin{aligned}
\mathcal{P}_i &= A_i^T P_i A_i - P_i + P_{\tau i} - C_i^T Q_i C_i,\\
\mathcal{P}_{\tau i} &= A_i^{\tau^T} \mathcal{P}_i A_i^{\tau} - P_{\tau i} - F_i^T Q_{\tau i} F_i,\\
\mathcal{B}_i &= [A_i^{\tau-1} B_{si} \ldots A_i B_{si} B_{si}]
\end{aligned}$$

Proof. From the state space model (2.1), we obtain

$$x_i = A_i^{\tau} x_i(-\tau) + \mathcal{B}_i \breve{u}_{ri}$$

where

$$\begin{aligned}
\mathcal{B}_i &= [A_i^{\tau-1} B_{si} \ldots A_i B_{si} \quad B_{si}], \quad B_{si} = [B_i \quad E_i \quad L_i],\\
\breve{u}_{ri} &= [u_{si}(k-\tau) \ldots u_{si}(k-\tau)], \quad u_{si} = [u_i \quad d_i \quad u_i]
\end{aligned}$$

Thus we have

$$\begin{aligned}
&[x_i^T(-\tau) A_i^{\tau^T} + \breve{u}_{\tau i}^T \mathcal{B}_i^T] \mathcal{P}_i [A_i^{\tau} x_i(-\tau) + \mathcal{B}_i \breve{u}_{\tau i}]\\
&\quad + u_i^T (B_i^T P_i B_i - R_i) u_i + u_i^T (E_i^T P_i E_i - R_i) u_i\\
&\quad + 2[x_i^T(-\tau) A_i^{\tau^T} + \breve{u}_{\tau i}^T \mathcal{B}_i^T](A_i^T P_i B_i - C_i^T S_i) u_i\\
&\quad + 2[x_i^T(-\tau) A_i^{\tau^T} + \breve{u}_{\tau i}^T \mathcal{B}_i^T] A_i^T P_i E_i u_i\\
&\quad + 2[x_i^T(-\tau) A_i^{\tau^T} + \breve{u}_{\tau i}^T \mathcal{B}_i^T] A_i^T P_i L_i d_i\\
&\quad + 2u_i^T B_i^T P_i E_i u_i + 2u_i^T B_i^T P_i L_i d_i + 2u_i^T E_i^T P_i L_i d_i\\
&\quad + x_i^T(-\tau)(-P_{\tau i} - F_i^T Q_{\tau i} F_i) x_i(-\tau) - 2x_i^T(-\tau) F_i^T S_{\tau i} u_i,\\
&\quad + d_i^T (L_i^T P_i L_i - R_i^d) d_i \le 0
\end{aligned}$$

This is equivalent to

$$\begin{aligned}
&\sigma^t \Pi_i \sigma \le 0\\
&\beta_i = [x_i(-\tau) \quad \breve{u}_{ri} v_i \quad v_i \quad d_i]^t
\end{aligned}$$

where Π_i is given by (2.10). This completes the proof. □

Remark 2.1: *It is expected that the delay-independent result in Theorem (2.1) is conservative compared to the result in Theorem (2.2), reasoned by the fact that two vectors $x_i(k)$ and $x_i(k-\tau)$ from the same state vector of $\mathcal{S}_i$ appear in the stacking vector of the delay-independent inequality (2.9) that is fulfilled by (2.8).*

2.2.3 Quadratic dissipativity

The interconnection stability condition for the networked system proposed in this chapter is based on the quadratic dissipativity of the controller trajectory. Here, the real-time controller $\mathcal{C}_i$ is considered as an uncertain system connected with subsystem $\mathcal{S}_i^s$ (2.4) in the stand-alone mode or with subsystem $\mathcal{S}_i$ in the distributed control scheme. Defining a quadratic function ξ^{ic} with respect to the output and input pair of $((u_i, d_i), y_i)$ by

$$\xi^{ic}((u_i, d_i), y_i) = u_i^T Q_{ic} u_i + 2u_i^T S_{ic}^T y_i + d_i^T Q_{ic}^d d_i \tag{2.11}$$

the trajectory of controller C_i is then said to be quadratically dissipative with respect to the supply rate ξ^{ic} (2.11), if the accumulation from time zero of $\xi^{ic}((u_i, d_i), y_i)$ over time, denoted as $\xi^{ic}(.)$, is nonnegative [152], that is

$$\begin{aligned} &\xi^{ic}(k) \geq 0 \ \forall k \geq 0, \\ &\xi^{ic}(k) = \sum_{k=0}^{k} \xi^{ic}((u_i(k), d_i(k)), y_i(k)) \end{aligned} \tag{2.12}$$

The matrices $Q_{ic}, Q_{ic}^d, S_{ic}, R_{ic}$ are called controller dissipativity parameter matrices. They will be determined in alliance with the dissipativity matrices of the corresponding open-loop subsystem from the LMIs derived in the stability conditions for interconnected systems.

2.2.4 Interconnection stability condition

To establish the stability condition for distributed control problems, we consider a block diagonal controller C, which can be implemented by h decoupled local controllers C_i. By defining the dissipativity matrices for the interconnected system Σ and the global diagonal control C as

$$\begin{aligned} Q &= \text{diag}[Q_i]_1^{ns}, \quad S = \text{diag}[S_i]_1^{ns}, \quad R = [R_i]_1^{ns}, \quad R^d = \text{diag}[R_i^d]_1^{ns} \\ Q_{\tau i} &= [Q_{\tau i}]_1^{ns}, \quad S_{\tau i} = [S_{\tau i}]_1^{ns}, \quad R_{\tau i} = [R_{\tau i}]_1^{ns} \\ R_c &= [R_{ic}]_1^{ns}, \quad S_c = [S_{ic}]_1^{ns}, \quad Q_c = [Q_{ic}]_1^{ns}, \quad Q_c^d = [Q_{ic}^d]_1^{ns} \end{aligned}$$

The interconnection stability condition can be stated in the next theorems. The LMI-based stability of the closed-loop large-scale system is established from the delay-robust dissipativity of all open-loop subsystems and their controller trajectories, as stated in Theorems 2.3 and 2.4. Theorem 2.3 is a delay-independent condition while Theorem 2.4 is a delay-dependent condition.

Theorem 2.3: *Consider h subsystems S_i (2.1) interconnected with each other by the coupling delay process H, and h nonnegative storage function of the Lyapunov–Krasovskii functional form (2.7). The closed-loop subsystem associated with controller C_i is stable if the controller trajectories are quadratically dissipative, as defined in (2.12), and the following LMIs are feasible in $P_i, P_{\tau i}, Q_i, S_i, R_i, R_i^d$, $Q_{\tau i}, S_{\tau i}, R_{\tau i}, Q_{ic}, Q_{ic}^d, S_{ic}, R_{ic}$:*

$$\begin{aligned} &F^T Q_\tau F + F^T S_\tau^T HF + F^T H^T S_\tau F + D' H^T R_\tau HF < 0 \\ &\begin{bmatrix} Q_i + R_{ic} & S_i S_{ic} \\ \bullet & R_i + Q_{ic} \end{bmatrix} < 0, \qquad R_i^d + Q_{ic}^d < 0 \end{aligned} \tag{2.13}$$

and (2.8) for $i = 1, \ldots, h$.

Proof. From (2.12), the dissipativity condition for the large-scale diagonal controller $\mathcal{C}$ is as follows:

$$\xi_{\Sigma}^{c}(\kappa) \geq 0 \quad \forall\, k \geq 0$$

where

$$\begin{aligned}\xi_{\Sigma}^{c}(\kappa) &= \sum_{k=0}^{\kappa} \xi^{c}((u_i(k), d_i(k)), y_i(k)),\\ \xi^{c}((u(k), d(k)), y(k)) &= y^T(k)R_c y(k) + 2y^T(k)S_c u(k)\\ &\quad + u^T(k)Q_c u(k) + d^T(k)Q_c^d d(k)\end{aligned}$$

□

Consider the large-scale vectors x, u, d, y, v and w integrated with the delayed interconnection process H in a quadratic function for the supply rate of the form

$$\begin{aligned}\xi((w(k), y(k)), (v(k), u(k), d(k))) &= y^T(k)Qy(k) + 2y^T(k)Su(k)\\ &\quad + u^T(k)R_u(k) + d^T(k)R^d d(k)\\ &\quad + w^T(-\tau)(k)Q_\tau w(-\tau)(k)\\ &\quad + 2w^T(-\tau)(k)S_\tau Hw(-\tau)(k)\\ &\quad + w^T(-\tau)(k)h^T R_\tau H_w(-\tau)(k)\end{aligned}$$

The sum of this function and the quadratic supply rate of $\mathcal{C}$ is given in the following equation:

$$\sigma(k) = \xi((w(k), y(k)), (v(k), u(k), d(k))) + \xi^{c}((u(k), d(k)), y(k)) \tag{2.14}$$

Omitting the time index k for conciseness, we have

$$\begin{aligned}\sigma &= y^T R_c y + y^T S_c u + u^T Q_c u + d^T Q_c^d d + y^T Qy + 2y^T Su\\ &\quad + u^T Ru + d^T R^d d + w^T(-\tau)Q_\tau w(-\tau) + 2w^T(-\tau)S_\tau H_w(-\tau)\\ &\quad + w^T(-\tau)H^T R_\tau H_w(-\tau)\\ &= x^T C^T R_c Cx + 2x^T C^T S_c u + u^T Q_c u + d^T Q_c^d d + x^T C^T QCx\\ &\quad + 2x^T C^T Su + u^T Ru + d^T R^d d\\ &\quad + x(-\tau)F^T Q_\tau w(-\tau) + 2x^T(-\tau)F^T S_\tau HC_x(-\tau)\\ &\quad + x^T(-\tau)F^T H^T R_\tau HCx(-\tau)\\ &= x^T C^T (R_c + Q)Cx + 2x^T C^T (S_c + S)u + u^T (Q_c + R)u\\ &\quad + d^T (Q_c^d + R^d)d + x^T(-\tau)F^T Q_\tau w(-\tau) + 2x^T(-\tau)F^T S_\tau HCx(-\tau)\\ &\quad + x^T(-\tau)F^T H^T R_\tau HCx(-\tau)\\ &= \begin{bmatrix} Cx \\ u \end{bmatrix}^T \begin{bmatrix} Q_i + R_{ic} & S_i + S_{ic} \\ \bullet & R_i + Q_{ic} \end{bmatrix} \begin{bmatrix} Cx \\ u \end{bmatrix} + d^T (Q_c^d + R^d)d\\ &\quad + x^T(-\tau)(F^T Q_\tau F + 2F^T S_\tau HF + F^T H^T R_\tau HF)x(-\tau)\end{aligned}$$

Due to (2.13), we have $\sigma(k) < 0$ for all $[x(k-\tau)Cx(k)u(k)d(k)]^T \neq 0$.

From the definitions of dissipation inequalities (2.6) and (2.12), by introducing

$$V_\Sigma(k) = x^T(k)Px(k) + \sum_{j=k-\tau}^{k-1} x^T(j)P_\tau x(j),$$
$$P = \mathrm{diag}[P_i]_1^{n_s}, \quad P_\tau = \mathrm{diag}[P_{\tau i}]_1^{n_s}$$

with $P_i > 0$ and $P_{\tau i} > 0$ as defined in (2.7), the following inequality can be obtained:

$$[V_\Sigma(k+1) + \xi_\Sigma(k)] - [V_\Sigma(k) + \xi_\Sigma(k-1)] \leq \sigma(k)$$

Therefore

$$[V_\Sigma(k+1) + \xi_\Sigma(k)] - [V_\Sigma(k) + \xi_\Sigma(k-1)] < 0$$

for all $[x(k) \quad u(k) \quad d(k)]^T \neq 0$ and equals to zero when $x(k) = 0$, $u(k) = 0$ and $d(k) = 0$.

With the assumption on the zero state observability, and $\xi_\Sigma(k) \geq 0 \; \forall \; k \in \mathbb{Z}^+$, $V_\Sigma(k) + \xi_\Sigma(k-1)$ can be chosen as a Lyapunov functional form for the closed-loop large-scale system. From the Lyapunov–Krasovskii condition for stability and the Krasovskii–La Salle principle, the closed-loop large-scale system is globally asymptotically stable.

Based on the proof of Theorem 2.3, the delay-dependent stability condition is derived in Theorem 2.4.

Theorem 2.4: *Consider h subsystems $S_i, i = 1, \ldots, h$, where S_i as in (2.1), interconnected with each other by the coupling delay process H, and h nonnegative storage function of the Lyapunov–Krasovskii functional form (2.7). The closed-loop subsystem associated with controller C_i is stable if the controller trajectories are quadratically dissipative as defined in (2.12) and the following LMIs are feasible in $P_i, P_{\tau i}, Q_i, S_i, R_i, R_i^d, Q_{\tau i}, S_{\tau i}, R_{\tau i}, Q_{ic}, Q_{ic}^d, S_{ic}, R_{ic}$:*

$$\Upsilon_i = \begin{bmatrix} A^{\tau^T} C^T(R_c + Q)CA^\tau + \mathcal{H} & A^{\tau^T} C^T(R_c + Q) & A^{\tau^T} C^T(S_c + S) \\ \bullet & R_c + Q & S_c + S \\ \bullet & \bullet & Q_c + R \end{bmatrix} \tag{2.15}$$
$$< 0$$
$$R^d + Q_c^d < 0$$

for $\tau = 1, \ldots, \tau_{\max}$ and (2.10) for $i = 1, \ldots, h$, where

$$\mathcal{H} = F^T Q_\tau F + F^T S_\tau^T HF + F^T H^T S_\tau F + F^T H^T R_\tau HF$$

Proof. From the definition of $\sigma(k)$ in (2.14) in the proof of Theorem 2.3 and its expansion, by writing the state vector as $x(k) = x^T(k-\tau)A^{\tau^T} + \breve{u}^T X^T$, where $\breve{u}$ is a vector sequence formed by stacking the historical state and control vectors, we then obtain

$$\begin{aligned}\sigma =\ & [x^T(-\tau)A^{\tau^T} + \breve{u}^T X^T]C^T(R_c + Q)Cx[A^\tau x(-\tau) + X\breve{u}] \\ & + 2[x^T(-\tau)A^{\tau^T} + \breve{u}^T X^T]C^T(S_c + S)u \\ & + u^T(Q_c + R)u + d^T(Q_c^d + R^d)d \\ & + x^T(-\tau)(F^T Q_\tau F + 2F^T S_\tau HF + F^T H^T R_\tau HF)x(-\tau)\end{aligned}$$

This is equivalent to

$$\omega^t \hat{\Upsilon}_i \omega < 0,$$

$$\omega = \begin{bmatrix} x(-\tau) \\ CX\breve{u} \\ u \\ d \end{bmatrix}, \quad \hat{\Upsilon}_i = \begin{bmatrix} \Upsilon_i & 0 \\ \bullet & R^d + Q_c^d \end{bmatrix}$$

Due to (2.15), we have $\sigma(k) < 0$ for all $[x(k-\tau) \quad CX\breve{u}(k) \quad u(k) \quad d(k)]^T \neq 0$. Similar to Theorem 2.3, the resultant stability follows. □

2.2.5 Semidefinite programming

In an MPC problem, the two main elements of interest are the predictive control sequence and initial state vector. The predictive control sequence $\{u_i(k), u_i(k+1), \ldots, u_i(k+N)\}$ is denoted by $\hat{u}_i(k, x_i^k, N)$ or simply $\hat{u}_i$, where N is the predictive horizon, $x_i(k)$ is an estimated state vector, and k is the current time step in the discrete-time domain. The predictive state sequence, denoted as

$$\hat{x}_i(k, \hat{u}_i, N) = \left\{x_i^{\hat{u}_i}(k+1, x_i(k)), \ldots, x_i^{\hat{u}_i}(k+N, x_i(k))\right\}$$

is obtained as a result of applying the predictive control sequence $\hat{u}$ to the open-loop subsystem $\mathcal{S}_i^s$. The control objective is to steer the state to the origin. For an initial state vector $x_i(k)$, the cost function is generally defined by

$$\mathcal{J}^i(\hat{u}_i(k, x_i^k, N)) = F_i(x_i(k+N)) + \sum_{j=k}^{k+N} \ell_i(x_i(j), u_i(j))$$

The terminal time $k+N$, which could vary with time k but should not be decreasing, is often referred to as a predictive horizon. $\ell_i(x_i(j), u_i(j))$ is the objective function. $F_i(x_i(k+N))$ is the terminal cost function which is used to emulate the infinite-horizon optimizing performance for the above finite-horizon problem. At every time step k, the open-loop optimization problem of minimizing $\mathcal{J}^i(\hat{u}_i(k, x_i^k, N))$ subject to $u_i(j) \in \mathbb{U}_i$ and $x_i(j) \in \mathcal{X}_i$ is solved to yield the minimizer of the form

$$\hat{u}_i^0(k, x_i^k, N) = \left\{u_i^0(k, xi(k)), u_i^0(k+1, xi(k)), \ldots, u_i^0(k+N, x_i(k))\right\}$$

The first control move $u_i^0(k, x_i(k))$ is then applied to control the plant. The whole process is repeated in a rolling or receding manner in the next iteration. Since N is finite, a global minimum of $\mathcal{J}^i(\hat{u}_i(k, x_i^k, N))$ exists if $\ell_i(x_i, u_i)$ are continuously differentiable, $\mathbb{U}_i$ is bounded, $\mathbb{U}_i$ is closed and bounded. A comprehensive literature review for MPC methods is given in [117].

Here, the MPC problem is formulated in semidefinite programming in association with a new stability constraint to be developed. Stability constraints are artificial constraints adding to the optimization problem of MPC solely for stability assurance purposes [153].

2.2.6 Objective function

For each stand-alone subsystem or subsystem in a distributed control scheme, a quadratic objective function of the states and control inputs is considered in association with the adequately chosen weighting matrices Q^x, Q^u as follows:

$$\mathcal{J}^i(\hat{u}_i(k,x_i^k,N)) = \sum_{j=1}^{N+1} x_i^T(k+j)Q_i^x x_i(k+j) + \sum_{j=0}^{N} u_i^T(k+j)Q_i^u u_i(k+j) \tag{2.16}$$

which is subject to the equality constraints of the state space model (2.1) and the inequality constraints deduced from physical limits of x_i and u_i. The problem is formulated by predicting the state vector x_i in N time steps using the current estimate of state vectors $x_i(k)$. The predictive state vector is obtained from the perfect model without the input disturbance vector d_i as follows:

$$\hat{x}_i(k,\hat{u}_i,N) = \Theta_i x_i(k) + \Gamma_i \hat{u}_i(k,x_i^k,N)$$

where

$$\begin{aligned}\hat{u}_i(k,x_i^k,N) &= [u_i(k)u_i(k+1)\dots u_i(k+N)]^T\\ \hat{x}_i(k,\hat{u}_i^k,N) &= [x_i(k+1)x_i(k+2)\dots x_i(k+N+1)]^T\end{aligned}$$

$$\Gamma_i = \begin{bmatrix} B_i & 0 & \dots & 0\\ A_iB_i & B_i & \dots & 0\\ \dots & \dots & \dots & \dots\\ A_i^N B_i & \dots & A_iB_i & B_i\end{bmatrix} \quad \text{and} \quad \Theta_i = \begin{bmatrix} A_i\\ A_i^2\\ \vdots\\ A_i^{N+1}\end{bmatrix}$$

By denoting $x_i(k,\hat{u}_i,N)$ as $\hat{x}_i(k)$ or just $\hat{x}_i$, and $u_i(k,x_i^k,N)$ as $\hat{u}_i(k)$ or just $\hat{u}_i$, we have

$$\hat{x}_i = r_i^k + \Gamma_i\hat{u}_i \tag{2.17}$$

where $r_i^k = \Theta_i x_i(k)$. The objective function is now rewritten as

$$\mathcal{J}^i(\hat{u}_i(k,x_i^k,N)) = [r_i^k + \Gamma_i\hat{u}_i]^T Q_{iN}^x [r_i^k + \Gamma_i\hat{u}_i] + \hat{u}_i^T Q_{iN}^u \hat{u}_i$$

where $Q_{iN}^x = \text{diag}[Q_i^x]_1^N$, $Q_{iN}^u = \text{diag}[Q_i^u]_1^N$, which can be cast in the form of

$$\mathcal{J}^i(\hat{u}_i(k,x_i^k,N)) = \hat{u}_i^T \Phi_i \hat{u}_i + 2\Upsilon_i^k \hat{u}_i + \delta_i^k \tag{2.18}$$

where

$$\Phi_i = \Gamma_i^T Q_{iN}^x \Gamma_i + Q_{iN}^u, \quad \Upsilon_i^k = r_i^{k^T} Q_{iN}^x \Gamma_i, \quad \delta_i^k = r_i^{k^T} Q_{iN}^x r_i^k$$

2.2.7 Constraint on decision variables

Let us consider a bounded constraint on decision variables of the following form:

$$\|u_i\|_2^2 \leq \eta_i \tag{2.19}$$

where η_i is the upper bound deduced from the physical limits of the control input vector u_i. Notably, the output constraints can also be incorporated into the optimization problem by converting them into constraints on vector u_i using the state predictive equation (2.17).

2.2.8 Semidefinite program

The problem of minimizing $\mathcal{J}^i(\hat{u}_i(k, x_i^k, N))$ subject to the required constraints can be written in semidefinite program (SDP) by applying the Schur complement to (2.18) and (2.19) as

$$\begin{gathered} \min_{\hat{u}_i} \gamma_i(k) \quad \text{subject to} \quad \gamma_i(k) > 0 \\ \begin{bmatrix} -\gamma_i(k) + \delta_i^k + 2\Upsilon_i^k \hat{u}_i & \hat{u}_i^T \\ \hat{u}_i & -\Phi_i^{-1} \end{bmatrix} < 0, \quad \begin{bmatrix} \eta_{\Sigma i} & \hat{u}_i^T \\ \hat{u}_i & I_{m_i \times N} \end{bmatrix} > 0 \end{gathered} \tag{2.20}$$

The parameters Φ_i, Υ_i^0, and δ_i^0 are found from the state space realization matrices A_i, B_i, the weighting matrices Q_{iN}^x, Q_{iN}^u of the objective function and the initial state vector $x_i(0)$. $\eta_{\Sigma i}$ is the bound of the control sequence $\hat{u}_i$ deduced from (2.19). Υ_i^k and δ_i^k are subsequently updated at every step using the updated state vector $x_i(k)$.

For each subsystem, the optimization problem (2.20) is solved by the local controller for the minimizing vector sequence $\hat{u}_i$ which comprises N elements of u_i. Only the first element $u_i(k)$ of the sequence $\hat{u}_i$ is applied to control the plant. This rolling process is repeated at the next time step, and continues thereon. Solving this problem does not, however, guarantee the stability of the closed-loop system according to the MPC literature [117]. In what follows, a dynamic stability constraint to tackle this issue will be developed.

2.2.9 Stability constraint

Among various MPC formulations, the stability-enforced MPC achieves the closed-loop stabilization by imposing an additional constraint to the optimization problem. In this context, a robust control Lyapunov function with Sontag's formula has been used in two inequality constraints [153] to guarantee the stability of the closed-loop system and approximate the optimal performance of the unconstrained predictive controller, regardless of the choice of the horizon lengths.

In this section, we proceed to add a point-wise constraint on the control vectors to the optimization problem of MPC. This constraint, being of a quadratic type, is based on the dissipativity inequality of controller trajectories subject to the satisfaction of the interconnection stability condition.

By rewriting (2.12) as

$$u_i^T Q_{ic} u_i + 2u_i^T S_{ic}^T y_i + y_i^T R_{ic} y_i + d_i^T Q_{ic}^d d_i + \xi_S^{ic}(k-1) \geq 0 \tag{2.21}$$

where $Q_{ic} < 0$, $Q_{ic}^d > 0$, $\|d_i\|_2 \leq 1$, $\xi_S^{ic}(k-1)$ is the historical data, y_i is the current measurement output vector, the dissipativity inequality (2.12) can be developed into a one-time-step quadratic constraint on the initial control vector $u_i^0(k, x_i(k))$ for the optimization problem of MPC. The convexity of the inequality constraint will be ensured if $Q_{ic} < 0$. The stability constraint on control vectors u_i that is not depend on the input disturbances can be derived when $Q_{ic}^d < 0$. Both $Q_{ic} < 0$ and $Q_{ic}^d < 0$ can be obtained from the LMIs of stability conditions in Theorem 2.3 or Theorem 2.4.

The one-time-step constraint may, however, lead to a conservative control performance. In Section 2.2.10, the state prediction in N steps will be calculated for the proposed receding horizon stability constraint, which is a structural constraint on the whole predictive control sequence $\hat{u}_i$. By employing the state predictions toward the end of the predictive horizon N, the controller trajectory is not required to satisfy a dissipative condition at the current time step. It will, therefore, create a much less conservative constraint than the one-time-step constraint since the controller trajectories would not normally reach a dissipative mode during transient periods. This is explained by the fact that the nonnegative condition on $\xi_S^{ic}(k)$ is much less conservative than the nonnegative condition of each of $\xi^{ic}(j)$.

2.2.10 Receding-horizon stability constraint

Consider the accumulative supply rate for the controller up until the predictive horizon N, $\xi_S^{ic}(k+N)$. The accumulation $\xi_S^{ic}(k+N)$ is split into two components encapsulating the output and input evolutions in the past and in the future as follows:

$$\begin{aligned} \xi_\Sigma^{ic}(k+N) &= \xi_\Sigma^{ic}(k-1) + \xi_\Sigma^{ic}(k+N \mid K) \\ \xi_\Sigma^{ic}(k+N \mid k) &= \sum_{j=k}^{k+N} \xi_\Sigma^{ic}(u_i(j), d_i(j), y_i(j)) \end{aligned}$$

From the state predictive equation (2.17), by denoting the disturbance sequence in N time steps as $d_i(k, x_i(k), N) = \hat{d}_i$, we have

$$\begin{aligned} \xi_\Sigma^{ic}(k+N \mid k) &= \xi_\Sigma^{ic}(k-1) + [r_i^k + \Gamma_i u_i]^T R_{icN} [r_i^k + \Gamma_i u_i] \\ &\quad + 2[r_i^k + \Gamma_i u_i]^T S_{icN} \hat{u}_i + \hat{u}_i^T Q_{icN} \hat{u}_i + \hat{d}_i^T Q_{icN}^d \hat{d}_i \end{aligned}$$

where

$$\begin{aligned} R_{icN} &= \mathrm{diag}[R_{ic}]_1^N, \quad S_{icN} = \mathrm{diag}[S_{ic}]_1^N, \quad Q_{icN} = \mathrm{diag}[Q_{ic}]_1^N, \\ Q_{icN}^d &= \mathrm{diag}[Q_{ic}^d]_1^N \end{aligned}$$

Then $\xi_\Sigma^{ic}(k+N) \geq 0$ is equivalent to

$$\hat{u}_i^T \Psi_{ic} \hat{u}_i + 2\Upsilon_{ic}^k \hat{u}_i + \delta_{ic}^k \geq 0 \tag{2.22}$$

where

$$\begin{aligned}\Psi_{ic} &= \Gamma_i^T R_{icN}\Gamma_i + 2\Gamma_i^T S_{icN} + Q_{icN} \\ \Upsilon_{ic}^k &= r_{ic}^{k^T}(R_{icN}\Gamma_i + S_{icN}) \\ \delta_{ic}^k &= r_{ic}^{k^T} R_{icN} r_{ic}^k + \xi_\Sigma^{ic}(k-1) - N\|Q_i c^d\| \\ \|d_i(k)\|_2 &\leq 1, \quad Q_{ic}^d < 0\end{aligned}$$

Since the stability constraint is updated at every time step using the state predictions toward the end of the horizon in the same rolling manner of receding horizon control, it is called receding horizon stability constraint herein. The parameters of stability constraints (Ψ_{ic}, Υ_{ic}^k, δ_{ic}^k) for DMPCs will be determined from the dissipativity matrices that meet the stability condition for the interconnected system as shown in the sequel. In this Chapter, only convex problems are considered, therefore, the condition of $\Psi_{ic} < 0$ must also be included.

2.2.11 Dissipativity-based algorithm

By applying the dissipativity parameter matrices that are found from the feasible solutions to the LMIs given in Theorem 2.3 or Theorem 2.4 to the stability constraint (2.22), and imposing the resulting stability constraint to the SDP (2.20) for each subsystem, we are now able to formulate the DMPC problem as follows:

$$\min_{\hat{u}_i} \gamma_i(k), \quad \gamma_i(k) > 0 \tag{2.23}$$

where

$$\begin{bmatrix} -\eta_{\Sigma^i} & \hat{u}_i^T \\ \hat{u}_i & I_{m_i \times N} \end{bmatrix} > 0, \quad \begin{bmatrix} -\delta_{ic}^k - 2\Upsilon_{ic}^k \hat{u}_i & \hat{u}_i^T \\ \hat{u}_i & \Psi_{ic}^{-1} \end{bmatrix} < 0,$$

$$\begin{bmatrix} -\gamma_i(k) + \delta_i^k + 2\Upsilon_i^k \hat{u}_i & \hat{u}_i^T \\ \hat{u}_i & -\phi_i^{-1} \end{bmatrix} < 0$$

where δ_i^k, Υ_i^k, Ψ_i are defined in the predictive objective function (2.18), δ_{ic}^k, Υ_{ic}^k, Ψ_{ic} are defined in the receding-horizon stability constraint (2.22), η_{Σ^i} is the control bound deduced from (2.19), Θ_i and Γ_{ic} are defined in the state predictive equation (2.17), and Q_{icN}, R_{icN}, S_{icN} are diagonal matrices formed by respective dissipativity parameter matrices Q_{ic}, R_{ic}, S_{ic} of the controller trajectory.

2.2.12 Robust feasibility

The feasibility set of the stability constraint (2.22) is an ellipsoid w.r.t. $u_i(k)$ due to the consideration of convex problems in this chapter, that is $\Psi_{ic} < 0$. The optimization problem (2.23) is feasible if and only if the intersections of the two feasibility set of (2.19) and (2.22) at every time step is nonempty. The recursive feasibility condition is thus stated in the next theorem.

Theorem 2.5: *Suppose that the state vector xi(k) at the current step k and the quadratic dissipativity matrices R_{ic}, S_{ic}, and Q_{ic} (2.11) of S_i are known. If $\hat{u}_i(k)$ is a feasible solution to the following LMIs:*

$$\begin{aligned}
&\begin{bmatrix} \tilde{\delta}_i^k + 2\tilde{\Upsilon}_i^k M_i \hat{u}_i(k) & \hat{u}_i^T(k) M_i^T B_i^T \\ \bullet & -\tilde{\Psi}_{ic}^{-1} \end{bmatrix} < 0, \\
&\Psi_{ic} < 0 \\
&\begin{bmatrix} \breve{\delta}_i^k + 2\breve{\Upsilon}_i^k M_i \hat{u}_i(k) & \hat{u}_i^T(k) M_i^T B_i^T \\ \bullet & -(Q_{ic} + \Delta_{ic})^{-1} \end{bmatrix} > 0, \\
&Q_{ic} + \Delta_{ic} < 0
\end{aligned} \tag{2.24}$$

where

$$\begin{aligned}
M_i &= [I_{m_i \times N}, 0 \ldots 0], \\
\tilde{\Psi} &= \Theta_i^T (R_{icN}\Gamma_i + S_{icN})^T (-\Psi_{ic}^{-1})(R_{icN}\Gamma_i + S_{icN}), \\
\tilde{\delta}_i^k &= x_i^T(k) A_i^T \tilde{\Psi}_i A_i x_i(k) - \eta_{\Sigma^i}, \quad \tilde{\Upsilon}_i^k = x_i^T(k) A_i^T \tilde{\Psi}_i B_i, \\
\Delta_{ic} &= \Theta_i^T (R_{icN} + (R_{icN}\Gamma_i + S_{icN})^T (R_{icN}\boldsymbol{G}_i + S_{icN}))\boldsymbol{Q}_i, \\
\breve{\delta}_i^k &= \xi_\Sigma^{ic}(k-1) - N\|Q_i c^d\| + x_i^T(k)(C_i^T R_{ic} C_i + A_i^T \Delta_{ic} A_i) x_i(k), \\
\breve{\Upsilon}_i^k &= x_i^T(k)(A_i^T \Delta_{ic} B_i + C_i^T S_{ic}).
\end{aligned}$$

in which Θ_i and Γ_i are defined in the state predictive equation (2.17), Ψ_{ic} is defined in (2.22), and R_{ic}, S_{ic}, and Q_{ic} are the quadratic dissipativity matrices of S_i trajectory defined in (2.11), then the optimization problem (2.23) is feasible at step $(k+1)$.

Proof. By rewriting (2.22) (with $\Psi_{ic} < 0$) for step $(k+1)$ as

$$(\hat{u}_i^T - \Upsilon_{ic}^{k+1} \Psi_{ic}^{-\frac{1}{2}})(-\Psi_{ic})(\hat{u}_i - \Psi_{ic}^{-\frac{1}{2}} \Upsilon_{ic}) - \delta_{ic}^{k+1} - \Upsilon_{ic}^{k+1} \Upsilon_{ic}^{k+1^T} \leq 0 \tag{2.25}$$

where

$$\begin{aligned}
\Psi_{ic} &= \Gamma_i^T R_{icN} \Gamma_i + 2\Gamma_i^T S_{icN} + Q_{icN} < 0 \\
\Upsilon_{ic}^{k+1} &= r_{ic}^{k+1^T} (R_{icN} \Gamma_{ic} + S_{icN}) \\
\delta_{ic}^{k+1} &= r_{ic}^{k+1^T} R_{icN} \tau_{ic}^{k+1} + \xi_\Sigma^{ic}(k-1) - N\|Q_{ic}^d\| \\
r_{ic}^{k+1} &= \Theta_i(A_i x_i(k) + B_i u_i(k))
\end{aligned}$$

it is apparent that the feasibility set of (2.25) is an ellipsoid region having the center at $\tilde{u}_i(k+1) = -\Psi_{ic}^{-\frac{1}{2}} \Upsilon_{ic}^{(k+1)}$ if and only if $\Upsilon_i < 0$ and

$$\delta_{ic}^{k+1} + \Upsilon_{ic}^{k+1} \Upsilon_{ic}^{k+1^T} \geq 0 \tag{2.26}$$

Then, if the center of the ellipsoid lies inside the constraint region of (2.19) the feasible region of (2.23) will be nonempty, that is

$$
\begin{aligned}
&\left(\Upsilon_{ic}^{k+1^T}\Psi_{ic}^{-\frac{1}{2}}\right)\left(\Psi_{ic}^{-\frac{1}{2}}\Upsilon_{ic}^{k+1}\right) \le \eta_{\Sigma^i} \\
&\quad \Leftrightarrow \Upsilon_{ic}^{k+1^T}(-\Psi_{ic}^{-1})\Upsilon_{ic}^{k+1} - \eta_{\Sigma^i} \le 0, \quad \Psi_{ic} < 0 \\
&\quad \Leftrightarrow (x_i^T(k)A_i^T + u_i^T(k)B_i^T)(\tilde{\Psi}_i)(A_ix_i(k) + B_iu_i(k)) - \eta_{\Sigma^i} \le 0
\end{aligned} \tag{2.27}
$$

where

$$
\tilde{\Psi}_i = \Theta_i^T(R_{icN}\Gamma_i + S_{icN})^T(-\Psi_i^{-1})(R_{icN}\Gamma_i + S_{icN})\Theta_i
$$

It is not difficult to see that (2.27) is equivalent to

$$
\begin{aligned}
&u_i^T(k)B_i^T\tilde{\Psi}_iB_iu_i(k) + 2x_i^T(k)A_i^T\tilde{\Psi}_iB_iu_i(k) \\
&\qquad + x_i^T(k)A_i^T\tilde{\Psi}_iA_ix_i(k) - \eta_{\Sigma^i} \le 0 \\
&\quad \Leftrightarrow x_i^T(k)A_i^T\tilde{\Psi}_iA_ix_i(k) - 2x_i^T(k)A_i^T\tilde{\Psi}_iB_iu_i(k) \\
&\qquad - u_i^T(k)B_i^T(-\tilde{\Psi}_i)B_iu_i(k) \le 0
\end{aligned}
$$

By Schur complement and $u_i(k) = M_i\hat{u}_i(k)$, (2.27) is equivalent to the first LMI of $\mathcal{F}(\hat{u}_i)$ in (2.24). Proceeding further, (2.26) is equivalent to

$$
\begin{aligned}
&\delta_{ic}^{k+1} + \Upsilon_{ic}^{k+1}\Upsilon_{ic}^{k+1^T} \ge 0 \\
&\quad \Leftrightarrow \xi_{\Sigma}^{ic}(k) - N\|Q_{ic}^d\| \\
&\qquad + r_{ic}^{k+1^T}\left(R_{icN} + (R_{icN}\Gamma_i + S_{icN})^T(R_{icN}\Gamma_i + S_{icN})\right)r_{ic}^{k+1} \ge 0 \\
&\quad \Leftrightarrow \xi_{\Sigma}^{ic}(k)(k-1) - N\|Q_{ic}^d\| \\
&\qquad + y_{ic}^T(k)R_{ic}y_{ic}(k) + 2y_{ic}^T(k)S_{ic}u_{ic}(k) + u_{ic}^T(k)Q_{ic}u_{ic}(k) \\
&\qquad + (u_i^TB_i^T + x_i^T(k)A_i^T)\Delta_{ic}(B_iu_i + A_ix_i(k)) \ge 0
\end{aligned}
$$

where

$$
\Delta_{ic} = \Theta_i^T((R_{icN}\Gamma_i + S_{icN})^T(R_{icN}\Gamma_i + S_{icN}))\Theta_i
$$

Finally, we have

$$
\begin{aligned}
&\Leftrightarrow \xi_{\Sigma}^{ic}(k) - N\|Q_{ic}^d\| + x_i^T(k)(C_i^TR_{ic}C_i + A_i^T\Delta_{ic}A_i)x_i(k) \\
&\quad + 2x_i^T(k)(C_i^TS_{ic} + A_i^T\Delta_{ic}B_i)u_i(k) - u_i^TB_i^T(-Q_{ic} - \Delta_{ic})B_iu_i \ge 0
\end{aligned}
$$

which is fulfilled by the second LMI of $\mathcal{F}(\hat{u}_i)$ in (2.24) by Schur complement. □

The LMIs in Theorem 2.5 will be used in association with other LMI constraints in the MPC optimization problem (2.23). With $\Psi_{ic} < 0$, $Q_{ic} + \Delta_{ic} \le 0$, and the current state vector $x_i(k)$ is known, the MPC optimization problem with guaranteed recursive feasibility is as follows:

$$
\begin{aligned}
&\min_{\hat{u}_i} \gamma_i(k), \quad \text{subject to } \gamma_i(k) > 0 \\
&\begin{bmatrix} -\eta_{\Sigma^i} & \hat{u}_i^T \\ \bullet & I_{m_i \times N} \end{bmatrix} > 0 \\
&\begin{bmatrix} -\delta_{ic}^k - 2\Upsilon_{ic}^k \hat{u}_i(k) & \hat{u}_i^T(k) \\ \bullet & \Psi_{ic}^{-1} \end{bmatrix} < 0 \\
&\begin{bmatrix} -\gamma_i(k) + \delta_i^k + 2\Upsilon_i^k \hat{u}_i & \hat{u}_i^T \\ \bullet & -\Phi_i^{-1} \end{bmatrix} < 0 \\
&\begin{bmatrix} \tilde{\delta}_i^k + 2\tilde{\Upsilon}_i^k M_i \hat{u}_i(k) & \hat{u}_i^T(k) M_i^T B_i^T \\ \bullet & -\tilde{\Psi}_{ic}^{-1} \end{bmatrix} < 0 \\
&\begin{bmatrix} \breve{\delta}_i^k + 2\breve{\Upsilon}_i^k M_i \hat{u}_i(k) & \hat{u}_i^T(k) M_i^T B_i^T \\ \bullet & -(Q_{ic} + \Delta_{ic})^{-1} \end{bmatrix} > 0
\end{aligned} \tag{2.28}
$$

where δ_i^k, Υ_i^k, Ψ_i are defined in the predictive objective function (2.18), δ_{ic}^k, Υ_{ic}^k, Ψ_{ic} are defined in the receding-horizon stability constraint (2.22), $\hat{\delta}_i^k$, $\hat{\Upsilon}_i^k$, Δ_{ic}, $\tilde{\delta}_i^k$, $\tilde{\Upsilon}_i^k$, $\tilde{\Psi}_i^k$ are defined in the recursive feasibility condition (2.24), η_{Σ^i} is the control bound deduced from (2.19), Θ_i and Γ_i are defined in the state predictive equation (2.17), and Q_{icN}, R_{icN}, S_{icN} are the diagonal matrices formed by respective dissipativity parameter matrices Q_{ic}, R_{ic}, S_{ic} of the controller trajectory.

2.2.13 Dissipativity matrices

The dissipativity parameter matrices in (2.28), including Q_{ic}, Q_{ic}^d, S_{ic}, R_{ic}, $i = 1, \ldots, h$, are determined in alliance with Q_i, S_i, R_i, R_i^d, $Q_{\tau i}$, $S_{\tau i}$, $R_{\tau i}$ from feasible solutions to (2.10) and (2.15) for the delay-dependent criterion, or alternatively (2.13) and (2.8) for the delay-independent criterion as stated in Theorems 2.3 and 2.4, together with $\Psi_{ic} < 0$ and $Q_{ic} + \Delta_{ic} \leq 0$ for recursive feasibility, as stated in the below corollary.

Corollary 2.1: *The dissipativity parameter matrices $Q_i, S_i, R_i, R_i^d, Q_{\tau i}, S_{\tau i}, R_{\tau i}, Q_{ic}$, Q_{ic}^d, S_{ic}, R_{ic} are found from feasible solutions to the following LMIs:*

1. *(2.10) and (2.15) (for delay-dependent criterion), or*
2. *(2.8) and (2.13) (for delay-independent criterion),*

and

$$
N(R_{ic}, S_{ic}, Q_{ic}) = \begin{cases} \Gamma_i^T R_{icN} \Gamma_i + 2\Gamma_i^T S_{icN} + Q_{icN} < 0, \\ \begin{bmatrix} Q_{ic} + \Theta_i^T R_{icN} \Theta_i & \Theta_i^T (R_{icN} \Gamma_i + S_{icN})^T \\ \bullet & -I \end{bmatrix} < 0 \end{cases} \tag{2.29}
$$

Proof. The first LMI in (2.29) is $\Psi_{ic} < 0$. The second LMI in (2.29) is equivalent to $Q_{ic} + \Delta_{ic} < 0$ due to Schur complement, that is

$$Q_{ic} + \Theta_i^T(R_{icN} + (R_{icN}\Gamma_i + S_{icN})^T(R_{icN}\Gamma_i + S_{icN}))\Theta_i < 0$$

Since $\mathcal{H} := F^TQ_\tau F + 2F^TS_\tau HF + F^TH^TR_\tau HF$ in (2.13) and (2.15) is applicable to the global large-scale system as a whole (i.e., a centralized matrix), the stability conditions have taken the coupling information into account. The proposed control strategy can, therefore, stabilize the large-scale networked systems in distributed architectures. □

2.2.14 Dissipativity-based DMPC algorithm

The distributed predictive control algorithm has both offline and online calculations involved as summarized in the below two-phase procedure:

Phase I: Parameter calculations

- Determine the dissipativity parameter matrices of Q_{ic}, Q_{ic}^d, S_{ic}, R_{ic} for all h local controllers $\mathcal{C}_i$ of h subsystems $\mathcal{S}_i$ using offline LMIs in Corollary 2.1.
- The offline parameters Γ_i, Θ_i, Φ_i, Ψ_{ic}, $\tilde{\Psi}_{ic}$, Δ_{ic} are found from the state space realization matrices A_i, B_i, the weighting matrices Q_{iN}^x and Q_{iN}^u of the objective function, and the dissipativity parameter matrices using (2.18), (2.22), and (2.24).

Phase II: At every synchronized step of the online controllers, formulate and solve the optimization problem (2.28) for each local controller using the predetermined parameters from the above offline calculations. Details are as follows:

1. *At step* $k = 1$, initiate parameters:

   - Initiate $\xi_\Sigma^{ic}(0)$, $x(1)$.
   - Υ_i^1 and δ_i^1 are found from Θ_i^1, Γ_i^1, the weighting matrices Q_{iN}^x, Q_{iN}^u of the objective function and the initial state vector $x_i(1)$ using (2.18).
   - Υ_{ic}^1 and δ_{ic}^1 are found from Θ_i^1, Γ_i^1, the dissipativity matrices Q_{ic}, S_{ic}, R_{ic}, Q_{ic}^d, the initial $\xi_\Sigma^{ic}(0)$, and the initial state vector $x_i(1)$ using (2.22).
   - $\hat{\Upsilon}_i^1$, $\hat{\delta}_i^1$, $\tilde{\Upsilon}_i^1$, and $\tilde{\delta}_i^1$ are found from Θ_i^1, Γ_i^1, the dissipativity matrices Q_{ic}, S_{ic}, R_{ic}, Q_{ic}^d, the initial $\xi_\Sigma^{ic}(0)$, and the initial state vector $x_i(1)$ using (2.24).

   Then, formulate and solve the optimization problem (2.28) with offline parameters Φ_i, Ψ_{ic}, $\tilde{\Psi}_{ic}$, Δ_{ic}, and the initial parameters Υ_i^1, δ_i^1, Υ_{ic}^1, δ_{ic}^1, $\hat{\Upsilon}_i^1$, $\hat{\delta}_i^1$, $\tilde{\Upsilon}_i^1$ and $\tilde{\delta}_i^1$ to yield the local control sequence $\hat{u}_i(1, x_i^1, N)$. The first vector element $u_i(1)$ is then applied to control the corresponding plant (subsystem $\mathcal{S}_i$).

2. *At step* $k \geq 2$, update online parameters with $x_i(k)$:

   - Calculate $\xi_\Sigma^{ic}(k-1)$.
   - Update Υ_i^k and δ_i^k using (2.18).
   - Update Υ_{ic}^k and δ_{ic}^k using (2.22).
   - Update $\hat{\Upsilon}_i^k$, $\hat{\delta}_i^k$, $\tilde{\Upsilon}_i^k$, and $\tilde{\delta}_i^k$ using (2.24).

Then, formulate and solve the optimization problem (2.28) with offline parameters Φ_i, Ψ_{ic}, $\tilde{\Psi}_{ic}$, Δ_{ic}, and the initial parameters Υ_i^k, δ_i^k, Υ_{ic}^k, δ_{ic}^k, $\hat{\Upsilon}_i^k$, $\hat{\delta}_i^k$, $\tilde{\Upsilon}_i^k$ and $\tilde{\delta}_i^k$ to yield the local control sequence $\hat{u}_i(k, x_i^k, N)$. The first vector element $u_i(k)$ is then applied to control the corresponding plant (subsystem S_i).

2.2.15 *Illustrative example 2.1*

In this section, simulation results for a networked system consisting of three coupled subsystems that suffer from an arbitrary coupling delay are included to demonstrate the effectiveness of the proposed partially DMPC strategy. Consider the following continuous time state space models of three subsystems with the random coupling delay occurring within seven time steps, that is, $\tau_{\max} = 7$ steps. The 0.2 second sampling time is applied to accommodate for synchronization between all three subsystems. The output constraints are ignored to emulate the system instability.

$$A_1 = \begin{bmatrix} -1 & 0.6 \\ 1 & -0.7 \end{bmatrix} B_1 = \begin{bmatrix} 0.8 \\ 0.1 \end{bmatrix} E_1 = \begin{bmatrix} 0.1 \\ 0 \end{bmatrix}, C_1 = \begin{bmatrix} 0.2 & -0.7 \\ 1.2 & 3.1 \end{bmatrix}$$

$$F_1 = \begin{bmatrix} 0.1 & 0.2 \end{bmatrix}$$

$$A_2 = \begin{bmatrix} -0.6 & 0 \\ 1 & -0.4 \end{bmatrix} B_2 = \begin{bmatrix} 1.2 \\ 0.2 \end{bmatrix} E_2 = \begin{bmatrix} 0.1 \\ 0 \end{bmatrix}, C_2 = \begin{bmatrix} 1.9 & 0.1 \\ -0.1 & -1.5 \end{bmatrix}$$

$$F_2 = \begin{bmatrix} 1 & 2 \end{bmatrix}, \quad H = \begin{bmatrix} 0 & 0 & 1 \\ 1 & 0 & 0 \\ 0 & 1 & 0 \end{bmatrix}$$

$$A_3 = \begin{bmatrix} -0.8 & 0 \\ 0.4 & -0.3 \end{bmatrix} B_3 = \begin{bmatrix} 1.5 \\ 0.1 \end{bmatrix} E_3 = \begin{bmatrix} 0.1 \\ 0 \end{bmatrix}, C_3 = \begin{bmatrix} -1.5 & -0.1 \\ 0.5 & 1.3 \end{bmatrix}$$

$$F_3 = \begin{bmatrix} 0.5 & 4 \end{bmatrix}$$

This networked system reflects typical structures of many washing circuits in the mineral processing plants, such as the alumina refinery in this example. The system consists of three washing filters interconnected to each other in a ring configuration. The controls are the condensate flow rates and slurry flow rates feeding into the filters in reverse directions. The measurement outputs are the levels of fine seed products and residue course seed products from the filters. The control and state weighting coefficients of the quadratic objective function, and constraints on the manipulation variables are as follows:

$$Q_1^u = 2, \quad Q_1^x = \begin{bmatrix} 1 & 0 \\ 0 & 1 \end{bmatrix}, \quad \eta_1 = 5$$

$$Q_2^u = 0.5, \quad Q_2^x = \begin{bmatrix} 0.1 & 0 \\ 0 & 0.1 \end{bmatrix}, \quad \eta_2 = 5$$

$$Q_3^u = 0.5, \quad Q_3^x = \begin{bmatrix} 0.1 & 0 \\ 0 & 0.1 \end{bmatrix}, \quad \eta_3 = 5$$

It is noted that, the weighting coefficients of the objective function are reserved for the users to choose. This is an advantageous feature of the present MPC strategy.

The simulation starts with the formulations of three model predictive controllers in SDP without any stability constraints (2.20). The predictive horizon of 16 time steps is implemented for all three subsystems, using the Matlab LMI toolbox. The output time responses are depicted by Figure 2.1, which clearly imply unstable subsystems.

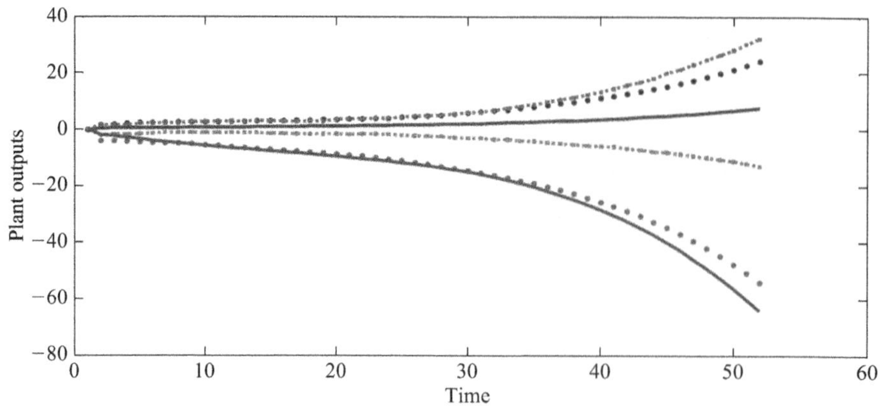

Figure 2.1 Distributed MPCs without stability constraints

In the next two simulations for delay-independent and delay-dependent approaches, the stability constraints are incorporated into the SDP (2.28) and the control algorithm of DMPC procedure is implemented. The delay-independent dissipativity parameter matrices are obtained from the feasible solutions to (2.8), (2.13), and (2.29), as detailed in the offline stage of DMPC procedure.

$$Q_1 = \begin{bmatrix} -1.1586 & 0.1363 \\ 0.1383 & -0.1896 \end{bmatrix}, \quad Q_2 = \begin{bmatrix} -0.1122 & -0.0695 \\ -0.0695 & -0.0526 \end{bmatrix},$$

$$Q_3 = \begin{bmatrix} -0.1828 & -0.0543 \\ -0.0543 & -0.0907 \end{bmatrix}$$

$$S_1 = \begin{bmatrix} 11.1081 \\ 1.5522 \end{bmatrix}, \quad S_2 = \begin{bmatrix} 0.2210 \\ 0.0238 \end{bmatrix}, \quad S_3 = \begin{bmatrix} -0.1860 \\ 0.1689 \end{bmatrix}$$

$$R_1 = [1.8854], \quad R_2 = [1.3104], \quad R_3 = [1.3808]$$

$$R_{1c} = \begin{bmatrix} 1.3787 & 0.1646 \\ 0.1646 & 0.2256 \end{bmatrix}, \quad R_{2c} = \begin{bmatrix} 0.1376 & 0.0912 \\ 0.0912 & 0.0645 \end{bmatrix},$$

$$R_{3c} = \begin{bmatrix} 0.1908 & 0.0657 \\ 0.0657 & 0.0797 \end{bmatrix},$$

$$S_{1c} = \begin{bmatrix} -10.9081 \\ -1.7342 \end{bmatrix}, \quad S_{2c} = \begin{bmatrix} -0.2650 \\ 0.171 \end{bmatrix}, \quad S_{3c} = \begin{bmatrix} 0.4160 \\ -0.0587 \end{bmatrix}$$

$$Q_{1c} = [-2.2437], \quad Q_{2c} = [-1.5189], \quad Q_{3c} = [-1.3808]$$

The initial control and state vectors of three subsystems are as follows:

$$u_1 = \begin{bmatrix}-1\end{bmatrix} x_1 = \begin{bmatrix}-1\\-1\end{bmatrix}, \quad u_2 = \begin{bmatrix}-0.1\end{bmatrix} x_2 = \begin{bmatrix}-1\\-1\end{bmatrix},$$

$$u_3 = \begin{bmatrix}-1\end{bmatrix} x_3 = \begin{bmatrix}-1\\-1\end{bmatrix}$$

The initial values of $\zeta_{ic} = +5, i = 1, 2, 3$ for all simulations. When the predictive horizon of 25 time steps is implemented for all subsystems, the output time responses from DMPC procedure are depicted by Figure 2.2.

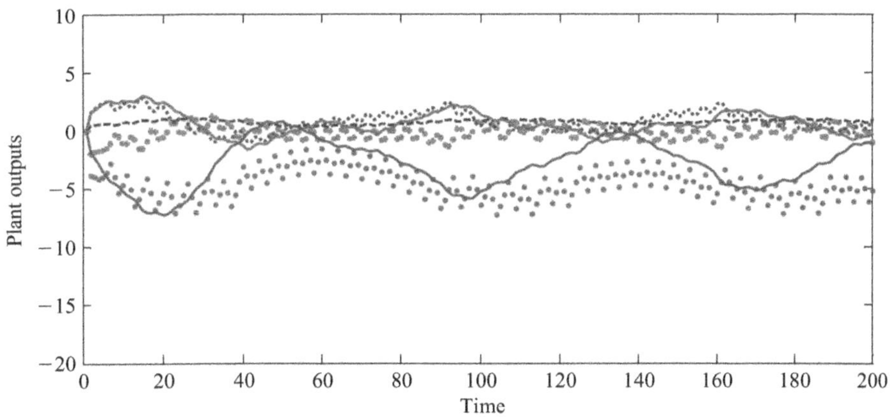

Figure 2.2 Distributed MPCs: delay-independent criteria

The delay-dependent dissipativity parameter matrices obtained from the feasible solutions to (2.10), (2.15), and (2.29), as detailed in the offline stage of DMPC procedure, are given below:

$$Q_1 = \begin{bmatrix}-1.2204 & 0.2117\\ 0.2117 & -0.2284\end{bmatrix}, \quad Q_2 = \begin{bmatrix}-1.1118 & -0.0737\\ -0.0737 & -0.0523\end{bmatrix},$$

$$Q_3 = \begin{bmatrix}-0.1927 & -0.0662\\ -0.0662 & -0.0805\end{bmatrix},$$

$$S_1 = \begin{bmatrix}0.3625\\ 0.0871\end{bmatrix}, \quad S_2 = \begin{bmatrix}0.2802\\ 0.0275\end{bmatrix}, \quad S_3 = \begin{bmatrix}-0.3954\\ 0.0695\end{bmatrix},$$

$$R_1 = \begin{bmatrix}1.1969\end{bmatrix}, \quad R_2 = \begin{bmatrix}1.3017\end{bmatrix}, \quad R_3 = \begin{bmatrix}1.3721\end{bmatrix}$$

$$R_{1c} = \begin{bmatrix}1.0984 & 0.1905\\ 0.1905 & 0.2056\end{bmatrix}, \quad R_{2c} = \begin{bmatrix}0.1018 & 0.0670\\ 0.0670 & 0.0476\end{bmatrix},$$

$$R_{3c} = \begin{bmatrix}0.1965 & 0.0675\\ 0.0675 & 0.0821\end{bmatrix}$$

$$S_{1c} = \begin{bmatrix} -0.3584 \\ -0.0957 \end{bmatrix}, \quad S_{2c} = \begin{bmatrix} -0.2521 \\ -0.0243 \end{bmatrix}, \quad S_{3c} = \begin{bmatrix} 0.3557 \\ -0.0622 \end{bmatrix}$$

$$Q_{1c} = \begin{bmatrix} -1.0773 \end{bmatrix}, \quad Q_{2c} = \begin{bmatrix} -1.4319 \end{bmatrix}, \quad Q_{3c} = \begin{bmatrix} -1.5093 \end{bmatrix}$$

The initial control and state vectors of three subsystems are as follows:

$$u_1 = \begin{bmatrix} -0.1 \end{bmatrix} x_1 = \begin{bmatrix} -3 \\ -4 \end{bmatrix}, \quad u_2 = \begin{bmatrix} -0.5 \end{bmatrix} x_2 = \begin{bmatrix} 0.2 \\ 0.3 \end{bmatrix},$$

$$u_3 = \begin{bmatrix} -1 \end{bmatrix} x_3 = \begin{bmatrix} 0.4 \\ 4 \end{bmatrix}$$

When the predictive horizon of 12 time steps is implemented for all subsystems, the output time responses from DMPC procedure are depicted by Figure 2.3. The average optimizing cost trend from the foregoing Phases I and II of the last simulation with input disturbances is provided in Figure 2.4. Simulation results have proved that the control algorithm of DMPC procedure successfully stabilized the networked system suffering from an arbitrary coupling delay. Both delay-dependent and delay-independent criteria have delivered stabilized responses. They clearly shown that the stability constraints derived from the delay-dependent criterion provide better control performances than those from the delay-independent criterion. The chosen predictive horizon plays a significant role in the proposed DMPC strategy implemented with the control algorithm of DMPC procedure.

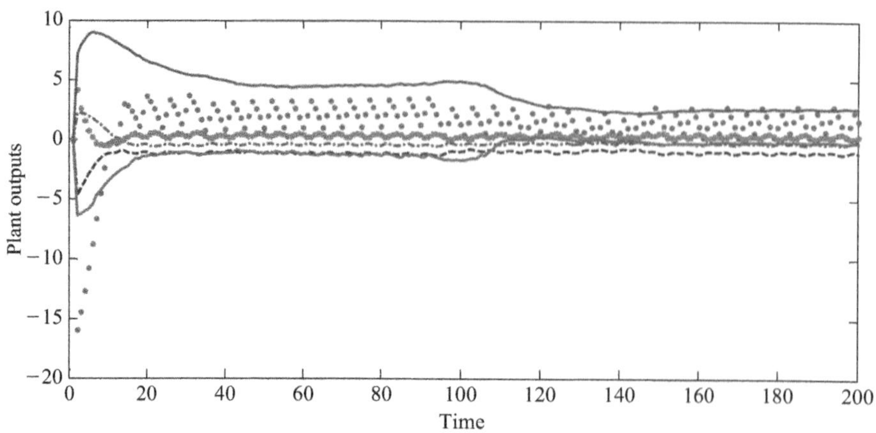

Figure 2.3 Distributed MPCs: delay-independent criteria

Remark 2.2: *Model predictive control subsystems have been widely implemented across the chemical industry sector, exploiting some of the latest theoretical developments in the area [154,155]. In cases of tightly coupled systems, the interplay between high-performance local control and subsystem–subsystem interactions leads to a deterioration in system-wide control performance. For most networked systems,*

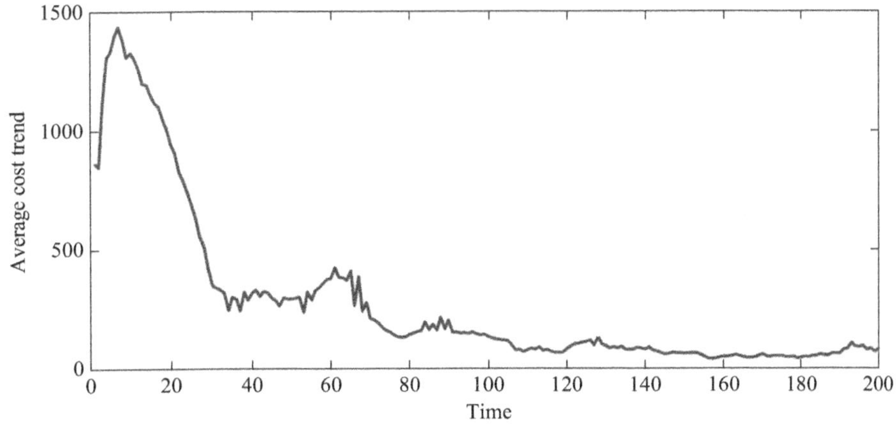

Figure 2.4 The average optimizing cost trend

the primary hurdles to centralized control are not computational but organizational. Operators of large, networked systems view centralized control as monolithic and inflexible. In many instances, different parts of the networked system are owned by different organizations making the comprehensive information sharing required by centralized control impractical. The opportunity presented for cross-integration within the MPC framework and potential requirements and benefits of such technology has been discussed in [156] and [157]. A DMPC formulation that considers coupling between various subsystem dynamics was proposed in [158] and further extended in [159].

2.3 Constrained model predictive control

In what follows, we explore an alternative distributed DMPC scheme. The controllers apply MPC policies to their local subsystems. They exchange their predictions by communication and incorporate the information from other controllers into their local MPC problem so as to coordinate with each other. For the full local state feedback and one-step delayed prediction exchange case, stability is ensured for controllable systems satisfying a matching condition by imposing stability constraints on the next state in the prediction. An example of multi-area load-frequency control is used as an example application for this DMPC scheme to show the performance of the scheme.

2.3.1 Introduction

We learned that model predictive control (MPC), also called receding horizon control, is a control scheme where the control input is obtained by solving an open-loop optimal control problem over a given horizon. Optimization produces an optimal control sequence. The first control in that sequence is applied to the physical system.

At the next control instant, the controller obtains new measurements and solves an updated optimal control problem. MPC is very popular in the process industries because it is an optimal control strategy that incorporates operating constraints explicitly. Many successful MPC applications have been reported in the last two decades [118,160]. In the MPC literature, various methods have been proposed to assure stability for MPC strategies. In the dominant scheme, stability constraints are applied to the end state in the prediction [115,117], but it is necessary for the prediction horizon to be long enough. Recently, Cheng and Krogh proposed a new scheme, stability-constrained model predictive control (SC-MPC) in which a constraint propagated from previous steps is imposed on the first state in the prediction in the current optimization problem [161]. This stability constraint becomes a contractive constraint that is always feasible for linear-time invariant (LTI) controllable systems. We use this idea to guarantee stability in our distributed multi-controller MPC strategy.

MPC is normally implemented in a centralized fashion. One controller has the full knowledge about the system and computes all the control inputs for the system. In large-scale interconnected systems, such as power systems, water distribution systems, and traffic systems such a centralized control scheme may not be suitable or even possible for technical or commercial reasons. Decentralized or distributed control is required [2,3,503].

In this section, we propose a scheme called distributed model predictive control with stability constraint (DMPC-SC) in which controllers coordinate with each other by exchanging their predictions. There is no centralized coordinator. In our scheme, the overall control problem is decomposed into several small ones, each of which is dealt with by a local MPC controller that uses local state measurements. The controllers exchange information about their measurements and predictions and incorporate this information in their local computations. The SC-MPC idea is used to guarantee the stability of the system. We show that if the system satisfies a structural property called a matching condition, the system controlled using the DMPC-SC policy with one-step delayed communication will be asymptotically stable.

2.3.2 Interconnected systems

In this chapter, we consider an LTI system composed of M interconnected subsystems described by the following equation:

$$z(k+1) = Az(k) + Bu(k) + Ew(k) \tag{2.30}$$

where

$$z(k) = \begin{bmatrix} z_1(k) \\ \vdots \\ z_M(k) \end{bmatrix} \in \Re^n, \quad u(k) = \begin{bmatrix} u_1(k) \\ \vdots \\ u_M(k) \end{bmatrix} \in \Re^m (n > m),$$

$$w(k) = \begin{bmatrix} w_1(k) \\ \vdots \\ w_M(k) \end{bmatrix} \in \Re^\ell$$

are the state, the control and the disturbance vectors of the system. The subvectors $z_i(k) \in \Re^{n_i}, u_i(k) \in \Re^{m_i}$, and $w_i(k) \in \Re^{l_i}$ are the state, control and disturbance vectors of the ith subsystem where

$$\sum_{i=1}^{M} n_i = n, \quad \sum_{i=1}^{M} m_i = m, \quad \sum_{i=1}^{M} l_i = l$$

$$A = \begin{bmatrix} A_{11} & \cdots & A_{1M} \\ \vdots & \ddots & \vdots \\ A_{M1} & \cdots & A_{MM} \end{bmatrix} \in \Re^{mn}, \quad A_{ij} \in \Re^{n_i \times n_j}$$

$$B = \begin{bmatrix} B_1 & & \\ & \ddots & \\ & & B_M \end{bmatrix} \in \Re^{mn}, \quad B_i \in \Re^{n_i \times m_i} (n_i > m_i)$$

$$E = \begin{bmatrix} E_1 & & \\ & \ddots & \\ & & E_M \end{bmatrix} \in \Re^{mn}, \quad E_i \in \Re^{n_i \times l_i}$$

If this system satisfies the structural condition in the following lemma, the control input to one subsystem does not have direct effect on other subsystems, that is, the coupling among subsystems is only through state interactions.

Lemma 2.1: *Consider systems in the form of (2.30). Suppose* (A_{ii}, B_i) $(i = 1, \ldots, M)$ *are controllable and* B_i *is of full rank. If*

$$rank(B_i \quad A_{i1} \quad \cdots \quad A_{i,i-1} \quad A_{i,i+1} \quad \cdots \quad A_{iM} \quad E_i) = m_i \ (i = 1, \ldots, M)$$

there exists a similarity transform matrix $P = \mathrm{diag}(P_1, P_2, \ldots, P_M)$ *where* $P_i \in \Re^{n_i \times n_j}$, *such that*

$$\begin{bmatrix} x_1(k+1) \\ \vdots \\ x_M(k+1) \end{bmatrix} = \begin{bmatrix} \bar{A}_{11} & \cdots & \bar{A}_{1M} \\ \vdots & \ddots & \vdots \\ \bar{A}_{M1} & \cdots & \bar{A}_{MM} \end{bmatrix} \begin{bmatrix} x_1(k) \\ \vdots \\ x_M(k) \end{bmatrix} + \begin{bmatrix} \bar{B}_1 & & \\ & \ddots & \\ & & \bar{B}_M \end{bmatrix} \begin{bmatrix} u_1(k) \\ \vdots \\ u_M(k) \end{bmatrix} + \begin{bmatrix} \bar{E}_{11} & & \\ & \ddots & \\ & & \bar{E}_M \end{bmatrix} \begin{bmatrix} w_1(k) \\ \vdots \\ w_M(k) \end{bmatrix} \tag{2.31}$$

where

$$P_z = \begin{bmatrix} x_1 \\ \vdots \\ x_M \end{bmatrix}, \quad x_i = \begin{bmatrix} x_i^1 \\ x_i^2 \end{bmatrix}, \quad x_i^1 \in \Re^{m_i}, \quad x_i^2 \in \Re^{n_i - m_i}$$

$$PAP^{-1} = \bar{A} = \begin{bmatrix} \bar{A}_{11} & \dots & \bar{A}_{1M} \\ \vdots & \ddots & \vdots \\ \bar{A}_{M1} & \dots & \bar{A}_{MM} \end{bmatrix}$$

$$\bar{A}_{ii} = \begin{bmatrix} 0 & I_{ii} \\ A_{ii}^1 & A_{ii}^2 \end{bmatrix}, \quad \bar{A}_{ij} = \begin{bmatrix} 0 & 0 \\ A_{ij}^1 & A_{ij}^2 \end{bmatrix} (j \neq i)$$

$$I_{ii} \in \Re^{(n_i - m_i)\times(n_i - m_i)}, \quad A_{ij}^1 \in \Re^{m_1 \times (n_j - m_j)}$$

$$A_{ij}^2 \in \Re^{m_i \times m_j} \ (i,j = 1,\dots,M)$$

$$PB = \bar{B} = \begin{bmatrix} \bar{B}_1 & & \\ & \ddots & \\ & & \bar{B}_M \end{bmatrix}, \quad \bar{B}_i = \begin{bmatrix} 0 \\ B_i^2 \end{bmatrix}, \quad B_i^2 \in \Re^{m_i \times m_j}$$

$$PE = \bar{E} = \begin{bmatrix} \bar{E}_1 & & \\ & \ddots & \\ & & \bar{E}_M \end{bmatrix}, \quad \bar{E}_i = \begin{bmatrix} 0 \\ E_i^2 \end{bmatrix}, \quad E_i^2 \in \Re^{m_i \times l_i}$$

The form (2.31) is called the decentralized controllable companion form.

This lemma follows directly from the controllable form in [2].

Remark 2.3: *It follows that the system discussed in Lemma 2.1 satisfies*

$$Span(A_{ij}) \subset span(B_i)(j \neq i)$$

which is frequently called the matching condition in decentralized control literature [2]. This condition on the structure of the system guarantees that the system can be stabilized by local state feedback. To design the feedback matrix for the distributed controller, only the knowledge of the local structure is needed. We note that the similarity transformation in Lemma 2.1 can be computed and implemented in a distributed fashion. That is, if each controller is designated to control a subsystem, the distributed controllers can do the transformation individually so that the model of the overall system is transformed to the decentralized controllable companion form (2.31).

Lemma 2.2: *Consider a system in decentralized controllable companion form (2.31). The state $x(k)$ can be represented in terms of components of the future states as*

$$x(k) = \begin{bmatrix} I_{n_i} & 0_{n_i \times (m_i \times l - n_i)} & & & \\ & & \ddots & & \\ & & & I_{n_M} & 0_{n_M \times (m_M \times l - n_M)} \end{bmatrix} \begin{bmatrix} x_1^l(k) \\ \vdots \\ x_1^l(k+l-1) \\ \vdots \\ x_M^l(k) \\ \vdots \\ x_M^l(k+l-1) \end{bmatrix} \quad (2.32)$$

where

$$\ell = \max_i \left(\left\lceil \frac{n_i}{m_i} \right\rceil \right)$$

$$\left\lceil \frac{n_i}{m_i} \right\rceil = \begin{cases} n_i/m_i & \text{if } n_i \text{ is a multiple of } m_i \\ int(n_i/m_i) & \text{otherwise} \end{cases}$$

We note that (2.32) follows from the definition of controllable for (A, B).

Lemma 2.3: *Consider a system in decentralized controllable companion form (2.31). Suppose the local disturbance $w_i(k)$ is measurable by the ith controller. For each subsystem, for any $x_i(k)$ and $0 < \beta_i < 1$, there exists $u_i(k)$ such that*

$$\|x_i(k+1)\|^2 \le \|x_i(k)\|^2 - \beta_i \|x_i^\ell(k)\|^2 \tag{2.33}$$

Moreover, if $u(k) = \begin{bmatrix} u_1(k) \\ \vdots \\ u_M(k) \end{bmatrix}$, *where $u_i(k)$ satisfies (2.33), and* $\beta = \min(\beta_1, \ldots, \beta_M)$, *then*

$$\|x(k+1)\|^2 \le \|x(k)\|^2 - \beta \|x^\ell(k)\|^2 \tag{2.34}$$

where $x^\ell(k) = \begin{bmatrix} x_1^\ell(k) \\ \vdots \\ x_2^\ell(k) \end{bmatrix}$.

Proof. From the definition of the decentralized controllable companion form for (A, B)

$$\begin{aligned} \|x_i(k+1)\|^2 &= \left\| A_{ii}x_i(k) + B_i u_i(k) + \sum_{j=1, j\neq i}^{M} A_{ij}x_j(k) + E_i W_i(k) \right\|^2 \\ &= \|x_i^2(k)\|^2 \\ &\quad + \left\| \sum_{j=1}^{M} [A_{ij}^1 x_j^1(k) + A_{ij}^2 x_j^2(k)] + B_i^2 u_i(k) + E_i^2 W_i(k) \right\|^2 \\ &= \|x_i(k)\|^2 - \|x_i^1(k)\|^2 \\ &\quad + \left\| \sum_{j=1}^{M} [A_{ij}^1 x_j^1(k) + A_{ij}^2 x_j^2(k)] + B_i^2 u_i(k) + E_i^2 W_i(k) \right\|^2 \end{aligned}$$

On substituting this into (2.33), it yields the following equivalent inequality:

$$\left\| \sum_{j=1}^{M} [A_{ij}^1 x_j^1(k) + A_{ij}^2 x_j^2(k)] + B_i^2 u_i(k) + E_i^2 W_i(k) \right\|^2 \le (1 - \beta_i)\|x_i^1(k)\|^2$$

It follows that $u_i(k) = (B_i^2)^{-1}\{\sum_{j=1}^{M} [A_{ij}^1 x_j^1(k) + A_{ij}^2 x_j^2(k)] + E_i^2 W_i(k)\}$ is a feasible solution.

Next consider the overall system. When (2.33) holds:

$$\begin{aligned}
\|x(k+1)\|^2 &= \sum_{i=1}^{M} \|x_i(k+1)\|^2 \\
&\leq \sum_{i=1}^{M} [\|x_i(k)\|^2 - \beta_i \|x_i^1(k)\|^2] \\
&\leq \sum_{i=1}^{M} [\|x_i(k)\|^2 - \beta_i \|x_i^1(k)\|^2] \\
&= \|x(k)\|^2 - \beta \|x^1(k)\|^2
\end{aligned}$$

Feasible solutions of subsystems consist a feasible solution to (2.34). □

Lemma 2.4: *When the constraint (2.34) is applied, the feasible control set is nonempty. It guarantees that, when distributed controllers are used to control the system with local constraint (2.33), the controls they find are feasible for the overall system.*

2.3.3 *MPC strategy*

In MPC, control decisions are made at discrete-time instants, which usually represent equally spaced time intervals. At each decision instant, the controller samples the current state of the system and then solves an optimization problem to find the control action. Currently, most applications of MPC are implemented in a centralized manner. The controller has the knowledge of the overall system and computes all the control inputs for the system. When there are multiple controllers, a DMPC scheme must be formulated. In the general problem formulation being considered, we propose that each controller follows the standard MPC strategy with some variation. Each controller must now include the behavior of other subsystems in its local optimization problem in order to coordinate with other controllers. So, for controller i, the basic optimal control problem must be modified as follows:

Performance index:

$$\min_{X_i(k),U_i(k)} J_i(X_i(k), U_i(k))$$

where

$$\begin{aligned}
X_i(k) &= \{\hat{x}_i(k+1 \mid k), \ldots, \hat{x}_i(k+N \mid k)\} \\
U_i(k) &= \{\hat{u}_i(k \mid k), \ldots, \hat{u}_i(k+N-1 \mid k)\}
\end{aligned}$$

Such that

Prediction:

$$\hat{x}_i(k+j+1 \mid k) = F_i(\hat{x}_i(k+j \mid k), \hat{u}_i(k+j \mid k), \hat{v}_i(k+j \mid k), \hat{w}_i(k+j \mid k),$$
$$(j = 0, \ldots, N-1)$$

Constraints:

$$G_i(\hat{x}_i(k+j \mid k), \hat{u}_i(k+j \mid k), \hat{v}_i(k+j \mid k)) \leq 0$$

Initial condition:

$$\hat{x}_i(k \mid k) = x_i(k)$$

Here $\hat{x}_i$ means the predicted value of the variable x_i and so on. The predicted values of the disturbance input have no effect on the stability of the control. But these predictions can affect the performance of the system.

In this formulation, the control prediction horizon and the state prediction horizon are assumed to be the same. We assume that the state variables x_i and the disturbance w_i can be measured or estimated directly by the controller i. The control variables u_i are controller i's control actions, $u_i(k) = \hat{u}_i(k \mid k)$. The variables v_i represent all variables outside the domain of the controller. In this section, the system with the decentralized controllable companion form (2.31) is considered so that

$$\hat{v}_j(k+i \mid k) = [\hat{x}_1^T(k+i \mid k-1) \cdots \hat{x}_{j-1}^T(k+i \mid k-1)$$
$$\hat{x}_{j+1}^T(k+i \mid k-1) \cdots \hat{x}_M^T(k+i \mid k-1)]^T$$

This is a one-step delayed prediction exchange. In this DMPC scheme, the stability constraints (2.33) are incorporated to ensure the stability of the system.

2.3.4 MPC with stability constraint

For the ith controller, the algorithm is described as follows:

Step 1. *Communication*: Send out its previous predictions $X_j(k-1)$ to other controllers and also receive information $V_j(k) = \{\hat{v}_j(k \mid k), \ldots, \hat{v}_j(k+N-1 \mid k)\}$ from other controllers.

Step 2. *Initialization*: Given measured $x_j(k)$ and $w_j(k)$ and $l_j(k)$ from the previous iteration (set $l_j(0)$ to be an arbitrary number), and $0 < \beta_j \leq 1$, define

$$\hat{l}_j(k) = \max\{l_j(k), \|x_j(k)\|^2\} - \beta_j \|x_j^1(k)\|^2$$

Set $\hat{x}_j(k \mid k) = x_j(k)$ and $\hat{w}_j(k \mid k) = w_j(k)$. Predict the future disturbances $\hat{w}_j(k+1 \mid k), \ldots, \hat{w}_j(k+N-1 \mid k)$. [*Note*: These predictions have no influence on the stability results.]

Step 3. *Optimization*: Solve the following optimal control problem:

$$\min J_j(X_j^P, U_j^P)$$

subject to

$$\begin{aligned}\hat{x}_j(k+i+1 \mid k) = A_{jj}\hat{x}_j(k+i \mid k) + B_j\hat{u}_j(k+i \mid k) \\ + K_j\hat{v}_j(k+i \mid k) + E_j\hat{w}_j(k+i \mid k), \\ (i = 0, 1, \ldots, N-1)\end{aligned}$$

$$\|\hat{x}_j(k+1 \mid k)\|^2 \leq \hat{l}_{j,k}$$

Step 4. *Assignment*: Let

$$u_j(k) = \hat{u}_j(k \mid k), l_j(k+1) = \|\hat{x}_j(k+1 \mid k)\|^2$$

Step 5. *Implementation*: Application control $u(k)$. Set $k = k + 1$ and return Step 1 at the next sample time.

End.

Remark 2.4: *In implementing this scheme, it is not necessary for each controller to communicate with all the other controllers. Controller i needs to send information to controller j only if the corresponding sub-matrix in the decentralized controllable companion form is nonzero. If the system is loosely coupled, the amount of communication will not be large.*

2.3.5 Stability consideration

Upon applying the DMPC-SC scheme, the stability of the system can be ensured if the coupling among the subsystems satisfies the condition in the following theorem.

Theorem 2.6: *Consider a system in decentralized controllable companion form (2.31). The control is computed at each control instant using DMPC-SC. The system is asymptotically stable if the following matrix is stable (i.e., has all eigenvalues inside the unit circle).*

$$\bar{A} = \begin{bmatrix} 0 & A_{12}^2 & \ldots & A_{1M}^2 \\ A_{21}^2 & 0 & \ldots & A_{2M}^2 \\ \vdots & \vdots & \ddots & \vdots \\ A_{M1}^2 & A_{M2}^2 & \ldots & 0 \end{bmatrix} \tag{2.35}$$

Proof. We first show that the error in the state predicted by each local controller (due to the delayed prediction information from the other controllers) goes to zero asymptotically. According to the DMPC-SC scheme, we have

$$x_i(k+1 \mid k) = \bar{A}_{ii}x_i(k) + \bar{B}_iu_i(k) + \sum_{\substack{j=1 \\ j\neq i}}^{M} \bar{A}_{ij}\hat{x}_j(k \mid k-1) + \bar{E}_iw_i(k)$$

while the state of the system is

$$x_i(k+1) = \bar{A}_{ii}x_i(k) + \bar{B}_i u_i(k) + \sum_{\substack{j=1 \\ j\neq i}}^{M} \bar{A}_{ij}x_j(k) + \bar{E}_i w_i(k)$$

So the error of the prediction $x_{e_i}(k+1 \mid k) = x_i(k+1) - \hat{x}_i(k+1 \mid k)$ is

$$x_{e_i}(k \mid k-1) = \sum_{\substack{j=1 \\ j\neq i}}^{M} \bar{A}_{ij}x_{e_j}(k-1 \mid k-2)$$

so that $x_e(k+l \mid k) = \hat{A}x_e(k \mid k-l)$, where $\hat{A} = \bar{A} - \text{diag}(\bar{A}_{11}, \ldots, \bar{A}_{MM})$. If $\hat{A}$ is stable, $x_e(k \mid k-1) \to 0$. Consider an eigenvalue λ and corresponding eigenvector ν, we have

$$\hat{A}\nu = \begin{bmatrix} 0 \\ \sum_{j\neq i}(A_{ij}^1\nu_j^1 + A_{ij}^2\nu_j^2) \\ \vdots \\ 0 \\ \sum_{j\neq M}(A_{Mj}^1\nu_j^1 + A_{Mj}^2\nu_j^2) \end{bmatrix}$$

$$\nu = \begin{bmatrix} \nu_1 \\ \vdots \\ \nu_M \end{bmatrix}, \nu_i = \begin{bmatrix} \nu_i^1 \\ \nu_i^2 \end{bmatrix}, \nu_i^1 \in \Re^{n_i-m_i}, \quad \text{and} \quad \nu_i^2 \in \Re^{m_i}$$

For nonzero eigenvalue, the corresponding eigenvector satisfies $\nu_i^1 = 0$. We have λ that is an eigenvalue of $\tilde{A}$ and $\begin{bmatrix} \nu_1^2 \\ \vdots \\ \nu_M^2 \end{bmatrix}$ is the corresponding eigenvector. So d is stable, which implies that $\hat{A}$ is stable. Therefore, $x_e(k \mid k-1) \to 0$.

We now show the system is asymptotically stable. By definition

$$\begin{aligned} \hat{\ell}_i(k) &= \max\{l_i(k), \|x_i(k)\|^2\} - \beta_i\|x_i^1(k)\|^2 \\ &= \max\{\|\hat{x}_i(k \mid k-1), \|\hat{x}_i(k \mid k-1) + x_{e_i}(k \mid k-1)\|^2\} - \beta_i\|x_i^1(k)\|^2 \\ &\leq \|\hat{x}_i(k \mid k-1)\|^2 + \|x_{e_i}(k \mid k-1)\|^2 - \beta_i\|x_i^1(k)\|^2 \\ &\leq \hat{\ell}_i(k-1) + \|\hat{x}_{e_i}(k \mid k-1)\|^2 - \beta_i\|x_i^1(k)\|^2 \end{aligned}$$

Suppose there exists a subsystem i for which the $x_i^1(k)$ does not converge to zero. Then, there must exist positive number a and an infinite sequence of positive integers $0 < k_1 < k_2 < \ldots$ such that $\|\hat{x}_i^1(k_j)\| > \alpha(j = 1, 2, \ldots)$. Since $x_e(k \mid k-1) \to 0, x_{e_i}(k \mid k-1) \to 0$ must hold. There must exist $K > 0$ such that $\|x_{e_i}(k \mid k-1)\|^2 < \beta_i\alpha^2 - \gamma$ for all $j \geq K$.

We have

$$\begin{aligned}\hat{\ell}_i(k_{K+s+1}) &< \hat{\ell}_i(k_{K+s}) + \beta_i\alpha^2 - \gamma - \beta_i\alpha^2 \\ &< \hat{\ell}_i(k_{K+s}) - \gamma < \hat{\ell}_i(k_K) - s\gamma\end{aligned}$$

When $s > \hat{\ell}_i(k_K)/\gamma, \hat{\ell}_i(k_{K+s}) < 0$. This contradicts the fact that $\hat{\ell}_{i,k} \geq 0$. So $x_i^1(k) \to 0$ for all $i = 1, \ldots, M$. By Lemma 2.2, we have $x(k) \to 0$. □

2.4 A cooperative game approach

In this section, we develop a DMPC scheme based on a cooperative game in which two different agents communicate in order to find a solution to the problem of controlling two constrained linear systems coupled through the inputs. We assume that each agent only has partial information of the model and the state of the system. In the proposed scheme, the agents communicate twice each sampling time in order to share enough information to take a cooperative decision. We provide sufficient conditions that guarantee practical stability of the closed-loop system as well as an optimization-based procedure to design the controller so that these conditions are satisfied. The theoretical results and the design procedure are illustrated using two different examples 2.2 and 2.3.

2.4.1 Introduction

So far, we learned that MPC is an optimization-based control strategy that has been successfully applied to process control, especially to constrained multivariable systems exhibiting dead times [116,502]. At each sampling time, a finite-horizon optimal control problem based on the model of the system is solved. Only the first control move of the optimal solution is applied and the rest of the sequence is discarded repeating the procedure when a new measurement is available (the so-called receding horizon scheme). In many cases, MPC cannot be applied to large-scale systems due to the computational requirements or to the impossibility of obtaining a centralized model of the whole system. Typical examples of large-scale systems are transportation systems such as traffic, water, or power networks [162]. In addition, there is an increasing interest in networked control systems, where dedicated local control networks can be augmented with additional networked (wired and/or wireless) actuator/sensor devices which have become cheap and easy-to-install [163,164].

Most large-scale and networked control systems are based on a decentralized architecture; that is, the system is divided into several subsystems, each controlled by a different agent that does not share information with the rest. Each of the agents implements an MPC based on a reduced model of the system and on partial state information, which in general results in an optimization problem with a lower computational burden. Figures 2.5 and 2.6 show a centralized and a decentralized MPC control scheme for two subsystems coupled through the inputs. For example, in [146],

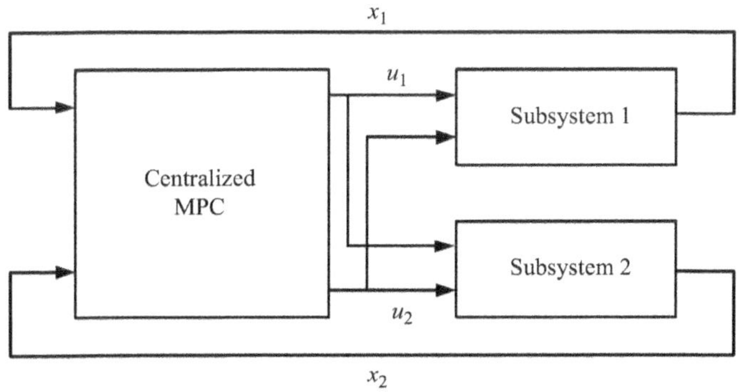

Figure 2.5 Centralized MPC scheme

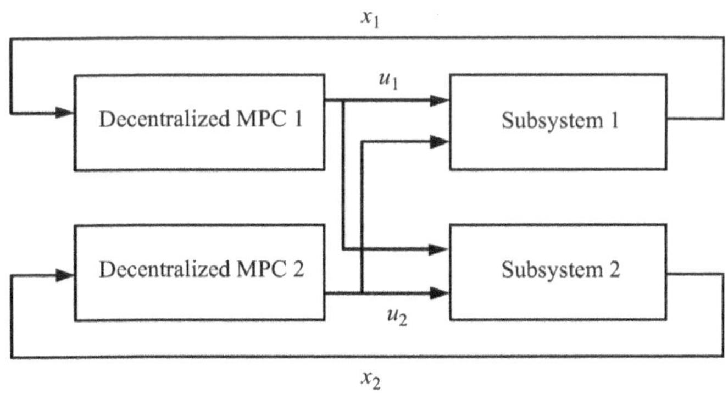

Figure 2.6 Decentralized MPC scheme

an MPC algorithm was proposed under the main assumptions that the system is non-linear, discrete-time and no information is exchanged between local controllers. The stability of this class of systems, from an input-to-state stability point of view, was studied in [165].

While this paradigm to process control has been successful, there is an increasing interest in developing DMPC schemes, where agents share information in order to improve closed-loop performance, robustness, and fault-tolerance; see Figure 2.7 for an example of this control scheme. The performance of the closed-loop system depends on the decisions that all the agents take; hence, cooperation and communication policies become very important issues. In this context, several DMPC schemes have been proposed in the literature that deal with the coordination of separate MPC controllers that communicate in order to obtain optimal input trajectories in a distributed manner; see [52] and [158] for reviews of results in this area. In [166], sufficient conditions that guarantee stability of a class of distributed controllers are given. In [167], the problem of distributed control of dynamically coupled non-linear

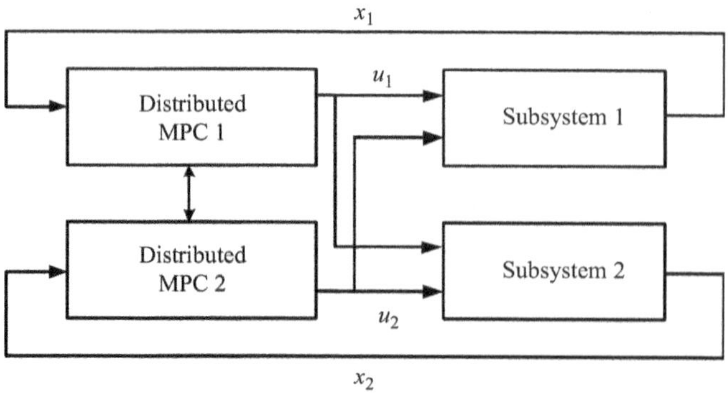

Figure 2.7 Distributed MPC scheme

systems that are subject to decoupled constraints was considered. In [168] and [169], the effect of the coupling was modeled as a bounded disturbance compensated using a robust MPC formulation. In [145], DMPC of decoupled systems (a class of systems of relevance in the framework of multi-agents systems) was studied. Several works in the literature assume that each agent has access to the model of the full system. For example, in [159], it was proved that through multiple communications between the distributed controllers and using system-wide control objective functions, stability of the closed-loop system can be guaranteed. In [170], a decentralized control architecture for non-linear systems with continuous and asynchronous measurements was presented. Following up on this work, in [171], a DMPC method for the design of networked control systems based on Lyapunov-based MPC was presented. In both cases, each agent had access to the full system model. However, sometimes it is not possible to obtain a model that describes a system globally, or even if it is possible, it cannot be computed in a reasonable time or it is very expensive, for example, in large-scale systems.

Motivated by these issues, we develop in the sequel a DMPC algorithm for two agents based on game theory in which two different agents communicate in order to find a solution to the problem of controlling two constrained linear systems coupled through the inputs. We assume that each agent only has partial information of the model and the state of the system. Game theory is a theoretical framework that allows one to study the problem of cooperation of different agents with, maybe, conflicting control goals, from a mathematical point of view [172,173]. In the proposed scheme, the coordination problem between the agents is reduced to a cooperative game where they have to choose one out of three options, and only two communication cycles are needed to reach an agreement. In addition, we provide sufficient conditions that guarantee practical stability of the closed-loop system as well as an optimization-based procedure to design the controller so that these conditions are satisfied. The theoretical results and the design procedure are illustrated using two different examples.

2.4.2 Problem formulation

Consider the following class of distributed linear systems in which two subsystems coupled with the neighbor subsystem through the inputs are defined:

$$\begin{aligned} x_1(t+1) &= A_1x_1(t) + B_{11}u_1(t) + B_{12}u_2(t) \\ x_2(t+1) &= A_2x_2(t) + B_{21}u_1(t) + B_{22}u_2(t) \end{aligned} \tag{2.36}$$

where $x_i \in \mathbb{R}^{n_i}, i = 1, 2$, are the states of each subsystem and $u_i \in \mathbb{R}^{m_i}, i = 1, 2$, are the different inputs. This class of systems is of relevance when identifications techniques are used to obtain the transfer function of a process. We consider the following linear constraints in the state and the inputs:

$$x_i \in \mathcal{X}_i, u_i \in \mathcal{U}_i, i = 1, 2$$

where $\mathcal{X}_i$ and $\mathcal{U}_i$ with $i = 1, 2$ are defined by a set of linear inequalities.

The control objective is to regulate the system to the origin while guaranteeing that the constraints are satisfied. Centralized MPC solves a single optimization problem to decide the optimal sequences of the inputs u_1 and u_2 with respect to a given performance index based on the full model of the system and on the measurements from all the sensors (see Figure 2.5). In distributed and decentralized schemes, two independent controllers (hereby denoted agents) are defined. Agent 1 has access to the model of subsystem 1, its state x_1, and decides the value of u_1. On the other hand, agent 2 has access to the model of subsystem 2, its state x_2, and decides the value of u_2. This implies that neither agent has access to the full model or state information and that in order to find a cooperative solution, they must communicate.

A control system is decentralized if there is no communication among the agents (see Figure 2.6). This is the worst scenario from the performance point of view, because each agent has to cope alone with its control problem with the risk that the absence of coordination in the agents' decisions may lead to the instability of the system. The control system is distributed if there is communication between agents (see Figure 2.3). The degree of communication depends on the control problem and the communication constraints. In this section, we present a DMPC controller based on a cooperative game scheme between two different agents.

The objective of the DMPC scheme is to minimize a performance index that depends on the future evolution of both states and inputs. At each sampling time, each agent solves a sequence of reduced dimension optimization problems based on the model of its subsystem and assuming a given fixed input trajectory for its neighbor. In order to describe the algorithm, we need to introduce the following definitions:

- U_i: Future input sequence of agent i: These are the decision variables of the optimization problems solved by both agents

$$U_1 = \begin{bmatrix} u_{1,0} \\ u_{1,1} \\ \vdots \\ u_{1,N-1} \end{bmatrix}, \quad U_2 = \begin{bmatrix} u_{2,0} \\ u_{2,1} \\ \vdots \\ u_{2,N-1} \end{bmatrix}$$

- n_i: Neighboring agent of agent i, that is, $U_{n_1} = U_2$ and $U_{n_2} = U_1$.
- J_i: Local cost function of agent i based on the predicted trajectories of its state defined as follows:

$$J_1(x_1, U_1, U_2) = \sum_{k=0}^{N-1} L_1(x_{1,k}, u_{1,k}) + F_1(x_{1,N})$$

$$J_2(x_2, U_2, U_1) = \sum_{k=0}^{N-1} L_2(x_{2,k}, u_{2,k}) + F_2(x_{2,N})$$

where $L_i(\cdot)$ and $F_i(\cdot)$ with $i = 1, 2$ are the stage and terminal cost functions, respectively, defined as

$$L_i(x, u) = x^T Q_i x + u^T R_i u$$
$$F_i(x) = x^T P_i x$$

with $Q_i, P_i > 0, Ri \geq 0$. The prediction horizon is N. We use the notation $x_{i,k}$ to denote the k-steps ahead future-predicted state obtained from the initial state x_i applying the input trajectories defined by U_1 and U_2. Note that the second and third parameters of functions J_1 and J_2 are switched. This will allow us to simplify the algorithm definition.
- $U_i^d(t)$: Optimal input sequence of agent i at time t, denoted as $U_i^d(t)$, defined as

$$U_1^d(t) = \begin{bmatrix} u_{1,0}^d \\ u_{1,1}^d \\ \vdots \\ u_{1,N-1}^d \end{bmatrix}, \quad U_2^d(t) = \begin{bmatrix} u_{2,0}^d \\ u_{2,1}^d \\ \vdots \\ u_{2,N-1}^d \end{bmatrix}$$

- $U_i^s(t)$: Shifted optimal input sequence of agent i obtained from the optimal input sequence of agent i at time $t-1$, denoted as $U_i^d(t-1)$, as follows:

$$U_1^s(t) = \begin{bmatrix} u_{1,1}^d \\ u_{1,2}^d \\ \vdots \\ u_{1,N-1}^d \\ K_1 x_{1,N} \end{bmatrix}, \quad U_2^d(t) = \begin{bmatrix} u_{2,1}^d \\ u_{2,2}^d \\ \vdots \\ u_{2,N-1}^d \\ K_2 x_{2,N} \end{bmatrix}$$

where $x_{1,N}, x_{2,N}$ are the N-steps ahead predicted state obtained from $x_1(t-1)$, $x_2(t-1)$, respectively, applying the input trajectories $U_1^d(t-1)$, $U_2^d(t-1)$ and K_1, K_2 are two known feedback gains.

The proposed DMPC algorithm is the following:

1. At time step t, each agent i receives its corresponding partial state measurement $x_i(t)$.

2. Each agent i minimizes J_i assuming that the neighbor keeps applying the optimal trajectory evaluated at the previous time step, that is, $U_{ni} = U^s_{ni}(t)$. Agent 1 solves the following optimization problem:

$$\begin{aligned} U_1^*(t) = \arg\min_{U_1} \; & J_1(x_1(t), U_1, U_2^s(t)) \\ & x_{1,k+1} = A_1 x_{1,k} + B_{11} u_{1,k} + B_{12} u_{2,k} \\ & x_{1,0} = x_1(t) \\ & x_{1,k} \in \mathcal{X}_1, \quad k = 0, \ldots, N \\ & u_{1,k} \in \mathcal{U}_1, \quad k = 0, \ldots, N-1 \\ & x_{1,N} \in \Omega_1 \end{aligned} \tag{2.37}$$

Agent 2 solves the following optimization problem:

$$\begin{aligned} U_2^*(t) = \arg\min_{U_2} \; & J_2(x_2(t), U_2, U_2^s(t)) \\ & x_{2,k+1} = A_2 x_{2,k} + B_{22} u_{2,k} + B_{21} u_{1,k} \\ & x_{2,0} = x_2(t) \\ & x_{2,k} \in \mathcal{X}_2, \quad k = 0, \ldots, N \\ & u_{2,k} \in \mathcal{U}_2, \quad k = 0, \ldots, N-1 \\ & x_{2,N} \in \Omega_2 \end{aligned} \tag{2.38}$$

The sets Ω_1 and Ω_2 define the terminal region constraints that are necessary to prove closed-loop practical stability following a terminal region/terminal cost approach. Note that in both optimization problems the free variable is U_i while the neighbor input trajectory U_{ni} is fixed.

3. Each agent i minimizes J_i optimizing the neighbor input assuming that it applies the input trajectory computed in the previous optimization problem U_i^*. Agent 1 solves the following optimization problem:

$$\begin{aligned} U_2^w(t) = \arg\min_{U_2} \; & J_1(x_1(t), U_1^*, U_2(t)) \\ & x_{1,k+1} = A_1 x_{1,k} + B_{11} u_{1,k} + B_{12} u_{2,k} \\ & x_{1,0} = x_1(t) \\ & x_{1,k} \in \mathcal{X}_1, \quad k = 0, \ldots, N \\ & u_{2,k} \in \mathcal{U}_2, \quad k = 0, \ldots, N-1 \\ & x_{1,N} \in \Omega_1 \end{aligned} \tag{2.39}$$

Agent 2 solves the following optimization problem:

$$\begin{aligned} U_1^w(t) = \arg\min_{U_1} \; & J_2(x_2(t), U_2^*, U_1(t)) \\ & x_{2,k+1} = A_2 x_{2,k} + B_{22} u_{2,k} + B_{21} u_{1,k} \\ & x_{2,0} = x_2(t) \\ & x_{2,k} \in \mathcal{X}_2, \quad k = 0, \ldots, N \\ & u_{1,k} \in \mathcal{U}_1, \quad k = 0, \ldots, N-1 \\ & x_{2,N} \in \Omega_2 \end{aligned} \tag{2.40}$$

In this optimization problem, the free variable is U_{ni} (the input trajectory U_i is fixed). Solving this optimization problem, agent i defines an input trajectory for its neighbor that optimizes its local cost function J_i.

4. Both agents communicate. Agent 1 sends $U_1^*(t)$ and $U_2^w(t)$ to agent 2 and receives $U_2^*(t)$ and $U_1^w(t)$.
5. Each agent evaluates the local cost function J_i for each of the nine different possible combinations of input trajectories, that is, $U_1 \in \{U_1^s(t), U_1^w(t), U_1^*(t)\}$ and $U_2 \in \{U_2^s(t), U_2^w(t), U_2^*(t)\}$.
6. Both agents communicate and share the information of the value of the local cost function for each possible combination of input trajectories. In this step, both agents receive enough information to take a cooperative decision.
7. Each agent applies the input trajectory that minimizes $J = J_1 + J_2$. Because both agents have access to the same information after the second communication cycle, both agents choose the same optimal input sets. We denote the chosen set of input trajectories as $U_1^d(t), U_2^d(t)$.
8. The first input of each optimal sequence is applied and the procedure is repeated the next sampling time.

From a game theory point of view, at each time step both agents are playing a cooperative game. This game can be synthesized in strategic form by a 3 × 3 matrix. Each row represents one of the three possible decisions of agent 1, and each column represents one of the three possible decisions of agent 2. The cells contain the sum of the cost functions of both agents for a particular choice of future inputs. At each time step, the option that yields a lower global cost is chosen. Note that both agents share this information, hence they both choose the same option. The nine possibilities are shown in Table 2.1.

Table 2.1 Cost function table used for the decision making

	$\boldsymbol{U_2^s(t)}$	$\boldsymbol{U_2^*(t)}$
$U_1^s(t)$	$J_1(x_1(t), U_1^s(t), U_2^s(t)) + J_2(x_2(t), U_2^s(t), U_1^s(t))$	$J_1(x_1(t), U_1^s(t), U_2^*(t)) + J_2(x_2(t), U_2^*(t), U_1^s(t))$
$U_1^*(t)$	$J_1(x_1(t), U_1^*(t), U_2^s(t)) + J_2(x_2(t), U_2^s(t), U_1^*(t))$	$J_1(x_1(t), U_1^*(t), U_2^*(t)) + J_2(x_2(t), U_2^*(t), U_1^*(t))$
$U_1^w(t)$	$J_1(x_1(t), U_1^w(t), U_2^s(t)) + J_2(x_2(t), U_2^s(t), U_1^w(t))$	$J_1(x_1(t), U_1^w(t), U_2^*(t)) + J_2(x_2(t), U_2^*(t), U_1^w(t))$
	$\boldsymbol{U_2^w(t)}$	
$U_1^s(t)$	$J_1(x_1(t), U_1^s(t), U_2^w(t)) + J_2(x_2(t), U_2^w(t), U_1^s(t))$	
$U_1^*(t)$	$J_1(x_1(t), U_1^*(t), U_2^w(t)) + J_2(x_2(t), U_2^w(t), U_1^*(t))$	
$U_1^w(t)$	$J_1(x_1(t), U_1^w(t), U_2^w(t)) + J_2(x_2(t), U_2^w(t), U_1^w(t))$	

Remark 2.5: *At each sampling time, the controllers decide among three different options. The shifted optimal input trajectory $U_i^s(t)$ keeps applying the latest optimal trajectory. The selfish option $U_i^*(t)$ provides the best improvement in J_i if the rest of the system's manipulated variables stay unchanged. The altruist option $U_i^w(t)$ provides the best improvement for the neighbor agent cost function J_2. In this case, the agent i sacrifices its own welfare in order to improve the overall performance.*

Remark 2.6: *Centralized MPC solves a single large-scale problem based on the model of the whole system (see Figure 2.1). In the example section, we will compare the performance of the proposed approach with a centralized MPC controller based on the following optimization problem:*

$$
\begin{aligned}
\{U_1^c(t), U_2^c(t)\} &= \arg\min_{U_1,U_2} \quad J_1(x_1(t), U_1, U_2) + J_2(x_1(t), U_2, U_1) \\
x_{1,k+1} &= A_1 x_{1,k} + B_{11} u_{1,k} + B_{12} u_{2,k} \\
x_{1,0} &= x_1(t) \\
x_{1,k} &\in \mathcal{X}_1, \quad k = 0, \ldots, N \\
u_{1,k} &\in \mathcal{U}_1, \quad k = 0, \ldots, N-1 \\
x_{1,N} &\in \Omega_1 \\
x_{2,k+1} &= A_2 x_{2,k} + B_{22} u_{2,k} + B_{21} u_{1,k} \\
x_{2,0} &= x_2(t) \\
x_{2,k} &\in \mathcal{X}_2, \quad k = 0, \ldots, N \\
u_{2,k} &\in \mathcal{U}_2, \quad k = 0, \ldots, N-1 \\
x_{2,N} &\in \Omega_2
\end{aligned}
\tag{2.41}
$$

The centralized MPC provides, in general, the best closed-loop performance, but can only be applied when it is possible to control the system with a single controller that has access to the full model and state of the same.

Remark 2.7: *In general, the minimum number of communication steps needed to implement a cooperative control scheme is two. In the first step each agent informs of its intentions to its neighbors and in the second it can confirm if it accepts its neighbors' intentions. In the best case, an agreement can be achieved in the second step, but in general an iterative procedure will be needed to reach an agreement.*

Remark 2.8: *The proposed controller scheme is cooperative from a game theory point of view because each agent chooses the solution that optimizes a cost function that depends on both subsystems, not only on the future trajectories of its subsystem. If the decision taken does not depend on a global performance index, the solution is not cooperative. In the simulation section, we will compare the proposed distributed controller with a distributed scheme in which the two agents communicate, but do not take a cooperative decision. They iterate until an agreement is obtained. In this case,* *the solution is a Nash equilibrium [59] of the multi-objective optimization problem defined by the cost functions of both agents. At each iteration, agent 1 solves the following optimization problem:*

$$
\begin{aligned}
U_1^{\ell+1} = \arg\min_{U_1} \; & J_1(x_1, U_1, U_2^l) \\
& x_{1,k+1} = A_1 x_{1,k} + B_{11} u_{1,k} + B_{12} u_{2,k} \\
& x_{1,0} = x_1 \\
& x_{1,k} \in \mathcal{X}_1, \quad k = 0, \ldots, N \\
& u_{1,k} \in \mathcal{U}_1, \quad k = 0, \ldots, N-1 \\
& x_{1,N} \in \Omega_1
\end{aligned}
\tag{2.42}
$$

and agent 2 solves the following optimization problem:

$$
\begin{aligned}
U_2^{l+1} = \arg\min_{U_2} \; & J_2(x_2, U_2, U_1^l) \\
& x_{2,k+1} = A_2 x_{2,k} + B_{22} u_{2,k} + B_{21} u_{1,k} \\
& x_{2,0} = x_2 \\
& x_{2,k} \in \mathcal{X}_2, \quad k = 0, \ldots, N \\
& u_{2,k} \in \mathcal{U}_2, \quad k = 0, \ldots, N-1 \\
& x_{2,N} \in \Omega_2
\end{aligned}
\tag{2.43}
$$

with $U_1^0 = U_1^s$ and $U_2^0 = U_2^s$, that is, the initial guess is given by the shifted trajectory. An agreement is reached when the difference between the proposed control vector by each agent at one iteration and its value at the previous iteration is below a threshold. This implies that they do not share information about the utility of each decision, they reach an agreement when neither of them can improve, hence reaching a Nash equilibrium. In the example, we will compare the proposed controller with different controllers based on this distributed scheme, each one carrying out a fixed number of iterations to demonstrate that the proposed cooperative scheme provides a better performance with a lower number of iterations.

Remark 2.9: *Although the option chosen by the algorithm is the Pareto optimum of the game that both agents are playing, in general, it is not a Pareto optimum of the multi-objective optimization problem defined by the cost functions J_1 and J_2.*

Remark 2.10: *The proposed scheme can be extended to deal with N agents; however, in order to build a global cost table to take a cooperative decision, the complexity grows exponentially. In order to reduce the complexity, the structure of the system may be exploited taking into account that an input may not affect all the outputs. Also, in general, not all the possible cooperation options are employed with the same frequency; hence, it is possible to reduce further the complexity by not taking into account the less frequent options. An interesting feature of the proposed algorithm is that both agents can operate in parallel, that is, the agents can compute U_i^* and U_i^w simultaneously (steps 2 and 3).*

2.4.3 Stability properties

Controlling a system between two independent agents may lead to an unstable system. The resulting closed-loop system is a multiprocess system and studying the stability of this class of systems is in general a difficult task. Following a terminal region/terminal constraint approach [10,93], in this section we provide sufficient conditions that guarantee practical stability of the closed-loop system as well as an optimization-based procedure to design the controller so that these conditions are satisfied. This result is stated in the following theorem.

Theorem 2.7: *Assume that there exists linear feedbacks*

$$u_1 = K_1 x_1, \quad u_2 = K_2 x_2$$

such that the following conditions hold:

$$F_1((A_1 + B_{11}K_1)x_1 + B_{12}K_2x_2) - F_1(x_1) + L_1(x_1, K_1x_1) - d_1 \leq 0 \qquad \forall x_2 \in \Omega_2 \tag{2.44}$$

$$F_2((A_2 + B_{22}K_2)x_2 + B_{21}K_1x_1) - F_1(x_2) + L_1(x_2, K_2x_2) - d_2 \leq 0 \qquad \forall x_1 \in \Omega_1 \tag{2.45}$$

$$x_1 \in \Omega_1 \rightarrow (A_1 + B_{11}K_1)x_1 + B_{12}K_2x_2 \quad \in \Omega_1 \qquad \forall x_2 \in \Omega_2 \tag{2.46}$$

$$x_2 \in \Omega_2 \rightarrow (A_2 + B_{22}K_2)x_2 + B_{21}K_1x_1 \quad \in \Omega_2 \qquad \forall x_1 \in \Omega_1 \tag{2.47}$$

$$K_1x_1 \in \mathcal{U}_1 \qquad \forall x_1 \in \Omega_1 \tag{2.48}$$

$$K_2x_2 \in \mathcal{U}_2 \qquad \forall x_2 \in \Omega_2 \tag{2.49}$$

$$\Omega_1 \in \mathcal{X}_1 \tag{2.50}$$

$$\Omega_2 \in \mathcal{X}_2 \tag{2.51}$$

Then, if at $t = 0$, $U_1^s(0)$, $U_2^s(0)$ are given such that problems (2.37) and (2.38) are feasible for $x_{1,0} = x_1(0), x_{2,0} = x_2(0)$, $U_1 = U_1^s(0)$, and $U_2 = U_2^s(0)$, then the proposed algorithm is feasible for all time steps $t \geq 0$, and system (2.36) in closed-loop with the proposed DMPC controller is ultimately bounded in a region that contains the origin in its interior.

Proof. The proof consists of two parts. We first prove recursive feasibility of problems (2.37) and (2.38) if at time t, $U_1^s(t)$, $U_2^s(t)$ are given such that (2.37) and (2.38) are feasible for $x_{1,0} = x_1(t), x_{2,0} = x_2(t)$, $U_1 = U_1^s(t)$, and $U_2 = U_2^s(t)$. Then we prove that under the stated assumptions

$$J(t) = J_1(x_1(t), U_1^d(t), U_2^d(t)) + J_2(x_2(t), U_2^d(t), U_1^d(t))$$

is a decreasing sequence of values with a lower bound. This implies that system (2.1) in closed-loop with the proposed DMPC controller is ultimately bounded in a region that contains the origin in its interior.

Part 1: We will prove this part by recursion. First, we prove that if the state and input trajectories obtained from $x_1(t-1)$ and $x_2(t-1)$ applying $U_1^d(t-1)$ and

$U_2^d(t-1)$ satisfy all the constraints of problems (2.37) and (2.38), then $U_1^d(t)$ and $U_2^d(t)$ also satisfy all the constraints. Recalling step 5 of the proposed algorithm, to prove this statement it is sufficient to prove that there exists at least a pair of input trajectories that satisfy all the constraints. To this end, we will prove that $U_1^s(t)$, $U_2^s(t)$ provide a feasible solution for $x_1(t)$ and $x_2(t)$. Note that, in general, it is not possible to guarantee that any of the other options are feasible.

Taking into account that by definition $U_1^d(t-1)$ and $U_2^d(t-1)$ satisfy the constraints of problems (2.37) and (2.38), the following statements hold:

$$\begin{aligned} &x_{1,k} \in \mathcal{X}_1, \quad k = 0, \ldots, N \\ &u_{1,k}^d \in \mathcal{U}_1, \quad k = 0, \ldots, N-1 \\ &x_{1,N} \in \Omega_1 \\ &x_{2,k} \in \mathcal{X}_2, \quad k = 0, \ldots, N \\ &u_{2,k}^d \in \mathcal{U}_2, \quad k = 0, \ldots, N-1 \\ &x_{2,N} \in \Omega_2 \end{aligned}$$

where $x_{1,k}, x_{2,k}$ are the k-steps ahead predicted state obtained from $x_1(t-1)$, $x_2(t-1)$, respectively, applying the input trajectories $U_1^d(t-1)$, $U_2^d(t-1)$ defined by $u_{1,k}^d, u_{2,k}^d$ with $k = 0, \ldots, N-1$. At time step $t-1$, the first input of the chosen trajectories $U_1^d(t-1)$, $U_2^d(t-1)$ is applied, that is, $u_1(t-1) = u_{1,0}^d$ and $u_2(t-1) = u_{2,0}^d$. This implies that

$$\begin{aligned} x_1(t) &= A_1 x_1(t-1) + B_{11} u_1(t-1) + B_{12} u_2(t-1) \\ &= A_1 x_{1,0} + B_{11} u_{1,0}^d + B_{12} u_{2,0}^d = x_{1,1} \end{aligned}$$

Taking into account the definitions of $U_1^s(t)$ and $U_2^s(t)$, it can be proved that the k-steps ahead predicted states obtained from $x_1(t), x_2(t)$, respectively, applying the input trajectories $U_1^s(t)$, $U_2^s(t)$ satisfy all the constraints from $k = 0$ to $N-1$. Moreover, as

$$x_{1,N} \in \Omega_1, \quad x_2 \in \Omega_2$$

it holds that

$$\begin{aligned} &(A_1 + B_{11} K_1) x_{1,N} + B_{12} K_2 x_{2,N} \in \Omega_1 \\ &(A_2 + B_{22} K_2) x_{2,N} + B_{21} K_1 x_{1,N} \in \Omega_2 \end{aligned}$$

and hence all the constraints of problems (2.37) and (2.38) are satisfied which implies that $U_1^s(t)$, $U_2^s(t)$ and hence $U_1^d(t)$, $U_2^d(t)$ provide a feasible solution for $x_1(t)$ and $x_2(t)$. Taking into account that by assumption, $U_1^s(0)$, $U_2^s(0)$ satisfy all the constraints for $x_1(0)$ and $x_2(0)$ and using the above result recursively, the statement of this part is proved.

Part 2: In this part we will prove that

$$J_1(x_1(t), U_1^s(t), U_2^s(t)) + J_2(x_2(t), U_2^s(t), U_1^s(t)) \leq J(t-1) + d_1 + d_2$$

where

$$
\begin{aligned}
J(t-1) = {} & J_1(x_1(t-1), U_1^d(t-1), U_2^d(t-1)) \\
& + J_2(x_2(t-1), U_2^d(t-1), U_1^d(t-1))
\end{aligned}
$$

Taking into account the definitions of $U_1^d(t-1)$ and $U_1^s(t)$, it follows that

$$
J_1(x_1(t), U_1^s(t), U_2^s(t)) - J_1(x_1(t-1), U_1^d(t-1), U_2^d(t-1))
$$

is equal to

$$
\begin{aligned}
& F_1((A_1 + B_{11}K_1)x_{1,N} + B_{12}K_2x_{2,N}) - F_1(x_{1,N}) \\
& \quad + L_1(x_{1,N}, K_1x_{1,N}) - L_1(x_{1,0}, K_1x_{1,0})
\end{aligned}
$$

As $L_1(x_{1,0}, K_1x_{1,0}) \geq 0$ and taking into account (2.44) and (2.45) that $x_{1,N} \in \Omega_1$ and that $x_{2,N} \in \Omega_2$, it follows that

$$
J_1(x_1(t), U_1^s(t), U_2^s(t)) - J_1(x_1(t-1), U_1^d(t-1), U_2^d(t-1)) - d_1 \leq 0
$$

Following the same steps for J_2 we obtain that

$$
J_2(x_2(t), U_2^s(t), U_1^s(t)) - J_2(x_2(t-1), U_2^d(t-1), U_1^d(t-1)) - d_2 \leq 0
$$

and hence

$$
J_1(x_1(t), U_1^s(t), U_2^s(t)) + J_2(x_2(t), U_2^s(t), U_1^s(t)) \leq J(t-1) + d_1 + d_2
$$

As the proposed algorithm chooses $U_1^d(t), U_2^d(t)$ as the pair of input trajectories that yield the minimum cost, it is easy to see that

$$
J(t) \leq J(t-1) + d_1 + d_2
$$

Following standard Lyapunov arguments and taking into account that recursive feasibility is guaranteed (see the first part of the proof), it is proved that system (2.36) in closed-loop with the proposed controller is ultimately bounded in a region that contains the origin in its interior. □

2.4.4 *Design procedure*

In the previous section, we have provided sufficient conditions to guarantee that the system in closed-loop with the proposed DMPC scheme is practically stable. In general, designing the controller parameters so that these conditions are satisfied is a hard problem because the design constraints are coupled; for example, the constraints that define the invariant set Ω_1 depend on the set Ω_2 and vice versa. For centralized MPC controllers, there are various methods described in the literature on how to design a stabilizing local controller, terminal cost function and terminal region [93,174] (e.g., the local controller and the terminal cost can be obtained solving an linear quadratic regulator (LQR) problem). These results, however, cannot be applied to the distributed case. In this section, we present an optimization-based procedure to find local controllers K_1, K_2, matrices P_1, P_2, and regions Ω_1, Ω_2 such that (2.44) holds for a given system.

The procedure determines first matrices K_1, K_2, P_1, and P_2 such that (2.44) and (2.45) hold for any given sets Ω_1 and Ω_2 solving a LMI optimization problem. Once the local feedbacks K_1 and K_2 are fixed, the invariant sets Ω_1 and Ω_2 are obtained. Note that constants d_1 and d_2 are determined *a posteriori*, once the local feedbacks, terminal costs, and terminal regions are fixed.

From the point of view of each agent, its neighbor's input can be viewed as a disturbance. This allows us to use well-known tools from control of linear uncertain systems in order to determine a local controller such that a given degree of robustness is guaranteed. In [175–177], several methods to solve this class of problems are presented. In particular, constraint (2.44) can be transformed into an LMI and solved using standard techniques; moreover, it is equivalent to designing an $\mathcal{H}$-infinity controller for subsystem 1 assuming that u_2 is the disturbance [177]. The same technique can be followed to design K_2 and P_2. The following theorem defines an LMI constraint that only depends on the system matrices that guarantee that there exists K_1, K_2, P_1, and P_2 such that (2.44) and (2.45) hold.

Theorem 2.8: *Consider system (2.36). If there exists matrices W_i, Y_i and a constant γ_i such that the following inequality holds:*

$$\begin{bmatrix} \gamma_i I & 0 & B_{i,ni}^T & 0 & 0 \\ * & W_i & W_i A_{ii}^T + Y_i^T B_{ii}^T & W_i Q_i^{1/2} & Y_i^T \Re_i^{1/2} \\ * & * & W_i & 0 & 0 \\ * & * & * & I & 0 \\ * & * & * & * & I \end{bmatrix} > 0 \tag{2.52}$$

then (2.44) (or (2.45), depending on the agent) is satisfied for

$$\begin{aligned} P_i &= W_i^{-1}, K_i = Y_i W_i^{-1} \\ d_i &= \gamma_i \max_{x_i \in \Omega_{ni}} (K_{ni} x)^T K_{ni} x \end{aligned}$$

Proof. In [177], it is proved that if (2.52) holds, then the following constraint is satisfied:

$$\begin{aligned} &F_i((A_i + B_{ii}K_i)x_i + B_{i,ni}v) - F_i(x_i) \\ &\quad + Li(x_i, K_i x_i) - \gamma_i v^T v \le 0 \quad \forall v \end{aligned} \tag{2.53}$$

It follows that (2.44) (or (2.45), depending on the agent) holds for

$$d_i = \gamma_i \max_{x \in \Omega_{ni}} (K_{ni} x)^T K_{ni} x$$

□

Once the local controllers and the terminal cost functions are fixed, in order to design a DMPC scheme that satisfies the assumptions of Theorem 2.7, one needs to find sets Ω_1, Ω_2 such that (2.46)–(2.50) hold. In general this is a difficult problem because each of the sets depends on the other. The size of the terminal region for agent 1 is determined by the magnitude of the disturbances induced by its neighbor agent 2 and vice versa. We provide next an optimization-based procedure to solve this problem. In order to present the algorithm we need the following definitions.

Definition 2.1: Given the following discrete-time linear system subject to bounded additive uncertainties:

$$x^+ = Ax + Bu + Dw$$

with $w \in \mathcal{W}$ subject to constraints in the state and the input $x \in \mathcal{X}, u \in \mathcal{U}$ and a linear feedback $u = Kx$; a set Ω is said to be a robust positive invariant set for the system if the following constraints hold:

$$\begin{aligned} x \in \Omega &\to (A + BK)x + BKx \in \Omega \quad \forall w \in \mathcal{W} \\ Kx &\in \mathcal{U} \\ \Omega &\in \mathcal{X} \end{aligned}$$

Given system matrices A, B, D, K and the sets $\mathcal{X}$, $\mathcal{U}$, $\mathcal{W}$, there exists several methods to find a set Ω that satisfies these constraints, see, for example, [178] for a procedure to find the maximal robust positive invariant and [179] for a procedure to find an approximation of the minimal robust positive invariant. We denote $\Omega(A, B, D, \mathcal{K}, \mathcal{X}, \mathcal{U}, \mathcal{W})$ the corresponding maximal robust positive invariant set.

Taking into account that the input of the neighbor agent can be considered as an unknown bounded disturbance, in order to decouple the computation of the sets Ω_1 and Ω_2, we use the following result based on finding a robust positive invariant set for each subsystem:

Theorem 2.9: *Given constants* $\lambda_1 \in (0, 1]$ *and* $\lambda_2 \in (0, 1]$*, if the sets defined as*

$$\begin{aligned} \Omega_1 &= \Omega(A_1, B_{11}, B_{12}, \mathcal{X}_1, K_1, \lambda_1\mathcal{U}_1, \lambda_2\mathcal{U}_2) \\ \Omega_2 &= \Omega(A_2, B_{22}, B_{21}, \mathcal{X}_2, K_2, \lambda_2\mathcal{U}_2, \lambda_1\mathcal{U}_1) \end{aligned}$$

are not empty, then constraints (2.46)–(2.50) are satisfied.

Proof. The theorem is easily proved taking into account the definition of the operator Ω and that $\lambda_1\mathcal{U}_1 \subseteq \mathcal{U}_1$ and $\lambda_2\mathcal{U}_2 \subseteq \mathcal{U}_2$.

The main idea is that to determine the invariant sets both agents limit their inputs by a factor $\lambda_i \in (0, 1]$ with $i = 1, 2$, hence the other agent can find the maximal robust positive invariant set with respect to a known bounded disturbance. For example, agent 1 finds the maximal robust positive invariant with respect to a disturbance bounded in $\lambda_2\mathcal{U}_2$ assuming that its input is bounded in $\lambda_1\mathcal{U}_1$. Agent 2 does the same. Note that these sets may be empty depending on the value of λ_1 and λ_2. If both sets exists, then they satisfy the stability constraints. In general, an infinite number of possible values of λ_1 and λ_2 such that both sets are nonempty may exist. In order to choose one, we propose to solve the following optimization problem to maximize the feasibility region of the DMPC controller:

$$\begin{aligned} \max_{\lambda_1 \in (0,1], \lambda_2 \in (0,1]} \quad & f(\Omega_1, \Omega_2) \\ & \Omega_1 = \Omega(A_1, B_{11}, B_{12}, K_1, X_1, \lambda_1 \mathcal{U}_1, \lambda_2 \mathcal{U}_2) \\ & \Omega_2 = \Omega(A_2, B_{22}, B_{21}, K_2, X_2, \lambda_2 \mathcal{U}_2, \lambda_1 \mathcal{U}_1) \end{aligned} \tag{2.54}$$

where $f(\cdot)$ is a measure of the size of a polyhedron (e.g., its Chebychev radius).

Once matrices K_1, K_2, P_1, P_2 and the sets Ω_1 and Ω_2 are determined, constants d_1 and d_2 can be calculated in order to obtain an estimation of the set in which the closed-loop system is ultimately bounded. □

Remark 2.11: *Theorem 2.8 guarantees that the closed-loop system is ultimately bounded in a closed region that contains the origin; however, taking into account that the local controller guarantees that each subsystem is input-to-state stable [55] with respect to the input of the neighbor subsystem, asymptotic stability can be proved under some additional assumptions regarding the decrease rate of the norm of the input with respect to the norm of the state as it approaches the origin.*

In the following section, the theoretical results and the design procedure are illustrated using two different examples. The first example is focused on the controller design procedure. The second controller shows the application of the proposed approach to a supply chain problem.

2.4.5 Illustrative example 2.2

The system considered is composed of two double integrators with coupled inputs. The first subsystem is defined by the following matrices:

$$A_1 = \begin{bmatrix} 1 & 1 \\ 0 & 1 \end{bmatrix}, \quad B_{11} = \begin{bmatrix} 0 \\ 1 \end{bmatrix}, \quad B_{12} = \begin{bmatrix} 0 \\ 0.4 \end{bmatrix}$$

and the second subsystems is defined by

$$A_2 = \begin{bmatrix} 1 & 1 \\ 0 & 1 \end{bmatrix}, \quad B_{22} = \begin{bmatrix} 0 \\ 1 \end{bmatrix}, \quad B_{21} = \begin{bmatrix} 0 \\ 1 \end{bmatrix}$$

The state and the input must satisfy the following constraints:

$$\|x_1\|_\infty \leq 1, \|x_2\|_\infty \leq 2, |u_1| \leq 1, |u_2| \leq 1$$

The stage cost functions of each agent are defined by $Q_i = I$ and $R_i = 1$ with $i = 1, 2$.

In order to determine the local controllers K_i and the corresponding weight matrices P_i that define the terminal cost function, an LMI problem based on (2.10) that minimizes the constant γ_i is solved for each agent. The following matrices are obtained:

$$K_i = [-0.2023 \quad -0.9254], \quad i = 1, 2$$

$$P_1 = \begin{bmatrix} 32.6719 & -17.5149 \\ -17.5149 & 54.6366 \end{bmatrix}, \quad P_2 = \begin{bmatrix} 38.4509 & -5.6447 \\ -5.6447 & 50.1686 \end{bmatrix}$$

The last step necessary to apply the proposed algorithm is to determine an invariant region for the two agents Ω_1 and Ω_2. Different approaches can be used to determine the values of λ_i that maximize the size of the terminal regions. In this example, the terminal region was calculated for a grid with different values of λ_i. The criterion to select the maximum invariant region was the Chebychev radius of the maximum ball inside the region. The results were $\lambda_1^* = 0.3$ and $\lambda_2^* = 0.5$. Figure 2.8 shows a 3D plot of the Chebychev radius as a function of λ_1 and λ_2.

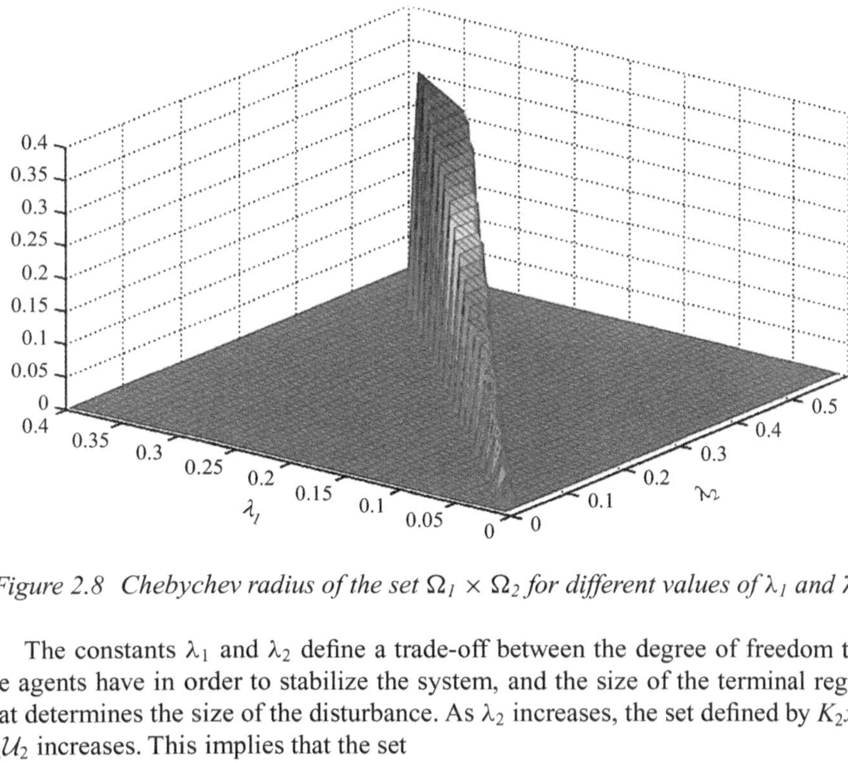

Figure 2.8 Chebychev radius of the set $\Omega_1 \times \Omega_2$ for different values of λ_1 and λ_2

The constants λ_1 and λ_2 define a trade-off between the degree of freedom that the agents have in order to stabilize the system, and the size of the terminal region that determines the size of the disturbance. As λ_2 increases, the set defined by $K_2x \in \lambda_2\mathcal{U}_2$ increases. This implies that the set

$$\Omega_2 = \omega(A_2, B_{22}, B_{21}, K_2, \mathcal{X}_2, \lambda_2\mathcal{U}_2, \lambda_1\mathcal{U}_1)$$

becomes larger because the feasibility region of the input is larger, while the set $\Omega_1 = \omega(A_1, B_{11}, B_{12}, K_1, \mathcal{X}_1, \lambda_1\mathcal{U}_1, \lambda_2\mathcal{U}_2)$ has to take into account bigger disturbances and may even cease to be defined (i.e., it is empty). This happens when the minimum positive robust invariant set for an uncertainty bounded in $\lambda_2\mathcal{U}_2$ is not included in the feasibility region defined by $\mathcal{X}_1$ and $K_1x_1 \in \lambda_1\mathcal{U}_1$. In Figure 2.9(a), inner approximations of the minimum positive invariant sets of subsystem 1 for different values of λ_2 and a fixed value of λ_1 are shown. It can be seen that for large values of λ_2, the inner approximation is not contained in the feasibility region of agent 2 (shown in dashed line), and hence, it is empty. In Figure 2.9(b) the maximum positive invariant set for the same values of λ_1 and a fixed value of λ_2 are shown. It can be seen how the size of the set always increases with λ_2.

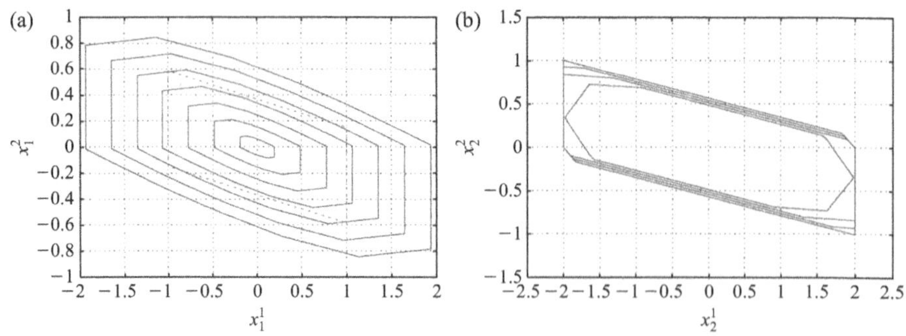

Figure 2.9 *(a) Minimum robust invariant set for agent 1 as a function of* λ_2 *with* λ_1 *fixed and (b) maximum robust invariant set for agent 2 as a function of* λ_2 *with* λ_1 *fixed*

In the first time step, a feasible solution for the centralized problem is used as the shifted trajectories. Simulations results are shown in the next figures for the following initial conditions:

$$x_1(0) = \begin{bmatrix} 0.7 \\ -1 \end{bmatrix}, \quad x_2(0) = \begin{bmatrix} 1 \\ 0.8 \end{bmatrix}$$

Figure 2.10 shows the trajectories of the states of each agent, the inputs, and the cost index. Figure 2.11(a,b) show the state trajectories of each agent along with its corresponding invariant set.

2.4.6 *Illustrative example 2.3*

In this section, we apply the proposed controller to a reduced version of the MIT beer game and compare the performance with other control schemes. The MIT beer game was developed by Jay Forrester in the late 1950s to show his management students how oscillations arise in a supply chain (see [180]). A supply chain is the set of structures and processes used by an organization to provide a service or a good to a consumer. Typically three phenomena take place in supply chains flows: oscillation, amplification, and phase lag. Owing to material or informational delays in the flows of the supply chain, they are prone to oscillation, that is, production and inventories overshoot and undershoot the optimal levels. The magnitude of the fluctuations increases as they propagate from the customer to the factory, with each upstream stage tending to lag behind the downstream one in what is commonly known as the bull whip effect.

The original MIT beer game is composed of four agents: retailer, wholesaler, distributor, and factory. Customers demand beer from the first one, who orders beer from the wholesaler, who orders and receives beer from the distributor, who finally orders and receives orders from the factory. There are shipping and processing delays at each stage. In [180], the original model and all the difficulties of the corresponding

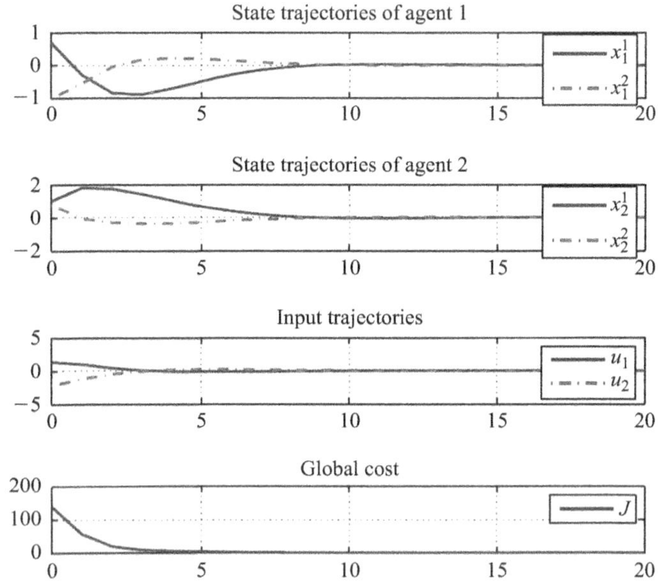

Figure 2.10 State, input and global cost $J(t)$ trajectories of the closed-loop double integrators

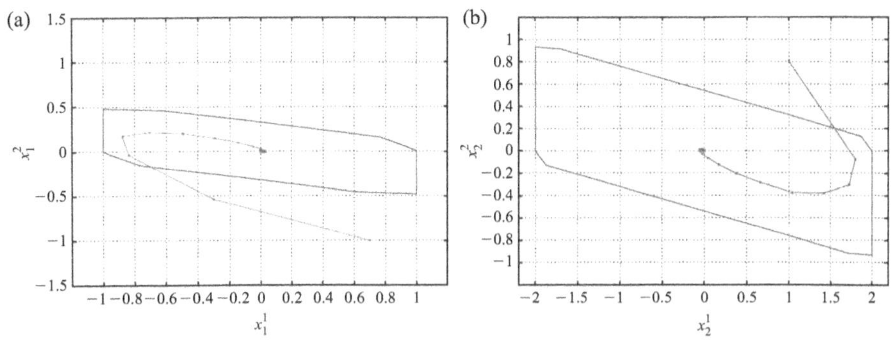

Figure 2.11 (a) Agent 1 state evolution and (b) agent 2 state evolution

stock management problem are explained in detail. This problem has been widely used in the literature. In particular, in [57] it has been used as application example for a DMPC scheme. The main difference between the proposed scheme and the DMPC proposed in [181] is that the agents only communicated once and the only information shared was the future input trajectories (a strategy similar to Iter 1).

In what follows, a reduced version of the problem with two agents is considered: the retailer and its supplier (see Figure 2.12). There is no loss of generality since the

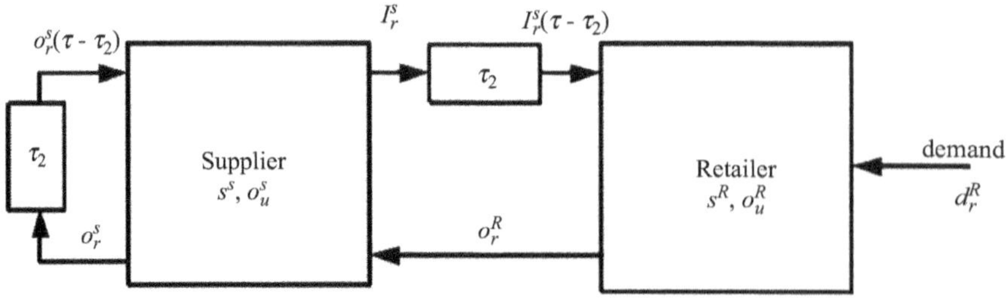

Figure 2.12 Reduced beer game

structure of the game is regular: there is a cascade of firms, each maintaining and controlling its stock. The continuous time equations for the supplier are [56]:

$$\begin{aligned}
\dot{s}^S(\tau) &= o_{\Re}^S(\tau - \tau_2) - o_{\Re}^R(\tau - \tau_1) - b^S(\tau)/t_b \\
\dot{o}^R(\tau) &= o_{\Re}^R(\tau) - o_{\Re}^R(\tau - \tau_2) \\
\dot{b}^S(\tau) &= o_{\Re}^R(\tau) - o_{\Re}^R(\tau - \tau_1) - b^S(\tau)/t_b
\end{aligned} \tag{2.55}$$

The equations for the retailer are

$$\begin{aligned}
\dot{s}^S(\tau) &= o_{\Re}^R(\tau - \tau_1 - \tau_2) + b^S(\tau - \tau_2)/t_b - d_r(\tau - \tau_1) - b^R(\tau)/t_b \\
\dot{o}_u^R(\tau) &= o_{\Re}^R(\tau) - o_{\Re}^R(\tau - \tau_1 - \tau_2) - b^S(\tau - \tau_2)/t_b \\
\dot{b}^R(\tau) &= d_{\Re}^R(\tau) - d_{\Re}^R(\tau - \tau_1) - b^R(\tau)/t_b
\end{aligned} \tag{2.56}$$

The superscripts R, S denote variables from the retailer and the supplier, respectively. Variable $s^i(\tau)$ is the stock level, that is, the number of items available in that stage for shipment downstream. The unfulfilled order of stock $o_u^i(\tau)$ stands for the number of ordered items that the agent is waiting to receive from the upstream stage. The backlog of unfulfilled orders $b^i(\tau)$ accounts for the number of committed items that have to be shipped to the downstream stage. The parameter t_b stands for the average backlog clearance time and introduces a first-order dynamic in the process. The customer demand $d_r^R(\tau)$ represents how many items are demanded by the customers. From a control point of view, it can be seen as a measurable perturbation that has to be rejected in order to maintain the stock and the production at the desired levels. The information flows are assumed to have no time delays and the material flows have a delay modeled by τ_2. A delay for processing the received orders is introduced by means of the parameter τ_1. The manipulated variable at each stage is the order rate o_r^i, that is, the number of items demanded upstream. The supplier demands directly to the factory, which is modeled here as a pure delay. This model is different from other supply chain models in which each agent has to decide not only what to order downstream but what to send upstream. In this model of the MIT beer game, items sent to the upstream agent are not a decision variable. They are fixed by the orders received. In particular, items sent and the orders received are related through a first-order system with a delay, that is, the shipment rate $l_r^i(\tau)$ is defined by the following equations:

$$
\begin{aligned}
l_r^S(\tau) &= o_r^R(\tau, \tau_1) + b^S(\tau)/t_b \\
l_r^R(\tau) &= d_r^R(\tau, \tau_1) + b^R(\tau)/t_b
\end{aligned}
$$

These relations have already been taken into account in the model.

The model of the system is defined by the parameters τ_1, τ_2, and t_b. In the simulations done in this chapter, we use $\tau_1 = 2d, \tau_2 = 1d$, and $t_b = 4d$. In order to obtain a discrete-time model of the system, the continuous time model of (2.55) and (2.56) is discretized with a sampling time $\Delta = 1d$. Auxiliary states are introduced to take into account the delays. The resulting discrete-time linear model is the one used in all the simulations carried out in this section.

In addition, an integrator is added to the controller, that is, the MPC controller decides the increment on the orders made downstream. This implies that the controller evaluates Δo_r^S and Δo_r^R defined as follows:

$$
\begin{aligned}
\Delta o_r^S(t) &= o_r^S(t) - o_r^S(t-1) \\
\Delta o_r^R(t) &= o_r^R(t) - o_r^R(t-1)
\end{aligned}
$$

The state of the model of the first subsystem (the retailer) is given by

$$
x_1(t) = \begin{bmatrix} s^R(t) \\ o_u^R(t) \\ b^R(t) \\ o_r^R(t-1) \\ o_r^R(t-2) \\ o_r^R(t-3) \\ b^S(t) \\ b^S(t-1) \\ d_r(t-1) \\ d_r(t-2) \end{bmatrix}
$$

The state of the model of the second subsystem (the supplier) is given by

$$
x_2(t) = \begin{bmatrix} s^S(t) \\ o_u^S(t) \\ b^S(t) \\ o_r^S(t-1) \\ o_r^R(t-1) \\ o_r^R(t-2) \end{bmatrix}
$$

It can be seen that both models share some information. In particular, the retailer model needs to keep track of the unfulfilled orders of the supplier, while the supplier model needs to keep track of the orders received of the retailer. The input of the first agent is $u_1 = \Delta o_r^R(t)$ and the input of the second agent is $u_2 = \Delta o_r^S(t)$.

The control objective is to regulate the stock levels and the orders placed by both agents to a desired value. The orders received by the retailer from the external demand forces him to send items upstream, and hence to lose stock. These orders can be seen as external disturbances that have to be rejected. To this end, the retailer sends and order for more items downstream. These orders can be seen as an external disturbance for the supplier, which in order to reject this disturbances generates new items. The retailer's stock has to be regulated to a reference value of $\Re_s^R(t)$. Analogously, the supplier's stock is regulated to a value $\Re_s^S(t)$. The reference signals for the orders are given by $\Re_o^R(t)$ for the retailer and by $\Re_o^S(t)$ for the supplier. Note that in general, the orders references signal should be chosen accordingly with the predicted demand.

To this end, we consider different MPC controllers based on the following cost functions:

$$J_1 = \sum_{k=0}^{N-1} (\Re_{s,k}^R - s_k^R)^2 W_s^R + (\Re_{o,k}^R - o_k^R)^2 W_o^R + (\Delta o_k^R)^2 W_\Delta^R$$

$$J_2 = \sum_{k=0}^{N-1} (\Re_{s,k}^S - s_k^S)^2 W_s^R + (\Re_{o,k}^S - o_k^S)^2 W_o^R + (\Delta o_k^S)^2 W_\Delta^S$$

where N is the prediction horizon, the subindex k denotes the k-steps predicted value of a signal and $W_s^R, W_o^R, W_\Delta^R, W_s^S, W_o^S, W_\Delta^S$ are constant weight matrices that define the stage cost. It is important to remark that no terminal cost function is considered in this example.

Note that in order to obtain predictions for the states of the retailer, an estimation of the future demand is needed. We denote the estimated demand as $d_r^R(t)$. This signal may be different from the actual demand $d_r^R(t)$ in a given simulation.

The following values were used for the controller parameters:

$$N = 6, W_s^R = 30, W_o^R = 30, W_\Delta^R = 1, W_s^S = 30, W_o^S = 30, W_\Delta^S = 1 \qquad (2.57)$$

For these simulations we have considered that the stocks and orders must be nonnegative.

The objective of this section is to compare the performance of different MPC schemes. To this end, we have carried out a set of simulations in four different scenarios for each controller. The first controller considered is the centralized MPC scheme defined by the optimization problem (2.41). This controller decides both inputs with a single optimization problem based on the full model of the system and the global cost function $J = J_1 + J_2$. In general, the centralized MPC provides the best performance and has the higher computational burden. The second control scheme considered is the proposed DMPC controller in which two different agents communicate to find a cooperative solution. In addition, we have considered several controllers based on the iterative controller defined by the optimization problems (2.42), (2.43). To avoid the case in which the agents do not reach an agreement, a maximum number of iterations is fixed. Different controllers with a maximum number of iterations of *1*, *2*, *5*, and *10* have been considered. We denote these controllers as *Iter 1*, *Iter 2*, *Iter 5*, and *Iter 10*, respectively. In the case of convergence in this bargaining process, the agents reach a Nash solution from a game theory point of view. None of them consider the cost function of the other agent. For any given input trajectory proposed by its neighbor, the agent evaluated the best possible input for his performance index. By definition, in a situation of equilibrium, this situation constitutes a Nash equilibrium. It is important to remark that in the controller defined by a single iteration (*Iter 1*), the agents do not reach an agreement. They just advise each other about their predicted inputs. Each agent uses this information to estimate the future behavior of the other one. This is not a cooperative scheme because agents do not have a chance to bargain. From the point of view of each agent, the other actions are simply measurable disturbances.

In order to compare the performance of the controllers, four different scenarios have been taken into account. Each scenario is defined by a different initial state, a different retailer demand, and a different demand forecast. All the simulations are done with the discrete-time model presented before.

Scenario 1: Both agents begin with 250 items in stock. The demand of the system $d_r^R(t)$ is defined in the following way: during the first 15 days its value is 70. After that, it is set to 130 during 10 days and finally it returns to its initial value for 70 days. The estimated demand $d_r^R(t)$ is equal to the real demand.

Scenario 2: Same initial state and estimated demand of the first scenario. In this case, the real demand differs from the estimated demand. At each time step, the real demand is obtained adding a random variable with mean 0 and standard deviation of 15 to the estimated demand.

Scenario 3: Same initial state and real demand of the first scenario. In this case, the estimated demand is supposed to be constant and equal to the latest demand received, that is, the instant demand is extended in time as a forecast.

Scenario 4: Same real and estimated demands of the first scenario. The initial state is below the reference. The retailer has an initial stock of 100 items while the supplier is out of stock.

The results obtained are shown in Tables 2.2–2.5. In these tables, the total accumulated cost and the total CPU time of each simulation is shown. The total CPU time includes not only the time of solving the different optimization problems but also all the additional computations such as evaluating the system model. In the simulations, the distributed schemes have not been implemented in parallel, and hence the centralized and the distributed controllers have the same computational power. The total simulation time provides an estimate of the computational burden of each of the controllers; in particular, it shows that for this particular example the centralized problem has a low computational burden and that the computational burden of

Table 2.2 Results for scenario 1

	J	T_{sim}
Centralized	3.6179e+006	1.4187
DMPC	4.9827e+006	0.8806
Iter 1	2.1866e+007	0.5362
Iter 2	5.6999e+006	0.7200
Iter 5	5.8449e+006	1.7651
Iter 10	4.1679e+006	2.2721

Table 2.3 Results for scenario 2

	J	T_{sim}
Centralized	4.3228e+006	1.44
DMPC	5.3558e+006	0.8890
Iter 1	2.1921e+007	0.5008
Iter 2	6.0941e+006	0.8771
Iter 5	6.2743e+006	1.8354
Iter 10	4.7223e+006	2.2581

Table 2.4 Results for scenario 3

	J	T_{sim}
Centralized	5.9327e+006	1.427
DMPC	9.6698e+006	0.8265
Iter 1	2.2047e+007	0.6887
Iter 2	9.0370e+006	0.8287
Iter 5	1.0595e+007	1.8209
Iter 10	6.2798e+006	1.2073

Table 2.5 Results for scenario 4

	J	T_{sim}
Centralized	7.2608e+006	1.55
DMPC	8.1302e+006	0.9521
Iter 1	2.9982e+007	0.6116
Iter 2	1.0397e+007	0.8542
Iter 5	1.0444e+007	1.8621
Iter 10	9.4679e+006	2.4107

the iterative controllers increases as the maximum number of iterations increases. In addition, for scenario 1 figures are shown for all the different controllers considered.

Some conclusions can be obtained from the preceding experiments. In general, the proposed algorithm provides a performance of the same order of magnitude than the one provided by the centralized MPC which, as expected, has the best results. Regarding the simulation time, it can be seen that for this particular case, the CPU time needed to solve in parallel the sequence of low-order optimization problems is very similar to the time needed to solve the large-scale problem. With respect to the non-cooperative DMPC controllers, the proposed distributed scheme provides a better performance than *Iter 1*, *Iter 2*, and *Iter 5*. The controller *Iter 10* provides a better performance but needs more communication cycles in order to achieve an agreement. Even in this case, the solution is still a Nash equilibrium, hence there is no guarantee that it will provide a good overall performance. Note that the iterative controllers' results show that increasing the number of iterations of the bargaining process does not guarantee an improvement in the performance. It can be seen that *Iter 5* sometimes is worse than *Iter 2* from the performance point of view.

The simulations demonstrate that the proposed distributed scheme provides a good performance with only two communication cycles because it obtains a cooperative solution, that is, the decision is taken in order to optimize a global performance index. The iterative controllers do not take a cooperative decision and this implies that, in general, the solutions provided are worst. This can be clearly seen in Figures 2.13–2.18, which show the closed-loop trajectories of each of the controllers considered for scenario 1. In this scenario, the centralized MPC is able to react in advance to the demand peak, maintaining the stocks close to the references. The trajectories of the proposed DMPC scheme show a larger deviation of the stocks from the references; however, these trajectories do not present oscillations as the trajectories corresponding to *Iter 1*, *Iter 2*, and *Iter 5*. Oscillations are a common result of non-cooperative bargaining processes. In this scenario, *Iter 10*,

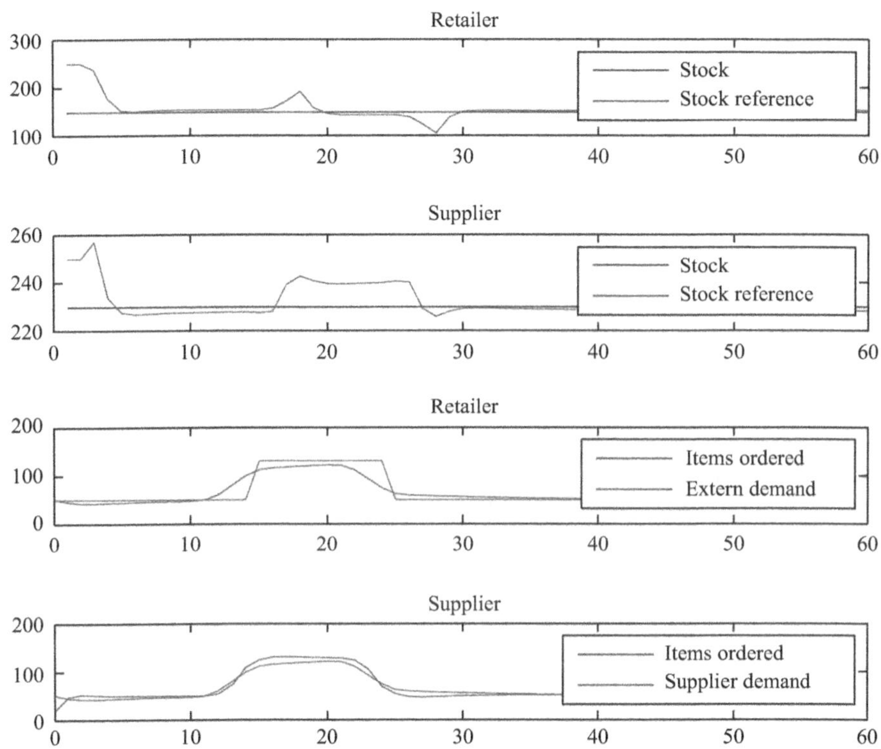

Figure 2.13 Centralized MPC closed-loop trajectories for scenario 1

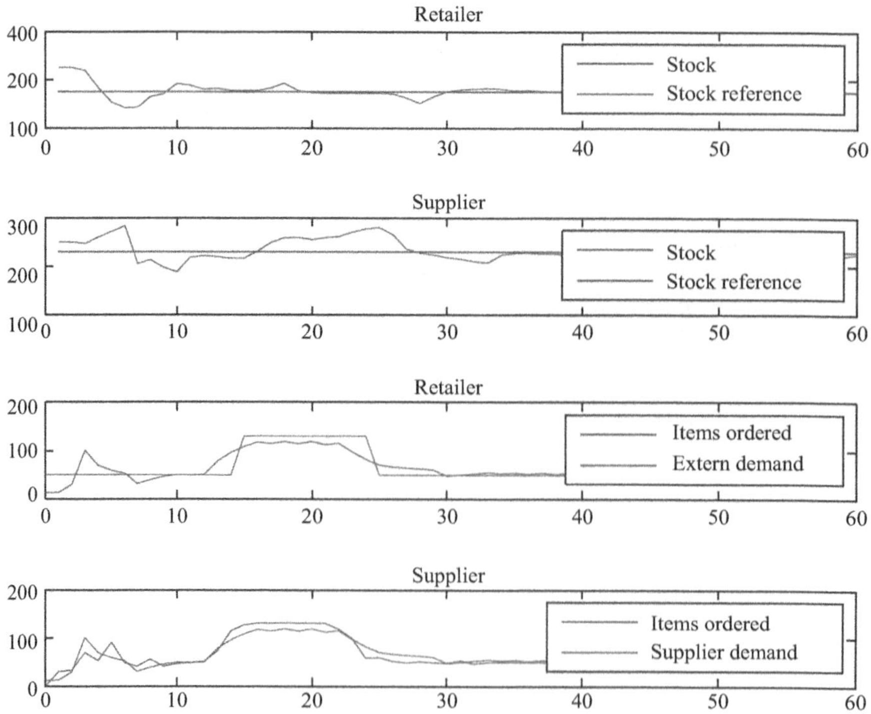

Figure 2.14 Proposed DMPC closed-loop trajectories for scenario 1

however, provides a better response than the proposed DMPC at the cost of a high computational burden and a large number of communication steps.

2.5 Feasible cooperation model predictive control

A DMPC framework was proposed in [131] for the class of systems that have independent subsystem dynamics but are linked through their cost functions. More recently in [167], an extension of the method described in [131] that handles systems with weakly interacting subsystem dynamics was proposed. Stability is proved through the use of a conservative, consistency constraint that forces the predicted and assumed input trajectories to be close to each other. Also, as pointed out by the author, the performance of the DMPC framework in [167] is, in most cases, different from that of centralized MPC.

A DMPC algorithm for unconstrained LTI systems was proposed in [158], where the evolution of the states of each subsystem is assumed to be influenced only by the states of interacting subsystems and local subsystem inputs. This choice of modeling framework can be restrictive. In many cases, such as the two area power network with FACTS device and most chemical plants, the evolution of the subsystem states is also influenced by the inputs of interconnected subsystems. More crucially for the DMPC

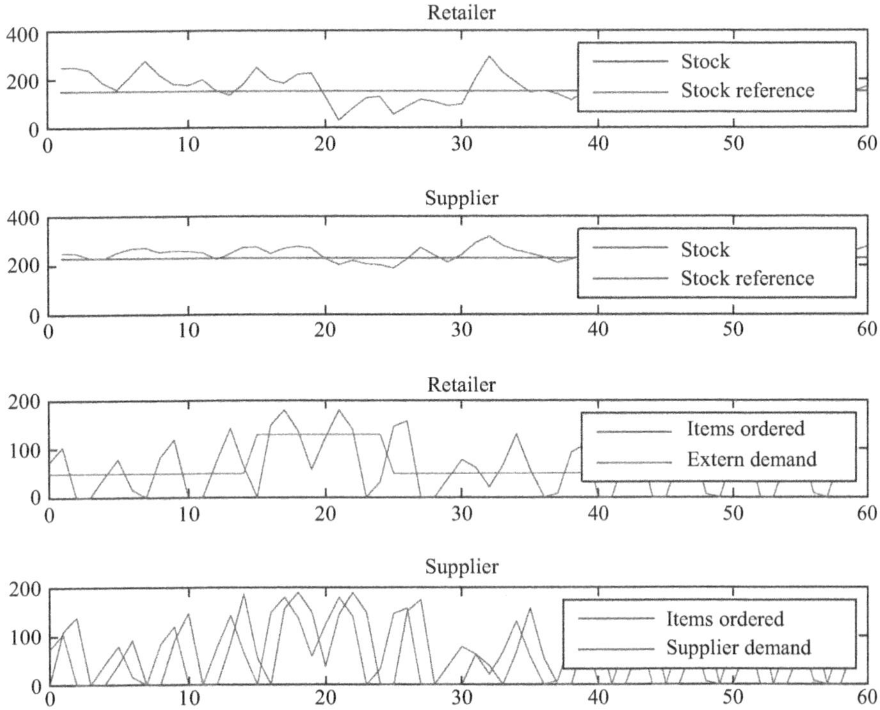

Figure 2.15 Iter 1 closed-loop trajectories for scenario 1

framework proposed in [158], the subsystem-based MPCs have no knowledge of each other's cost/utility functions.

2.5.1 System models

As we learned before, DMPC relies on decomposing the overall system model into appropriate subsystem models. A system composed of interconnected subsystems will be used to establish these concepts.

Centralized model: The overall system model is represented as a discrete, LTI model of the form

$$x(k+1) = Ax(k) + Bu(k), \quad y(k) = Cx(k)$$

in which k denotes discrete time and

$$A = \begin{bmatrix} A_{11} & A_{12} & \cdots & A_{1M} \\ \vdots & \vdots & \ddots & \vdots \\ A_{i1} & A_{i2} & \cdots & A_{iM} \\ \vdots & \vdots & \ddots & \vdots \\ A_{M1} & A_{i2} & \cdots & A_{MM} \end{bmatrix}$$

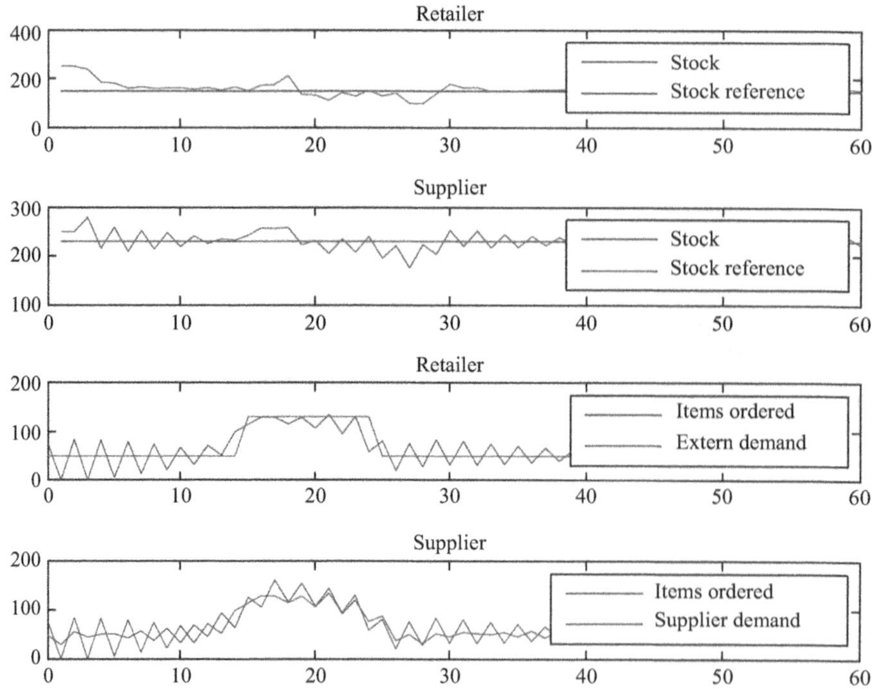

Figure 2.16 Iter 2 closed-loop trajectories for scenario 1

$$B = \begin{bmatrix} B_{11} & B_{12} & \cdots & B_{1M} \\ \vdots & \vdots & \ddots & \vdots \\ B_{i1} & B_{i2} & \cdots & B_{iM} \\ \vdots & \vdots & \ddots & \vdots \\ B_{M1} & B_{i2} & \cdots & B_{MM} \end{bmatrix}$$

$$C = \begin{bmatrix} C_{11} & 0 & \cdots & 0 \\ 0 & C_{22} & \cdots & 0 \\ \vdots & \vdots & \ddots & \vdots \\ 0 & \cdots & \cdots & C_{MM} \end{bmatrix}$$

$$u = \begin{bmatrix} u_1' & u_2' & \cdots & u_M' \end{bmatrix}' \in \mathbb{R}^m$$

$$x = \begin{bmatrix} x_1' & x_2' & \cdots & x_M' \end{bmatrix}' \in \mathbb{R}^n$$

$$y = \begin{bmatrix} y_1' & y_2' & \cdots & y_M' \end{bmatrix}' \in \mathbb{R}^z$$

For each subsystem $i = 1, 2, \ldots, M$, the triplet (u_i, x_i, y_i) represents the subsystem input, state, and output vector, respectively. The centralized model pair is assumed to be stabilizable and is detectable.

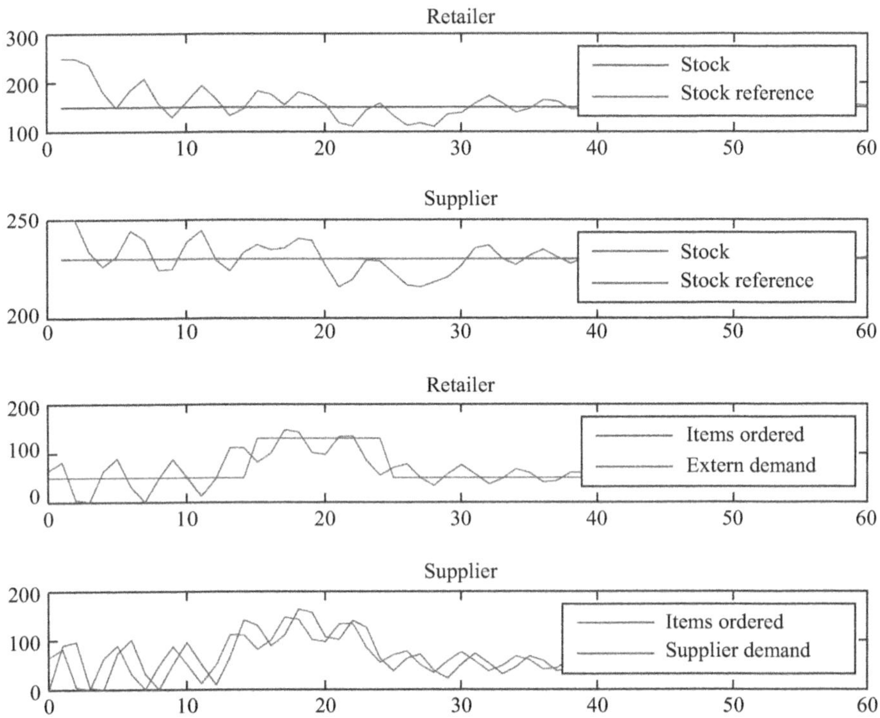

Figure 2.17 Iter 5 closed-loop trajectories for scenario 1

Decentralized model: In the decentralized modeling framework, it is assumed that the interaction between the subsystems is negligible. Subsequently, the effect of the external subsystems on the local subsystem is ignored in this modeling framework. The decentralized model for subsystem $i = 1, 2, \ldots, M$ is

$$x_i(k+1) = A_{ii}x_i(k) + B_{ii}u_i(k), \quad y_i(k) = C_{ii}x_i(k)$$

Partitioned model (PM): The PM for subsystem combines the effect of the local subsystem variables and the effect of the states and inputs of the interconnected subsystems. The PM for subsystem is obtained by considering the relevant partition of the centralized model and can be explicitly written as

$$x_i(k+1) = A_{ii}x_i(k) + B_{ii}u_i(k) + \sum_{j \neq i}(A_{ij}x_j(k) + B_{ij}u_j(k)) \tag{2.58a}$$

$$y_i(k) = C_{ii}x_i(k). \tag{2.58b}$$

2.5.2 Information structures

For any subsystem $i = 1, 2, \ldots, M$, let the predicted state and input at time instant $k+j, j \geq 0$, based on data at time be denoted by $x_i(k+j \mid k) \in \mathbb{R}^{n_i}$ and $u_i(k+j \mid k) \in \Omega_i \subset \mathbb{R}^{m_i}$, respectively, where Ω_i is the set of admissible controls for subsystems i.

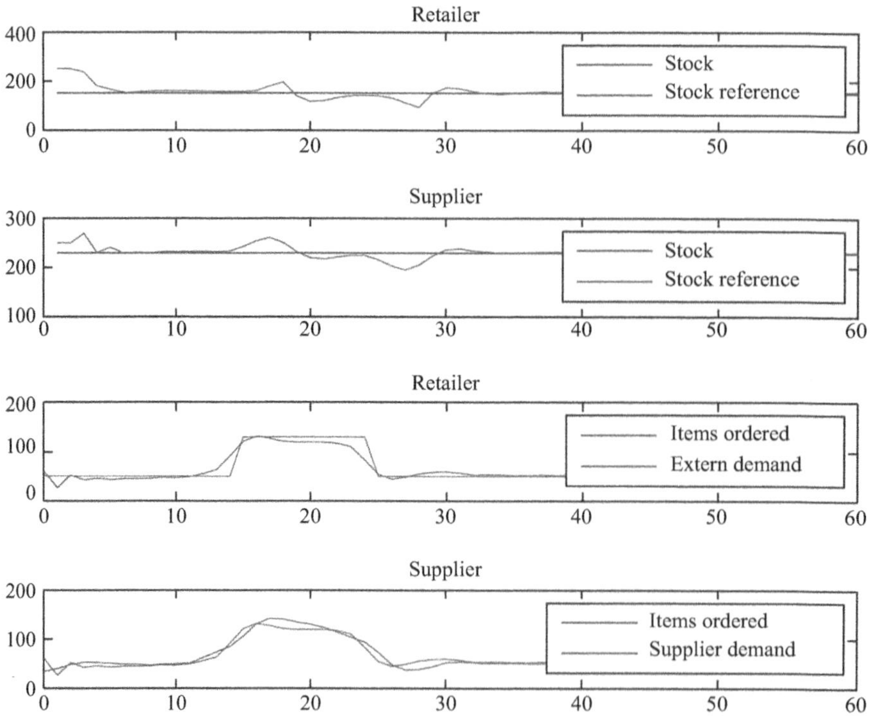

Figure 2.18 Iter10 closed-loop trajectories for scenario 1

We have the following information structures for the infinite horizon predicted state and input trajectory vectors in the different MPC frameworks:

Centralized state trajectory

$$\mathbf{x}^T(k) = [x(k+1 \mid k)', x(k+2 \mid k)^T, \ldots]$$

Centralized input trajectory

$$\mathbf{u}^T(k) = [u(k \mid k)', u(k+1 \mid k)^T, \ldots]$$

State trajectory

$$\mathbf{x}_i^T(k) = [x_i(k+1 \mid k)', x_i(k+2 \mid k)^T, \ldots]$$

Input trajectory

$$\mathbf{u}_i^T(k) = [u_i(k \mid k)', u_i(k+1 \mid k)^T, \ldots]$$

Let N denote the control horizon. Define $\mathcal{U}_i = \Omega_i \times \Omega_i \times \cdots \times \Omega_i \subset \mathbb{R}^{m_i N}$. The following notation is used to represent the finite horizon predicted state and input trajectory vectors in the different MPC frameworks:

Centralized state trajectory

$$\bar{\mathbf{x}}^T(k) = [x(k+1 \mid k)', x(k+2 \mid k)', \ldots, x(k+N \mid k)^T]$$

Centralized input trajectory

$$\bar{\mathbf{u}}^T(k) = [u(k \mid k)', u(k+1 \mid k)', \ldots, u(k+N-1 \mid k)^T]$$

State trajectory

$$\bar{\mathbf{x}}_i^T(k) = [x_i(k+1 \mid k)', x_i(k+2 \mid k)', \ldots, x_i(k+N \mid k)^T]$$

Input trajectory

$$\bar{\mathbf{u}}_i^T(k) = [u_i(k \mid k)', u_i(k+1 \mid k)', \ldots, u_i(k+N-1 \mid k)^T]$$

2.5.3 *MPC frameworks*

Let each set of admissible controls Ω_i be a nonempty, compact, convex set with $0 \in \text{int}(\Omega_i)$. The set of admissible controls for the whole plant Ω is defined to be the Cartesian product of the admissible control sets $\Omega_i, \forall i = 1, 2, \ldots, M$.

The *stage cost* at stage along the prediction horizon is defined as

$$L_i(x_i(t \mid k), u_i(t \mid k)) = \frac{1}{2}[x_i(t \mid k)'Q_i x_i(t \mid k) + u_i(t \mid k)'R_i u_i(t \mid k)] \tag{2.59}$$

in which $Q_i \geq 0$, $R_i > 0$ are symmetric weighting matrices and $(A_i, Q_i^{1/2})$ is detectable. The *cost function* for subsystem is defined over an infinite horizon and is written as

$$\phi_i(x_i, u_i; x_i(k)) = \sum_{t=k}^{\infty} L_i(x_i(t \mid k), u_i(t \mid k)) \tag{2.60}$$

with $x_i(k \mid k) \equiv x_i(k)$. For any system, the constrained stabilizable set (also termed *Null controllable domain*) $\mathcal{X}$ is the set of all initial states $x \subseteq \mathbb{R}^n$ that can be steered to the origin by applying a sequence of admissible controls. It is assumed throughout that the initial system state vector $x(k) \in \mathcal{X}$, in which $\mathcal{X}$ denotes the constrained stabilizable set for the overall system. A feasible solution to the corresponding optimization problem, therefore, exists. For notational simplicity, we drop the time dependence of the state and input trajectories in each MPC framework. For instance, in the centralized MPC framework, we write $\mathbf{x} \leftarrow \mathbf{x}(k)$ and $\mathbf{u} \leftarrow \mathbf{u}(k)$. In the DMPC framework, we use $\mathbf{x}_i \leftarrow \mathbf{x}_i(k)$ and $\mathbf{u}_i \leftarrow \mathbf{u}_i(k)$, $\forall i = 1, 2, \ldots, M$.

Five MPC-based control frameworks are described in the following. In each MPC framework, the controller is defined by implementing the first input of the solution to the corresponding optimization problem.

Centralized MPC: In the centralized MPC framework, the MPC for the overall system solves the following optimization problem:

$$\begin{aligned} &\min_{\mathbf{x},\mathbf{u}} \phi(\mathbf{x}, \mathbf{u}; x(k)) = \sum_i w_i \phi_i(\mathbf{x}_i, \mathbf{u}_i; x_i(k)) \\ &\text{subject to } x(l+1 \mid k) = Ax(l \mid k) + Bu(l \mid k), \quad k \leq l \\ &\qquad u_i(l \mid k) \in \Omega_i, k \leq l, \quad i = 1, 2, \ldots, M \end{aligned} \tag{2.61}$$

where $w_i > 0, \sum w_i = 1$.

For any system, centralized MPC achieves the best attainable performance (Pareto optimal) as the effect of interconnections among subsystems are accounted for exactly. Furthermore, any conflicts among controller objectives are resolved optimally.

Decentralized MPC: In the decentralized MPC, each subsystem-based MPC solves the following optimization problem:

$$\begin{aligned} &\min_{\mathbf{x}_i,\mathbf{u}_i} \phi_i(\mathbf{x}_i, \mathbf{u}_i; x_i(k)) \\ &\text{subject to}\; x_i(l+1 \mid k) = A_{ii}x_i(l \mid k) + B_{ii}u_i(l \mid k), \quad k \leq l \\ &\qquad u_i(l \mid k) \in \Omega_i, \quad k \leq l \end{aligned} \tag{2.62}$$

Each decentralized MPC solves an optimization problem to minimize its (local) cost function. The effects of the interconnected subsystems are assumed to be negligible and are ignored. In many situations, however, the previous assumption is not valid and leads to reduced control performance.

Distributed MPC: The partitioned model for each subsystem $i = 1, 2, \ldots, M$ is assumed to be available. Two formulations for DMPC, namely communication-based MPC and cooperation-based MPC, are considered. Communication based strategies form the basis for the DMPC formulations in [65] and [92]. In the sequel, the suitability of pure communication-based MPC, as a candidate systemwide control formulation, is assessed. For both communication and cooperation-based MPC, several subsystem optimizations and exchanges of variables between subsystems are performed during a sample time. An optimization and exchange of variables is termed an *iterate*. We may choose not to iterate to convergence. The iteration number is denoted by p.

Communication-based MPC: For communication-based MPC,[1] the optimal state-input trajectory $(\mathbf{x}_i^p, \mathbf{u}_i^p)$ for subsystem i, $i = 1, 2, \ldots, M$ at iterate is obtained as the solution to the optimization problem

$$\begin{aligned} \min_{\mathbf{x}_i,\mathbf{u}_i} \quad & \phi(\mathbf{x}_i, \mathbf{u}_i; x_i(k)) \\ \text{subject to} \quad & x_i(l+1 \mid k) = A_{ii}x_i(l \mid k) + B_{ii}u_i(l \mid k) \\ & \qquad + \sum_{j \neq i} \left[A_{ij}x_j^{p-1}(l \mid k) + B_{ij}u_j^{p-1}(l \mid k)\right], \quad k \leq l \\ & u_i(l \mid k) \in \Omega_i, \quad k \leq l \end{aligned}$$

Each communication-based MPC utilizes the objective function for that subsystem only. For each subsystem i at iteration p, only that subsystem input sequence $\mathbf{u}_i$ is optimized and updated. The other subsystems' inputs remain at $\mathbf{u}_j^{p-1}$, $\forall j = 1, 2, \ldots, M$, $j \neq i$. If the communication-based iterates converge, then at convergence, a Nash equilibrium is achieved. In this work, the term *communication-based MPC* alludes to the previous framework at convergence of the exchanged trajectories. Examples are presented in Section 2.5.7 for which communication-based MPC leads to either unacceptable closed-loop performance or closed-loop instability.

[1] Similar strategies have been proposed by [8] and [20].

Feasible cooperation-based MPC (FC-MPC): To arrive at a reliable DMPC framework, we need to ensure that the subsystems' MPCs cooperate, rather than compete, with each other in achieving systemwide objectives. The local controller objective $\phi_i(\cdot)$ is replaced by an objective that measures the systemwide impact of local control actions. The simplest choice for such an objective is a strict convex combination of the controller objectives, that is, $\phi(\cdot) = \sum_i w_i \phi_i(\cdot)$, $w_i > 0$, $\sum_i w_i = 1$.

In large-scale implementations, the system sampling interval may be insufficient to allow convergence of an iterative, cooperation-based algorithm. In such cases, the cooperation-based algorithm has to be terminated prior to convergence of exchanged trajectories. The final calculated input trajectories are used to define a suitable DMPC control law. To enable intermediate termination, it is necessary that all iterates generated by the cooperation-based algorithm are strictly *systemwide feasible* (i.e., satisfy all model and inequality constraints) and the resulting nominal distributed control law is closed-loop stable. Such a DMPC algorithm is presented in Section 2.5.

For notational convenience, we drop the $k$dependence of $\bar{\mathbf{x}}_i(k)$, $\bar{\mathbf{u}}_i(k)$, $i = 1, 2, \ldots, M$. It is shown in [3] that each $\bar{\mathbf{x}}_i$ can be expressed as

$$\bar{\mathbf{x}}_i = E_{ii}\bar{\mathbf{u}}_i + f_{ii}x_i(k) + \sum_{j \neq i} [E_{ij}\bar{\mathbf{u}}_j + f_{ij}x_j(k)] \tag{2.63}$$

We consider the more practical case of open-loop stable systems first. A DMPC methodology capable of handling large, open-loop unstable systems is described in Section 2.6.

For open-loop stable systems, the FC-MPC optimization problem for subsystem i, denoted as $\mathcal{F}_i$, is defined as

$$\mathcal{F}_i \triangleq \min_{\mathbf{u}_i} \sum_{r=1}^{M} w_r \Phi_r(\mathbf{u}_1^{p-1}, \ldots, \mathbf{u}_i^{p-1} - 1, \mathbf{u}_i, \mathbf{u}_{i+1}^{p-1}, \ldots, \mathbf{u}_M^{p-1}; x_r(k)) \tag{2.64a}$$

subject to

$$u_i(t \mid k) \in \Omega_i, \quad k \leq t \leq k + N - 1 \tag{2.64b}$$

$$u_i(t \mid k) = 0, \quad k + N \leq t \tag{2.64c}$$

The infinite horizon input trajectory $\mathbf{u}_i$ is obtained by augmenting $\bar{\mathbf{u}}_i$ with the input sequence $u_i(k \mid k) = 0, k + N \leq t$. The infinite horizon state trajectory $\mathbf{x}_i$ is derived from $\bar{\mathbf{x}}_i$ by propagating the terminal state $x_i(k + N \mid k)$ using (2.1) and $u_i(t \mid k) = 0$, $k + N \leq t$, $\forall i = 1, 2, \ldots, M$. The cost function $\Phi_i(\cdot)$ is obtained by eliminating the state trajectory $\mathbf{x}_i$ from (2.3) using (2.6) and the input, state parameterization described before. The solution to the optimization problem $\mathcal{F}_i$ is denoted by $u_i^{*(p)}$. By definition

$$\mathbf{u}_i^{*(p)} = [u_i^{*(p)}(k \mid k)^T, u_i^{*(p)}(k+1 \mid k)^T, \ldots]^T$$

$$\bar{\mathbf{u}}_i^{*(p)} = [u_i^{*(p)}(k \mid k)^T, u_i^{*(p)}(k+1 \mid k)^T, \ldots, u_i^{*(p)}(k+N-1 \mid k)^T]^T$$

2.5.4 Terminal penalty method

For the quadratic form of $\phi_i(\,\cdot\,)$ given by (2.60), the FC-MPC optimization problem (2.64a)–(2.64c), for each subsystem $i = 1, 2, \ldots, M$, can be written as

$$\mathcal{F}_i \triangleq \min_{\bar{\mathbf{u}}_i} \frac{1}{2}\bar{\mathbf{u}}_{i'}\mathfrak{R}_i\bar{\mathbf{u}}_i + \left(\mathbf{r}_i(x(k)) + \sum_{j\neq i}\mathcal{H}_{ij}\bar{\mathbf{u}}_j^{p-1}\right)\bar{\mathbf{u}}_i \tag{2.65a}$$

subject to

$$\bar{\mathbf{u}}_i \in \mathcal{U}_i \tag{2.65b}$$

in which

$$\begin{aligned}
\mathfrak{R}_i &= \mathbb{R}_i + \sum_{j=1}^{M} E'_{ji}\mathbb{Q}_j E_{ji} + \sum_{j=1}^{M} E'_{ji}\sum_{l\neq j}\mathbb{T}_{jl}E_{li}\\
\mathbb{Q}_i &= \operatorname{diag}(w_iQ_i(1), \ldots, w_iQ_i(N-1), P_{ii})\\
\mathbb{T}_{ij} &= \operatorname{diag}(0, \ldots, 0, P_{ij})\\
\mathbb{R}_i &= \operatorname{diag}(w_iR_i(0), w_iR_i(1), \ldots, w_iR_i(N-1))\\
\mathbf{r}_i(x(k)) &= \sum_{j=1}^{M} E'_{ji}\mathbb{Q}_j\mathbf{g}_j(x(k)) + \sum_{j=1}^{M} E'_{ji}\sum_{l\neq j}\mathbb{T}_{jl}\mathbf{g}_l(x(k))\\
\mathcal{H}_{ij} &= \sum_{l=1}^{M} E'_{li}\mathbb{Q}_l E_{lj} + \sum_{l=1}^{M} E'_{li}\sum_{s\neq l}\mathbb{T}_{ls}E_{sj}\\
\mathbf{g}_i(x(k)) &= \sum_{j=1} f_{ij}x_j(k)
\end{aligned}$$

and

$$P = \begin{bmatrix} P_{11} & P_{12} & \cdots & \cdots & P_{1M}\\ P_{21} & P_{22} & \cdots & \cdots & P_{2M}\\ \vdots & \vdots & \ddots & \ddots & \vdots\\ P_{M1} & P_{M2} & \cdots & \cdots & P_{MM}\end{bmatrix} \tag{2.66}$$

is a suitable terminal penalty matrix. Restricting attention (for now) to open-loop stable systems simplifies the choice of P. For each $i = 1, 2, \ldots, M$, let $Q_i(0) = Q_i(1) = \cdots = Q_i(N-1) = Q_i$. The terminal penalty P can be obtained as the solution to the centralized Lyapunov equation

$$A'PA - P = -\mathcal{Q} \tag{2.67}$$

in which $\mathcal{Q} = \operatorname{diag}(w_1Q_1, w_2Q_2, \ldots, w_MQ_M)$. The centralized Lyapunov equation (2.67) is solved offline. The solution to (2.10) has to be recomputed if the subsystems' models and/or cost functions are altered.

2.5.5 Algorithm and properties

At time k, let $p_{\max}(k)$ represent the maximum number of permissible iterates for the sampling interval. The following algorithm is employed for cooperation-based DMPC.

Algorithm 2.1:

Terminal penalty algorithm
Given $\bar{\mathbf{u}}_i^0 k, x_i(k), \mathbb{Q}_i, \mathbb{R}_i, i = 1, 2, \ldots, M$
$p_{\max}(k) \geq 0$ and $\varepsilon > 0$
$p \leftarrow 1, \rho_i \leftarrow \Gamma_\varepsilon, \Gamma \gg 1$
while $\rho_i > \varepsilon$ for some $i = 1, 2, \ldots, M$ and $p \leq p_{\max}$
do $\forall i = 1, 2, \ldots, M$
$\bar{\mathbf{u}}_i^{*(p)} = \arg(\mathcal{F}_i)$, (see (2.8))
end (do)
for each $i = 1, 2, \ldots, M$
$\bar{\mathbf{u}}_i^p \leftarrow w_i \bar{\mathbf{u}}_i^{*(p)} + (1 - w_i)\bar{\mathbf{u}}_i^{p-1}$
$\rho_i \leftarrow \|\bar{\mathbf{u}}_i^p - \bar{\mathbf{u}}_i^{p-1}\|$
Transmit $\bar{\mathbf{u}}_i^p$ to each interconnected subsystem
$j = 1, 2, \ldots, M, j \neq i$.
end (for)
$p \leftarrow p + 1$
end (while)

The state trajectory for subsystem i generated by the input trajectories $\bar{\mathbf{u}}_1, \bar{\mathbf{u}}_2, \ldots, \bar{\mathbf{u}}_M$ and initial state z is represented as $\bar{\mathbf{x}}_i(\bar{\mathbf{u}}_1, \bar{\mathbf{u}}_2, \ldots, \bar{\mathbf{u}}_M; z)$. At each iterate p in Algorithm 2.1, the state trajectory for subsystem can be calculated as $\bar{\mathbf{x}}_i^p(\bar{\mathbf{u}}_1^p, \bar{\mathbf{u}}_2^p, \ldots, \bar{\mathbf{u}}_M^p; x(k))$. At each k, $p_{\max}(k)$ represents a design limit on the number of iterates; the user may choose to terminate Algorithm 2.1 prior to this limit.

The infinite horizon input and state trajectories $\mathbf{x}_i^p, \mathbf{u}_i^p$ can be readily obtained. Denote the cooperation-based cost function after iterates by

$$\Phi(\mathbf{u}_1^p, \mathbf{u}_2^p, \ldots, \mathbf{u}_M^p; x(k)) = \sum_{r=1}^{M} w - r\Phi_r(\mathbf{u}_1^p, \mathbf{u}_2^p, \ldots, \mathbf{u}_M^p; x_r(k))$$

To proceed further, we recall the following result on minimum principle for constrained, convex optimization:

Lemma 2.5 [182]: *Let $\mathcal{X}$ be a convex set and let f be a convex function over $\mathcal{X}$. A necessary and sufficient condition for x^* to be a global minimum of over $\mathcal{X}$ is*

$$\nabla f(x^*)'(x - x^*) \geq 0, \quad \forall x \in \mathcal{X}.$$

The following properties can be established for the FC-MPC formulation employing the terminal control algorithm:

Lemma 2.6: *Given the DMPC formulation $\mathcal{F}_i$ defined in (2.64a) through (2.65), $\forall i = 1, 2, \ldots, M$, the sequence of cost functions $\{\Phi(\mathbf{u}_1^p, \mathbf{u}_2^p, \ldots, \mathbf{u}_M^p; x(k))\}$ generated by Algorithm 2.1 is nonincreasing with iteration number p.*

Proof. From the terminal penalty algorithm, we know that

$$\begin{aligned}&\Phi(\mathbf{u}_1^{p-1}, \ldots, \mathbf{u}_{i-1}^{p-1}, \mathbf{u}_i^{*(p)}, \mathbf{u}_{i+1}^{p-1}, \ldots, \mathbf{u}_M^{p-1}; x(k))\\ &\quad \leq \Phi(\mathbf{u}_1^{p-1}, \mathbf{u}_2^{p-1}, \ldots, \mathbf{u}_M^{p-1}; x(k)), \quad \forall i = 1, 2, \ldots, M\end{aligned} \tag{2.68}$$

Therefore, from the definition of $\mathbf{u}_i^p$ in the terminal penalty algorithm, we have

$$\begin{aligned}&\Phi(\mathbf{u}_1^p, \mathbf{u}_2^p, \ldots, \mathbf{u}_M^p; x(k))\\ &\quad = \Phi(w_1\mathbf{u}_1^{*(p)} + (1 - w_1)\mathbf{u}_1^{p-1}, \ldots, w_M\mathbf{u}_M^{*(p)} + (1 - w_M)\mathbf{u}_M^{p-1}; x(k))\\ &\quad = \Phi(w_1\mathbf{u}_1^{*(p)} + w_2\mathbf{u}_1^{p-1} + \cdots + w_M\mathbf{u}_1^{p-1}, w_1\mathbf{u}_2^{p-1} + w_2\mathbf{u}_2^{*(p)} + \cdots\\ &\qquad + w_M\mathbf{u}_2^{p-1}, \ldots, w_1\mathbf{u}_M^{p-1} + w_2\mathbf{u}_M^{p-1} + \cdots + w_M\mathbf{u}_M^{*(p)}; x(k))\end{aligned}$$

By convexity of $\Phi(\,\cdot\,)$

$$\begin{aligned}&\leq \sum_{r=1}^{M} w_r\Phi(u_1^{p-1}, \ldots, u_{r-1}^{p-1}, u_r^{*(p)}, u_{r+1}^{p-1}, \ldots, u_M^{p-1}; x(k))\\ &\leq \sum_{r=1}^{M} w_r\Phi(u_1^{p-1}, \ldots, u_{r-1}^{p-1}, u_r^{p-1}, u_{r+1}^{p-1}, \ldots, u_M^{p-1}; x(k))\\ &= \Phi(u_1^{p-1}, u_2^{p-1}, \ldots, u_M^{p-1}; x(k))\end{aligned} \tag{2.69}$$

in which equality is obtained if $\mathbf{u}_i^p = \mathbf{u}_i^{p-1}$, $\forall i = 1, 2, \ldots, M$. □

Using Lemma 2.6 and the fact that $\Phi(\,\cdot\,)$ is bounded below assures convergence of the sequence of cost functions with iteration number.

Consider the centralized MPC optimization problem obtained by eliminating the subsystem states using the PM equations (1), $\forall i = 1, 2, \ldots, M$

$$\min_{\mathbf{u}_1, \mathbf{u}_2, \ldots, \mathbf{u}_M} \Phi(\mathbf{u}_1, \mathbf{u}_2, \ldots, \mathbf{u}_M; x(k)) = \sum_{i=1}^{M} w_i\Phi_i(\mathbf{u}_1, \mathbf{u}_2, \ldots, \mathbf{u}_M; x_i(k)) \tag{2.70a}$$

subject to

$$u_i(l \mid k) \in \Omega_i, \quad k \leq l \leq k + N - 1, \tag{2.70b}$$

$$u_i(l \mid k) = 0, \quad k + N \leq l \tag{2.70c}$$

$$\forall i = 1, 2, \ldots, M$$

From the definition of $\phi_i(\,\cdot\,)$ given by (2.60), we have $\mathbb{R}_i > 0$. Hence, $\mathfrak{R}_i > 0$, $\forall i = 1, 2, \ldots, M$ in (2.65a)–(2.27). It follows that $\Phi_i(\,\cdot\,)$ is strictly convex. Using convexity of $\Omega = \Omega_1 \times \Omega_2 \times \cdots \times \Omega_M$ and strict convexity of $\Phi(\,\cdot\,)$, the solution

$(\mathbf{u}_1^*, \ldots, \mathbf{u}_M)$ to the centralized MPC optimization problem (2.11) exists and is unique. By definition, $\mathbf{u}_i^* = [\bar{\mathbf{u}}_i^{*'}, 0, 0, \ldots]$.

Lemma 2.7: *Consider $\Phi(\cdot)$ positive definite quadratic and let $\Omega_i, \forall i = 1, 2, \ldots, M$ be convex, compact. Assume the solution to Algorithm 2.1 after p iterates is $(\mathbf{u}_1^p, \ldots, \mathbf{u}_M^p)$ with an associated cost function value $\Phi(\mathbf{u}_1^p, \ldots, \mathbf{u}_M^p; x(k))$, in which $\mathbf{u}_i^p = [\bar{\mathbf{u}}_i^{p'}, 0, 0, \ldots]$. Denote the unique solution to (2.11) by $(\mathbf{u}_1^*, \mathbf{u}_2^*, \ldots, \mathbf{u}_M^*)$, in which $u_i^* = [\bar{u}_i^{*'}, 0, 0, \ldots]'$, and let $\Phi(\mathbf{u}_1^*, \mathbf{u}_2^*, \ldots, \mathbf{u}_M^*; x(k))$ represent the optimal cost function value. The solution obtained at convergence of Algorithm 2.1 satisfies*

$$\lim_{p \to \infty} \Phi(\mathbf{u}_1^p, \mathbf{u}_2^p, \ldots, \mathbf{u}_M^p; x(k)) = \Phi(\mathbf{u}_1^*, \mathbf{u}_2^*, \ldots, \mathbf{u}_M^*; x(k))$$

$$\lim_{p \to \infty} (\mathbf{u}_1^p, \mathbf{u}_2^p, \ldots, \mathbf{u}_M^p) = (\mathbf{u}_1^*, \mathbf{u}_2^*, \ldots, \mathbf{u}_M^*).$$

Proof. Since the level set

$$\begin{aligned} S_0 = &\{(\bar{\mathbf{u}}_1, \bar{\mathbf{u}}_2, \ldots, \bar{\mathbf{u}}_M) \mid \Phi(\mathbf{u}_1, \mathbf{u}_2, \ldots, \mathbf{u}_M; x(k)) \\ &\leq \Phi(\mathbf{u}_1^0, \mathbf{u}_2^0, \ldots, \mathbf{u}_M^0; x(k))\} \end{aligned}$$

is closed and bounded (hence compact), a limit point for the terminal penalty algorithm exists. We know that $(\mathbf{u}_1^*, \mathbf{u}_2^*, \ldots, \mathbf{u}_M^*)$ is the unique solution for the centralized MPC optimization problem (2.11). Let $\Phi^* = \Phi(\mathbf{u}_1^*, \mathbf{u}_2^*, \ldots, \mathbf{u}_M^*; x(k))$. Define $\mathbf{u}_i^\infty = [\bar{\mathbf{u}}_i^{\infty'}, 0, 0, \ldots]$. Assume that the sequence $(\bar{\mathbf{u}}_1^p, \bar{\mathbf{u}}_2^p), \ldots, \bar{\mathbf{u}}_M^p$, generated by terminal penalty algorithm, converges to a feasible subset of the non-optimal level set

$$S_\infty = \{(\bar{\mathbf{u}}_1, \mathbf{u}_2, \ldots, \mathbf{u}_M,) \mid \Phi(\mathbf{u}_1, \mathbf{u}_2, \ldots, \mathbf{u}_M; x(k)) = \Phi^\infty\}$$

Since $\Phi(\cdot)$ is strictly convex and by assumption of non-optimality $\Phi^\infty > \Phi^*$. Let $(\bar{\mathbf{u}}_1^\infty, \ldots, \bar{\mathbf{u}}_M^\infty) \in S_\infty$ be generated by the terminal penalty algorithm for p large. To establish convergence of the terminal penalty algorithm to a point rather than a limit set, we assume the contrary and show a contradiction. Suppose that the terminal penalty algorithm does not converge to a point. Our assumption here implies that there exists $(\bar{\mathbf{v}}_1, \ldots, \bar{\mathbf{v}}_M) \in S_\infty$ generated by the next iterate of Algorithm 2.1 with $((\bar{\mathbf{v}}_1, \ldots, \bar{\mathbf{v}}_M) \in S_\infty) \neq (\bar{\mathbf{u}}_1^\infty, \ldots, \bar{\mathbf{u}}_M^\infty)$. Consider the set of optimization problems

$$\mathbf{z}_i^\infty = \arg\min_{\mathbf{u}_i} \Phi(\mathbf{u}_1^\infty, \ldots, \mathbf{u}_{i-1}^\infty, \mathbf{u}_i, \mathbf{u}_{i+1}^\infty, \ldots, \mathbf{u}_M^\infty; x(k)) \tag{2.71a}$$

$$u_i(l \mid k) \in \Omega_i, \quad 0 \leq l \leq N-1 \tag{2.71b}$$

$$u_i(l \mid k) = 0, \quad N \leq l \tag{2.71c}$$

$$\forall i = 1, 2, \ldots, M.$$

We have $\mathbf{z}_i^\infty = [\bar{\mathbf{z}}_i^{\infty'}, 0, 0, \ldots]$ in which $\bar{\mathbf{z}}_i^\infty = [z_i^\infty(0)', \ldots, z_i^\infty(N-1)']$. By assumption, there exists at least one i for which $\bar{\mathbf{z}}_i^\infty \neq \mathbf{u}_i^\infty$. We let $\bar{\mathbf{z}}_1^\infty \neq \mathbf{u}_1^\infty$. By definition,

$\bar{\mathbf{v}}_i = w_i\bar{\mathbf{z}}_i^\infty + (1-w_i)\bar{\mathbf{u}}_i^\infty$, $\forall i = 1,2,\ldots,M$. It follows that $\mathbf{v}_i = [\bar{\mathbf{v}}_{i'}, 0, 0, \ldots]$. Since $(\bar{\mathbf{v}}_1, \bar{\mathbf{v}}_2, \ldots, \bar{\mathbf{v}}_M) \in S_\infty$, $\Phi(\mathbf{v}_1, \ldots, \mathbf{v}_M; x(x)) = \Phi^\infty$. Using convexity of Φ^∞, we have

$$\begin{aligned}
\Phi^\infty &= \Phi(\mathbf{v}_1, \ldots, \mathbf{v}_M; x(x)) \\
&= \Phi(w_1\mathbf{z}_1^\infty + (1-w_1)\mathbf{u}_1^\infty, \ldots, w_M\mathbf{z}_M^\infty + (1-w_M)\mathbf{u}_M^\infty; x(k)) \\
&< w_1\Phi(\mathbf{z}_1^\infty, \mathbf{u}_2^\infty, \ldots, \mathbf{u}_M^\infty; x(k)) + \cdots + w_M\Phi(\mathbf{u}_1^\infty, \ldots, \mathbf{u}_{M-1}^\infty, \mathbf{z}_M^\infty; x(k)) \\
&< w_1\Phi^\infty + \cdots + w_M\Phi^\infty \\
&= \Phi^\infty
\end{aligned}$$

in which the strict inequality follows from $\mathbf{z}_i^\infty$ for at least one $i = 1,2,\ldots,M$. Hence, a contradiction. Suppose now that $(\bar{\mathbf{u}}_1^p, \bar{\mathbf{u}}_2^p, \ldots, \bar{\mathbf{u}}_M^p) \to (\bar{\mathbf{u}}_1^\infty, \bar{\mathbf{u}}_2^\infty, \ldots, \bar{\mathbf{u}}_M^\infty) \neq (\bar{\mathbf{u}}_1^*, \bar{\mathbf{u}}_2^*, \ldots, \bar{\mathbf{u}}_M^*)$. From uniqueness of the optimizer, $\Phi(\bar{\mathbf{u}}_1^*, \bar{\mathbf{u}}_2^*, \ldots, \bar{\mathbf{u}}_M^*; x(k)) < \Phi(\bar{\mathbf{u}}_1^\infty, \bar{\mathbf{u}}_2^\infty, \ldots, \bar{\mathbf{u}}_M^\infty; x(k))$. Since $(\bar{\mathbf{u}}_1^p, \bar{\mathbf{u}}_2^p), \ldots, \bar{\mathbf{u}}_M^p$, generated using Algorithm 2.1, converges to $(\bar{\mathbf{u}}_1^\infty, \bar{\mathbf{u}}_2^\infty, \ldots, \bar{\mathbf{u}}_M^\infty)$, we have

$$\mathbf{u}_i^\infty = \arg\min_{\mathbf{u}_i} \Phi(\mathbf{u}_1^\infty, \ldots, \mathbf{u}_{i-1}^\infty, \mathbf{u}_i, \mathbf{u}_{i+1}^\infty, \ldots, \mathbf{u}_M^\infty; x(k)) \tag{2.72a}$$

$$u_i(l \mid k) \in \Omega_i, \quad 0 \leq N-1 \tag{2.72b}$$

$$u_i(l \mid k) = 0, \quad N \leq l$$

$$\forall i = 1,2,\ldots,M \tag{2.72c}$$

Recall that

$$\nabla\bar{\mathbf{u}}_j\Phi(\mathbf{u}_1^\infty, \ldots, \mathbf{u}_M^\infty; x(k))'(\bar{\mathbf{u}}_j^* - \bar{\mathbf{u}}_j^\infty) \geq 0, \quad \forall j = 1,2,\ldots,M.$$

Define $\Delta\bar{\mathbf{u}}_j = \bar{\mathbf{u}}_j^* - \bar{\mathbf{u}}_j^\infty$ and $\Delta\mathbf{u}_j = \mathbf{u}_j^* - \mathbf{u}_j^\infty = [\Delta\bar{\mathbf{u}}_{j'}, 0, 0, \ldots]'$, $\forall j = 1,2,\ldots,M$. We have, from our assumption $(\bar{\mathbf{u}}_1^\infty, \bar{\mathbf{u}}_2^\infty, \ldots, \bar{\mathbf{u}}_M^\infty) \neq (\bar{\mathbf{u}}_1^*, \bar{\mathbf{u}}_2^*, \ldots, \bar{\mathbf{u}}_M^*)$, that $\Delta\bar{\mathbf{u}}_i \neq 0$ for at least one index i, $1 \leq i \leq M$.

A second-order Taylor series expansion around $(\mathbf{u}_1^\infty, \mathbf{u}_2^\infty, \ldots, \mathbf{u}_M^\infty)$ gives

$$\begin{aligned}
&\Phi(\mathbf{u}_1^*, \mathbf{u}_2^*, \ldots, \mathbf{u}_M^*; x(k)) \\
&\quad = \Phi(\mathbf{u}_1^\infty + \Delta\mathbf{u}_1, \mathbf{u}_2^\infty + \Delta\mathbf{u}_2, \ldots, \mathbf{u}_M^\infty + \Delta\mathbf{u}_M; x(k)) \\
&\quad = \Phi(\mathbf{u}_1^\infty, \mathbf{u}_2^\infty, \ldots, \mathbf{u}_M^\infty; x(k)) \\
&\qquad \underbrace{+\sum_{j=1}^{M} \nabla_{\bar{\mathbf{u}}_j}\Phi(\mathbf{u}_1^\infty, \mathbf{u}_2^\infty, \ldots, \mathbf{u}_M^\infty; x(k))'\Delta\mathbf{u}_j}_{\geq 0,\ \text{Lemma 3}} \\
&\qquad \underbrace{+\frac{1}{2}\begin{bmatrix}\Delta\bar{\mathbf{u}}_1 \\ \vdots \\ \Delta\bar{\mathbf{u}}_M\end{bmatrix}' \nabla^2\Phi(\mathbf{u}_1^\infty, \mathbf{u}_2^\infty, \ldots, \mathbf{u}_M^\infty; x(k)) \begin{bmatrix}\Delta\bar{\mathbf{u}}_1 \\ \vdots \\ \Delta\bar{\mathbf{u}}_M\end{bmatrix}}_{\geq 0,\ \text{since } \Phi(\,\cdot\,)\ \text{p.d. quadratic}}.
\end{aligned} \tag{2.73}$$

Using (2.73) and optimality of $(\mathbf{u}_1^*, \mathbf{u}_2^*, \ldots, \mathbf{u}_M^*)$ gives

$$\begin{aligned}\Phi(\mathbf{u}_1^*, \ldots, \mathbf{u}_M^*; x(k)) &= \Phi(\mathbf{u}_1^{infty}, \ldots, \mathbf{u}_M^{infty}; x(k)) + \beta(\Delta\mathbf{u}_1, \ldots, \Delta\mathbf{u}_M) \\ &\leq \Phi(\mathbf{u}_1^{infty}, \ldots, \mathbf{u}_M^{infty}; x(k))\end{aligned} \tag{2.74}$$

in which $\beta(\,\cdot\,)$ is a positive definite function from (2.73). We have from (2.26) that $\beta(\,\cdot\,) \leq 0$, which implies $\beta(\Delta\mathbf{u}_1, \ldots, \Delta\mathbf{u}_M) = 0$. It follows, therefore, that $\bar{\mathbf{u}}_j^\infty = \bar{\mathbf{u}}_j^*$, $\forall j = 1, 2, \ldots, M$. Using the previous relation gives $\mathbf{u}_j^\infty = [\bar{\mathbf{u}}_j^{\infty'}, 0, 0, \ldots]' = \mathbf{u}_j^*$, $\forall j = 1, 2, \ldots, M$. Hence, $\Phi(\mathbf{u}_1^*, \mathbf{u}_2^*, \ldots, \mathbf{u}_M^*; x(k)) = \Phi(\mathbf{u}_1^\infty, \mathbf{u}_2^\infty, \ldots, \mathbf{u}_M^\infty; x(k))$. □

2.5.6 Distributed MPC control law

At time k, let the FC-MPC algorithm (Algorithm 2.1) be terminated after $p(k)$ iterates, with

$$\mathbf{u}_i^{p(k)}(k; x(k)) = [u_i^{p(k)}(k; x(k))', u_i^{p(k)}(k+1; x(k))', \ldots]', \quad i = 1, 2, \ldots, M$$

representing the solution to Algorithm 2.1 after $p(k)$ cooperation-based iterates. The DMPC control law is obtained through a receding horizon implementation of optimal control whereby the input applied to subsystem i is

$$u_i(k) = u_i^{p(k)}(k; x(k)) \tag{2.75}$$

2.5.7 Feasibility of FC-MPC optimizations

Since $x(0) \in \mathcal{X}$, there exists a set of feasible, open-loop input trajectories $\mathbf{u}_1$, $\mathbf{u}_2, \ldots, \mathbf{u}_M$, such that $x_i(k) \to 0$, $\forall i = 1, 2, \ldots, M$ and k sufficiently large. Convexity of Ω_i, $\forall i = 1, 2, \ldots, M$ and Algorithm 2.1 guarantee that given a feasible input sequence at time $k = 0$, a feasible input sequence exists for all future times. One trivial choice for a feasible input sequence at $k = 0$ is $u_i(k + l \mid k) = 0, l \geq 0$, $\forall i = 1, 2, \ldots, M$. This choice follows from our assumption that each Ω_i is nonempty and $0 \in \text{int}\Omega_i$. Existence of a feasible input sequence for each subsystem i at $k = 0$ ensures that the FC-MPC optimization problem (2.64a) through (2.65) has a solution for each $i = 1, 2, \ldots, M$ and all $k \geq 0$.

2.5.8 Initialization

At discrete time $k + 1$, define $\forall i = 1, 2, \ldots, M$

$$\begin{aligned}\mathbf{u}_i^0(k+1)' = &[u_i^{p(k)}(k+1; x(k))', u_i^{p(k)}(k+2; x(k))', \ldots, \\ &u_i^{p(k)}(k+N-1; x(k))', 0, 0, \ldots]\end{aligned} \tag{2.76}$$

It follows that $\mathbf{u}_1^0(k+1), \mathbf{u}_2^0(k+1), \ldots, \mathbf{u}_M^0(k+1)$ constitute feasible subsystem input trajectories with an associated cost function $\Phi(\mathbf{u}_1^0(k+1), \mathbf{u}_2^0(k+1), \ldots, \mathbf{u}_M^0(k+1); x(k+1))$.

2.5.9 Nominal closed-loop stability

Given the set of initial subsystem states $x_i(0)$, $\forall i = 1, 2, \ldots, M$. Define $\tilde{J}_N(x(0))$ to be the value of the cooperation-based cost function with the set of zero input trajectories $u_i(k+j \mid k) = 0$, $j \geq 0$, $\forall i = 1, 2, \ldots, M$. At time k, let $J_N^0(x(k))$ represent the value of the cooperation-based cost function with the input trajectory initialization described in (2.13). For notational convenience we drop the function dependence of the generated state trajectories and write $\mathbf{x}_i \equiv \mathbf{x}_i(\mathbf{u}_1, \mathbf{u}_2, \ldots, \mathbf{u}_M; z)$, $\forall i = 1, 2, \ldots, M$. The value of the cooperation-based cost function after iterates is denoted by $J_N^{p(k)}(x(k))$. Thus

$$J_N^{p(k)}(x(k)) = \sum_{i=1}^{M} w_i \phi_i(\mathbf{x}_i^{p(k)}, \mathbf{u}_i^{p(k)}; x(k)) \tag{2.77a}$$

$$= \sum_{i=1}^{M} w_i \sum_{j=0}^{\infty} L_i(x_i^{p(k)}(k+j \mid k), u_i^{p(k)}(k+j \mid k)) \tag{2.77b}$$

At $k = 0$, we have, using Lemma 2.6, that $J_N^{p(0)} \leq J_N^0(x(0)) = \tilde{J}_N(x(0))$. It follows from (2.76) and Lemma 2.6 that

$$\begin{aligned} 0 \leq J_N^{p(k)}(x(k)) \leq J_N^0(x(k)) = J_N^{p(k-1)}&(x(k-1)) \\ &- \sum_{i=1}^{M} w_i L_i(x_i(k-1), u_i^{p(k-1)}(k-1)), \\ &\forall k > 0 \end{aligned} \tag{2.78}$$

Using the previous relationship recursively from time k to time 0 gives

$$\begin{aligned} J_N^{p(k)}(x(k)) &\leq \tilde{J}_N(x(0)) - \sum_{j=0}^{k-1} \sum_{i=1}^{M} w_i L_i(x_i(j), u_i^{p(j)}(j)) \\ &\leq \tilde{J}_N(x(0)) \end{aligned} \tag{2.79}$$

From (2.77a) and (2.77b), we have $\frac{1}{2}\lambda_{\min}(Q)\|x(k)\|^2 \leq J_N^{p(k)}(x(k))$. Using (2.79) gives $J_N^{p(k)}(x(k)) \leq \tilde{J}_N(x(0)) = \frac{1}{2}x(0)'Px(0) \leq \frac{1}{2}\lambda_{\max}(P)\|x(0)\|^2$. From the previous two cost relationships, we obtain $\|x(k)\| \leq \sqrt{(\lambda_{\max}(P))/(\lambda_{\min}(Q))}\|x(0)\|$, which shows that the closed-loop system is Lyapunov stable. In fact, using the cost convergence relationship (2.78) the closed-loop system is also attractive, which proves asymptotic stability under the DMPC control law.

Before embarking on the main result, we recall the following result:

Lemma 2.8 [183]: *Let the input constraints in (2.65)–(2.65) be specified in terms of a collection of linear inequalities. Consider the closed ball $B_\varepsilon(0)$, in which $\varepsilon > 0$ is chosen such that the input constraints in each FC-MPC optimization problem (2.65)–(2.65) are inactive for each $x(k) \in B_\varepsilon(0)$. The DMPC control law defined by* *the FC-MPC formulation of Theorem 2.10 is a Lipschitz continuous function of $x(k)$, for all $x(k) \in B_\varepsilon(0)$.*

Lemmas 2.6 and 2.7 can now be used to establish the following (stronger) exponential closed-loop stability result.

Theorem 2.10: *Given Algorithm 2.1 using the DMPC optimization problem (2.65)–(2.65) with $N \geq 1$. In Algorithm 2.1, let $0 < p_{\max}(k) \leq p^* < \infty, \forall k \geq 0$. If A is stable, P is obtained from (2.67), and*

$$\begin{aligned} Q_i(0) = Q_i(1) = \cdots = Q_i(N-1) = Q_i &> 0 \\ R_1(0) = R_i(1) = \cdots = R_i(N-1) = R_i &> 0 \\ \forall i &= 1, 2, \ldots, M \end{aligned}$$

then the origin is an exponentially stable equilibrium for the closed-loop system

$$x(k+1) = Ax(k) + Bu(k)$$

in which

$$u(k) = [u_1^{p(k)}(k; x(k))', \ldots, u_M^{p(k)}(k; x(k))']'$$

for all $x(k) \in \mathbb{R}^n$ and any $p(k) = 1, 2, \ldots, p_{\max}(k)$.

Proof. Since $Q > 0$ and A is stable, $P > 0$. The constrained stabilizable set $\mathcal{X}$ for the system is $\mathbb{R}^n$. To prove exponential stability, we use the value function $J_N^{p(k)}(x(k))$ as a candidate Lyapunov function. We need to show [94, p. 267] that there exists constants $a, b, c > 0$, such that

$$a\|x(k)\|^2 \leq J_N^p(x(k)) \leq b\|x(k)\|^2 \tag{2.80a}$$

$$\Delta J_N^p(x(k)) \leq -c\|x(k)\|^2 \tag{2.80b}$$

in which $\Delta J_N^{k(k)}(x(k)) = J_N^{p(k+1)}(x(k+1)) - J_N^{p(k)}(x(k))$.

Let ε be chosen such that the input constraints remain inactive for $x \in B_\varepsilon(0)$. Such an ε exists because the origin is Lyapunov stable and $0 \in \text{int}(\Omega_1 \times \cdots \times \Omega_M)$. Since Ω_i is compact $\forall i = 1, 2, \ldots, M$, there exists $\sigma > 0$ such that $\|\bar{\mathbf{u}}_i\| \leq \sigma$. For any x satisfying $\|x\| > \varepsilon$, $\|\bar{\mathbf{u}}_i\| < \sigma/\varepsilon\|x\|$, $\forall i = 1, 2, \ldots, M$. For $x(k) \in B_\varepsilon(0)$, we have from Lemma 2.8 that $\bar{\mathbf{u}}_i^{p(k)}(x(k))$ is a Lipschitz continuous function of $x(k)$. There exists, therefore, a constant $\rho > 0$, such that $\|\bar{\mathbf{u}}_i^{p(k)}(x(k))\| \leq \rho\|x(k)\|$, $\forall 0 < p(k) \leq p^*$. Define $K_u = \max(\sigma/\varepsilon, \rho)^2$, in which $K_u > 0$ and independent of $x(k)$. The previous definition gives $\|u_i^{p(k)}(k+j; x(k))\| \leq \sqrt{K_u M}\|x(k)\|$, $\forall i = 1, 2, \ldots, M$, $k \geq 0$ and all $0 < p(k) \leq p_{\max}(k)$. For $j \geq 0$, define $u(k+j \mid k) = [u_1^{p(k)}(k+j; x(k))', \ldots, u_M^{p(k)}(k+j; x(k))']'$. By definition, $u(k \mid k) \equiv u(k)$. We have $\|u(k+j \mid k)\| = \sqrt{\sum_{i=1}^{M} \|u_i^{p(k)}(k+j; x(k))\|^2} \leq \sqrt{K_u M}\|x(k)\|$. Similarly, define $x(k+j \mid k) = [x_1^{p(k)}(k+j \mid k)', \ldots, x_M^{p(k)}(k+j \mid k)']'$, $\forall j \geq 0$. By definition $x(k \mid k) \equiv x(k)$. Since A is stable, there exists $\bar{c}$, such that $\|A^j\| \leq \bar{c}\lambda^j$ [91, Corollary 5.6.13, p. 199], in which $\lambda_{\max}(A) \leq \lambda < 1$. Hence

$$
\begin{aligned}
\|x(k+j \mid k)\| &\leq \|A^j\| \|x(k)\| + \sum_{l=0}^{j-1} \|A^{j-1-l}\| \|B\| \|u(k+l \mid k)\| \\
&\leq \bar{c}\lambda^j \|x(k)\| + \sum_{l=0}^{j-1} \bar{c}\lambda^{j-1-l} \|B\| \|u(k+l \mid k)\| \\
&\leq \bar{c}\left(1 + \frac{\|B\|}{1-\lambda}\sqrt{MK_u}\right) \|x(k)\| \forall j \geq 0
\end{aligned}
$$

since $\sum_{l=0}^{j} \lambda^l \leq \sum_{l=0}^{\infty} = 1/(1-\lambda)$, $\forall j \geq 0$. Let $\mathcal{R} = \text{diag}(w_1 R_1, w_2 R_2, \ldots, w_M R_M)$ and $\Gamma = [\bar{c}(1 + \|B\|/(1-\lambda)\sqrt{MK_u})]^2$. Then

$$
\begin{aligned}
J_N^{p(k)}(x(k)) &= \frac{1}{2}\sum_{i=1}^{M} w_i \sum_{j=0}^{\infty} \left[\|x_i^{p(k)}(k+j \mid k)\|_{Q_i}^2 + \|u_i^{p(k)}(k+j \mid k)\|_{R_i}^2\right] \\
&= \frac{1}{2}\sum_{j=0}^{N-1} [x(k+j \mid k)' \mathcal{Q} x(k+j \mid k) + u(k_j \mid k)' \mathcal{R} u(k+j \mid k)] \\
&\quad + \frac{1}{2} x(k+N \mid k)' P x(k+N \mid k) \\
&\leq \frac{1}{2}\left[\sum_{j=0}^{N-1} (\lambda_{\max}(\mathcal{Q}) \|x(k+j \mid k)\|^2 \right. \\
&\quad \left. + \lambda_{\max}(\mathcal{R}) \|u(k+j \mid k)\|^2) + \lambda_{\max}(P) \|x(k+N \mid k)\|^2 \right] \\
&\leq \frac{1}{2}[N\lambda_{\max}(\mathcal{Q})\Gamma + N\lambda_{\max}(\mathcal{R})K_u M + \lambda_{\max}(P)\Gamma] \|x(k)\|^2 \\
&\leq b\|x(k)\|^2
\end{aligned}
$$

in which $0 < \frac{1}{2}[N\lambda_{\max}(\mathcal{Q})\Gamma + N\lambda_{\max}(\mathcal{R})K_u M + \lambda_{\max}(P)\Gamma] \leq b$.

Also, $\frac{1}{2}\lambda_{\min}(\mathcal{Q})\|x(k)\|^2 \leq J_N^{p(k)}(x(k))$. Furthermore

$$
\begin{aligned}
J_N^{p(k+1)}(x(k+1)) - J_N^{p(k)}(x(k)) &\leq J_N^0(x(k+1)) - J_N^{p(k)}(x(k)) \\
&= -\sum_{i=1}^{M} w_i L_i(x_i(k), u_i^{p(k)}(k; x(k))) \\
&\leq -\sum_{i=1}^{M} w_i L_i(x_i(k), 0) \\
&= -\frac{1}{2} x(k)' \mathcal{Q} x(k) \\
&\leq -\frac{1}{2}\lambda_{\min}(\mathcal{Q}) \|x(k)\|^2
\end{aligned}
\tag{2.81}
$$

which proves the theorem. □

Remark 2.12: *If* $(A, Q^{1/2})$ *is detectable, then the weaker requirement* $Q_i \geq 0$, $R_i > 0, \forall i = 1, \ldots, M$ *is sufficient to ensure exponential stability of the closed-loop system under the DMPC control law.*

2.6 Application to power plant control

In this section, we illustrate the performance of a decentralized MPC controller by applying it to a combined cycle power train shown in Figure 2.19. The power-train state-space model was 28th-order, comprising 8th-order, 11th-order and 9th-order gas turbine, boiler and steam turbine models respectively. In each case, the states were associated with physical mass and energy process variables, sensor, actuation, and low-level PID control dynamics. The decentralized ground predictive control applied to the power train is illustrated in Figure 2.20. In this application, the decentralized

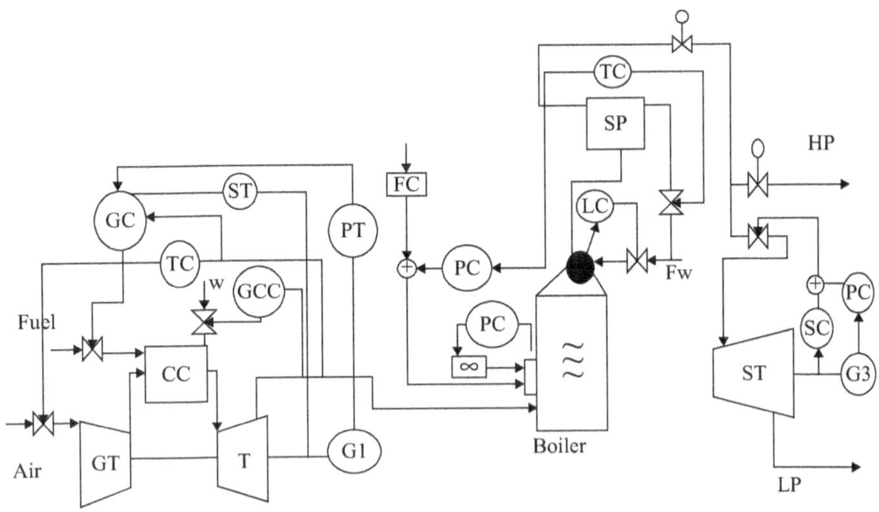

Figure 2.19 The power-plant train

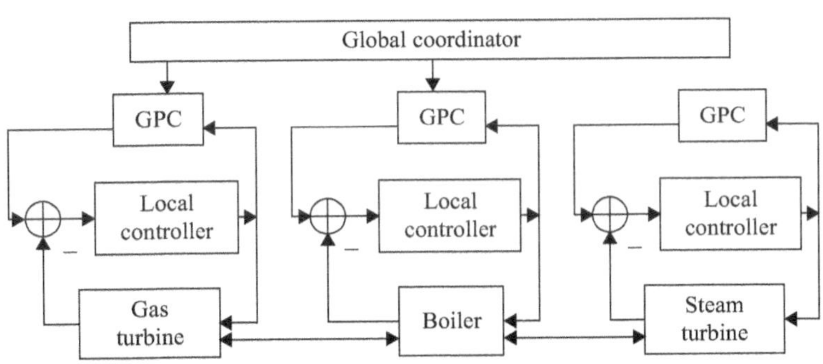

Figure 2.20 Power plant and decentralized MPC

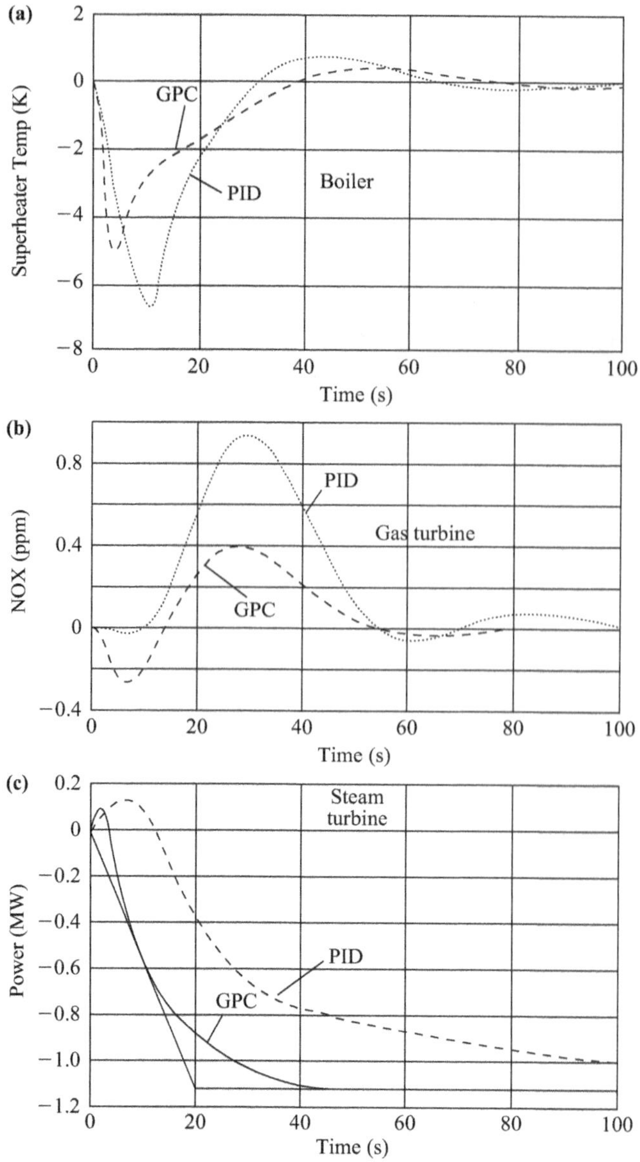

Figure 2.21 (a) The boiler superheater temperature; (b) the NOX response of the gas turbine; (c) the steam turbine output power

MPC control exists as an upper dynamic control layer above the low-level PID regulatory loops. The PID controls are usually integrated in the system units supplied by the vendors. In this regards, the task of the decentralized ground predictive controller (GPC) is to optimize the power train dynamic operating point transitions by manipulation of the low-level PID controller set points. The PID loop set points manipulated by the decentralized GPC, the initial steady-state setting, and the reference change are as follows:

Gas turbine: power 33.9 MW + 10%, exhaust temperature 101.6 K + 10%, nitrous oxide 90 ppm
Boiler: superheat steam pressure 40 bar + 0%, furnace pressure 1 bar, superheated steam temperature 700 K + 0%
Steam turbine: power 11.13 MW − 10%.

The decentralized GPC parameter settings were as follows: $t_u = 1$ s, $t_2 - t_1 = 30$ s and $\lambda = 0$. The simulation results are presented in Figure 2.21. It is quite clear that the decentralized MPC controller reduces the transition error compared with PID loops acting alone. This was achieved by utilizing future reference and predicted error information in the decentralized GPD controller, which is not available to the low-level PID controllers.

2.7 Application to quadruple-tank process

A quadruple-tank process consisting of four interconnected water tanks and two pumps is considered for simulation. Its manipulated variables are voltages to the pumps and the controlled variables (CVs) are the water levels in the two lower tanks. The quadruple-tank process is being built by considering the concept of two double-tank processes. The quadruple tank system presents a multi-input multi-output (MIMO) system. This system is a real-life control problem prototyped to experiment on, and try to solve in the most efficient way, since it deals with multiple variables, thus it gives a reflection for the large systems in industry. The schematic description of the four tank system can be visualized by Figure 2.22. The system has two control inputs (pump throughputs) that can be manipulated to control the water level in the tanks. The two pumps are used to transfer water from a sump into four overhead tanks. By adjusting the bypass valves of the system, the proportion of the water pumped into different tanks can be changed to adjust the degree of interaction between the pump throughputs and the water levels. Thus, each pump output goes to two tanks, one lower and another upper, diagonally opposite and the ratio of the split up is controlled by the position of the valve. Because of the large water distribution load, the pumps have been supplied 12 V each. The mathematical modeling of the quadruple-tank process can be obtained by using Bernoulli's law. Combining all the equations for the interconnected four-tank system we obtain the physical system. A fault model can then be constructed by adding extra holes to each tank. The mathematical model of the faulty quadruple tank system can be given as

$$
\begin{aligned}
\frac{dh_1}{dt} &= -\frac{a_1}{A_1}\sqrt{2gh_1} + \frac{a_3}{A_1}\sqrt{2gh_3} + \frac{\gamma_1 k_1}{A_1}v_1 + \frac{d}{A_1} \\
\frac{dh_2}{dt} &= -\frac{a_2}{A_2}\sqrt{2gh_2} + \frac{a_4}{A_2}\sqrt{2gh_4} + \frac{\gamma_2 k_2}{A_2}v_2 - \frac{d}{A_2}
\end{aligned}
$$

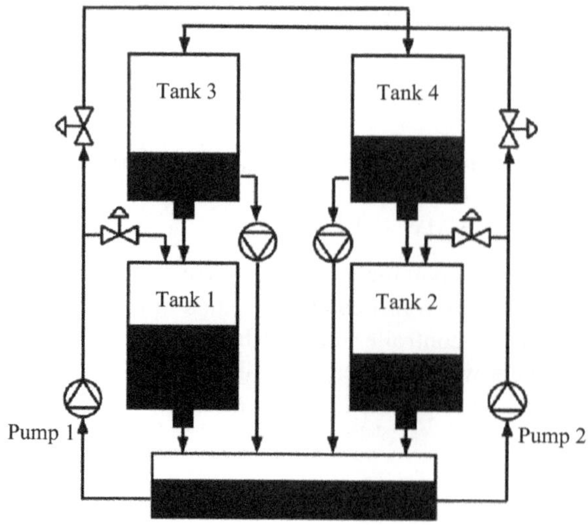

Figure 2.22 Schematic diagram of quadruple tank system

$$
\begin{aligned}
\frac{dh_3}{dt} &= -\frac{a_3}{A_3}\sqrt{2gh_3} + \frac{(1-\gamma_2)k_2}{A_3}v_2 \\
\frac{dh_4}{dt} &= -\frac{a_4}{A_4}\sqrt{2gh_4} + \frac{(1-\gamma_1)k_1}{A_4}v_1 \\
\frac{dv_1}{dt} &= -\frac{v_1}{\tau_1} + \frac{1}{\tau_1}u_1 \\
\frac{dv_2}{dt} &= -\frac{v_2}{\tau_2} + \frac{2}{\tau_2}u_2
\end{aligned}
\tag{2.82}
$$

Barring the information about the flow disturbance and invoking Nomenclature, the non-linear model (2.82) with parameter values can be linearized at the nominal operating conditions to yield a discrete-time system with a sampling frequency 1 s^{-1}. The four-tank state model is given by

$$
\begin{aligned}
x_p(k) &= \begin{bmatrix} h_1 & h_3 & v_3 & v_1 & h_2 & h_4 \end{bmatrix}^t \\
x_p(k+1) &= Ax_p(k) + Bu_p(k), \quad y_p(k) = Cx_p(k) \\
A &= \begin{bmatrix} 0.982 & 0.028 & 0 & 0.001 & 0 & 0 \\ 0 & 0.972 & 0.002 & 0 & 0 & 0 \\ 0 & 0 & 0.607 & 0 & 0 & 0 \\ 0 & 0 & 0 & 0.607 & 0 & 0 \\ 0 & 0 & 0.001 & 0 & 0.981 & 0.034 \\ 0 & 0 & 0 & 0.002 & 0 & 0.965 \end{bmatrix}
\end{aligned}
$$

$$B = \begin{bmatrix} 0 & 0 \\ 0 & 0.001 \\ 0 & 0.787 \\ 0.787 & 0 \\ 0 & 0 \\ 0.001 & 1 \end{bmatrix}, \quad C^t = \begin{bmatrix} 1 & 0 \\ 0 & 0 \\ 0 & 0 \\ 0 & 0 \\ 0 & 1 \\ 0 & 0 \end{bmatrix}$$

The control objective in the four-tank system is to keep the levels at tanks 1 and 2 at specified reference values in the face of flow disturbances. This objective and an examination of the process lead to a clear division of the system into two physical subsystems with each subsystem containing a controlled tank 1 or tank 2 and their matching overhead tank 3 or tank 4. However, allocating the process inputs (control signals to pumps) into these two groups is not a trivial task, because of the multivariable interactions in the system. For this purpose a relative gain array analysis is conducted on the linearized process model.

This analysis revealed that pairing tank 1 level with pump 2 and tank 2 level with pump 1 is more effective. Based on this result, the four-tank system is partitioned into two subsystems $[h_1, h_3, u_2]$ and $[h_2, h_4, u_1]$ for a DMPC application.

The states of the centralized model in (2.82) are prearranged for the demonstration of the suggested mathematical decomposition. The measured states h_1 and h_2 are the focal points for the distribution of states among the two subsystems and they form block diagonal groups on the two corners of the state transition matrix. This arrangement forms a fairly disconnected system, except for states v2 and v1 (corresponding to pump speeds 2 and 1), which are shared between the two subsystems leading to nodal models given in (2.22) and (2.23). These models correspond to (2.14) and (2.15) of the DMPC framework. In this distribution, x_1 contains states h_1, h_3, v_2, v_1 and x_2 contains states v_2, v_1, h_2, h_4. The observability and controllability matrices for the two nodes are of full rank, which indicates that this decomposition also satisfies the self-sufficiency requirement of a DMPC network.

The next stage in the construction of a DMPC network is to augment the anticipated disturbance and noise models on the different subsystems. In this case, a single unmeasured disturbance and two input and a single-output noise channels are added to both nodes. Kalman filters, based on these models, are then designed with expected covariance values for the process and measurement noises.

The overlapping states between the subsystems were determined during the nodal decomposition step; however, for the DMPC design, the reliability factors for these states at each node has to be specified. Since both sensors and the model fidelity was equivalent in the two nodes, the reliability factors are the same and can be determined in the course of simulation.

The performance of the proposed DMPC network for the four-tank system is initially evaluated in a simulation study. The non-linear model in (2.20) with the parameter set given in Nomenclature is used to simulate the four-tank system with the three different control strategies outlined in the previous section. The size and shape of the disturbance to be used in the simulations are determined by performing on-off tests with the constant-speed submersible pump as a 20 cm^3 s^{-1} step. For the evaluation of the MPC-based algorithms the disturbance is introduced after 100 s and the system is simulated for 2000 s. A comparison of the distributed and decentralized estimation and control (DDEC) algorithm with the centralized MPC is also provided with the same input and output weights in the regulator design. However, in these simulations the disturbance is removed after 5 s to avoid the steady-state offset suffered by the LQG-based DDEC methodology. The simulation results are shown in Figures 2.6 and 2.7.

The simulation study showed that the DDEC methodology performs very similar to a centralized MPC for pulsatile disturbances. The simulation results also showed that the proposed DMPC framework could be successfully applied to the four-tank system. The performance comparison between the different MPC-based systems demonstrated expected results, where the performance of DMPC was better than the fully decentralized MPCs, and the former approached the performance of the centralized system.

2.8 Application to automatic generation control

Most interconnected power systems rely on automatic generation control (AGC) for regulating system frequency and tie-line interchange [184]. These objectives are achieved by controlling the real power output of generators throughout the system, taking into account restrictions on the amount and rate of generator power deviations. To cope with the expansive nature of power systems, a distributed control structure has been adopted for AGC. Distributed MPC offers an effective means of achieving the desired controller coordination and performance improvements, while alleviating the organizational and computational burden associated with centralized control. AGC therefore provides a very relevant example for illustrating the performance of DMPC in a power system setting.

Power system terminology and control area model: For the purposes of AGC, power systems are decomposed into control areas, with tie-lines providing interconnections between areas [26]. Each area typically consists of numerous generators and loads. It is common, although, for all generators in an area to be lumped as a single equivalent generator, and likewise for loads. Furthermore, because AGC operation is limited to relatively small system disturbances, use of linearized models is standard [26]. Those modeling simplifications are adopted in all subsequent examples. Some basic power systems terminology is provided in Nomenclature. The notation is used to indicate a deviation from steady state. For example, it represents a deviation in the angular frequency from its nominal operating value (60 Hz).

Consider any control area $i = 1, 2, \ldots, M$, interconnected to control area $j, j \neq i$ through a tie-line. A simplified model for such a control area i is given by

Area i

$$\frac{d\Delta w_i}{dt} + \frac{1}{M_i^a} D_i \Delta w_i + \frac{1}{M_i^a} \Delta P_{\text{tie}}^{ij} - \frac{1}{M_i^a} \Delta P_{\text{mech}_i} = -\frac{1}{M_i^a} \Delta P_{L_i} \tag{2.83a}$$

$$\frac{d\Delta P_{\text{mech}_i}}{dt} + \frac{1}{T_{\text{CH}_i}} \Delta P_{\text{mech}_i} - \frac{1}{T_{\text{CH}_i}} \Delta P_{\text{v}_i} = 0 \tag{2.83b}$$

$$\frac{d\Delta P_{v_i}}{dt} + \frac{1}{T_{G_i}}\Delta P_{v_i} - \frac{1}{T_{G_i}}\Delta P_{\text{ref}_i} + \frac{1}{R_i^f T_{G_i}} = 0 \tag{2.83c}$$

Tie-line power flow between areas i and j

$$\frac{d\Delta P_{\text{tie}}^{ij}}{dt} = T_{ij}(\Delta w_i - \Delta w_j) \tag{2.83d}$$

$$\Delta P_{\text{tie}}^{ij} = -\Delta P_{\text{tie}}^{ij}. \tag{2.83e}$$

Performance comparison: The cumulative stage cost Λ is used as an index for comparing the performance of different MPC frameworks. Define

$$\Lambda = \frac{1}{t}\sum_{k=0}^{t-1}\sum_{i=1}^{M} L_i(x_i(k), u_i(k)) \tag{2.84}$$

where t is the simulation horizon.

1. *Two-area power system network*

An example with two control areas interconnected through a tie-line is considered initially. A control horizon $N = 15$ is used for each MPC. The CV for area 1 is the frequency deviation $\Delta\omega_1$ and the CV for area 2 is the deviation in the tie-line power flow between the two control areas $\Delta P_{\text{tie}}^{12}$. From the control area model (17), if $\Delta\omega_1 \to 0$ and $\Delta P_{\text{tie}}^{12} \to 0$ then $\Delta\omega_2 \to 0$.

For a 25% load increase in area 2, the load disturbance rejection performance of the FC-MPC formulation is evaluated and compared against the performance of centralized MPC (cent-MPC), communication-based MPC (comm-MPC), and standard AGC with anti-reset windup. The load reference setpoint in each area is constrained between ± 0.3. In practice, a large load change, such as the one considered above, would result in curtailment of AGC and initiation of emergency control measures such as load shedding. The purpose of this exaggerated load disturbance is to illustrate the influence of input constraints on the different control frameworks.

The relative performance of standard AGC, cent-MPC, and FC-MPC (terminated after one iterate) rejecting the load disturbance in area 2 is depicted in Figure 2.1. The closed-loop trajectory of the FC-MPC controller, obtained by terminating Algorithm 2.1 after one iterate, is almost indistinguishable from the closed-loop trajectory of cent-MPC. Standard AGC performs nearly as well as cent-MPC and FC-MPC in driving the local frequency changes to zero. Under standard AGC, however, the system takes in excess of 400 s to drive the deviational tie-line power flow to zero. With the cent-MPC or the FC-MPC framework, the tie-line power flow disturbance is rejected in about 100 s. A closed-loop performance comparison of the different control frameworks is given in Table 2.6. The comm-MPC framework stabilizes the system but incurs a control cost that is nearly 18% greater than that incurred by FC-MPC (one iterate). If five iterates per sampling interval are allowed, the performance of FC-MPC is almost identical to that of cent-MPC.

Note from Figure 2.23 that the initial response of AGC is to increase generation in both areas. This causes a large deviation in the tie-line power flow. On the other

Table 2.6 Performance of different control formulations w.r.t. cent-MPC $\Delta\Lambda\% = (\Lambda_{\text{config}}) - \Lambda_{\text{cent}} \times 100$

	Λ	ΔΛ %
Standard AGC	39.26	189
Comm-MPC	15.82	18.26
FC-MPC (1 iterate)	13.42	0.26
FC-MPC (5 iterates)	~13.38	~0
Cent-MPC	13.38	0

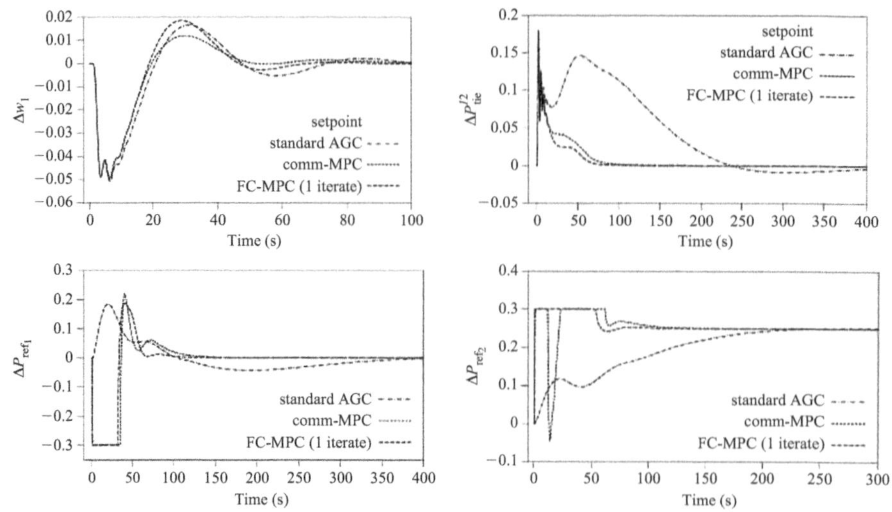

Figure 2.23 Performance of different control frameworks rejecting a load disturbance in area 2

hand, under comm-MPC and FC-MPC, MPC_1 initially reduces area 1 generation and MPC_2 orders a large increase in area 2 generation (the area where the load disturbance occurred). This strategy enables a much more rapid restoration of tie-line power flow.

2. *Four-area power system network*

Consider the four area power system shown in Figure 2.24. The model for each control area follows from (2.17). In each control area, a change in local power demand (load) alters the nominal operating frequency. The MPC in each control area i manipulates the load reference setpoint P_{ref_i} to drive the frequency deviations $\Delta\omega_i$ and tie-line power flow deviations $\Delta P_{\text{tie}}^{ij}$ to zero. Power flow through the tie-lines gives rise to interactions among the control areas. Hence, a load change in area 1, for instance, causes a transient frequency change in all control areas.

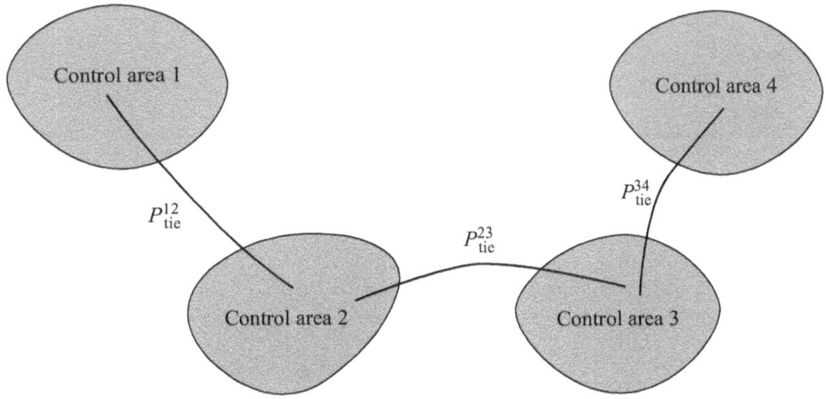

Figure 2.24 Four-area power system

The relative performance of cent-MPC, comm-MPC, and FC-MPC is analyzed for a 25% load increase in area 2 and a simultaneous 25% load drop in area 3. This load disturbance occurs at 5 s. For each MPC, we choose a control horizon of $N = 20$. In the comm-MPC and FC-MPC formulations, the load reference setpoint ΔP_{ref} in each area is manipulated to reject the load disturbance and drive the change in local frequencies $\Delta\omega_i$ and tie-line power flows $\Delta P_{\mathrm{tie}}^{ij}$ to zero. In the cent-MPC framework, a single MPC manipulates all four $\Delta P_{\mathrm{ref}_i}$. The load reference setpoint for each area is constrained between ± 0.5.

The performances of cent-MPC, comm-MPC, and FC-MPC (one iterate) are shown in Figure 2.25. Only $\Delta\omega_2$ and $\Delta P_{\mathrm{tie}}^{23}$ are shown as the frequency and tie-line

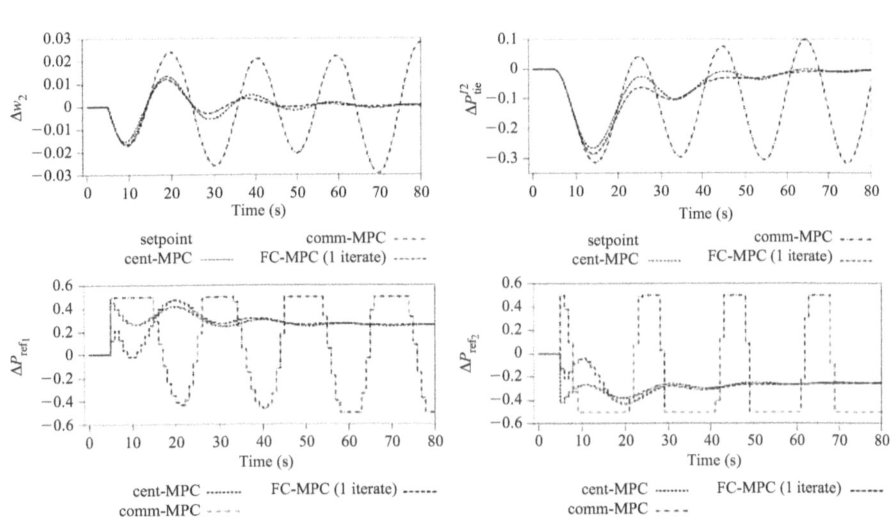

Figure 2.25 Performance of different control frameworks rejecting a load disturbance in areas 2 and 3

Table 2.7 Performance of different MPC frameworks relative to cent-MPC $\Delta\Lambda\% = (\Lambda_{\text{config}} - \Lambda_{\text{cent}})/(\Lambda_{\text{cent}}) \times 100$

	$\Lambda \times 10^{-12}$	$\Delta\Lambda$ %
Cent-MPC	7.6	0
Comm-MPC	∞	∞
FC-MPC (1 iterate)	9.6	26
FC-MPC (5 iterates)	7.87	3.7

power flow deviations and the other areas display similar qualitative behavior. Likewise, only ΔP_{ref_2} and ΔP_{ref_3} show other load reference setpoints behave similarly. The control costs are given in Table 2.7. Under comm-MPC, the load reference setpoints for areas 2 and 3 switch repeatedly between their upper and lower saturation limits. Consequently, the power system network is unstable under comm-MPC. The closed-loop performance of the FC-MPC formulation, terminated after just one iterate, is within 26% of cent-MPC performance. If the FC-MPC algorithm is terminated after five iterates, the performance of FC-MPC is within 4% of cent-MPC performance. By allowing the cooperation-based iterative process to converge, the closed-loop performance of FC-MPC can be driven to within any prespecified tolerance of cent-MPC performance.

3. *Two-area power system with FACTS device*

In this example, we revisit the two area network considered in Section 2.5.7. In this case, although, a FACTS device is employed by area 1 to manipulate the effective impedance of the tie-line and control power flow between the two interconnected control areas. The control area models follow from (2.83a)–(2.83e). In order to incorporate the FACTS device (2.83a) in area 1 is replaced by

$$\begin{aligned}
\frac{d\Delta\delta_{12}}{dt} &= (\Delta\omega_1 - \Delta\omega_2) \\
\frac{d\Delta\omega_1}{dt} &= -\frac{1}{M_1^a} D_1 \Delta\omega_1 - \frac{1}{M_1^a} T_{12}\Delta\delta_{12} + \frac{1}{M_1^a} K_{12}\Delta X_{12} \\
&\quad + \frac{1}{M_1^a}\Delta P_{\text{mech}_1} - \frac{1}{M_1^a}\Delta P_{L_1}
\end{aligned}$$

and in area 2 by

$$\begin{aligned}
\frac{d\Delta\omega_2}{dt} &= -\frac{1}{M_2^a} D_2 1\Delta\omega_2 - \frac{1}{M_2^a} T_{12}\Delta\delta_{12} + \frac{1}{M_2^a} K_{12}\Delta X_{12} \\
&\quad + \frac{1}{M_2^a}\Delta P_{\text{mech}_2} - \frac{1}{M_2^a}\Delta P_{L_2}
\end{aligned}$$

where ΔX_{12} is the impedance deviation induced by the FACTS device. The tie-line power flow deviation becomes

$$\Delta P_{tie}^{12} = -\Delta P_{tie}^{21} = T_{12}\Delta\delta_{12} - K_{12}\Delta X_{12}$$

Note that if ΔX_{12}, the model reverts to (2.17). The MPC for area 1 manipulates ΔP_{ref_1} and ΔX_{12} to drive $\Delta\omega_1$ and the relative phase difference $\delta_{12} = \Delta\delta_1 - \Delta\delta_2$ to zero. The MPC for area 2 manipulates ΔP_{ref_2} to drive $\Delta\omega_2$ to zero.

The relative performance of cent-MPC, comm-MPC, and FC-MPC rejecting a simultaneous 25% increase in the load of areas 1 and 2 is investigated. The closed-loop performance of the different MPC frameworks is shown in Figure 2.26. The associated control costs are given in Table 2.8. The performance of FC-MPC (one iterate) is within 28% of cent-MPC performance. The performance of comm-MPC,

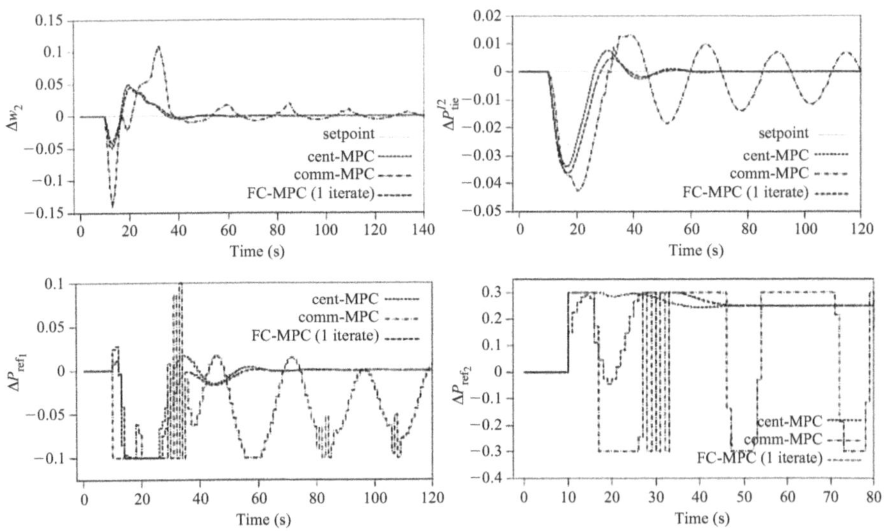

Figure 2.26 Performance of different control frameworks rejecting a load disturbance in area 2

Table 2.8 Performance of different MPC frameworks relative to cent-MPC $\Delta\Lambda\% = (\Lambda_{\text{config}} - \Lambda_{\text{cent}})/(\Lambda_{\text{cent}}) \times 100$

	$\Lambda \times 10^{-2}$	$\Delta\Lambda\%$
Cent-MPC	3.06	0
comm-MPC	9.53	211
FC-MPC (1 iterate)	3.92	28
FC-MPC (5 iterates)	3.13	2.3

on the other hand, is highly oscillatory and significantly worse than that of FC-MPC (one iterate). While comm-MPC is stabilizing, the system takes nearly 400 s to reject the load disturbance. With FC-MPC (one iterate), the load disturbance is rejected in less than 80 s. If five iterates per sampling interval are possible, the FC-MPC framework achieves performance that is within 2.5% of cent-MPC performance.

The terminal penalty-based FC-MPC framework considered earlier utilizes a suboptimal parameterization of the postulated input trajectories. Accordingly, performance is infinite horizon optimal only in the limit as $N \to \infty$. Otherwise, convergence achieves performance that is within a prespecified tolerance of a modified infinite horizon optimal control problem (2.11). The motivation behind terminal control-based FC-MPC is to achieve infinite horizon optimal (centralized, constrained, LQR) performance at convergence using finite values of N.

For terminal control FC-MPC, the unconstrained centralized feedback law is employed as the terminal feedback law. The idea is to force the collection of subsystem-based MPCs to drive the system state to a neighborhood of the origin in which the unconstrained centralized feedback law is feasible. From [59], we know that such a neighborhood of the origin is well defined and can be computed offline. Following the description in [59], we use $\mathcal{O}_\infty(A, C)$ to denote the maximal output admissible set for the overall system (A, B, C). Since Ω_i, $\forall i = 1, 2, \ldots, M$ and Ω are convex, we have from Theorem 2.1 in [59] that $\mathcal{O}_\infty$ is convex. We assume that each Ω_i is a polytope, that is, $\Omega_i = \{u_i \mid D_i u_i \leq d_i, d_i > 0\}$. The determination of $\mathcal{O}_\infty$, in this case, involves the solution to a set of linear programs. Because (A, C) is detectable only (and not observable), $\mathcal{O}_\infty(A, C)$ is a cylinder with infinite extent along directions in the unobservable subspace.

Let K denote the optimal, centralized LQR gain and let Π denote the solution to the corresponding centralized discrete steady-state Riccati equation, that is

$$\Pi = \mathcal{Q} + A'\Pi A - A'\Pi B(\mathcal{R} + B'\Pi B)^{-1}B'\Pi A \tag{2.85a}$$

$$K = -(\mathcal{R} + \mathcal{B}'\Pi\mathcal{B})^{-1}\mathcal{B}'\Pi A \tag{2.85b}$$

in which $\mathcal{Q} = \text{diag}(w_1 Q_1, w_2 Q_2, \ldots, w_M Q_M)$ and $\mathcal{R} = \text{diag}(w_1 R_1, w_2 R_2, \ldots, w_M R_M)$. Conditions for existence of a solution to (2.19) are well known [29,36]. Using a subsystem-wise partitioning for K and Φ gives

$$K = \begin{bmatrix} K_{11} & K_{12} & \cdots & K_{1M} \\ K_{21} & K_{22} & \cdots & K_{2M} \\ \vdots & \ddots & \ddots & \vdots \\ K_{M1} & K_{M2} & \cdots & K_{MM} \end{bmatrix}$$

$$\Pi = \begin{bmatrix} \Pi_{11} & \Pi_{12} & \cdots & \Pi_{1M} \\ \Pi_{21} & \Pi_{22} & \cdots & \Pi_{2M} \\ \vdots & \ddots & \ddots & \vdots \\ \Pi_{M1} & \Pi_{M2} & \cdots & \Pi_{MM} \end{bmatrix}$$

The terminal control law for subsystem $i = 1, 2, \ldots, M$ at time k is, therefore, $u_i(t \mid k) = K_{ii}x_i(t \mid k) + \sum_{j \neq i}^{M} K_{ij}x_j(t \mid k)$, $k + N \leq t$. To arrive at the terminal control FC-MPC optimization problem, we use existing definitions in (2.8) and redefine

$$\begin{aligned} \mathbb{Q}_i &= \text{diag}(w_i Q_i(1), \ldots, w_i Q_i(N-1), \Pi_{ii}) \\ \mathbb{T}_{ij} &= \text{diag}(0, \ldots, 0, \Pi_{ij}) \end{aligned}$$

The terminal control FC-MPC optimization problem is then given by (2.8), with these modifications. Algorithm 2.1 is again utilized for terminal control FC-MPC.

Initialization: To initialize Algorithm 2.1 for terminal control FC-MPC, it is necessary to calculate a set of subsystem input trajectories that steers the terminal system state (i.e., the predicted state at the end of the control horizon of each subsystem-based MPC) inside $\mathcal{O}_\infty(A, C)$. For the initial system state $x(0) \notin \mathcal{O}_\infty(A, C)$, such a set of subsystem input trajectories can be computed by solving a simple quadratic program (QP). One formulation for this initialization QP is described as follows:

$$\mathcal{L}^N(x(k)) = \arg\min_{\tilde{\mathbf{u}}(k)} \|\tilde{\mathbf{u}}(k)\|^2 \tag{2.86a}$$

$$\text{subject to } \mathcal{T}'(\mathbb{E}\tilde{\mathbf{u}}(k) + \mathbf{g}(x(k))) \in \mathcal{O}_\infty(A, C) \tag{2.86b}$$

$$\tilde{\mathbf{u}}(k) \in \mathcal{U}_i, i = 1, \ldots, M \tag{2.86c}$$

in which

$$\mathbb{E} = \begin{bmatrix} E_{11} & E_{11} & \cdots & E_{1M} \\ E_{21} & E_{21} & \cdots & E_{2M} \\ \vdots & \ddots & \ddots & \vdots \\ E_{M1} & E_{M1} & \cdots & E_{MM} \end{bmatrix}$$

$$\mathcal{T}_i = \begin{bmatrix} 0 \\ \vdots \\ \vdots \\ I \end{bmatrix}, \quad \tilde{\mathbf{u}}(k) = \begin{bmatrix} \tilde{\mathbf{u}}_1(k) \\ \tilde{\mathbf{u}}_2(k) \\ \vdots \tilde{\mathbf{u}}_M(k) \end{bmatrix}, \quad \mathbf{g}(x) = \begin{bmatrix} \mathbf{g}_1(x) \\ \mathbf{g}_2(x) \\ \vdots \mathbf{g}_M(x) \end{bmatrix}$$

$\mathcal{T} = \text{diag}(\mathcal{T}_1, \mathcal{T}_2, \ldots, \mathcal{T}_M)$, with E_{ij} defined in (2.6), and $\mathbf{g}_i(x(k))$ defined in (2.8). The definition of $\mathcal{T}_i$ is such that $x_i(k + N \mid k) = \mathcal{T}_{i'}\bar{\mathbf{x}}_i(k)$. The QP (2.20) is a centralized calculation; distributed versions for this initialization QP can be derived using techniques similar to those presented here, but are not pursued in this chapter.

Define the steerable set

$$\mathbb{S}_k = \{x(k) \mid \exists \bar{\mathbf{x}}_i(k) \in \mathcal{U}_i, i = 1, \ldots, M,$$

such that

$$x(k + N \mid k) = \mathcal{T}'\bar{\mathbf{x}}(k) \in \mathcal{O}_\infty(A, C)\}.$$

The set $\mathbb{S}_k$ denotes the set of all $x(k)$ for which the initialization QP (2.20) is feasible for a given N. We have $\mathbb{S} \subseteq \mathcal{X}$. Constrained stabilizability, therefore, follows.

At each iterate of the terminal control FC-MPC algorithm, the validity of the terminal set constraint $x^p(k+N \mid k) \in \mathcal{O}_\infty(A, C)$ must be verified. Two approaches are available for ensuring the validity of the postulated terminal control law without explicitly enforcing a terminal set constraint. In the first approach, the value of N is altered online to ensure validity of the terminal set constraint. At each iterate, a subsystem-based procedure is used to verify the validity of the postulated terminal control law. If the selected control horizon is not sufficient to ensure feasibility of the terminal control law, N is increased and the subsystems' terminal control FC-MPC optimizations are resolved using the new value of N.

Rather than increasing N online, a second approach may be adopted. The idea in this case is to restrict the set of permissible initial states to a positively invariant set in which the terminal set constraint is feasible for each subsystem $i = 1, \ldots, M$. This positively invariant set depends on the choice of N. For a given $N > 0$, we first construct the steerable set $\mathbb{S}$. Next, we determine the set of all possible combinations of system states and assumed subsystem input trajectories for which the solution to the terminal control FC-MPC optimization problem for each subsystem satisfies the terminal set constraint. Finally, the domain of the controller, which is the largest positively invariant set for which the terminal control FC-MPC control law is stabilizing, is constructed. To construct this invariant set, one may employ standard techniques available in the literature for backward construction of polytopic invariant sets under constraints.

For the nominal case, the set of shifted input trajectories (2.13), obtained using the solution to Algorithm 2.1 for terminal control FC-MPC at time, is a feasible set of input trajectories at time $k+1$. For this case, therefore, the initialization QP (2.20) has to be solved only once at $k = 0$. Lemmas 2.1 and 2.2 established for terminal penalty FC-MPC (see Section 2.5) are also valid for terminal control FC-MPC. At convergence of the exchanged input trajectories, the performance of terminal control FC-MPC is within a prespecified tolerance of the centralized constrained LQR (CLQR) [56,185] performance. If (A, B) is stabilizable, (A, C) and $(A, Q^{1/2})$ are detectable, and $Q_i \geq 0$, $R_i > 0$, $\forall i = 1, \ldots, M$, the terminal control FC-MPC control law is nominally asymptotically stable for all values of the iteration number $p(k) > 0$.

Unstable four-area power network: Consider the four-area power network described in Section 2.5.7. In this case, the value of M_4^a was increased to force the system to be open-loop unstable. At time 10 s, the load in area 2 increases by 15% and simultaneously the load in area 3 decreases by 15%. The load disturbance rejection performance of terminal control FC-MPC [FC-MPC (tc)] is investigated and compared to the performance of the benchmark centralized CLQR.

Figure 2.27 depicts the stabilizing and disturbance rejection performance of FC-MPC (tc) and CLQR. Only quantities relating to area 2 are shown as variables in other areas that displayed similar qualitative behavior. The associated control costs are given in Table 2.9. For terminal control FC-MPC terminated after one iterate, the load disturbance rejection performance is within 13% of CLQR performance. If five iterates per sampling interval are possible, the incurred performance loss drops to <1.5%.

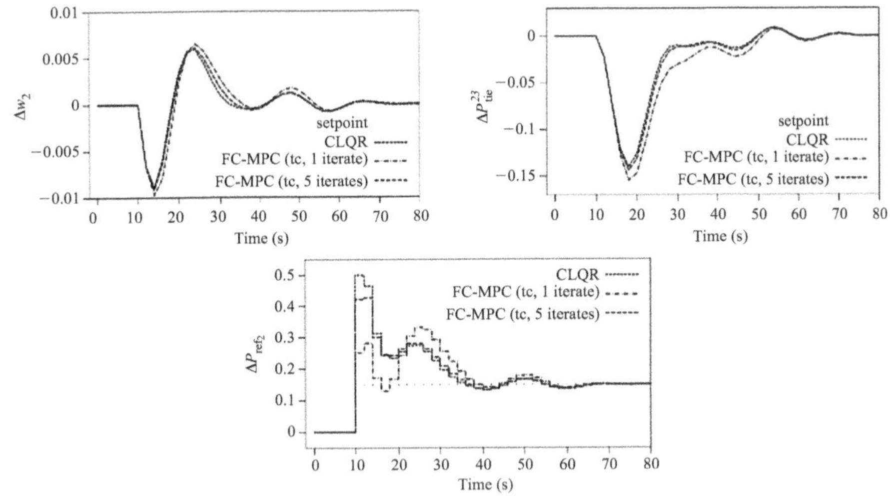

Figure 2.27 Performance of FC-MPC (tc) and CLQR, rejecting a load disturbance in areas 2 and 3

Table 2.9 Performance of terminal control FC-MPC relative to centralized contrained LQR (CLQR) for control of unstable four area network $\Delta\Gamma\% = (\Gamma_{\text{config}-\Gamma_{\text{cent}}}) \times 100$

	$\Gamma \times 10^{-2}$	$\Delta\gamma\%$
CLQR	4.91	0
FC-MPC (tc, 1 iterate)	5.52	12.4
FC-MPC (tc, 5 iterates)	4.97	1.2

2.9 Notes

In this chapter, we have presented a DMPC algorithm based on game theory for a class of systems controlled by two agents. The developed controller only needs two communication steps in order to obtain a cooperative solution to the centralized optimization problem. Each agent solves an optimization problem that only depends on its local model and partial state information. After sharing information about the local cost, the agents choose the solution that yields the best global performance among a set of suboptimal possibilities. The options are suboptimal because each agent has an incomplete view of the system and they propose the best solutions from their point of view. The proposed algorithm has low communication and computational burdens and provides a feasible solution to the centralized problem. In addition, we provided sufficient conditions that guarantee practical stability of the closed-loop system as well as an optimization-based procedure to design the controller so that these conditions are satisfied.

Although centralized MPC (CMPC) is not well suited for control of interconnected, geographically separated systems, performance benefits obtained with CMPC can be realized through distributed MPC (DMPC) strategies. For DMPC, the overall system is decomposed into interconnected subsystems. Iterative optimization and exchange of information among the subsystems is performed. An MPC optimization problem is solved within each subsystem, using local measurements and the latest available external information from the previous iterate. It is shown that FC-MPC guarantees smooth controller behavior by forcing the MPCs to cooperate toward achieving the overall control objectives. It is further demonstrated that the solution obtained at convergence of a terminal penalty version of FC-MPC is identical to the CMPC solution (and therefore, Pareto optimal). In addition, the FC-MPC algorithm can be terminated prior to convergence without compromising feasibility or closed-loop stability of the resulting distributed controller. Moreover, the FC-MPC framework allows smooth transitioning from completely decentralized control to completely centralized control.

Several extensions for the terminal penalty DMPC framework are possible. The proposed DMPC framework can be extended to penalize and constrain the rate of change of inputs. The state for subsystem is augmented with the input from the previous time step. Incorporation of the rate of change of input penalty results in additional terms in the FC-MPC cost function and additional input constraints. An important distinction, which arises due to the presence of the coupled input constraint, is that the limit points of the computational algorithm need not be optimal. The DMPC control law based on any intermediate iterate is feasible and closed-loop stable, but may not achieve centralized MPC performance at convergence of the iterates.

Nomenclature

Symbol	State/parameters
Nominal operating conditions and parameter values	
h^0	Nominal levels (11.4, 11.6, 5.3, 4 cm)
ν^0	Nominal pump settings (50%, 40%)
a_i	Drain area in tank i (2.1, 2.14, 2.2, 2.3 cm^2)
A_i	Areas of the tanks (730 cm^2)
γ_1	Ratio of flows in tank 1 to tank 4 (0.3)
γ_2	Ratio of flows in tank 2 to tank 3 (0.35)
k_j	Pump proportionality constants (7.45, 7.3 cm$^3/(s\%)$)
g	Gravitational acceleration (981 cm/s^2)
τ_i	Pump response time constants (2, 0, 2.1 s)

Symbol	State/parameters
Basic power systems terminology	
ω	Angular frequency of rotating mass
δ	Phase angle of rotating mass
M^a	Angular momentum
D	$\dfrac{\text{Percent change in load}}{\text{Percent change in frequency}}$
P_{mech}	Mechanical power
P_L	Nonfrequency sensitive load
T_{CH}	Charging time constant (prime mover)
P_{v}	Steam valve position
P_{ref}	Load reference setpoint
R	$\dfrac{\text{Percent change in frequency}}{\text{Percent change in unit output}}$
T_{G}	Governor time constant
P_{tie}^{ij}	Tie-line power flow between areas *i* and *j*
T_{ij}	Tie-line (between areas *i* and *j*) stiffness coefficient
K_{ij}	FACTS device coefficient (regulating impedance between areas *i* and *j*)

Chapter 3
Distributed linear quadratic control

In this chapter, we examine methods for distributed control (DC) based on linear quadratic theory. We begin with some general material, information structures, and notations to pave the way to subsequent sections. In one section, we deal with the case of identical decoupled dynamical systems. This is followed by analytical treatment of both semistable linear and nonnegative systems, respectively.

3.1 Introduction

Although DC is very widely used in system engineering practice, yet it is an ill-defined term. In what follows, we will consider one possible way of defining such systems. Looking at the conventional controller design problem, it assumes that all the controllers present in the system have access to the same information. The objective is to design a controller K for a plant P such that some performance specification $\min \|f(P, K)\|$ is met. Indeed, there is usually also an additional specification that K should stabilize the plant. As an example, the classical linear quadratic Gaussian (LQG) problem can be stated as follows. Given a plant P of the form

$$x(k+1) = Ax(k) + Bu(k) + w(k)$$

design a controller that generates control inputs $u(k)$ as a causal function of the measurements

$$y(k) = Cx(k) + v(k)$$

and minimizes a quadratic cost function of the form

$$J = \mathbb{E}\sum_{k=0}^{K} [x(k)^T Qx(k) + u(k)^T Ru(k)]$$

The noises $w(k)$ and $v(k)$ are assumed white and Gaussian.

In the corresponding DC problem, multiple plants of the form

$$x_j(k+1) = A_j x_j(k) + \sum_{i \neq j} A_{ij} x_j(k) + B_j u_j(k) + w_j(k)$$

are present. If the terms A_{ij} are all zero, the plants are said to be *dynamically uncoupled*. Each plant i has access to (possibly noisy) observations about the states of a set of other agents. We refer to this set as the out-neighbors of the agent i and denote it as $\mathcal{N}_j$.[1] For simplicity, throughout this lecture we will assume that each agent can access the states of all its out-neighbors perfectly. Denote by $x(k)$ the vector formed by stacking the states of all the individual agents $x_j(k)$ and define vectors $u(k)$ and $w(k)$ similarly. The aim is to design the control laws of the individual agents to minimize (say) a quadratic cost function of the form

$$J = E\sum_{k=0}^{K}[x(k)^T Qx(k) + u(k)^T Ru(k)]$$

where in general Q and R are full. The additional constraint is that each control input $u_j(k)$ can only depend on the states of agents in the set $\mathcal{N}_j$. If we try to minimize the cost function directly, we will come up with a control law of the form $u(k) = F(k)x(k)$ where the matrix $F(k)$ is full in general and thus does not satisfy this *topology constraint*. Solving the problem in the presence of this constraint is a much harder problem.

Thus, in general, a DC problem can be stated in the form [186]

$$\begin{aligned}&\text{minimize } \|f(P,K)\|\\&\text{subject to } K \text{ stabilizes } P\\&K \in S\end{aligned} \tag{3.1}$$

where S is a subspace. Note the way we have defined the problem makes it very similar to the problem of finding a *structured controller* for a plant. For a general linear time-invariant plant P and subspace S, there is no known tractable algorithm for computing the optimal K. We will restrict ourselves to the case of linear plants and quadratic costs from now on. In the sequel, we try to see why this problem is hard.

3.1.1 Information pattern

While the problem stated in (3.1) is tractable (at least for the special linear quadratic case we are concentrating on) if the subspace constraint is not present, imposing the constraint that K lie only in the subspace S render the problem open in general. One of the earliest works that pointed out that just the assumptions of a linear plant, quadratic cost, and Gaussian noises are not sufficient to obtain the solution was the famous counterexample provided by Witsenhausen [114,187–188]. The problem was originally posed in terms of two stages. We can view them as two agents in our setting. Consider $x(0)$ and v to be two independent scalar random variables. At the first stage, the random variable $x(0)$ is viewed. Thus, the output equation is

$$y(0) = x(0)$$

Based on this observation, a control input $u(0)$ is calculated and applied. The state then evolves to

$$x(1) = x(0) + u(0)$$

[1] By convention we assume that $i \in \mathcal{N}_j$.

At the next stage, the output equation is

$$y(1) = x(1) + v$$

A control input $u(1)$ that depends on $y(1)$ is then calculated and applied to obtain

$$x(2) = x(1) - u(1)$$

The objective is to minimize the cost function given by

$$J = k^2 u(0)^2 + x(2)^2$$

The admissible controllers are

$$\begin{aligned} u(0) &= \gamma_0(y(0)) \\ u(1) &= \gamma_1(y(1)) \end{aligned}$$

where γ_0 and γ_1 are Borel functions.

Note that if $u(1)$ were allowed the knowledge of $u(0)$, the problem can be solved using LQG like methods. However, in the present case, there is information to signal and the observation $y(1)$ can be used to signal that information. There is a trade-off between maximizing the information available (signaling) and minimizing the use of control at the first stage. Note the form of the cost function. At the second stage, all we are penalizing is $x(2)$, which is calculated as

$$x(2) = x(1) - u(1)$$

The controller needs to estimate $x(1)$ from $y(1)$ as best as it can, so that it can set $u(1)$ close to $x(1)$. On the other hand, at the first stage, we do not penalize the state $x(1)$ and hence the controller can choose $u(0)$ arbitrarily without worrying about $x(1)$. Thus, we are asking for $x(1)$ to be

1. low entropy, so that it can be easily predicted;
2. high energy, so that the noise v does not affect it much.

Affine controllers would mean Gaussian random variables and for Gaussian variables these two aims are in direct opposition. Nonlinear controllers thus can achieve better performance. While it is known that a non-linear controller can achieve much better performance than any linear controller, the optimal controller of this problem is still unknown. As an instance, for $k = 0.1$, the best possible affine control law gives a cost of 0.99, while non-linear controllers are possible that drive the cost as close to zero as desired. It can also be shown that the cost function is no longer convex in the controller variables, hence the problem is hard to solve numerically.

This simple counterexample is important since it shows that even for linear plants, quadratic costs and Gaussian noises, linear controls may not be optimal and the problem may be very difficult to solve. The additional piece that makes the conventional control problem simple is that of the information pattern. Informally, the information pattern is a representation of the information set that each decision maker in the problem (the controller) has access to at every time step when it makes the decision (e.g., calculates the control input). As an example, in the conventional LQG control problem, the controller at time step k has access to all the measurements $y(0), y(1), \ldots, y(k-1)$ as well as all the previous control inputs $u(0), u(1), \ldots, u(k-1)$. This is called a *classical information pattern*. As Witsenhausen's counterexample shows, a nonclassical information pattern can render a control problem intractable (Figure 3.1). Since in a DC problem, different controllers have access to different information sets, the information pattern is not classical and hence the problem is inherently difficult. It can be shown [189,190] that the problem of finding a stabilizing decentralized static output feedback is NP-complete.

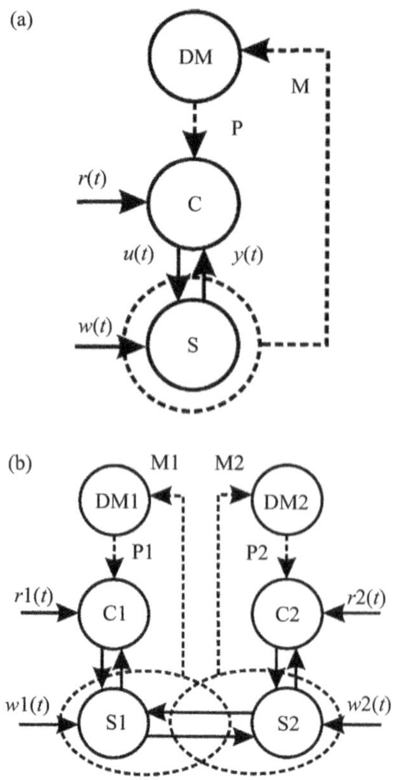

Figure 3.1 Information structures: (a) classical, (b) nonclassical

Since the general problem is difficult, there are two main approaches that have been proposed:

1. Identifying suboptimal solutions
2. Identifying special conditions or information patterns under which the problem can be solved

We now look at these approaches in a bit more detail.

3.1.2 Suboptimal controller synthesis

In this section, we will take a look at some of the approaches that have been suggested to implement suboptimal controllers for arbitrary interconnection topology (and hence arbitrary subspace constraints) on the controller.

Perhaps the approach that is most easy to understand is the one inspired by the design of reduced-order controllers [191]. This approach was used to obtain numerical algorithms for solving the optimal linear control with arbitrary number of free parameters for the infinite horizon case in [192–194].

Consider N dynamically uncoupled agents evolving as

$$x_j(k+1) = A_j x_j(k) + B_j u_j(k)$$

where the control of the ith agent can depend linearly on its own state value and the states of a specified set of other agents. On stacking the states of all the agents, the system evolves as

$$\begin{aligned} x(k+1) &= Ax(k) + Bu(k) \\ u(k) &= Fx(k) \end{aligned}$$

where F is a matrix that incorporates the interconnection information. In particular, F has a block structure, with the (i,j)th block zero if agent i cannot obtain the information about agent j to calculate its state value. Thus, F is constrained to lie in a particular space. Assume that the initial condition $x(0)$ is random and Gaussian with mean zero and covariance $R(0)$. We wish to find the constrained control law F that minimizes the cost function

$$J = E\left[\sum_{k=0}^{\infty}\{x^T(k)Qx(k) + u^T(k)Ru(k)\}\right]$$

Assume that a F exists in the required space, such that $A + BF$ is stable. Then, for that F, the cost function is given by

$$J = \mathbb{E}[x^T(0)Px(0)]$$

where P satisfies the discrete algebraic Lyapunov equation

$$P = (Q + F^T RF) + (A + BF)^T P(A + BF)$$

Thus, the cost is given by $J = \text{trace}(PR(0))$ with $R(0)$ as the initial covariance.

The case when noise is present can also be expressed similarly. Suppose that the system evolves as

$$\begin{aligned} x(k+1) &= Ax(k) + Bu(k) + w(k) \\ u(k) &= Fx(k) \end{aligned}$$

where F is chosen to minimize the cost function

$$J = \lim_{k\to\infty} E[x^T(k)Qx(k) + u^T(k)Ru(k)]$$

As an exercise, prove that the cost can now be written as $J = \text{trace}(PR_w)$ where R_w is the covariance of noise $w(k)$. Note that because F is stable, the initial condition $R(0)$ would not affect the cost function.

3.1.3 Stabilizability

Two questions arise immediately:

1. Is it possible to stabilize the system using information from other agents when the agents are individually not stable. In other words, if an agent is unstable, can the system be stabilized by the exchange of information between different agents?
2. Are some topologies inherently unstable in that even if the agents are stable, the information flow will always make it impossible to stabilize the formation?

Adequate answers to these questions are found in [194] and [195] and summarized below:

Proposition 3.1: *Consider a system of interconnected dynamically uncoupled agents as defined above.*

1. *The system is controllable if and only if each individual agent is controllable.*
2. *The system is stabilizable if and only if each individual agent is stabilizable.*

Without loss of generality, the proof is presented for the case of identical agents.

Proof. Suppose there are N agents each with state-space dimension m with state matrices Φ and Γ. Thus, the entire system has state-space dimension Nm and system matrices

$$A = I \otimes \Phi, \quad B = I \otimes \Gamma$$

where I is the identity matrix of suitable dimensions and $\otimes$ represents the Kronecker product. For controllability of the system, we thus want the following matrix to have rank Nm

$$M_1 = [I \otimes \Gamma \quad (I \otimes \Phi\Gamma) \quad (I \otimes \Phi^2\Gamma) \quad \cdots \quad (I \otimes \Phi^{Nm-1}\Gamma)]$$

Using the standard property of Kronecker product

$$(a \otimes b)(c \otimes d) = ac \otimes bd$$

we can rewrite M_1 as

$$M_1 = [I \otimes \Gamma \quad (I \otimes \Phi\Gamma) \quad (I \otimes \Phi^2\Gamma) \quad \cdots \quad (I \otimes \Phi^{Nm-1}\Gamma)]$$

This matrix has rank Nm if and only if the following matrix has rank m

$$M_2 = [\Gamma \quad \Phi\Gamma \quad \Phi^2\Gamma \quad \cdots \quad \Phi^{Nm-1}\Gamma]$$

Since Φ is an $m \times m$ matrix, the equivalent condition is that the matrix

$$M_3 = [\Gamma \quad \Phi\Gamma \quad \Phi^2\Gamma \quad \cdots \quad \Phi^{m-1}\Gamma]$$

has rank m. But M_3 being rank m is simply the condition for the individual agent being controllable. Thus, the system is controllable if and only if each individual agent is controllable. This proves the first part. The proof of the second part is similar. The subspace not spanned by the columns of M_1 is stable if and only if the subspace not spanned by the columns of M_3 is stable. □

3.1.4 Numerical algorithms

In the sequel, we obtain necessary conditions for the optimal solution that we can numerically solve. The objective is to find

$$F = \sum_{i=1}^{n} \alpha_j \Phi_j$$

such that trace($PR(0)$) is minimized, where

$$P = (Q + F^T RF) + (A + BF)^T P(A + BF) \tag{3.2}$$

For a critical point

$$\text{trace}\left(\frac{\partial P}{\partial \alpha_j} R(0)\right) = 0, \quad \forall i = 1, 2, \ldots, n$$

Let us define

$$\Sigma_j = \Phi_j^T [RF + B^T P(A + BF)] \tag{3.3}$$

Differentiating (3.2) with respect to α_j, we obtain

$$\frac{\partial P}{\partial \alpha_j} = (A + BF)^T \frac{\partial P}{\partial \alpha_j}(A + BF) + \Sigma_j + \Sigma_j^T.$$

Thus, the cost is given by

$$\text{trace}\left(\frac{\partial P}{\partial \alpha_j} R(0)\right) = \text{trace}\left(\left((A + BF)^T \frac{\partial P}{\partial \alpha_j}(A + BF) + \Sigma_j + \Sigma_j^T\right) R(0)\right)$$

Now the covariance of the state at time k evolves as

$$R(k + 1) = (A + BF)R(k)(A + BF)^T$$

Thus,

$$\text{trace}\left((A + BF)^T \frac{\partial P}{\partial \alpha_j}(A + BF)R(0)\right) = \text{trace}\left(\frac{\partial P}{\partial \alpha_j} R(1)\right)$$

Using this relation k times, we obtain

$$\text{trace}\left(\frac{\partial P}{\partial \alpha_j}R(0)\right) = \text{trace}\left(\left(\frac{\partial P}{\partial \alpha_j}R(k) + \Sigma_j X(k) + \Sigma_j^K X(k)\right) R(0)\right)$$

where

$$X(k) = R(0) + R(1) + \cdots + R(k)$$

But if $(A + BF)$ is stable, $R(k)$ would be approximately a zero matrix for sufficiently large values of k. Thus, if we denote

$$X = R(0) + R(1) + \cdots$$

we see that X satisfies the Lyapunov equation

$$X = R(0) + (A + BF)X(A + BF)^T \tag{3.4}$$

we obtain the following necessary condition for a critical point. We want

$$\text{trace}(\Sigma_j X + \Sigma_j^T X) = 0 \quad \forall i = 1, \ldots, n$$

where

$$F = \sum_{i=1}^{n} \alpha_j \Phi_j$$

P satisfies (3.2), Σ_j is defined by (3.3), and X satisfies (3.4). Note that (3.4) can either be solved iteratively or a gradient descent method can be used to obtain the control law.

As an exercise, show the following:

1. For the case when F has no restrictions on its structure, we obtain the usual condition

$$B^T P(A + BF) + RF = 0$$

2. If the initial conditions of the agents are independent, then for the completely decentralized case (when the control law of each agent can depend only on its own state value), only the diagonal terms of cost matrices Q and R are important. Note that this is not true in general. Even if agent i cannot access the state of agent j, the (i,j)th block of matrices Q and R are still important.

The algorithm we have discussed is for the infinite horizon case. A similar approach was addressed in [196] and [197]. For the finite horizon case, a similar algorithm can be applied as described, that is, in [198]. However, as [199] pointed out, there are computational difficulties arising out of solving a large number of coupled matrix equations. A suboptimal algorithm to get around this difficulty was proposed in [195] in which as opposed to NT coupled matrix equations (where T is the time horizon and N agents are present), N equations need to be solved T times.

3.1.5 Other approaches

We have described one particular approach toward obtaining suboptimal algorithms for the DC problem. Many other approaches have been proposed in the literature. (We cannot through them in any detail because of space constraint.) However, we summarize a couple of approaches here.

The problem of synthesizing a constrained controller while minimizing a $\mathcal{H}_2$ performance criterion was considered in [200]. The analysis problem (for a given controller) was shown to be convex. However, it was shown that for the synthesis problem, enforcing the topology constraint typically destroys convexity. A method to retain convexity at the expense of suboptimality was presented.

The problem of synthesizing a distributed controller achieving $\mathcal{H}_\infty$ performance was considered in [201]. They used tools inspired by dissipativity theory and derive sufficient *linear matrix inequality* (LMI) conditions on which performance constraints can be imposed. The controller structure that they come up with has the same interconnection topology as the plant interconnection topology. The tools have been extended to the case of lossy communication links in [202].

3.1.6 Identifying solvable information patterns

As we saw, the general problem of DC is very difficult and the optimal controller is not known for arbitrary information patterns. In particular, optimal controllers are not linear or even numerically easy to calculate in general. There have been numerous efforts to classify what information patterns lead to linear controllers being optimal and in what cases can the optimal linear controller be cast as a convex optimization problem. Witsenhausen [187] in a survey paper summarized several important results. He gave sufficient conditions under which the standard LQG theory could be applied and thus the optimal controller would be linear. Another important early contribution was [203] that showed the optimal controller to be linear for a class of information structures that they called *partially nested.* A partially nested information structure is one in which the *memory communication structure* is the same as the precedence relation in the *information structure diagram.* Informally, this means that a controller A has access to all the information that another controller B has access to, if the decision that B makes can affect the information set of A. Thus, once the control laws are fixed, any controller can deduce the action of all the controllers *precedent* to it. The only random effects are due to the structure of the external disturbances that are not control-law dependent.

As an example, consider a system where two agents evolve according to

$$
\begin{aligned}
x_1(k+1) &= A_1 x_1(k) + B_1 u_1(k) + w_1(k) \\
x_2(k+1) &= A_2 x_2(k) + A_{12} x_1(k) + B_2 u_2(k) + w_2(k)
\end{aligned}
$$

where $w_1(k)$ and $w_2(k)$ are white uncorrelated zero mean Gaussian noises. Further, let the initial conditions $x_1(0)$ and $x_2(0)$ be independent. Suppose the cost function to be minimized is

$$J = \mathbb{E}\left[\sum_{k=0}^{K}\{x_1^T(k+1)Q_1x_1(k+1) + x_2^T(k+1)Q_2x_2(k+1) + u_1^T(k)R_1u_1(k) + u_2^T(k)R_2u_2(k)\}\right]$$

The agents are being observed through measurements of the form

$$\begin{aligned} y_1(k) &= C_1x_1(k) + v_1(k) \\ y_2(k) &= C_2x_2(k) + v_2(k) \end{aligned}$$

with the usual assumptions on the noises. Obviously if both the agents have access to all previous control inputs $u_j(0), u_j(1), \ldots, u_j(k-1)$ and the measurements $y_j(0), y_j(1), \ldots, y_j(k)$ at any time step k, the information structure is classical. The problem then admits of unique optimal control inputs $u_j(k)$. Further they are linear in the measurements and can be obtained, for example, using the LQG theory. However, now consider an information pattern in which agent 1 has access to its own previous controls $u_1(0), u_1(1), \ldots, u_1(k-1)$ and its own measurements $y_1(0), y_1(1), \ldots, y_1(k)$. The agent 2 has access to its own control inputs but measurements from both agents. In this case, the information pattern is partially nested. Even for this information structure, the optimal control inputs are unique and linear in the measurements. This is so because agent 1 can choose its control input without worrying about agent 2's decision. Agent 2 can reconstruct agent 1's control input if it knows the control law followed by agent 1 even if it does not have access to the control input directly. Thus, it can also solve for the same control input as in the classical information pattern case.

There has been a lot of work on particular information patterns. For instance the one-step delayed information sharing pattern assumes that each controller has, at the current time, all the previously implemented control values, all the observations made anywhere in the system through, and including the previous time, and its own observation at the current time. Hence, current observations are not shared. Recursive solutions for this problem with a quadratic cost were provided using dynamic programming in [204], an exponential cost by [205] and with $\mathcal{H}_2$, $\mathcal{H}_\infty$, and L_1 costs by [206]. Some other structures that are tractable have been identified, for example, in [207] and [208]. A property called quadratic invariance was defined in [186] and it was shown that it is necessary and sufficient for the constraint set to be preserved under feedback, and that this allows optimal stabilizing decentralized controllers to be synthesized via convex programming.

3.2 Identical decoupled dynamical systems

In what follows, we consider a set of 'identical' decoupled dynamical subsystems and a control problem where the performance index couples the behavior of the systems. The interaction among subsystems is described through a communication graph where each subsystem is a node and the control action at each node is only function of its state and the states of its neighbors. A DC design method is presented that requires the solution of a single linear quadratic regulator (LQR) problem. The size of the LQR problem is equal to the maximum vertex degree of the communication graph plus one. The proposed design procedure illustrates how stability of the large-scale system is related to the robustness of local controllers and the spectrum of a matrix representing the desired sparsity pattern of the distributed controller design problem.

3.2.1 Introduction

Distributed control techniques today can be found in a broad spectrum of applications ranging from robotics and formation flight to civil engineering. Approaches to DC design differ from each other in the assumptions they make on:

1. the kind of interaction between different systems or different components of the same system (dynamics, constraints, objective);
2. the model of the system (linear, non-linear, constrained, continuous-time, discrete-time);
3. the model of information exchange between the systems; and
4. The control design technique used.

In what follows, we focus on *identical decoupled linear time-invariant systems*. Our interest in DC for such systems arises from the abundance of networks of independently actuated systems and the necessity of avoiding centralized design when this becomes computationally prohibitive. Networks of vehicles in formation, production units in a power plant, cameras at an airport, and an array of mechanical actuators for deforming a surface are just a few examples.

In a descriptive way, the problem of DC for decoupled systems can be formulated as follows. A dynamical system is composed of (or can be decomposed into) distinct dynamical subsystems that can be independently actuated. The subsystems are dynamically decoupled but have common objectives, which make them interact with each other. Typically the *interaction* is local, that is, the goal of a subsystem is a function of only a subset of other subsystems' states. The interaction will be represented by an *interaction graph*, where the nodes represent the subsystems and an edge between two nodes denotes a coupling term in the controller associated with the nodes. Also, typically it is assumed that the *exchange of information* has a special structure, that is, it is assumed that each subsystem can sense and/or exchange information with only a subset of other subsystems. We will assume that the *interaction graph* and the *information exchange graph* coincide. A DC scheme consists of distinct controllers, one for each subsystem, where the inputs to each subsystem are computed only based on local information, that is, on the states of the subsystem and its neighbors.

Over the past few years, there has been a renewal of interest in systems composed of a large number of interacting and cooperating interconnected units [145,194, 195,201,209–220]. Recent advances in stability analysis of such systems yielded greater insight into the relationship between the spectrum of the interconnection graph and global stability of the overall system. Research efforts in this area are epitomized by the pioneering work of Zhang et al. [8], which provided stability analysis tools for an interconnection of identical linear dynamical systems each using the same control law that operates on the average information obtained from neighbors.

Recent results are pertaining to the study of distributed parameter systems where the underlying dynamics are spatially invariant, and where the controls and measurements are spatially distributed. The fundamental work [210,221] in this field discusses distributed LQR design purely for infinite dimensional spatially invariant systems, where the problem diagonalizes exactly into a parameterized family of finite dimensional LQR problems. It is established that the corresponding *algebraic Riccati equation* (ARE) solutions are translation-invariant operators, and the optimal controller is a spatially invariant system. The authors show that quadratically optimal controllers for spatially invariant systems are themselves spatially invariant. Another recent result for spatially invariant systems and unbounded domains in [212] considers the problem of distributed controller design, when communication among the sites is limited. In particular, the controller is assumed to be constrained so that information is propagated with a delay that depends on the distance between subsystems. The authors show that the problem of optimal design can be cast as a convex problem provided that the propagation speeds in the controller are at least as fast as those in the plant. Special spatially invariant structures composed of an *infinite* string of linear systems have also received much attention. The work in [213] analyzes inverse optimality of localized distributed controllers in such systems.

Another significant group of recently explored strategies involves relaxations to the LMI versions of these spatially invariant problems. In [214], the authors consider heterogeneous subsystems and a relaxation, which is used to derive sufficient conditions for the existence of controllers that stabilize the system and provide a guaranteed level of performance. The work in [201] is concerned with finding distributed controllers for a set of arbitrarily connected, finite, and possibly heterogeneous linear-time invariant systems. The authors consider nonoriented edges in the graph interconnection and formulate convex conditions for the existence of output-feedback controllers, which achieve a certain $\mathcal{H}_\infty$ performance. The resulting linear matrix inequalities grow in size with the number of systems in the interconnection.

The complexity associated with the computation of distributed optimal controllers is well exemplified by the study in [195]. There, a finite-time LQR synthesis problem in discrete time is considered, where the matrix describing the control law is constrained to lie in a particular vector space. This vector space represents a pre-specified DC structure between a network of autonomous agents. The computationally intensive optimal solution for the DC problem is presented along with a computationally more tractable suboptimal one.

One of the most notable recent result in characterizing the complexity of optimal distributed controller design subject to constraints on the controller structure was presented in [186]. The authors establish that in general, efficient solution of such problems require a special property of the sparsity pattern or interconnection structure to hold. Specifically, it is shown that quadratic invariance of the assumed controller structure implies that the distributed minimum-norm problem may be solved efficiently via convex programming.

In the sequel, a *simple* distributed controller *design* approach is proposed and focuses on a class of systems, for which existing methods outlined above are either not efficient or would not even be directly applicable. Our method applies to large-scale systems composed of *finite* number of *identical subsystems where the interconnection structure or sparsity pattern is not required to have any special invariance properties*. The philosophy of our approach builds on the recent works [145,215,216], where at each node, the model of its neighbors are used to predict their behavior. We show that in absence of state and input constraints, and for identical linear system dynamics, such an approach leads to an extremely powerful result: the synthesis of stabilizing DC laws can be obtained by using a simple local LQR design, whose size is limited by the maximum vertex degree of the interconnection graph plus one. Furthermore, the design procedure proposed in this chapter illustrates how stability of the overall large-scale system is related to the robustness of local controllers and the spectrum of a matrix representing the desired sparsity pattern. In addition, the constructed distributed controller is stabilizing independent of the tuning parameters in the local LQR cost function. This leads to a method for designing distributed controllers for a finite number of dynamically decoupled systems, where the local tuning parameters can be chosen to obtain a desirable global performance. Such result can be immediately used to improve current stability analysis and controller synthesis in the field of distributed receding horizon control for dynamically decoupled systems [158,194,215,217,218]. We emphasize that the aforementioned advantages are the results of two main simplifying assumptions compared to a general structured optimal design: identical and decoupled subsystem dynamics and suboptimality of the global controller.

3.2.2 LQR properties

Consider a set of N_s identical, decoupled linear time-invariant dynamical systems, the ith system being described by the continuous-time state equation

$$\dot{x}_j = Ax_j + Bu_j, \quad x_j(0) = x_{i0} \tag{3.5}$$

where $x_j(t) \in \Re^n$, $u_j(t) \in \Re^m$ are states and inputs of the ith system at time t, respectively. Let $\tilde{x} \in \Re^{nN_s}$ and $\tilde{u}(t) \in \Re^{mN_s}$ be the vectors that collect the states and inputs of the N systems at time t:

$$\begin{aligned} \dot{\tilde{x}} &= \tilde{A}\tilde{x} + \tilde{B}\tilde{u}, \\ \tilde{x}(0) &= \tilde{x}_0 \triangleq [x_{10}, \ldots, x_{N_s 0}]^T \end{aligned} \tag{3.6}$$

with

$$\tilde{A} = I_{N_s} \otimes A, \quad \tilde{B} = I_{N_s} \otimes B \tag{3.7}$$

We consider an LQR control problem for the set of N systems where the cost function couples the dynamic behavior of individual subsystems:

$$J(\tilde{u},\tilde{x}_0) = \int_0^{\infty} \left[\sum_{i=1}^{N} (x_j(\tau)^T Q_{ii} x_j(\tau) + u_j(\tau)^T R_{ii} u_j(\tau)) \right. \\ \left. + \sum_{i=1}^{N} \sum_{j \neq i}^{N} (x_j(\tau) - x_j(\tau))^T Q_{ij} (x_j(\tau) - x_j(\tau)) \right] d\tau \tag{3.8}$$

with

$$R_{ii} = R_{ii}^T = R > 0, \quad Q_{ii} = Q_{ii}^T \geq 0 \quad \forall i \tag{3.9a}$$

$$Q_{ij} = Q_{ij}^T = Q_{ji} \geq 0, \quad \forall i \neq j \tag{3.9b}$$

The cost function (3.8) contains terms that weigh the ith system states and inputs, as well as the difference between the ith and the jth system states and can be rewritten using the following compact notation:

$$J(\tilde{u}(t),\tilde{x}_0) = \int_0^{\infty} [\tilde{x}(\tau)^T \tilde{Q} \tilde{x}(\tau) + \tilde{u}(\tau)^T \tilde{R} \tilde{u}(\tau)] d\tau \tag{3.10}$$

where the matrices $\tilde{Q}$ and $\tilde{R}$ have a special structure defined next. $\tilde{Q}$ and $\tilde{R}$ can be decomposed into N_L^2 blocks of dimension $n \times n$ and $m \times m$, respectively:

$$\tilde{Q} = \begin{bmatrix} \tilde{Q}_{11} & \tilde{Q}_{12} & \cdots & \tilde{Q}_{1N} \\ \vdots & \ddots & \vdots & \vdots \\ \tilde{Q}_{N1} & \cdots & \cdots & \tilde{Q}_{NN} \end{bmatrix} \\ \tilde{R} = \begin{bmatrix} R & 0 & \cdots & 0 \\ \vdots & \ddots & \vdots & \vdots \\ 0 & \cdots & \cdots & R \end{bmatrix} \tag{3.11}$$

with

$$\tilde{Q}_{ii} = Q_{ii} + \sum_{k=1,k\neq i}^{N} Q_{ik}, \quad i = 1,\ldots,N \\ \tilde{Q}_{ij} = -Q_{ij}, \quad i,j = 1,\ldots,N, \quad i \neq j \\ \tilde{R} = I_N \otimes R \tag{3.12}$$

Remark 3.1: *The cost function structure (3.8) can be used to describe several practical applications including formation flight, paper machine control, and monitoring networks of cameras [219,222].*

Let $\tilde{K}$ and $\tilde{x}_0^T \tilde{P} \tilde{x}_0$ be the optimal controller and the value function corresponding to the following LQR problem:

$$\min_{\tilde{u}} J(\tilde{u},\tilde{x}_0) \text{ subject to } \dot{\tilde{x}} = \tilde{A}\tilde{x} + \tilde{B}\tilde{u} \quad \tilde{x}(0) = \tilde{x}_0 \tag{3.13}$$

From now onward, we will assume that a stabilizing solution to the LQR problem (3.10) with finite performance index exists and is unique (see [223], p. 52, and references therein):

Assumption 3.1: *System $\tilde{A}, \tilde{B}$ is stabilizable and system $\tilde{A}, \tilde{C}$ is observable, where $\tilde{C}$ is any matrix such that $\tilde{C}^T\tilde{C} = \tilde{Q}$. We will also assume local stabilizability and observability.*

Assumption 3.2: *System A, B is stabilizable and systems A, C are observable, where C is any matrix such that $C^TC = Q$.*

It is well known that

$$\tilde{K} = -\tilde{R}^{-1}\tilde{B}^T\tilde{P} \tag{3.14}$$

where $\tilde{P}$ is the symmetric positive definite solution to the following ARE:

$$\tilde{A}^T\tilde{P} + \tilde{P}\tilde{A} - \tilde{P}\tilde{B}\tilde{R}^{-1}\tilde{B}^T\tilde{P} + \tilde{Q} = 0 \tag{3.15}$$

We decompose $\tilde{K}$ and $\tilde{P}$ into blocks of dimension $m \times n$ and $n \times n$, respectively. Denote by $\tilde{K}_{ij}$ and $\tilde{P}_{ij}$ the (i,j) block of the matrix $\tilde{K}$ and $\tilde{P}$, respectively. In the following theorems, we show that $\tilde{K}_{ij}$ and $\tilde{P}_{ij}$ satisfy certain properties that will be critical for the design of stabilizing distributed controllers in Section 3.4. These properties stem from the special structure of the LQR problem (3.10). Next, the matrix X is defined as $X = BR^{-1}B^T$.

Theorem 3.1: *Let $\tilde{K}$ and $\tilde{x}_0^T\tilde{P}\tilde{x}_0$ be the optimal controller and the value function solution to the LQR problem (3.10). Let $\tilde{K}_{ij} = \tilde{K}[(i-1)m : im, (j-1)n : jn]$ and $\tilde{P}_{ij} = \tilde{P}[(i-1)n : in, (j-1)n : jn]$ with $i = 1, \ldots, N$, $j = 1, \ldots, N$. Then*

$$\sum_{i=1}^{N}\left[A^T\sum_{j=1}^{N}\tilde{P}_{ij} + \sum_{j=1}^{N}\tilde{P}_{ij}A - \sum_{j=1}^{N}\tilde{P}_{ij}X\sum_{j=1}^{N}\tilde{P}_{ij} + Q_{ii}\right] = 0 \tag{3.16}$$

Proof. Consider the ARE (3.15) associated with the LQR problem (3.13). Without loss of generality, the diagonal blocks $\tilde{P}_{ii}$ of $\tilde{P}$ can be written as

$$\tilde{P}_{ii} = \tilde{F}_{ii} - \sum_{j=1, j\neq i}^{N}\tilde{P}_{ij} \tag{3.17}$$

for a certain $\tilde{F}_{ii}$. In general, $\tilde{F}_{ii}$ would be a function of $\sum_{j=1, j\neq i}^{N}\tilde{P}_{ij}$. We will show next that for the LQR problem (3.10), $\tilde{F}_{ii} = P$ and that it is not a function of $\sum_{j=1, j\neq i}^{N}\tilde{P}_{ij}$.

The equations for the diagonal blocks in the ARE (3.15) are

$$A^T\tilde{P}_{ii} + \tilde{P}_{ii}A - \sum_{k=1}^{N}(\tilde{P}_{ik}X\tilde{P}_{ik}) + \tilde{Q}_{ii} = 0 \tag{3.18}$$

for $i = 1, \ldots, N$. Note also that $\tilde{P}_{ij} = \tilde{P}_{ji}$. Substituting (3.18) in (3.15) leads to

$$\underbrace{A^T\tilde{F}_{ii} + \tilde{F}_{ii}A - A^T\left(\sum_{k=1,k\neq i}^{N}\tilde{P}_{ik}\right) - \left(\sum_{k=1,k\neq i}^{N}\tilde{P}_{ik}\right)A}_{1}$$

$$-\underbrace{\sum_{k=1,k\neq i}^{N}(\tilde{P}_{ik}X\tilde{P}_{ki})}_{2} - \tilde{F}_{ii}X\tilde{F}_{ii} - \underbrace{\left(\sum_{k=1,k\neq i}^{N}\tilde{P}_{ik}\right)X\left(\sum_{l=1,l\neq i}^{N}\tilde{P}_{il}\right)}_{3}$$

$$+\sum_{k=1,k\neq i}^{N}\tilde{P}_{ik}X\tilde{F}_{ii} + \underbrace{\tilde{F}_{ii}X\left(\sum_{k-1,k\neq i}^{N}\tilde{P}_{ik}\right)}_{4} + \tilde{Q}_{ii} = 0 \tag{3.19}$$

The equations for the off-diagonal blocks $\tilde{P}_{ij}$, $i \neq j$ of $\tilde{P}$ in the ARE (3.19) are

$$A^T\tilde{P}_{ij} + \tilde{P}_{ij}A - \sum_{k=1}^{N}(\tilde{P}_{ik}X\tilde{P}_{kj}) + \tilde{Q}_{ij} = 0 \tag{3.20}$$

for $i,j = 1, \ldots, N, i \neq j$. Substituting (3.18) into (3.20) leads to

$$A^T\tilde{P}_{ij} + \tilde{P}_{ij}A - \tilde{F}_{ii}X\tilde{P}_{ij} + \left(\sum_{k=1,k\neq i}^{N}\tilde{P}_{ij}\right)X\tilde{P}_{ij} - \tilde{P}_{ij}X\tilde{F}_{jj}$$

$$+\tilde{P}_{ij}X\left(\sum_{k=1,k\neq j}^{N}\tilde{P}_{jk}\right) - \sum_{k=1,k\neq i,k\neq j}^{N}(\tilde{P}_{ik}X\tilde{P}_{kj}) + \tilde{Q}_{ij} = 0 \tag{3.21}$$

Summing up (3.21) for all $j \neq i$ corresponds to a block-wise row sum of off-diagonal terms in the ith block row of the ARE (3.15). This summation leads to

$$\underbrace{A^T\left(\sum_{k=1,k\neq i}^{N}\tilde{P}_{ik}\right) + \left(\sum_{k=1,k\neq i}^{N}\tilde{P}_{ik}\right)A}_{1} - \underbrace{\tilde{F}\left(\sum_{k=1,k\neq i}^{N}\tilde{P}_{ik}\right)}_{4}$$

$$+\underbrace{\sum_{l=1,l\neq i}^{N}\left(\left(\sum_{k=1,k\neq i}^{N}\tilde{P}_{ik}\right)X\tilde{P}_{il}\right)}_{3}$$

$$+\underbrace{\sum_{k=1,k\neq i}^{N}\left(\tilde{P}_{ik}X\left(\sum_{l=1,l\neq k}^{N}\tilde{P}_{kl}\right)\right) - \sum_{k=1,k\neq i}^{N}\left(\sum_{l=1,l\neq i,l\neq k}^{N}(\tilde{P}_{il}X\tilde{P}_{lk})\right)}_{2}$$

$$-\sum_{k=1,k\neq i}^{N}\tilde{P}_{ik}X\tilde{F}_{kk} + \sum_{k=1,k\neq i}^{N}\tilde{Q}_{ik} = 0 \tag{3.22}$$

Note that

$$\sum_{k=1,k\neq i}^{N}\left(\tilde{P}_{ik}X\left(\sum_{l=1,l\neq k}^{N}\tilde{P}_{kl}\right)\right)-\sum_{k=1,k\neq i}^{N}\left(\sum_{l=1,l\neq i,l\neq k}^{N}(\tilde{P}_{il}X\tilde{P}_{lk})\right)$$
$$=\sum_{k=1,k\neq i}^{N}\sum_{l=1,l\neq k}^{N}\tilde{P}_{ik}X\tilde{P}_{kl}-\sum_{k=1,k\neq i}^{N}\sum_{l=1,l\neq k}^{N}\tilde{P}_{ik}X\tilde{P}_{kl}$$
$$=\sum_{k=1,k\neq i}^{N}\tilde{P}_{ik}X\tilde{P}_{ki} \tag{3.23}$$

where we used symmetry, switching of sum operators and renaming the indices. Adding (3.23) to (3.19) and using properties of $\tilde{Q}$ defined in (3.11), equally numbered terms will cancel each other out leading to

$$A^T\tilde{F}_{ii}+\tilde{F}_{ii}A-\tilde{F}_jiX\tilde{F}_{ii}+\sum_{k=1,k\neq i}^{N}(\tilde{P}_{ik}X(\tilde{F}_{ii}-\tilde{F}_{kk}))+Q_{ii}=0 \tag{3.24}$$

Summing up (3.24) over all $i=1,\ldots,N$ we obtain

$$\sum_{i=1}^{N}(A^T\tilde{F}_{ii}+\tilde{F}_{ii}A-\tilde{F}_{ii}X\tilde{F}_{ii}+Q_{ii})=0 \tag{3.25}$$

which proves the theorem. □

Next we particularize Theorem 3.1 to the case of identical weights and prove an additional property of LQR for decoupled systems.

Theorem 3.2: *Assume the weighting matrices (3.11) of the LQR problem (3.13) are chosen as*

$$\begin{aligned} Q_{ii}&=Q_1 \quad \forall i=1,\ldots,N\\ Q_{ij}&=Q_2 \quad \forall j=1,\ldots,N, \quad i\neq j \end{aligned} \tag{3.26}$$

Let $\tilde{x}_0^T\tilde{P}\tilde{x}_0$ be the value function of the LQR problem (3.13) with weights (3.26), and the blocks of the matrix $\tilde{P}$ be denoted by $\tilde{P}_{ij}=\tilde{P}[(i-1)n:in,(j-1)n:jn]$ with $i,j=1,\ldots,N$. Then

I. $\sum_{j=1}^{N}\tilde{P}_{ij}=P$ *for all $i=1,\ldots,N$, where P is the symmetric positive definite solution of the ARE associated with a single node local problem:*

$$A^TP+PA-PBR^{-1}B^TP+Q_1=0 \tag{3.27}$$

II. $\sum_{j=1}^{N}\tilde{K}_{ij}=K$ *for all $i=1,\ldots,N$, where $K=-R^{-1}B^TP$.*

III. $\tilde{P}_{ij}=\tilde{P}_{lm}:=\tilde{P}_2\ \forall i\neq j,\ \forall l\neq m$ *is a symmetric negative semidefinite matrix.*

Proof. The assumption in (3.26) requires that the weight Q_1 used for absolute states and the weight Q_2 used for neighboring state differences are *equal* for all nodes and for all neighbors of a node, respectively. Such an assumption and the fact that $\tilde{A}$

and $\tilde{B}$ are block-diagonal with identical blocks imply that the ARE in (3.13) is a set of N identical equations where the matrices $\tilde{P}_{ij}$ are all identical and symmetric for all $i \neq j$. We denote by $\tilde{P}_2$, the generic block $\tilde{P}_{ij}$ for $i \neq j$. The matrices $F_{ii} = \sum_{j=1}^{N} \tilde{P}_{ij}$ defined in (3.17) are all identical and therefore (3.16) in Theorem 3.1 becomes

$$N(A^T\tilde{F}_{ii} + \tilde{F}_{ii}A - \tilde{F}_{ii}X\tilde{F}_{ii} + Q_1) = 0 \tag{3.28}$$

which proves property I with $F_{ii} = P$. Property II follows from property I and from (3.14), which implies that $\tilde{K}_{ij} = -R^{-1}B^T\tilde{P}_{ij}$. Next we prove property III.

The ARE (3.20) for the block $\tilde{P}_{ij}$ with $i \neq j$ becomes

$$A^T\tilde{P}_2 + \tilde{P}_2A - \tilde{P}_2X\tilde{P}_1 - \tilde{P}_1X\tilde{P}_2 - (N-2)\tilde{P}_2X\tilde{P}_2 - Q_2 = 0 \tag{3.29}$$

which can be rewritten as follows in virtue of property I:

$$(A - XP)^T(-N\tilde{P})^T\tilde{P}_2 + \tilde{P}_2(A - XP) + (N)\tilde{P}_2X\tilde{P} - Q_2 = 0 \tag{3.30}$$

where P is the symmetric positive definite solution of the ARE (3.27) associated with a single node local problem.

Rewriting (3.30) as

$$\begin{aligned}&(A - XP)^T(-N\tilde{P}_2) + (-N\tilde{P}_2)(A - XP)\\ &\quad -(-N\tilde{P}_2)^TX(-N\tilde{P}_2) + NQ_2 = 0\end{aligned} \tag{3.31}$$

Since $X > 0$ and $Q_2 \neq 0$, (3.31) can be seen as an ARE associated with an LQR problem for the stable system $(A - XP, B)$ with weights NQ_2 and R. Let the matrix $-N\tilde{P}_2$ be its positive semidefinite solution. Then, the matrix

$$\tilde{P} = \begin{bmatrix} P-(N-1)\tilde{P}_2 & \tilde{P}_2 & \cdots & \tilde{P}_2 \\ \tilde{P}_2 & P-(N-1)\tilde{P}_2 & \cdots & \tilde{P}_2 \\ \vdots & \ddots & \ddots & \vdots \\ \tilde{P}_2 & \cdots & \cdots & P-(N-1)\tilde{P}_2 \end{bmatrix} \tag{3.32}$$

is a symmetric positive definite matrix since

$$\tilde{x}^T\tilde{P}\tilde{x} = x_j^TPx_j + \sum_{i=1}^{N}\sum_{j\neq i}^{N}(x_j - x_j)^T\tilde{P}_2(x_j - x_j) \tag{3.33}$$

and it is the unique symmetric positive definite matrix solution to the ARE (3.27). This proves the theorem. □

Under the hypothesis of Theorem 3.2, because of symmetry and equal weights Q_2 on the neighboring state differences and equal weights Q_1 on absolute states, the LQR optimal controller will have the following structure:

$$\tilde{K} = \begin{bmatrix} K_1 & K_2 & \cdots & K_2 \\ K_2 & K_1 & \cdots & K_2 \\ \vdots & \ddots & \ddots & \vdots \\ K_2 & \cdots & \cdots & K_1 \end{bmatrix} \tag{3.34}$$

with K_1 and K_2 functions of N, A, B, Q_1, Q_2, and R.

Remark 3.2: *We conjecture that Theorem 3.2 is true under much milder assumptions on the tuning parameters (even with different weights on neighboring state differences) as long as the cost function is defined as in (3.11). The authors are currently studying this conjecture.*

The following corollaries of Theorem 3.2 follow from the stability and the robustness of the LQR controller $-R^{-1}B^T(-N\tilde{P}_2)$ for system $A - XP$ in (3.31).

Corollary 3.1: *The matrix $A - XP + NX\tilde{P}_2$ is Hurwitz.*

From the gain margin properties [224], we have:

Corollary 3.2: *$A - XP + \alpha NX\tilde{P}_2$ is a Hurwitz matrix for all $\alpha > 1/2$, with $\alpha \in \mathbb{R}$.*

Remark 3.3: *$A - XP$ is a Hurwitz matrix, thus the system in Corollary 3.2 is stable for $\alpha = 0$ as well. ($A - XP = A + BK$ with K being the LQR gain for system A, B with weights (Q_1, R).)*

The following condition defines a class of systems and LQR weighting matrices that will be used in later sections to extend the set of stabilizing distributed controller structures.

Condition 3.1: *$A - XP + \alpha NX\tilde{P}_2$ is a Hurwitz matrix for all $\alpha \in [0, 1/2]$, with $\alpha \in \mathbb{R}$.*

Essentially, Condition 3.1 characterizes systems for which the LQR gain stability margin described in Corollary 3.2 is extended to any positive α. The fact that $A - XP$ is already stable does not necessarily guarantee this property.

Checking the validity of Condition 3.1 for a given tuning of P and $\tilde{P}_2$ may be performed as a stability test for a simple affine parameter-dependent model

$$\dot{x} = \underbrace{(A_0 + \alpha A_1)}_{A(\alpha)} x \tag{3.35}$$

where $A_0 = A - XP$, $A_1 = NX\tilde{P}_2$, and $0 \leq \alpha \leq 1/2$. This test can be posed as an LMI problem (Proposition 5.9 in [225]) searching for quadratic parameter–dependent Lyapunov functions.

3.2.3 Distributed control design

We consider a set of N linear, identical, and decoupled dynamical systems, described by the continuous-time time-invariant state (3.2), rewritten below as

$$\dot{x}_j = Ax_j + Bu_j, \quad x_j(0) = x_{i0} \tag{3.36}$$

where $x_j(t) \in \mathbb{R}^n$, $u_j(t) \in \mathbb{R}^m$ are states and inputs of the ith system at time t, respectively. Let $\hat{x}(t) \in \mathbb{R}^{Nn}$ and $\hat{u}(t) \in \mathbb{R}^{Nm}$ be the vectors that collect the states and inputs of the N systems at time t, then

$$\begin{aligned}\dot{\hat{x}} &= \hat{A}\hat{x} + \hat{B}\hat{u} \\ \hat{x}(0) &= \hat{x}_0 \triangleq [x_{10}, \ldots, x_{N0}]^T\end{aligned} \tag{3.37}$$

with

$$\hat{A} = I_N \otimes A, \quad \hat{B}I_N \otimes B$$

Remark 3.4: *Systems (3.5) and (3.37) differ only in the number of subsystems. We will use system (3.5) with N_s subsystems when referring to local problems, and system (3.37) with N subsystems when referring to the global problem. Accordingly, tilded matrices will refer to local problems and hat matrices will refer to the global problem.*

With reference to the appendix, we use a graph to represent the coupling in the control objective and the communication in the following way. We associate the ith system with the ith node of a graph $\mathcal{G} = (\mathcal{V}, \mathcal{A})$. If an edge (i,j) connecting the ith and jth node is present, then

1. the ith system has full information about the state of the jth system and
2. the ith system control law minimizes a weighted distance between the ith and the jth system states.

The class of $\mathcal{K}^N_{n,m}(\mathcal{G})$ matrices is defined as follows:

Definition 3.1: $\mathcal{K}^N_{n,m}(\mathcal{G}) = \{M \in \mathbb{R}^{nN \times mN} \mid M_{ij} = 0 \text{ if } (i,j) \notin \mathcal{A}, M_{ij} = M[(i-1)n : in, (j-1)m : jm], i,j = 1, \ldots, N\}$.

The distributed optimal control problem is defined as follows:

$$\min_{\hat{K}} \hat{J}(\hat{u}, \hat{x}_0) = \int_0^{\infty} (\hat{x}(\tau)^T \hat{Q}\hat{x}(\tau) + \hat{u}(\tau)^T \hat{R}\hat{u}(\tau))d\tau \tag{3.38a}$$

$$\text{subj. to } \dot{\hat{x}} = \hat{A}\hat{x} + \hat{B}\hat{u} \tag{3.38b}$$

$$\hat{u}(t) = \hat{K}\hat{x}(t) \tag{3.38c}$$

$$\hat{K} \in \mathcal{K}^N_{m,n}(\mathcal{G}) \tag{3.38d}$$

$$\hat{Q} \in \mathcal{K}^N_{n,n}(\mathcal{G}), \quad \hat{R} = I_N \otimes R \tag{3.38e}$$

$$\hat{x}(0) = \hat{x}_0 \tag{3.38f}$$

with $\hat{Q} = \hat{Q}^T \geq 0$ and $\hat{R} = \hat{R}^T > 0$. We also refer to problem (3.38a)–(3.38f) without (3.38d) as a centralized optimal control problem. In general, computing the solution to problem (3.38a)–(3.38f) is an NP-hard problem. Next, we propose a suboptimal control design leading to a controller $\hat{K}$ with the following properties:

1. $\hat{K} \in \mathcal{K}^N_{m,n}(\mathcal{G})$ (3.39a)
2. *$\hat{A} + \hat{B}\hat{K}$ is Hurwitz* (3.39b)
3. *Simple tuning of absolute and relative state errors and control effort within $\hat{K}$* (3.39c)

Such controller will be referred to as distributed suboptimal controller. The following theorem will be used to propose a distributed suboptimal control design procedure.

Theorem 3.3: *Consider the LQR problem (3.10) with $N = d_{\max}(\mathcal{G}) + 1$ and weights chosen as in (3.26) and its solution (3.32), (3.34).*

Let $M \in \mathbb{R}^{N \times N}$ be a symmetric matrix with the following property:

$$\lambda_j(M) > \frac{N}{2}, \forall \lambda_j(M) \in S(M) \backslash \{0\} \tag{3.40}$$

and construct the feedback controller

$$\hat{K} = -I_N \otimes R^{-1} B^T P + M \otimes R^{-1} B^T \tilde{P}_2 \tag{3.41}$$

Then, the closed loop system

$$A_{cl} = I_N \otimes A + (I_N \otimes B)\hat{K} \tag{3.42}$$

is asymptotically stable.

Proof. Consider the eigenvalues of the closed-loop system A_{cl}

$$S(I_N \otimes (A - XP) + M \otimes (X\tilde{P}_2))$$

By Proposition A.2 in the appendix:

$$\mathcal{S}(I_N \otimes (A - XP) + M \otimes (X\tilde{P}_2)) = \bigcup_{i=1}^{N} \mathcal{S}(A - XP + \lambda_j(M) X\tilde{P}_2) \tag{3.43}$$

We will prove that $(A - XP + \lambda_j(M)X\tilde{P}_2)$ is a Hurwitz matrix $\forall i = 1, \ldots, N$, and thus prove the theorem. If $\lambda_j(M) = 0$, then $A - XP + \lambda_j(M)X\tilde{P}_2$ is Hurwitz based on Remark 3.3. If $\lambda_j(M) \neq 0$, then from Corollary 3.1 and from condition (3.40), we conclude that $A - XP + \lambda_j(M)X\tilde{P}_2$ is Hurwitz. □

Theorem 3.3 has several main consequences:

1. If $M \in \mathcal{K}_{1,1}^{N}(\mathcal{G})$, then $\hat{K}$ in (3.41) is an asymptotically stable *distributed* controller.
2. We can use *one* local LQR controller to compose distributed stabilizing controllers for a collection of identical dynamically decoupled subsystems.
3. The first two consequences imply that we can not only find a stabilizing distributed controller with a desired sparsity pattern (which is in general a formidable task by itself), but it is enough to solve a low-dimensional problem characterized by $d_{\max}(\mathcal{G})$ compared to the full problem size (3.36). This attractive feature of our approach relies on the specific problem structure defined in Sections 3.3 and 3.5.
4. The eigenvalues of the closed-loop large-scale system $\mathcal{S}(A_{cl})$ can be computed through N smaller eigenvalue computations as $\bigcup_{i=1}^{N} \mathcal{S}(A - XP + \lambda_j(M)X\tilde{P}_2)$.

5. The result is independent of the local LQR tuning. Thus, Q_1, Q_2, and R in (3.26) can be used in order to influence the compromise between minimization of absolute and relative terms, and the control effort in the global performance.

For the special class of systems defined by Condition 3.1, the hypothesis of Theorem 3.3 can be relaxed as follows:

Theorem 3.4: *Consider the LQR problem (3.10) with weights chosen as in (3.26) and its solution (3.32), (3.34). Assume that Condition 3.1 holds.*

Let $M \in \mathbb{R}^{N\times N}$ be a symmetric matrix with the following property:

$$\lambda_j(M) \geq 0, \forall \lambda_j \in \mathcal{S}(M) \tag{3.44}$$

Then, the closed loop system (3.42) is asymptotically stable when $\hat{K}$ is constructed as in (3.41).

Proof. Note that if Condition 3.1 holds, then $\mathcal{S}(A - XP + \lambda_j X\tilde{P}_2)$ is Hurwitz for all $\lambda_j(M) \geq 0$ (from Corollaries 2.1 and 3.1). By Proposition A.2 in the appendix

$$\mathcal{S}(I_N \otimes (A - XP) + M \otimes (X\tilde{P}_2)) = \bigcup_{i=1}^{N} \mathcal{S}(A - XP + \lambda_j(M)X\tilde{P}_2)$$

which together with condition (3.44) proves the theorem. □

In the next sections, we show how to choose M in Theorems 3.3 and 3.4 in order to construct distributed suboptimal controllers. The matrix M will

1. reflect the structure of the graph $\mathcal{G}$;
2. satisfy (3.40) or (3.44); and
3. be computed by using the graph adjacency matrix or the Laplacian matrix.

3.2.4 Finite strings

Let $\mathcal{G}$ represent a string interconnection of N systems with $d_{\max}(\mathcal{G})$ and the following Laplacian:

$$L(\mathcal{G}) = \begin{bmatrix} 1 & -1 & 0 & \cdots & 0 \\ -1 & 2 & -1 & \cdots & 0 \\ \vdots & \ddots & \ddots & \ddots & \vdots \\ 0 & \cdots & -1 & 2 & -1 \\ 0 & \cdots & 0 & -1 & 1 \end{bmatrix} \tag{3.45}$$

Construct a distributed controller as follows. Solve the LQR problem (3.10), (3.26) with $N = 2 + 1 + 3$ and desired Q_1, Q_2, R. Consider the decomposition (3.32) and (3.34) of local controller and cost function, respectively. Take $b \geq 0$ and construct the global controller gain matrix $\hat{K}$ as

$$\hat{K} = \begin{bmatrix} K_1 & 0 & 0 & \cdots & 0 \\ 0 & K_1 & 0 & \cdots & 0 \\ \vdots & \ddots & \ddots & \ddots & \vdots \\ 0 & \cdots & 0 & K_1 & 0 \\ 0 & \cdots & 0 & 0 & K_1 \end{bmatrix} + \begin{bmatrix} aK_2 & bK_2 & 0 & \cdots & 0 \\ bK_2 & aK_2 & bK_2 & \cdots & 0 \\ \vdots & \ddots & \ddots & \ddots & \vdots \\ 0 & \cdots & bK_2 & aK_2 & bK_2 \\ 0 & \cdots & 0 & bK_2 & aK_2 \end{bmatrix} \tag{3.46}$$

The closed-loop system under the DC law $\hat{K}$ can be written as $\dot{\hat{x}} = A_{cl}\hat{x}$ with

$$A_{cl} = I_N \otimes A + (I_N \otimes B)\hat{K} \tag{3.47}$$

Rewrite $\hat{K}$ as

$$\begin{aligned} \hat{K} &= -I_N \otimes (R^{-1}B^T\tilde{P}_1) - aI_N \otimes (R^{-1}B^T\tilde{P}_2) \\ &= -bA(G) \otimes (R^{-1}B^T\tilde{P}_2) \end{aligned}$$

From Theorem 3.3, $\hat{K}$ is equal to

$$\hat{K} = -I_N \otimes (R^{-1}B^TP) + M \otimes (R^{-1}B^T\tilde{P}_2)$$

and

$$M = (N-1-a)I_N - bA(G), \quad N = 3 \tag{3.48}$$

The following corollary of Theorem 3.3 defines the range of a and b, which leads to a stable distributed controller.

Corollary 3.3: *Consider controller (3.46). If $a + 2b < 1/2$, then the closed loop system (3.42) is asymptotically stable. If Condition 3.1 holds then the closed loop system (3.42) is asymptotically stable for all $a + 2b \le 2$.*

Proof. From *Property 1*, $\lambda_{\max}(\mathbf{A}(\mathcal{G})) \le d_{\max}(\mathcal{G}) = 2$. Since $b \ge 0$, by using Proposition A.1 in the appendix applied to M we derive

$$\begin{aligned} \lambda_{\min}(M) &= N - 1 - a - b\lambda_{\max}(\mathbf{A}(\mathcal{G})) \\ &\ge N - 1 - a - 2b, \quad N = 3 \end{aligned} \tag{3.49}$$

Condition (3.40) of Theorem 3.3 holds if $a + 2b < 1/2$, which proves the first part of the theorem.

Assume that Condition 3.1 holds and consider (3.49). Then, condition (3.44) of Theorem 3.4 holds if $a + 2b \le 2$. □

3.2.5 Illustrative example 3.1

With $a = 0$, the distributed controller (3.46) becomes

$$\hat{K} = \begin{bmatrix} K_1 & bK_2 & 0 & \cdots & 0 \\ bK_2 & K_1 & bK_2 & \cdots & 0 \\ \vdots & \ddots & \ddots & \ddots & \vdots \\ 0 & \cdots & bK_2 & K_1 & bK_2 \\ 0 & \cdots & 0 & bK_2 & K_1 \end{bmatrix} \tag{3.50}$$

The closed loop system (3.42) is asymptotically stable if $b < 1/4$. If Condition 3.1 holds, then the closed loop system (3.42) is asymptotically stable if $b \leq 1$.

3.2.6 Illustrative example 3.2

Assume that Condition 3.1 holds. Since the graph Laplacian $L(\mathcal{G})$ is a symmetric positive semidefinite matrix, $M = aL(\mathcal{G})$, $a \geq 0$ satisfies condition (3.44) of Theorem 3.4. Thus, any distributed controller $\hat{K} = -I_N \otimes (R^{-1}B^T P) + aL(\mathcal{G}) \otimes (R^{-1}B^T \tilde{P}_2)$ having the following structure:

$$\hat{K} = \begin{bmatrix} 0 & K_2 & 0 & \cdots & 0 \\ K_2 & 0 & K_2 & \cdots & 0 \\ \vdots & \ddots & \ddots & \ddots & \vdots \\ 0 & \cdots & K_2 & 0 & K_2 \\ 0 & \cdots & 0 & K_2 & 0 \end{bmatrix} + I_N \otimes (K_1 - (2-a)K_2) \tag{3.51}$$

stabilizes system (3.42) for all $a \geq 0$.

3.2.7 Arbitrary graph structures

We consider a generic graph $\mathcal{G}$ for N nodes with an associated Laplacian $L(\mathcal{G})$ and maximum vertex degree $d_{\max}$. Let $0 = \lambda_1(G < \lambda_2(\mathcal{G}) \cdots \leq \lambda_N(\mathcal{G}))$ be the eigenvalues of the Laplacian $L(\mathcal{G})$. In Corollaries 3.3, 3.4, and 3.5, we present three ways of choosing M in (3.41) that lead to distributed suboptimal controllers.

Corollary 3.4: *Compute M in (3.41) as*

$$M = aL(\mathcal{G}) \tag{3.52}$$

If

$$a > \frac{N}{2\lambda_2(\mathcal{G})} \tag{3.53}$$

then the closed-loop system (3.42) is asymptotically stable when $\hat{K}$ is constructed as in (3.41). In addition, if Condition 3.1 holds, then the closed-loop system (3.42) is asymptotically stable for all $a \geq 0$.

Proof. Apply condition (3.40) in Theorem 3.3 to the case $M = aL(\mathcal{G})$. Note that $\lambda_j(\mathcal{G}) \geq \lambda_2(\mathcal{G})$ for all $i = 3,\ldots,N$ and $\lambda_2(\mathcal{G}) \neq 0$ by property 3 of the Laplacian matrix. Therefore, condition (3.40) becomes $\lambda_2(aL(\mathcal{G})) > N/2$. This implies $a\lambda_2(\mathcal{G}) > N/2$ and thus $a > N/2\lambda_2(\mathcal{G})$. The first part of the theorem is proven. If $a \geq 0$ then $\lambda_j(M) \geq 0 \forall i$. Therefore, the application of Theorem 3.4 to the matrix $M = aL(G)$ proves the second part of the theorem. □

Remark 3.5: *By using the third formula in property 4, condition (3.53) can be linked to the edge connectivity as follows:*

$$a > \frac{N}{2\eta(\mathcal{G})(1 - \cos(\frac{\pi}{N}))} \tag{3.54}$$

Remark 3.6: *Corollary 3.4 links the stability of the distributed controller to the size of the second smallest eigenvalues of the graph Laplacian. It is well known that graphs with large λ_2 (with respect to the maximal degree) have some properties that make them very useful in several applications such as computer science [226–229]. Interesting enough, this property is shown here to be crucial also for the design of distributed controllers. We refer the reader to [230] for a more detailed discussion on the importance of the second smallest eigenvalue of a Laplacian.*

Corollary 3.5: *Compute M in (3.41) as*

$$M = aI_N - b\mathbf{A}(\mathcal{G}), \quad b \geq 0 \tag{3.55}$$

If $a - bd_{\max} > N/2$, then the closed-loop system (3.42) is asymptotically stable when $\hat{K}$ is constructed as in (3.41). In addition, if Condition 3.1 holds, then the closed-loop system (3.42) is asymptotically stable if $a - bd_{\max} \geq 0$.

Proof. Note that $\lambda_{\min}(M) = a - b\lambda_{\max}(\mathbf{A}(\mathcal{G})) \geq a - bd_{\max}$. The proof is a direct consequence of Theorems 3.3 and 3.4 and property 1 of the adjacency matrix. □

Consider a weighted adjacency matrix $\mathbf{A}^w = \mathbf{A}^w(\mathcal{G})$, defined as follows. Denote by $\mathbf{A}^w_{i,j} \in \mathbb{R}$ its i,j element, then $\mathbf{A}^w_{i,j} = 0$, if $i = j$ and $(i,j) \notin \mathcal{A}$ and $\mathbf{A}^w_{i,j} = w_{ij}$ if $(i,j) \in A$, $\forall i,j = 1,\ldots,N$, $i \neq j$. Assume $w_{ij} = w_{ji} > 0$. Define $w_{\max}$ in the light of the following corollary:

Corollary 3.6: *Compute M in (3.41) as*

$$M = aI_N - \mathbf{A}^w(\mathcal{G}) \tag{3.56}$$

If $a > w_{\max} - N/2$, then the closed-loop system (3.42) is asymptotically stable when $\hat{K}$ is constructed as in (3.41). In addition, if Condition 3.1 holds, then the closed-loop system (3.42) is asymptotically stable if $a \geq w_{\max}$.

Proof. $\mathbf{A}^w\mathbf{1} \leq w_{\max}\mathbf{1}$ and by Perron–Frobenius theorem $\lambda_{\max}(\mathbf{A}^w) \leq w_{\max}$. Note that $\lambda_{\min}(M) = a - \lambda_{\max}(\mathbf{A}(\mathcal{G})) \geq a - w_{\max}$, then the proof is a direct consequence of Theorems 3.3 and 3.4. □

Table 3.1 Summary of stability conditions for systems (3.41)–(3.42)

Choice of M	**Stability condition**	**Stability condition if Condition 3.1 holds**
$M = aL(\mathcal{G})$	$a > \dfrac{N}{2\lambda_2(\mathcal{G})}$	$a \geq 0$
$M = aI_N - b\mathbf{A}(\mathcal{G}), b \geq 0$	$a - bd_{\max} > \dfrac{N}{2}$	$a - bd_{\max} \geq 0$
$aI_N - \mathbf{A}^w(\mathcal{G})$	$a > w_{\max} - \dfrac{N}{2}$	$a \geq w_{\max}$

The results of Corollaries 3.4–3.6 are summarized in Table 3.1.

Corollaries 3.4–3.6 present three choices of DC design with increasing degrees of freedom. In fact, a, b, and w_{ij} are additional parameters that, together with Q_1, Q_2, and R, can be used to tune the closed-loop system behavior. We recall here that from Theorem 3.3, the eigenvalues of the closed-loop large-scale system are related to the eigenvalues of M through the simple relation (3.43). Thus, as long as the stability conditions defined in Table 3.1 are satisfied, the overall system architecture can be modified arbitrarily by adding or removing subsystems and interconnection links. This leads to a very powerful modular approach for designing DC systems. The difference between the performance of the DC law and the performance of a centralized optimal solution can also be computed in a simple way. This is discussed in the next section.

3.2.8 Measure of suboptimality

Consider system (3.38f) and let K^* and P^* be an LQR controller and the corresponding ARE solution associated with the weights $\hat{Q} > 0$ and $\hat{R} > 0$. Therefore, $\hat{u}(t) = -K^*\hat{x}(t)$ minimizes the following cost function for any $\hat{x}_0$:

$$J(\hat{u}(t), \hat{x}_0) = \int_0^{\infty} \left(\hat{x}(\tau)^T \hat{Q}\hat{x}(\tau) + \hat{u}\tau^T \hat{R}\hat{u}(\tau)\right) d\tau \tag{3.57}$$

that is

$$J^*(\hat{x}_0) \triangleq \min_{\hat{u}(t)} J(\hat{u}(t), \hat{x}_0) = J(-K^*\hat{x}(t), \hat{x}_0) = \hat{x}_0^T P^* \hat{x}_0 \tag{3.58}$$

The next results aim at comparing the optimal centralized controller K^* and the distributed suboptimal controller $\hat{K}$ presented in the previous section. The comparison will be made assuming measure (3.57) reflects the desired performance index and computing the loss of performance introduced by the DC law (3.41) with chosen as in Corollaries 3.4–3.6.

Proposition 3.2: *Consider the distributed controller $\hat{K}$ in (3.41) and the closed-loop system (3.42). If (3.42) is stable, then*

$$J(-\hat{K}\hat{x}(t), \hat{x}_0) = \hat{x}_0^T \hat{P} \hat{x}_0 \tag{3.59}$$

where $\hat{P}$ is the positive definite solution of the following Lyapunov equation:

$$(I_N \otimes A + (I_N \otimes B)\hat{K})^T \hat{P} + \hat{P}(I_N \otimes A + (I_N \otimes B)\hat{K}) + (\hat{Q} + \hat{K}^T \hat{R} \hat{K}) = 0 \tag{3.60}$$

Proof. By assumption, system (3.42) is stable and thus the positive definite solution to the Lyapunov equation (3.60) can be written as

$$\hat{P} = \int_0^\infty (e^{(\hat{A}+\hat{B}\hat{K})\tau^T} (\hat{Q} + \hat{K}^T \hat{R} \hat{K}) e^{(\hat{A}+\hat{B}\hat{K})\tau}) d\tau \tag{3.61}$$

Consider $J(-\hat{K}\hat{x}(t), \hat{x}_0)$. From direct computation

$$J(-\hat{K}\hat{x}(t), \hat{x}_0) = \int_0^\infty (\hat{x}(\tau)^T (\hat{Q} + \hat{K}^T \hat{R} \hat{K}) \hat{x}(\tau)) d\tau \tag{3.62}$$

By substituting $\hat{x}$ with $e^{(\hat{A}+\hat{B}\hat{K})\tau}\hat{x}_0$ we obtain (3.59). □

Proposition 3.2 shows that the cost associated with the distributed controller $\hat{K}$ according to the performance index (3.57) can be computed by solving a Lyapunov equation for the closed-loop system with weight equal to $(\hat{Q} + \hat{K}^T \hat{R} \hat{K})$.

Since P^* is optimal, $J(-\hat{K}\hat{x}(t), \hat{x}_0) \geq J(-K^*\hat{x}(t), \hat{x}_0)$ for all $\hat{x}_0$ and thus $\Delta P = \hat{P} - P^*$ is a positive semidefinite matrix that is equal to zero if $\hat{K} = K^*$. Any norm of ΔP is a measure of the suboptimality of the distributed controller $\hat{K}$.

The 'best' linear state-feedback distributed controller $K_d^* \in \mathcal{K}_{m,n}^N(\mathcal{G})$ could be computed by solving

$$\begin{aligned} J_d^*(\hat{x}_0) &= \min_{K_d} \|\Delta P\| \\ \text{subj. to } \dot{\hat{x}} &= \hat{A}\hat{x} + \hat{B}\hat{u} \\ \hat{u}(t) &= K_d \hat{x}(t) \\ K_d &\in \mathcal{K}_{m,n}^N(G) \\ \hat{x}(0) &= \hat{x}_0 \end{aligned} \tag{3.63}$$

As discussed in the previous sections, computing the solution to problem (3.63) is a difficult problem in general without further assumptions. Its efficient solution is the topic of current research.

3.2.9 Illustrative example 3.3

Consider a 10×10 mesh interconnection of $N = 100$ identical, dynamically decoupled, and independently actuated linear systems moving in a plane with double integrator dynamics in both spatial dimensions

$$\ddot{x}_j = u_{x,i} \quad \ddot{y}_j = u_{y,i} \quad i = 1, \ldots, 100 \tag{3.64}$$

The interconnection structure is depicted in Figure 3.2. The control objective is to move each subsystem to a desired absolute position corresponding to its location in the pre-specified rectangular grid, which has equal separation distances defined between each orthogonal neighbor.

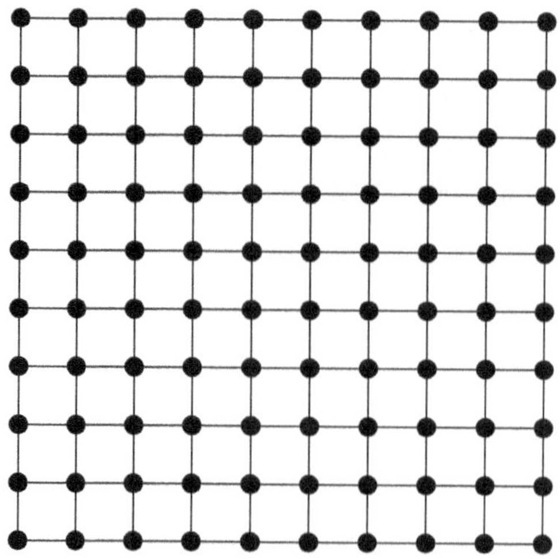

Figure 3.2 A 10 × 10 finite mesh grid interconnection structure

The maximum vertex degree of such an interconnection graph is 4, as the nodes located at the 'corners' of the rectangular grid have two neighbors, those along the 'edges' of the rectangle have three neighbors and the ones in the 'middle' have four each.

A stabilizing distributed controller is designed for the mesh interconnection of systems by solving the following simple LQR problem involving only $N = 5$ nodes ($d_{\max} + 1$)

$$\min_{\tilde{u}} J(\tilde{u}, \tilde{x}_0) \text{ subj. to } \dot{\tilde{x}} = \tilde{A}\tilde{x} + \tilde{B}\tilde{u} \quad \tilde{x}(0) = \tilde{x}_0 \tag{3.65}$$

where

$$\tilde{A} = I_5 \otimes A, \quad \tilde{B} = I_5 \otimes B$$
$$A = \begin{bmatrix} 0 & 1 & 0 & 0 \\ 0 & 0 & 0 & 0 \\ 0 & 0 & 0 & 1 \\ 0 & 0 & 0 & 0 \end{bmatrix}, \quad B = \begin{bmatrix} 0 & 0 \\ 1 & 0 \\ 0 & 0 \\ 0 & 1 \end{bmatrix} \tag{3.66}$$

The cost function defined in (3.8) uses the following weights for absolute and relative state information

$$Q_{ii} = Q_a, \quad Q_{ij} = Q_r, \quad \forall i \neq j \tag{3.67}$$

The weight on the control effort was kept constant in the following examples (for each control input in both spatial dimensions). The subsystem model and chosen weights satisfy Condition 3.1 (we tested the stability of (3.35) by solving an LMI problem searching for quadratic parameter–dependent Lyapunov functions), thus we can use less conservative stability ranges in Table 3.1. Alternatively, one can choose the more conservative stability ranges in Table 3.1 that can be *always* satisfied.

The solution of the above local LQR problem will yield a controller matrix of the following form in this specific example:

$$\tilde{K} = \begin{bmatrix} K_a & K_r & K_r & K_r & K_r \\ K_r & K_a & K_r & K_r & K_r \\ K_r & K_r & K_a & K_r & K_r \\ K_r & K_r & K_r & K_a & K_r \\ K_r & K_r & K_r & K_r & K_a \end{bmatrix} \tag{3.68}$$

The distributed controller for the grid of $N = 100$ interconnected systems has the same structure as the underlying graph and is constructed in the following simple way based on the forgoing results:

$$\hat{K} = I_N \otimes K_a + \mathbf{A} \otimes K_r \tag{3.69}$$

where $\mathbf{A}$ denotes the adjacency matrix of the mesh grid interconnection of systems. The distributed controller in (3.69) can be rewritten as $\hat{K} = I_N \otimes K - (4I_N - \mathbf{A}) \otimes K_r$, where $K = K_a + 4K_r$. This corresponds to $M = 4I_N - \mathbf{A}$ in Table 3.1 with stability condition $a - d_{\max} \geq 0$, which is satisfied in this example since $a = d_{\max} = 4$.

Two simulation results, starting from the same initial conditions, are presented with different choices for the parameters Q_a and Q_r in the local LQR problem tuning (3.67). The first simulation weighs the absolute state information in the local LQR solution more heavily than the second simulation and uses $Q_a(1, 1) = 1$ and $Q_r(1, 1) = 1$ and values for position states in both spatial dimensions. Velocity states were not weighted in these simulation examples. Snapshots of the simulation are shown in Figure 3.3, indicating that the subsystems converge to their desired absolute positions along fairly straight lines starting from their initial conditions. The second simulation illustrates the behavior of the large-scale DC system, when much more emphasis is put on the relative state information in the local LQR cost function by selecting $Q_a(1, 1) = 0.001$ and $Q_r(1, 1) = 1000$. As the snapshots in Figure 3.4 demonstrate, the behavior of the overall interconnected system has changed fundamentally. Although the distributed controller is still stabilizing, the dynamic behavior has changed drastically and shows wave-like oscillations due to the high weights on relative state information.

The purpose of these examples was to show with a numerical exercise that the stabilizing effect of the proposed DC design is independent of the local LQR weighting parameter selection; thus, these parameters are freely available for tuning and achieving different global performance objectives.

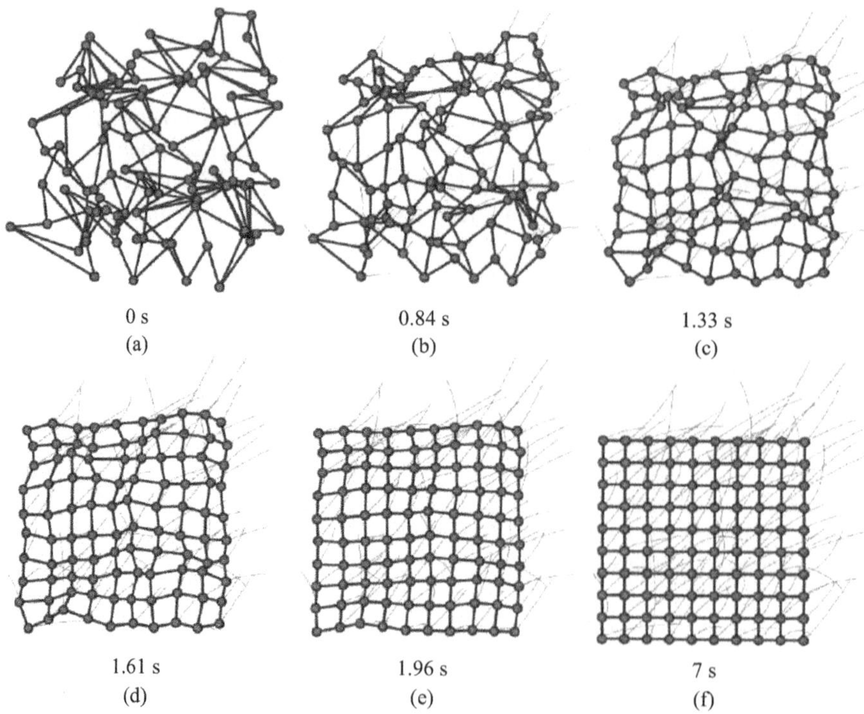

Figure 3.3 Snapshots of the mesh grid simulation with larger weight on absolute position references

3.3 Distributed semistable LQR control

A distributed LQR semistability theory for discrete-time systems is developed for designing optimal semistable controllers for discrete-time coupled systems. Unlike the standard LQR control problem, a unique feature of the proposed optimal control problem is that the closed-loop generalized discrete-time semistable Lyapunov equation can admit multiple solutions. Necessary and sufficient conditions for the existence of solutions to the generalized discrete-time semistable Lyapunov equation are derived and an optimization-based design framework for distributed optimal controllers is presented. The material covered in this section is based on the work of Hui [231].

3.3.1 Introduction

Recent years have witnessed a considerable interest in control of networks and control over networks. The merit of such network systems involving distributed decision-making for coordination of multiple agents is that they can enhance operational effectiveness via cooperative control. A unique feature of the closed-loop dynamics under many cooperative control algorithms in dynamical networks is the

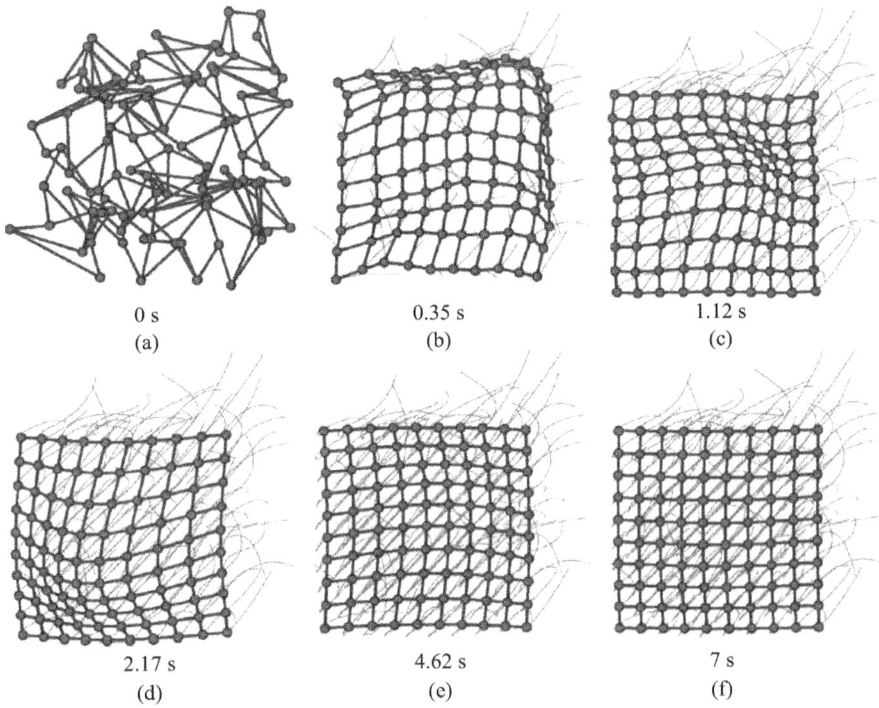

Figure 3.4 Snapshots of the mesh grid simulation with larger weight on relative position references

existence of a continuum of equilibria representing a desired state of convergence. Under such dynamics, the desired limiting state is not determined completely by the system dynamics, but depends on the initial system state as well [232,233]. In such network systems, *semistability*, and not asymptotic stability, is the relevant notion of stability [232,234]. Roughly speaking, semistability is the property that every trajectory starting in a neighborhood of a manifold consisting of Lyapunov stable equilibria will converge to another Lyapunov stable equilibrium on this manifold.

Even though many network coordination algorithms have been developed over the last several years (see [232,233,235–241], and the numerous references therein), LQR-type optimality properties of these algorithms have been largely ignored. LQR-type optimality here refers to minimization of an LQR-type quadratic cost functional for coordination algorithms subject to stability and network connectivity constraints. The physical meaning of this LQR-type quadratic cost functional is the trade-off between the algorithm convergence rate and control effort devoted to the closed-loop system.

It is important to note that although this optimality is close to the standard LQR problem, there are some differences between them. First, in the standard LQR the closed-loop system is asymptotically stable, not semistable. Second, the equilibrium point for the standard LQR is the origin, which is a fixed point. However, in the semistable LQR, the equilibrium point is a function of the initial condition, which is not fixed a priori. Next, the proposed control problem is also different from the ones in [242–244] and [245]. In particular, the authors in [242] and [245] consider the eigenvalue optimization for *undirected* connected graphs, while our problem considers the quadratic optimization for linear systems over arbitrary *directed* graphs. Although a least squares type minimization is taken into account in [243] and [244] to minimize the cost associated with the trajectories of the linear averaging algorithm, this minimization problem only considers a very special case of convergence-linear averaging and it does not consider a general semistable LQR problem.

A quite general, distributed semistable LQR control framework for discrete-time linear dynamical systems is pursued in the sequel following [231]. Unlike the standard QR control problem, a unique feature of the distributed semistable LQR control problem is that the closed-loop generalized Lyapunov equation guaranteeing discrete-time semistability can admit multiple solutions. We present necessary and sufficient conditions for the existence of solutions to the generalized discrete-time semistable Lyapunov equation. The distributed semistable LQR control problem is then converted into a constrained optimization problem. This optimization problem can be used to solve the original distributed semistable LQR control problem. Using a computationally feasible framework, the LQR optimal network coordination algorithm can be addressed to balance both the convergence rate and control energy for the network system.

3.3.2 Problem formulation

In the sequel, we consider q discrete-time linear systems $\mathcal{G}_j$, $i = 1, 2, \ldots, q$, given by

$$x_j(k+1) = A_j x_j(k) + B_j u_j(k), \quad x_j(0) = x_{i0}, \quad k \in \bar{\mathbb{Z}}_+ \tag{3.70}$$

where for every $k \in \bar{\mathbb{Z}}_+$, $x_j(k) \in \mathfrak{R}^n$ denotes states of the ith system, $u_j(k) \in \mathfrak{R}^m$ denotes inputs of the ith system, $A_j \in \mathfrak{R}^{n\times n}$, and $B_j \in \mathfrak{R}^{n\times m}$. Note that this model includes both homogeneous and heterogeneous systems for multiagent coordination. We can rewrite the overall system as a compact form

$$x(k+1) = Ax(k) + Bu(k), \quad x(0) = x_0, \quad k \in \bar{\mathbb{Z}}_+ \tag{3.71}$$

where $x \triangleq [x_1^T, \ldots, x_q^T]^T \in \mathfrak{R}^{nq}$, $u = [u_1^T, \ldots, u_q^T]^T \in \mathfrak{R}^{mq}$, $A \triangleq$ block-diag $[A_1, \ldots, A_q]$, and $B \triangleq$ block-diag $[B_1, \ldots, B_q]$.

The following definition of discrete-time semistability is given in [246]:

Definition 3.2: The discrete-time linear system $x(k+1) = Ax(k)$ is discrete-time semistable if for every $\lambda \in \operatorname{spec}(A), |\lambda| < 1$ or $\lambda = 1$ and, if $\lambda = 1$, then l is semisimple.

A direct consequence of Definition 3.2 is that $x(k+1) = Ax(k)$ is discrete-time semistable if and only if $\lim_{k\to\infty} x(k)$ exists [246]. In this section, we consider the distributed controller design $u_j = K_j x$ so that the closed-loop system is discrete-time semistable, that is

$$\lim_{k\to\infty} x_j(k) = \alpha_j, \quad i = 1, 2, \ldots, q \tag{3.72}$$

where α_j is a constant and determined by the initial condition, the cost function

$$J(u, x_0) = \sum_{k=0}^{\infty} [(x(k) - \alpha)^T Q(x(k) - \alpha) + (u(k) - \beta)^T R(u(k) - \beta)] \tag{3.73}$$

is minimized, where $Q = Q^T \geq 0$, $R = R^T > O$, $\alpha = \lim_{k\to\infty} x(k)$, and $\beta = \lim_{k\to\infty} u(k)$. For *consensus* [247] in multiagent coordination, $\alpha_1 = \cdots = \alpha_q$. Here, we assume that the communication topology is not determined yet, that is, the controller gain K_j does not have a fixed structure. We called this *problem P1*. The first question on this optimal control problem is about well-posedness. Is this optimal control problem a well-defined problem? Under what conditions is the optimal control problem well defined? To answer those questions, let us look at the following example.

3.3.3 Illustrative example 3.4

Consider a free discrete-time linear system of the type (3.72):

$$x_1(k + l) = u_1(k) \tag{3.74}$$

$$x_2(k + 1) = u_2(k) \tag{3.75}$$

Assume that the distributed controllers are given by

$$u_1(k) = k_{11}x_1(k) + k_{12}x_2(k) \tag{3.76}$$

$$u_2(k) = k_{21}x_1(k) + k_{22}x_2(k) \tag{3.77}$$

where $k_{ij} \in \Re$, $i, j = 1, 2$. The closed-loop system is given by

$$\begin{bmatrix} x_1(k+1) \\ x_2(k+1) \end{bmatrix} = \begin{bmatrix} k_{11} & k_{12} \\ k_{21} & k_{22} \end{bmatrix} \begin{bmatrix} x_1(k) \\ x_2(k) \end{bmatrix} \tag{3.78}$$

To achieve $\lim_{t\to\infty} x_1(t) = \lim_{t\to\infty} x_2(t) = \frac{1}{2}(x_1(0) + x_2(0))$, we first need to guarantee that the closed-loop system (3.78) is discrete-time semistable so that $\lim_{t\to\infty} x_j(t)$ exists for every $i = 1, 2$. Note that the matrix $\left[\begin{smallmatrix} k_{11} & k_{12} \\ k_{21} & k_{22} \end{smallmatrix}\right]$ has two eigenvalues λ_l and λ_2 satisfying $\lambda_1| + \lambda_2 = k_{11} + k_{22}$. It follows from Definition 3.2 that the discrete-time semistability of (3.78) implies that one of the eigenvalues for

$$\begin{bmatrix} k_{11} & k_{12} \\ k_{21} & k_{22} \end{bmatrix}$$

has to be 1. We assume, without loss of generality, that $\lambda_1 = 1$. Then $\lambda_2 = 1k_{11} - k_{22}$. It also follows from Definition 3.2 that $|\lambda_2| < 1$, which indicates that

$$|k_{11} + k_{22} - 1| < 1 \tag{3.79}$$

On the other hand, since $\lim_{t\to\infty} x_1(t) = \lim_{t\to\infty} x_2(t) = \frac{1}{2}(x_l(0) + x_2(0))$, it follows from [5] that $[1, 1]^T$ has to be a left and right eigenvector associated with the eigenvalue 1. Thus

$$k_{11} + k_{12} = 1 \tag{3.80}$$

$$k_{21} + k_{22} = 1 \tag{3.81}$$

In this case, we rewrite (3.78) as

$$\begin{bmatrix} x_1(k+1) \\ x_2(k+1) \end{bmatrix} = \begin{bmatrix} 1 - k_{12} & k_{12} \\ k_{21} & 1 - k_{21} \end{bmatrix} \begin{bmatrix} x_1(k) \\ x_2(k) \end{bmatrix} \tag{3.82}$$

and

$$0 < k_{12} + k_{21} < 2 \tag{3.83}$$

Then we have

$$k(k) = \begin{bmatrix} \dfrac{k_{21} + k_{12}(1 - k_{12} - k_{21})^k}{k_{12} + k_{21}} & \dfrac{k_{12} - k_{12}(1 - k_{12} - k_{21})^k}{k_{12} + k_{21}} \\ \dfrac{k_{21} - k_{21}(1 - k_{12} - k_{21})^k}{k_{12} + k_{21}} & \dfrac{k_{12} + k_{21}(1 - k_{12} - k_{21})^k}{k_{12} + k_{21}} \end{bmatrix} x(0) \tag{3.84}$$

and

$$\lim_{k\to\infty} x(k) = \begin{bmatrix} \dfrac{k_{21}}{k_{12} + k_{21}} & \dfrac{k_{12}}{k_{12} + k_{21}} \\ \dfrac{k_{21}}{k_{12} + k_{21}} & \dfrac{k_{12}}{k_{12} + k_{21}} \end{bmatrix} x(0) \tag{3.85}$$

Next, the cost function is chosen to be

$$J(u, x_0) = \sum_{k=0}^{\infty} [(u_1(k) - u_{e1})^2 + (u_2(k) - u_{e2})^2] \tag{3.86}$$

where $u_{ei} = \lim_{k\to\infty} u_j(k)$, $i = 1, 2$. Substituting (3.84) and (3.85) into (3.86) yields

$$J(u, x_0) = x^T(0) \begin{bmatrix} a_{11} & a_{12} \\ a_{21} & a_{22} \end{bmatrix} \tag{3.87}$$

where

$$\begin{aligned} a_{11} &= \frac{(k_{12}^2 + k_{21}^2)(1 - k_{12} - k_{21})^2}{(k_{12} + k_{21})^2[1 - (1 - k_{12} - k_{21})^2]} \\ a_{12} &= -a_{11}, \quad a_{21} = a_{12}, \quad a_{22} = a_{11} \end{aligned} \tag{3.88}$$

Thus

$$J(u, x_0) = a_{11}(x_1(0) - x_2(0))^2 \tag{3.89}$$

Note that a_{11} can be rewritten as

$$a_{11} = \frac{k_{12}^2 + k_{21}^2}{(k_{12} + k_{21})^2} \left[\frac{1}{1 - (1 - k_{12} - k_{21})^2} \right] \tag{3.90}$$

Since $(k_{12}^2 + k_{21}^2)/(k_{12} + k_{21})^2 \geq 2$ and $1/[1 - (1 - k_{12} - k_{21})^2] - 1 \geq 1$, it follows that $a_{11} \geq 0$, where the equality holds if and only if $k_{12} = k_{21} = 1/2$. Hence, $\min_u J(u, x_0) = 0$ and the optimal controller gain K is unique and given by

$$K = \frac{1}{2}\begin{bmatrix} 1 & 1 \\ 1 & 1 \end{bmatrix} \tag{3.91}$$

Alternatively, if we choose the cost function given by

$$\begin{aligned} J(u, x_0) &= \sum_{k=0}^{\infty} [(x_1(k) - x_{el})^2 + (x_2(k) - x_{e2})^2] \\ &\quad + \sum_{k=0}^{\infty} [(u_1(k) - u_{el})^2 + (u_2(k) - u_{e2})^2] \end{aligned} \tag{3.92}$$

where $x_{el} = \lim_{k\to\infty} x_j(k)$, $i = 1, 2$, then $J(u, x_0)$ is given by (3.89) with

$$\begin{aligned} a_{11} &= \frac{k_{12}^2 + k_{21}^2[1 + (1 - k_{12} - k_{21})^2]}{(k_{12} + k_{21})^2[1 - (1 - k_{12} - k_{21})^2]} \\ &= \frac{k_{12}^2 + k_{21}^2}{(k_{12} + k_{21})^2}\left[\frac{2}{1 - (1 - k_{12} - k_{21})^2} - 1\right] \end{aligned} \tag{3.93}$$

Since $(k_{12}^2 + k_{2l}^2)/(k_{12} + k_{2l})^2 \geq 1/2$ and $2/[1(1k_{l2}k_{21})^2]1 \geq 1$, it follows that $a_{11} = 1/2$, where the equality holds if and only if $k_{12} = k_{21} = 1/2$. Hence, $\min_u J(u, x_0) = (\frac{1}{2})(x_1(0)x_2(0))^2$ and the optimal controller gain K is unique and given by (3.91).

3.3.4 *Illustrative example 3.5*

Consider the stable discrete-time linear system of the type (3.71) given by

$$x_1(k+1) = \frac{1}{2}x_1(k) + u_1(k) \tag{3.94}$$

$$x_2(k+1) = \frac{1}{2}x_2(k) + u_2(k) \tag{3.95}$$

Assume that the distributed controllers are given by (3.76) and (3.77). The closed-loop system is given by

$$\begin{bmatrix} x_1(k+1) \\ x_2(k+1) \end{bmatrix} = \begin{bmatrix} k_{11} + 1/2 & k_{12} \\ k_{21} & k_{22} + 1/2 \end{bmatrix}\begin{bmatrix} x_1(k) \\ x_2(k) \end{bmatrix} \tag{3.96}$$

Similar to Illustrative example 3.4, to achieve $\lim_{t\to\infty} x_1(t) = \lim_{t\to\infty} x_2(t) = \frac{1}{2}(x_1(0) + x_2(0))$, we have

$$\begin{aligned} k_{11} + k_{12} &= 1/2 \\ k_{21} + k_{22} &= 1/2 \\ |k_{11} + k_{22}| &< 1 \end{aligned} \tag{3.97}$$

Hence, (3.82) and (3.83) hold in addition to (3.84) and (3.85). Next, let the cost function be (3.86). Substituting (3.84) and (3.85) into (3.86) yields (3.87) and (3.89), where

$$
\begin{aligned}
a_{11} &= \frac{(k_{12}^2 + k_{21}^2)(1/2 - k_{12} - k_{21})^2}{(k_{12} + k_{21})^2[1 - (1 - k_{12} - k_{21})^2]} \\
&= \frac{k_{12}^2 + k_{21}^2}{(k_{12} + k_{21})^2} \cdot \frac{(1/2 - k_{12} - k_{21})^2}{1 - (1 - k_{12} - k_{21})^2}
\end{aligned}
\tag{3.98}
$$

Since

$$(k_{12}^2 + k_{21}^2)/(k_{12} + k_{21})^2 \geq 1/2$$

and

$$(1/2 - k_{12} - k_{21})^2/[1 - (1 - k_{12} - k_{21})^2] \geq 0$$

it follows that $a_{11} \geq 0$, where the equality holds if and only if $k_{12} = k_{21} = l/4$. Hence, $\min_u J(u, x_0) = 0$ and the optimal controller gain K is unique and given by

$$K = \frac{1}{4}\begin{bmatrix} 1 & 1 \\ 1 & 1 \end{bmatrix}$$

3.3.5 Illustrative example 3.6

Consider the unstable discrete-time linear system of the type (3.71) given by

$$x_1(k+1) = 2x_1(k) + u_1(k) \tag{3.99}$$

$$x_2(k+1) = 2x_2(k) + u_2(k) \tag{3.100}$$

Assume that the distributed controllers are given by (3.76) and (3.77). The closed-loop system is given by

$$\begin{bmatrix} x_1(k+1) \\ x_2(k+1) \end{bmatrix} = \begin{bmatrix} k_{11} + 2 & k_{12} \\ k_{21} & k_{22} + 2 \end{bmatrix} \begin{bmatrix} x_1(k) \\ x_2(k) \end{bmatrix} \tag{3.101}$$

Similar to Illustrative example 3.4, to achieve

$$\lim_{t\to\infty} x_1(t) = \lim_{t\to\infty} x_2(t) = \frac{1}{2}(x_1(0) + x_2(0))$$

we have

$$
\begin{aligned}
k_{11} + k_{12} &= -1 \\
k_{21} + k_{22} &= -1 \\
|k_{11} + k_{22} + 3| &< 1
\end{aligned}
\tag{3.102}
$$

Hence, (3.82) and (3.83) hold in addition to (3.84) and (3.85). Next, let the cost function be (3.86). Equation (3.84) and (3.85) into (3.86) yields (3.87) and (3.89), where

$$a_{11} = \frac{(k_{12}^2 + k_{21}^2)(-1 - k_{12} - k_{21})^2}{(k_{12} + k_{21})^2[1 - (1 - k_{12} - k_{21})^2]}$$
$$= \frac{k_{12}^2 + k_{21}^2}{(k_{12} + k_{21})^2} \cdot \frac{[-2 + (1 - k_{12} - k_{21})]^2}{1 - (1 - k_{12} - k_{21})^2} \tag{3.103}$$

Observe that $(k_{12}^2 + k_{21}^2)/(k_{12} + k_{21})^2 \geq 1/2$. Now, letting

$$f(x) = \frac{(x-2)^2}{1-x^2}, \quad -1 < x < 1 \tag{3.104}$$

it follows that

$$f^T(x) = \frac{(2-x)(4x-2)}{(1-x^2)^2},$$
$$f''(x) = \frac{2(5-4x)(1-x^2) + 8x(2-x)(2x-1)}{(1-x^2)^3} \tag{3.105}$$

Setting $f(x) = 0$ yields $x = 1/2$. Since $f''(1/2) = 32/3 > 0$, it follows that $f(x)$ achieves the minimum at $x = 1/2$. Thus

$$(-1 - k_{12} - k_{21}^2)/[1 - (1 - k_{12} - k_{21})^2] \geq 3$$

and hence, $a_{11} \geq 3/2$, where the equality holds if and only if $k_{12} = k_{21} = 1/4$. Hence, $\min_u J(u, x_0) = (3/2)(x_1(0) - x_2(0))^2$ and the optimal controller gain K is unique and given by

$$K = \frac{1}{4}\begin{bmatrix} -5 & 1 \\ 1 & -5 \end{bmatrix}$$

Thus, it follows from Illustrative examples 3.4–3.6 that our problem is well defined. Moreover, these examples show that even for the simplest 2×2 matrix case, the direct computation of the optimal controller is a daunting task. In [248] and [249], the authors proposed a new characterization of the semistable LQR control problem for linear systems using linear matrix inequalities (LMIs) and bilinear matrix inequalities (BMIs). This new method requires the existence of the solutions to a Lyapunov equation. However, the Lyapunov equation approach based on the notion of *semiobservability* developed in [90] fails to work in some cases. To elucidate this point, let us look at the following example.

3.3.6 Illustrative example 3.7

Consider illustrative example 3.5 once again. Then the derivation in Illustrative example 3.6 gives us the following optimal solution:

$$K = \frac{1}{4}, \quad A + BK = \frac{1}{4}\begin{bmatrix} 3 & 1 \\ 1 & 3 \end{bmatrix}, \quad Q + K^T RK = \frac{1}{8}\begin{bmatrix} 1 & 1 \\ 1 & 1 \end{bmatrix} \tag{3.106}$$

Next, we show that the method proposed in [249] cannot be used to obtain the above solution. To see this, first note that the pair $(A + BK, Q + K^T RK)$ is not semiobservable in the sense of [249]. It is easy to verify that there does not exist $P \geq 0$

such that $P = (A + BK)^T P(A + BK) + Q + K^T RK$ holds. However, the method proposed in [249] to solve the semistable LQR problem does require the existence of $P \geq 0$ such that $P = (A + BK)^T P(A + BK) + Q + K^T RK$ holds. Thus, this method cannot be used to solve the optimal controller $u = Kx$ for the proposed semistable LQR problem. A new method is needed to address such a deficiency in [249].

Hence, all the existing methods in the literature cannot be used to solve the proposed semistable LQR problem. It is worthwhile developing a feasible technique to solve this issue. In the sequel, we will investigate this problem by generalizing [244] to develop distributed semistable LQR control theory for discrete-time linear dynamical systems.

Now we go back to the general optimal control problem. The closed-loop system can be written as

$$x(k+1) = \tilde{A}x(k), \quad x(0) = x_0, \quad k \in \bar{\mathbb{Z}}_+ \tag{3.107}$$

where $\tilde{A} \triangleq A + BK$ and $K = [K_1^T, \ldots, K_q^T]^T \in \mathfrak{R}^{mq \times nq}$.

Suppose now that the closed-loop system (3.107) is discrete-time semistable. Then it follows from [246, p. 447] that $\lim_{k\to\infty} x(k) = x_e$, where $x_e = [I_{nq} - (I_{nq} - \tilde{A})(I_{nq} - \tilde{A})^{\#}]x(0)$. Considering the cost functional (3.73), it follows that

$$\begin{aligned}
J(u, x_0) &= \sum_{k=0}^{\infty} [x(k) - x_e]^T (Q + K^T RK)[x(k) - x_e] \\
&= \sum_{k=0}^{\infty} [x(0) - x_e]^T (\tilde{A}^k)^T [x(0) - x_e] \\
&= \sum_{k=0}^{\infty} x^T(0)[(I_{nq} - \tilde{A})(I_{nq} - \tilde{A})^{\#}]^T \\
&\quad \times (\tilde{A}^k)^T \tilde{R} \tilde{A}^k (I_{nq} - \tilde{A})(I_{nq} - \tilde{A})^{\#} x(0) \\
&= Tr x^T(0) \left[\sum_{k=0}^{\infty} [(I_{nq} - \tilde{A})(I_{nq} - \tilde{A})^{\#}]^T \right. \\
&\quad \left. \times [(\tilde{A}^k)^T \tilde{R} \tilde{A}^k (I_{nq} - \tilde{A})(I_{nq} - \tilde{A})^{\#}] x(0) \right] \\
&= Tr \left[\sum_{k=0}^{\infty} [(I_{nq} - \tilde{A})(I_{nq} - \tilde{A})^{\#}]^T \right. \\
&\quad \left. \times [(\tilde{A}^k)^T \tilde{R} \tilde{A}^k (I_{nq} - \tilde{A})(I_{nq} - \tilde{A})^{\#}] x(0) x^T(0) \right] \\
&= TrPV
\end{aligned} \tag{3.108}$$

where $\tilde{R} \triangleq Q + K^T RK$ and

$$P \triangleq \sum_{k=0}^{\infty} [(I_{nq} - \tilde{A})(I_{nq} - \tilde{A})^{\#}]^T (\tilde{A}^k)^T \tilde{R} \tilde{A}^k (I_{nq} - \tilde{A})(I_{nq} - \tilde{A})^{\#} \tag{3.109}$$

Here we assume that the initial state x_0 is a random variable such that $\mathbb{E}[x_0] = 0$ and $\mathbb{E}[x_0 x_0^T] = V$ and we used the fact that $x(k) - x_e = \tilde{A}^k [x(0) - x_e]$.

The following lemma establishes the well-posedness of the semistable $\mathcal{H}_2$ form.

Lemma 3.1: *Consider the discrete-time linear dynamical system (3.107). If (3.107) is discrete-time semistable, then for every $nq \times nq$ nonnegative definite matrix R*

$$\sum_{k=0}^{\infty}(x(k)-x_e)^T R(x(k)-x_e) < \infty \tag{3.110}$$

where $x_e = [I_{nq} - (I_{nq} - \tilde{A})(I_{nq} - \tilde{A})^{\#}]x(0)$.

The following result is concerned with the well-posedness of (3.109).

Lemma 3.2: *If (3.107) is discrete-time semistable, then P given by (3.109) is well defined, that is, $\|P\| < \infty$.*

Proof. The proof is similar to that in [249]. For completeness, we include it here. Note that for every $x_0 \in \Re^{nq}$

$$\begin{aligned} x_0^T P x_0 &= \sum_{k=0}^{\infty}\sum_{k=0}^{\infty} x_0^T[(I_{nq}-\tilde{A})(I_{nq}-\tilde{A})^{\#}]^T \\ &\quad \times (\tilde{A}^k)^T\tilde{R}\tilde{A}^k(I_{nq}-\tilde{A})^{\#}x_0 \\ &= \sum_{k=0}^{\infty}(x_0-x_e)^T(\tilde{A}^k)^T\tilde{R}\tilde{A}^k(x_0-x_e) \\ &= \sum_{k=0}^{\infty}(x(k)-x_e)^T\tilde{R}(x(k)-x_e) \end{aligned} \tag{3.111}$$

where we used the fact that $x(k) - x_e = \tilde{A}^k(x_0 - x_e)$. It follows from Lemma 3.1 that P is well defined. □

The next result gives an alternative characterization of (3.109) using a Lyapunov-like equation.

Lemma 3.3: *If (3.107) is discrete-time semistable, then for any $k = l, 2, \ldots, P$ given by (3.109) satisfies*

$$((I_{nq}-\tilde{A})^k)^T P(I_{nq}-\tilde{A})^k = ((I_{nq}-\tilde{A})^k)^T(\tilde{A}^T P\tilde{A}+\tilde{R})(I_{nq}-\tilde{A})^k \tag{3.112}$$

Proof. Note that $(I_{nq} - \tilde{A})(I_{nq} - \tilde{A})^{\#}(I_{nq} - \tilde{A}) = I_{nq} - \tilde{A}$ and $\tilde{A}(I_{nq} - \tilde{A}) = (I_{nq} - \tilde{A})\tilde{A}$. Then it follows that

$$\begin{aligned} ((I_{nq}-\tilde{A})^k)^T P(I_{nq}-\tilde{A})^k &= \sum_{i=0}^{\infty}((I_{nq}-\tilde{A})^k)^T(\tilde{A}^i)^T\tilde{R}\tilde{A}^i(I_{nq}-\tilde{A})^k \\ &= \sum_{i=0}^{\infty}(\tilde{A}^i)^T((I_{nq}-\tilde{A})^k)^T\tilde{R}(I_{nq}-\tilde{A})^k\tilde{A}^i \end{aligned} \tag{3.113}$$

and

$$((I_{nq}-\tilde{A})^k)^T\tilde{A}^TP\tilde{A}(I_{nq}-\tilde{A})^k = \tilde{A}^T((I_{nq}-\tilde{A})^k)^TP(I_{nq}-\tilde{A})^k\tilde{A}$$
$$= \sum_{i=0}^{\infty}(\tilde{A}^{i+1})^T((I_{nq}-\tilde{A})^k)^T\tilde{R}(I_{nq}-\tilde{A})^k\tilde{A}^{i+1} \tag{3.114}$$

Hence $((I_{nq}-\tilde{A})^k)^TP(I_{nq}-\tilde{A})^k - ((I_{nq}-\tilde{A})^k)^T\tilde{A}^TP\tilde{A}(I_{nq}-\tilde{A})^k = ((I_{nq}-\tilde{A})^k)^T \tilde{R}(I_{nq}-\tilde{A})^k$. □

Observe that (3.112) is a generalized Lyapunov equation for discrete-time semistability of (3.107) and hence it is called the *generalized discrete-time semistable Lyapunov equation*. If (3.107) is discrete-time asymptotically stable, then the generalized discrete-time semistable Lyapunov equation (3.112) reduces to the standard Lyapunov equation $P = \tilde{A}^TP\tilde{A} + \tilde{R}$. The following result is immediate.

Lemma 3.4: *If (3.107) is discrete-time semistable, then for every nonnegative-definite matrix $\tilde{R} = \tilde{R}^T \in \Re^{nq\times nq}$ and every $k = 1, 2, \ldots$, there exists a nonnegative-definite matrix $P = P^T \in \Re^{nq\times nq}$ such that (3.112) holds.*

The following definition is a generalization of weak semiobservability in [244] and semiobservability [249].

Definition 3.3: Let $A \in \Re^{n\times n}$ and $C \in \Re^{m\times n}$. The pair (A, C) is weakly k-semiobservable if

$$k = \min\left\{m \in \mathbb{Z}_+ : \bigcap_{i=1}^{n} N(C(A-I_n)^{i+m-1}) = N(A-I_n)\right\} \tag{3.115}$$

Recall from [249] that the pair (A, C) is semiobservable if and only if

$$\bigcap_{i=1}^{n} \mathcal{N}(C(A-I_n)^{i-1}) = \mathcal{N}(A-I_n) \tag{3.116}$$

It is easy to see from the definition of weak k-semiobservability that (A, C) is weakly k-semiobservable if and only if (A, CA^k) is semiobservable. Motivated by Lemma 2.1 of [249], we have the following lemma:

Lemma 3.5 [249]: *If there exist matrices $P \geq 0$ and $\tilde{R} \geq 0$ in $\Re^{nq\times nq}$ and $k \in \mathbb{Z}_+$, such that (3.52) holds and the pair $(\tilde{A}, \tilde{R})$ is weakly k-semiobservable, then (i) $\mathcal{N}(P(\tilde{A}-I_{nq})^k) \subseteq \mathcal{N}(\tilde{A}-I_{nq}) \subseteq \mathcal{N}(\tilde{R}(\tilde{A}-I_{nq})^k)$ and (ii) $\mathcal{N}(\tilde{A}-I_{nq}) \cap \mathcal{R}(\tilde{A}-I_{nq}) = \{0\}$.*

Proof. Note that $(\tilde{A}, \tilde{R})$ is weakly k-semiobservable if

$$\bigcap_{i=1}^{nq} \mathcal{N}(\tilde{R}(\tilde{A}-I_{nq})^{i+k-1}) = \mathcal{N}(\tilde{A}-I_{nq}) \tag{3.117}$$

Since $\mathcal{N}(((I_{nq}-\tilde{A})^k)^T\tilde{R}(I_{nq}-\tilde{A})^k(I_{nq}-\tilde{A})^{i-1}) = \mathcal{N}(\tilde{R}(I_{nq}-\tilde{A})^k(I_{nq}-\tilde{A})^{i-1}) = \mathcal{N}(\tilde{R}(I_{nq}-\tilde{A})^{i+k-1})$, it follows that

$$\bigcap_{i=1}^{nq} \mathcal{N}(((I_{nq} - \tilde{A})^k)^T \tilde{R}(I_{nq} - \tilde{A})^{i+k-1}) = \bigcap_{i=1}^{nq} \mathcal{N}(\tilde{R}(I_{nq} - \tilde{A})^{i+k-1}) = \mathcal{N}(I_{nq} - \tilde{A}) \tag{3.118}$$

Let $\tilde{P} \triangleq ((I_{nq} - \tilde{A})^k)^T P(I_{nq} - \tilde{A})^k$ and $\tilde{Q} \triangleq ((I_{nq} - \tilde{A})^k)^T \tilde{R}(I_{nq} - \tilde{A})^k$. Then (3.112) becomes $\tilde{P} = \tilde{A}^T \tilde{P}\tilde{A} + \tilde{Q}$ and (3.118) becomes $\bigcap_{i=1}^{nq} \mathcal{N}(\tilde{Q}(I_{nq} - \tilde{A})^{i-1}) = \mathcal{N}(I_{nq} - \tilde{A})$. Now the result directly follows from Lemma 2.1 of [249] by noting that $\mathcal{N}(\tilde{P}) = \mathcal{N}(P(I_{nq} - \tilde{A})^k)$ and $\mathcal{N}(\tilde{Q}) = \mathcal{N}(\tilde{R}(I_{nq} - \tilde{A})^k)$. □

The partial converse result for Lemma 3.5 can be stated, as a direct consequence of Theorem 2.2 of [249], as follows.

Lemma 3.6: *Assume that for every nonnegative-definite matrix $\tilde{R} = \tilde{R}^T \in \Re^{nq \times nq}$ satisfying $(\tilde{A}, \tilde{R})$ is weakly k-semiobservable, there exists a nonnegative-definite matrix $P = P^T \in \Re^{nq \times nq}$ such that (3.112) holds. Then (3.107) is discrete-time semistable.*

Now we have our first necessary and sufficient condition for discrete-time semistability of (3.112) by using the generalized discrete-time semistable Lyapunov equation (3.107).

Lemma 3.7: *Consider the discrete-time linear dynamical system (3.107). Then (3.107) is discrete-time semistable if and only if for every weakly k-semiobservable pair (A, R) with nonnegative-definite R, there exists an $nq \times nq$ matrix P such that (3.112) holds.*

The result is a combination of Lemmas 3.4 and 3.6.

Next, we present a semistability property for Kronecker products, which will be needed for the main result of this section.

Lemma 3.8 [249]: *Let $A \in \Re^{n \times n}$ and $B \in \Re^{m \times m}$. If A and B are discrete-time semistable, then $A \otimes B$ is discrete-time semistable.*

The next result is needed for the main result of this section.

Lemma 3.9: *Let $x \in \Re^n$ and $A \in \Re^{n \times n}$, and assume A is discrete-time semistable. Then $\sum_{k=0}^{\infty} A^k x$ exists if and only if $x \in \mathcal{R}(A - I_n)$. In this case, $\sum_{k=0}^{\infty} A^k x = -(A - I_n)^{\#} x$.*

The proof is similar to the proofs of (viii) and (ix) of Lemma 5.2 of [250].

Now we come down to the main result of this section.

Theorem 3.5: *Consider the discrete-time linear dynamical system (3.107). Then (3.107) is discrete-time semistable if and only if for every weakly k-semiobservable pair $(\tilde{A}, \tilde{R})$ with nonnegative-definite $\tilde{R}$, there exists an $nq \times nq$ matrix $\hat{P} > 0$ such that*

$$\hat{P} = \tilde{A}^T \hat{P} \tilde{A} + ((I_{nq} - \tilde{A})^k)^T \tilde{R}(I_{nq} - \tilde{A})^k \tag{3.119}$$

Such a $\hat{P}$ is not unique. Furthermore, if $(\tilde{A}, \tilde{R})$ is weakly k-semiobservable and $\hat{P}$ satisfies (3.119), then

$$\hat{P} = \sum_{i=0}^{\infty} (\tilde{A}^i)^T((I_{nq} - \tilde{A})^k)^T \tilde{R}(I_{nq} - \tilde{A})^k \tilde{A}^i + \alpha x x^T, \quad x \in N(I_{nq} - \tilde{A}^T), \quad \alpha > 0 \tag{3.120}$$

Proof. Suppose $(\tilde{A}, \tilde{R})$ is weakly k-semiobservable. Then it follows from Lemma 3.7 that there exists an $nq \times nq$ matrix $P > 0$ such that (3.112) holds. Since, by Lemma 3.5, $\mathcal{N}(\tilde{A}I_{nq}) \cap R(\tilde{A} - I_{nq}) = \{0\}$, it follows from [251] (p. 119) that $\tilde{A}I_{nq}$ is group invertible. Thus, let $L \triangleq I_{nq} - (\tilde{A} - I_{nq})(\tilde{A} - I_{nq})^{\#}$ and note that $L^2 = L$. Hence, L is the unique $nq \times nq$ matrix satisfying $\mathcal{N}(L) = \mathcal{R}(\tilde{A} - I_{nq})$, $\mathcal{R}(L) = \mathcal{N}(\tilde{A} - I_{nq})$, and $Lx = x$ for all $x \in \mathcal{N}(\tilde{A} - I_{nq})$. Now, define

$$\hat{P} \triangleq ((\tilde{A} - I_{nq})^k)^T P(\tilde{A} - I_{nq})^k + L^T L \tag{3.121}$$

Next, we show that $\hat{P}$ is positive definite.

Consider the function $V(x) = x^T \hat{P} x, x \in \Re^{nq}$. If $V(x) = 0$ for some $x \in \Re^{nq}$, then $P(\tilde{A} - I_{nq})^k x = 0$ and $Lx = 0$. It follows from (i) of Lemma 3.5 that $x \in N(\tilde{A} - I_{nq})$, and $Lx = 0$ implies that $x \in R(\tilde{A} - I_{nq})$. Now, it follows from (ii) of Lemma 3.5 that $x = 0$. Hence, $\hat{P}$ is positive definite. Next, since $L(\tilde{A} - I_{nq}) = \tilde{A} - I_{nq} - (\tilde{A} - I_{nq})(\tilde{A} - I_{nq})^{\#}(\tilde{A} - I_{nq}) = 0$, it follows that

$$\begin{aligned}
\hat{P} - \tilde{A}^T \hat{P} \tilde{A} &= ((\tilde{A} - I_{nq})^k)^T P(\tilde{A} - I_{nq})^k + L^T L \\
&\quad - \tilde{A}^T((\tilde{A} - I_{nq})^k)^T P(\tilde{A} - I_{nq})^k \tilde{A} - \tilde{A}^T L^T L \tilde{A} \\
&= ((\tilde{A} - I_{nq})^k)^T P(\tilde{A} - I_{nq})^k + L^T L \\
&\quad - ((\tilde{A} - I_{nq})^k)^T \tilde{A}^T P \tilde{A}(\tilde{A} - I_{nq})^k - L^T L \\
&= ((\tilde{A} - I_{nq})^k)^T \tilde{R}(\tilde{A} - I_{nq})^k
\end{aligned} \tag{3.122}$$

Conversely, if there exists $\hat{P} > 0$ such that (3.119) holds, consider the function $U(x) = x^T \hat{P} x, x \in \Re^{nq}$. Then $U(x(t+1)) - U(x(t)) = -x^T(t)((I_{nq} - \tilde{A})^k)^T \tilde{R}(I_{nq}\tilde{A})^k x(t) \leq 0$, $t \in \bar{\mathbb{Z}}_+$, and $\{x \in \Re^{nq} : x^T((I_{nq} - \tilde{A})^k)^T \tilde{R}(I_{nq} - \tilde{A})^k x = 0\} = N(\tilde{R}(I_{nq} - \tilde{A})^k)$. To obtain the largest invariant set $\mathcal{M}$ contained in $\mathcal{N}(\tilde{R}(I_{nq} - \tilde{A})^k)$, consider a solution $x(t)$ of (3.107) such that $\tilde{R}(I_{nq} - \tilde{A}) = 0$ for $t \in \bar{\mathbb{Z}}_+$. Then, $\tilde{R}(I_{nq} - \tilde{A})^k x(t+1) - \tilde{R}(I_{nq} - \tilde{A})^k x(t) = 0$, that is, $\tilde{R}(I_{nq} - \tilde{A})^{k+1} x(t) = 0$. This implies $\tilde{R}(I_{nq} - \tilde{A})^{i+k-1} = 0$ for all $t \in \bar{\mathbb{Z}}_+$ and $i = 1, 2, \ldots$. Now, it follows from (3.117) that $x(t) \in \mathcal{N}(I_{nq} - \tilde{A})$ for all $t \in \bar{\mathbb{Z}}_+$. Thus, $\mathcal{M} \subseteq \mathcal{N}(I_{nq} - \tilde{A})$. Since $\mathcal{N}(I_{nq} - \tilde{A})$ consists of only equilibrium points, it follows that $\mathcal{M} = \mathcal{N}(I_{nq} - \tilde{A})$. For $x_e \in \mathcal{N}(I_{nq} - \tilde{A})$, Lyapunov stability of x_e now follows by considering the Lyapunov function $U(xx_e)$.

For any $\hat{P} > 0$ satisfying (3.119) and $M \geq 0$, let $\bar{P} \triangleq \hat{P} + L^T M L$. Clearly, $\bar{P} \geq \hat{P} > 0$. It is easy to verify that $\bar{P}$ is a solution to (3.119), and hence such a $\hat{P}$ is not unique.

Finally, since $\tilde{A}$ is discrete-time semistable, it follows from the above result that there exists an $n \times n$ positive-definite matrix $\hat{P}$ such that (3.119) holds or, equivalently, $\text{vec}\hat{P} = (\tilde{A} \otimes \tilde{A})^T \text{vec}\hat{P} + \text{vec}((I_{nq} - \tilde{A})^k)^T \tilde{R}(I_{nq} - \tilde{A})^k$, that is, $[I_{n^2q^2} - (\tilde{A} \otimes \tilde{A})^T]\text{vec}\hat{P} = \text{vec}((I_{nq} - \tilde{A})^k)^T \tilde{R}(I_{nq} - \tilde{A})^k$. Hence, $\text{vec}((I_{nq} - \tilde{A})^k)^T \tilde{R}(I_{nq} - \tilde{A})^k \in \mathcal{R}(I_{n^2q^2} - (\tilde{A} \otimes \tilde{A})^T)$. Next, it follows from Lemma 3.8 that $\tilde{A} \otimes \tilde{A}$ is discrete-time semistable, and hence, by Lemma 3.9

$$
\begin{aligned}
&\text{vec}^{-1}((I_{n^2q^2} - (\tilde{A} \otimes \tilde{A})^T)^{\#}\text{vec}((I_{nq} - \tilde{A})^k)^T \tilde{R}(I_{nq} - \tilde{A})) \\
&= \sum_{i=0}^{\infty} \text{vec}^{-1}(((\tilde{A} \otimes \tilde{A})^T)^i \text{vec}((I_{nq} - \tilde{A})^k)^T \tilde{R}(I_{nq} - \tilde{A})^k) \\
&= \sum_{i=0}^{\infty} \text{vec}^{-1}(((\tilde{A}^i)^T \otimes (\tilde{A}^i)^T)\text{vec}((I_{nq} - \tilde{A})^k)^T \tilde{R}(I_{nq} - \tilde{A})^k) \\
&= \sum_{i=0}^{\infty} (\tilde{A}^i)^T ((I_{nq} - \tilde{A})^k)^T \tilde{R}(I_{nq} - \tilde{A})^k \tilde{A}^i
\end{aligned} \tag{3.123}
$$

where in (3.123) we used the facts that $(X \otimes Y)^T = X^T \otimes Y^T$, $(X \otimes Y)(Z \otimes W) = XZ \otimes YW$, and $\text{vec}(XYZ) = (Z^T \otimes X)\text{vec}Y$ ([246], Chapter 7). Hence

$$
\hat{P} = \sum_{i=0}^{\infty} (\tilde{A}^i)^T ((I_{nq} - \tilde{A})^k)^T \tilde{R}(I_{nq} - \tilde{A})^k \tilde{A}^i + \text{vec}^{-1}(z) \tag{3.124}
$$

where z satisfies $z \in \mathcal{N}[I_{n^2q^2} - (\tilde{A} \otimes \tilde{A})^T]$ and $\text{vec}^{-1}(z) = (\text{vec}^{-1}(z))^T \geq 0$. Since $(\tilde{A} \otimes \tilde{A})^T$ is discrete-time semistable, it follows that $I_{n^2q^2} - (\tilde{A} \otimes \tilde{A})^T$ is discrete-time semistable. Hence, the general solution to the equation $(\tilde{A} \otimes \tilde{A})^T z = z$ is given by $z = \lambda x \otimes x$, where $x \in N(I_{nq} - \tilde{A}^T)$ and $\lambda \in \Re$. Hence, $\text{vec}^{-1}(z) = \text{vec}^{-1}(\alpha x \otimes x) = \alpha x x^T$, $\alpha > 0$, where we used the fact that $xy^T = \text{vec}^{-1}(y \otimes x)$ ([246], Proposition 7.1.8).

Theorem 3.5 gives a sharp necessary and sufficient condition for discrete-time semistability of (3.107) by using the refined Lyapunov equation (3.119), which is more applicable than the one given by Lemma 3.7 due to the involvement of k. □

3.3.7 Control design

In this section, we use distributed LQR semistability theory developed in Section 3.3.3 to explore a semistable optimal controller design by converting the original optimal control problem P1 into the following equivalent optimization problem.

Theorem 3.6: *Consider the discrete-time linear dynamical system (3.107). Assume $(\tilde{A}, \tilde{R})$ is weakly k-semiobservable for nonnegative definite $\tilde{R}$. Let $S_{\min}$ be a solution to the minimization problem*

$$
\min\{\text{tr}SV : S > 0 \text{ and } S = \tilde{A}^T S \tilde{A} + ((I_{nq} - \tilde{A})^k)^T \tilde{R}(I_{nq} - \tilde{A})^k\} \tag{3.125}
$$

where $V \in \Re^{nq \times nq}$, $V \geq 0$. Then for P given by (3.109)

$$
\text{tr}((I_{nq} - \tilde{A})^{\#})^T S_{\min}(I_{nq} - \tilde{A})^{\#} V = \text{tr}((I_{nq} - \tilde{A})^{k-1})^T P (I_{nq} - \tilde{A})^{k-1} V \tag{3.126}
$$

Furthermore, such an $S_{\min}$ is not unique. Finally, all the solutions to the minimization problem (3.125) can be parameterized as

$$P_{\min} = S_{\min} + \alpha x x^T, \quad x \in N(I_{nq} - \tilde{A}^T) \cap N(((I_{nq} - \tilde{A})^{\#})^T), \quad \alpha > 0 \quad (3.127)$$

Proof. The existence of $(I_{nq} - \tilde{A})^{\#}$ is guaranteed by a positive definite solution to (3.12). Since $(\tilde{A}, \tilde{R})$ is weakly k-semiobservable, it follows from Theorem 3.5 that all the solutions S of $S = \tilde{A}^T S \tilde{A} + ((I_{nq} - \tilde{A})^k)^T \tilde{R}(I_{nq} - \tilde{A})^k = 0$ have the form

$$S = \sum_{i=0}^{\infty} (\tilde{A}^i)^T ((I_{nq} - \tilde{A})^k)^T \tilde{R}(I_{nq} - \tilde{A})^k \tilde{A}^i + P_0 \quad (3.128)$$

where $P_0 \triangleq \alpha x x^T$, $\alpha > 0$, and $x \in \mathcal{N}(I_{nq} - \tilde{A}^T)$. Hence

$$\begin{aligned}
&\mathrm{tr}((I_{nq} - \tilde{A})^{\#})^T S (I_{nq} - \tilde{A})^{\#} V \\
&= \mathrm{tr} \sum_{i=0}^{\infty} ((I_{nq} - \tilde{A})^{\#})^T (\tilde{A}^i)^T ((I_{nq} - \tilde{A})^k)^T \tilde{R} \\
&\quad \times (I_{nq} - \tilde{A})^k \tilde{A}^i (I_{nq} - \tilde{A})^{\#} V + \mathrm{tr}((I_{nq} - \tilde{A})^{\#})^T P_0 (I_{nq} - \tilde{A})^{\#} V \\
&= \mathrm{tr} \sum_{i=0}^{\infty} ((I_{nq} - \tilde{A})^{k-1})^T ((I_{nq} - \tilde{A})(I_{nq} - \tilde{A})^{\#})^T (\tilde{A}^i)^T \tilde{R} \tilde{A}^i (I_{nq} - \tilde{A}) \\
&\quad \times (I_{nq} - \tilde{A})^{\#} (I_{nq} - \tilde{A})^{k-1} V + \mathrm{tr}((I_{nq} - \tilde{A})^{\#})^T P_0 (I_{nq} - \tilde{A})^{\#} V \\
&= \mathrm{tr}((I_{nq} - \tilde{A})^{k-1})^T P (I_{nq} - \tilde{A})^{k-1} V \\
&\quad + \mathrm{tr} V^{1/2} ((I_{nq} - \tilde{A})^{\#})^T P_0 (I_{nq} - \tilde{A})^{\#} V^{1/2} \\
&\geq \mathrm{tr}((I_{nq} - \tilde{A})^{k-1})^T P (I_{nq} - \tilde{A})^{k-1} V
\end{aligned} \quad (3.129)$$

Now, let $S_{\min} = ((\tilde{A} - I_{nq})^k)^T P (\tilde{A} - I_{nq})^k + L^T L$. It follows from Theorem 3.5 that $S_{\min}$ satisfies (3.119). Next

$$\begin{aligned}
S_{\min} &= \sum_{i=0}^{\infty} ((I_{nq} - \tilde{A})^k)^T ((I_{nq} - \tilde{A})(I_{nq} - \tilde{A})^{\#})^T (\tilde{A}^i)^T \\
&\quad \times \tilde{R} \tilde{A}^i (I_{nq} - \tilde{A})(I_{nq} - \tilde{A})^{\#} (I_{nq} - \tilde{A})^k + L^T L \\
&= \sum_{i=0}^{\infty} ((I_{nq} - \tilde{A})^k)^T (\tilde{A}^i)^T \tilde{R} \tilde{A}^i (I_{nq} - \tilde{A})^k + L^T L
\end{aligned} \quad (3.130)$$

Hence

$$\begin{aligned}
&\mathrm{tr}((I_{nq} - \tilde{A})^{\#})^T S_{\min} (I_{nq} - \tilde{A})^{\#} V \\
&= \mathrm{tr} \sum_{i=0}^{\infty} ((I_{nq} - \tilde{A})^{k-1})^T ((I_{nq} - \tilde{A})(I_{nq} - \tilde{A})^{\#})^T (\tilde{A}^i)^T \tilde{R} \tilde{A}^i \\
&\quad \times (I_{nq} - \tilde{A})(I_{nq} - \tilde{A})^{\#} (I_{nq} - \tilde{A})^{k-1} V \\
&\quad + \mathrm{tr}((I_{nq} - \tilde{A})^{\#})^T L^T L (I_{nq} - \tilde{A})^{\#} V \\
&= \mathrm{tr} P V + \mathrm{tr}((I_{nq} - \tilde{A})^{\#})^T L^T L (I_{nq} - \tilde{A})^{\#} V
\end{aligned} \quad (3.131)$$

Note that $L(I_{nq}-\tilde{A})^{\#} = (I_{nq}-(I_{nq}-\tilde{A})(I_{nq}-\tilde{A})^{\#})(I_{nq}-\tilde{A})^{\#} = (I_{nq}-\tilde{A})^{\#} - (I_{nq}-\tilde{A})^{\#}(I_{nq}-\tilde{A}) \times (I_{nq}-\tilde{A})^{\#} = 0$. Hence, $\mathrm{tr}((I_{nq}-\tilde{A})^{\#})^T L^T L(I_{nq}-\tilde{A})^{\#}V = 0$. Thus

$$\mathrm{tr}((I_{nq}-\tilde{A})^{\#})^T S_{\min}(I_{nq}-\tilde{A})^{\#}V = \mathrm{tr}((I_{nq}-\tilde{A})^{k-1})^T P(I_{nq}-\tilde{A})^{k-1}V \qquad (3.132)$$

To show nonuniqueness of $S_{\min}$, let $\hat{S} \triangleq S_{\min} + L^T ML$ for any $M \geq 0$. Clearly, $\hat{S} \geq S_{\min} > 0$. It is easy to verify that $\mathrm{tr}((I_{nq}-\tilde{A})^{\#})^T \hat{S}(I_{nq}-\tilde{A})^{\#}V = \mathrm{tr}((I_{nq}-\tilde{A})^{\#})^T S_{\min}(I_{nq}-\tilde{A})^{\#}V$, and hence such an $S_{\min}$ is not unique.

Finally, it follows from Theorem 3.5 that all the solutions satisfying (3.119) can be parameterized as

$$P_{\min} = S_{\min} + \alpha xx^T, \quad x \in \mathcal{N}(I_{nq}-\tilde{A}^T), \quad \alpha > 0 \qquad (3.133)$$

Next, if $P_{\min}$ is a solution to the minimization problem (3.125), then

$$((I_{nq}-\tilde{A})^{\#})^T xx^T(I_{nq}-\tilde{A})^{\#} = 0 \qquad (3.134)$$

which implies that $x \in \mathcal{N}(((I_{nq}-\tilde{A})^{\#})^T)$. □

Remark 3.7: *It is important to note that Theorem 3.6 is more general than Theorem 5.1 in [244]. Moreover, the nonuniqueness of the solutions to the generalized Lyapunov equation (3.119) in Theorem 3.5 leads to the nonuniqueness of the solutions to the optimization problem (3.125) in Theorem 3.6. Although the optimization problem (3.125) has infinitely many solutions, practically we are only interested in obtaining one of the optimal solutions, not all of the optimal solutions, for the purpose of optimal semistable control design of (3.107). Furthermore, if one can obtain one of the optimal solutions, all the other optimal solutions can be easily characterized by using (3.107) in Theorem 3.6, and hence the nonuniqueness of* $\hat{S}_{\min}$ *would not create a problem for the optimal semistable control design of (3.107).*

Theorem 3.5 states that the optimal solutions to *problem P1* are equivalent to the optimal solutions to (3.125). Thus, by solving (3.125), we obtain the solutions to *problem P1*.

3.3.8 Illustrative example 3.8

To illustrate Theorem 3.6, consider the general optimal control problem posed earlier. Assume $m = n = 1$, $A = 0$, $Q = 0$, and $B = R = I_q$. Now, it follows from Theorem 3.6 that the optimal control problem P1 is equivalent to the minimization problem (3.125). In this case, $S = K^T SK + (I_q K)^T K^T K(I_q - K)$. To simplify our discussion, we assume that K is idempotent and symmetric. Then it follows that K is discrete-time semistable. Furthermore, $S = K^S K$, or equivalently, $S - K = K^T(S-K)K$. Since K is discrete-time semistable, it follows from Theorem 3.5 that $S - K = xx^T$, where $x \in \mathcal{N}(I_q - K^T)$ and $K + xx^T > 0$. Hence, $S_{\min} = K + xx^T$. Note that $(I_q K)^2 = I_q - 2K + K^2 = I_q K$. Then it follows from (iii) of Proposition 6.2.2 of [246, p. 229] that $(I_q K)^{\#} = I_q K$, and hence the cost function is given by

$\text{tr}(I_qK)S_{\min}(I_qK)V = 0$. For $q = 2$, we solve $K^2 = K = K^T$ to obtain three nonzero matrices

$$K_1 = \begin{bmatrix} 1 & 0 \\ 0 & 1 \end{bmatrix}, \quad K_2 = \begin{bmatrix} a & \sqrt{a(1-a)} \\ \sqrt{a(1-a)} & 1-a \end{bmatrix},$$

$$K_3 = \begin{bmatrix} a & -\sqrt{a(1-a)} \\ -\sqrt{a(1-a)} & 1-a \end{bmatrix}$$

where $0 < a < 1$. The second matrix K_2 with $a = 1/2$ corresponds to distributed semistable state agreement design, which is consistent with the result in Illustrative example 3.3.

3.4 Distributed control of nonnegative systems

The study of matrices with nonnegative coefficients has a long history dating back to the Perron–Frobenius theorem. A classic book on the topic is [252]. The theory has been used in Leontief economics, where the states denote nonnegative quantities of commodities, in the study of Markov chains, where the states denote nonnegative probabilities, and compartment models, where the states could denote quantities of chemical species in an organism.

Positive systems have received increasing attention in the control literature during the last decade. In particular, stabilization of positive linear systems was studied in [253] and [254]. Basic control theory for monotone systems, the non-linear counterpart of positive systems, was developed in [255]. The importance of nonnegative matrices for consensus algorithms has been widely recognized (see [256]) and can be traced back to the fact that stability of a nonnegative matrix can be verified using a linear Lyapunov function [257].

A recent remarkable result in [258] shows that decentralized controllers can be optimized for positive systems using semidefinite programming. The criterion is the closed loop $\mathcal{H}_\infty$ norm and the authors show that diagonal quadratic storage functions can be used without conservatism.

In the sequel, we let $\Re_+$ denote the set of nonnegative real numbers. The inequality $X > 0$ ($X \geq 0$) means that all elements of the matrix (or vector) X are positive (nonnegative). For a symmetric matrix X, the inequality $X > 0$ means that the matrix is positive definite. The matrix $A \in \Re^{n\times n}$ is said to be *Schur* if all eigenvalues are strictly inside the unit circle. It is *Hurwitz* if all eigenvalues have positive real part. Finally, the matrix is said to be *Metzler* if all off-diagonal elements are nonnegative.

Proposition 3.3: *Let $A \in \Re_+^{n\times n}$. Then the following statements are equivalent:*

1. *The matrix A is Schur.*
2. *There is a $x \in \Re_+^n$ such that $Ax < x$.*
3. *There is a diagonal matrix $P > 0$ such that $A^tPA < P$.*

Remark 3.8: *If A is Schur, then the same is true for A', so the proposition guarantees existence of $p \in \Re^n_+$ with $A'p < p$. Hence, the linear function $V(x) = p'x$ is a Lyapunov function for the dynamics $x(k+1) = Ax$.*

Proof. If A is Schur, take any vector $\alpha > 0$ and define $x(I - A)^{-1}\alpha$. Then

$$x = \alpha + A\alpha + A^2\alpha + A^3\alpha + \cdots > 0$$

and $x - Ax = \alpha > 0$, so item 2 holds. On the other hand, if item 2 holds, choose $\varepsilon > 0$ such that $Ax < (1-\varepsilon)x$. Then for every $z \in \Re^n$ with $0 < z < x$ we have

$$0 \leq A^k z \leq A^k x < (1-\varepsilon)^k x, \quad \textit{for } k = 1, 2, 3, \ldots$$

In particular, $\lim_{k\to\infty} A^k z$, so A must be Schur.

The implication from item 3 to item 1 is standard, so it remains to prove the opposite direction. Hence, assuming that A (and therefore A') is Schur, we use item 2 to pick $x, y \in \Re^{n\times 1}_+$ with

$$Ax < x \quad A'y < y$$

Define the matrix $P = diag\{p_1, p_2, \ldots, p_n\}, p_k = y_k/x_k$ and the vector $z, \in \Re^n, z_k = \sqrt{x_k y_k}, k = 1, \ldots, n$. Then

$$\begin{aligned} P^{-1/2}A'PAP^{1/2}z &= P^{-1/2}A'PAx \\ &< P^{-1/2}A'Px = P^{-1/2}A'y < P^{-1/2}y = z \end{aligned}$$

which means that the symmetric matrix $P^{-1/2}A'PAP^{1/2}$ is Schur. This leads to

$$P^{-1/2}A'PAP^{1/2} < I$$

and thus item 3 follows. □

The corresponding continuous-time result can be stated as follows:

Proposition 3.4: *Let $A \in \Re^{n\times n}$ be Metzler. Then the following statements are equivalent:*

1. *The matrix A is Hurwitz.*
2. *There is a $x \in \Re^{n\times 1}_+$ such that $Ax < x$.*
3. *There is a diagonal matrix $P > 0$ such that $PA + A'P < 0$.*

3.4.1 Illustrative example 3.9

For purpose of illustration, we consider a linear model of platoon of four vehicles using distance measurements for position adjustment. This represents a dynamical system interconnected according to the graph illustrated in Figure 3.5.

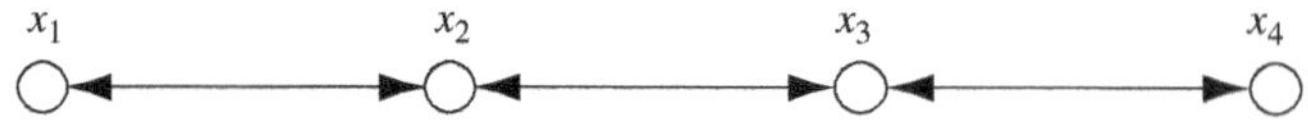

Figure 3.5 A one-dimensional graph of interconnected systems

$$\begin{aligned}
\dot{x}_1 &= \beta_1 x_1 + \omega_{12}(x_2 - x_1) \\
\dot{x}_2 &= \beta_2 x_2 + \omega_{21}(x_1 - x_2) + \omega_{23}(x_3 - x_2) \\
\dot{x}_3 &= \beta_3 x_3 + \omega_{32}(x_2 - x_3) + \omega_{34}(x_4 - x_3) \\
\dot{x}_4 &= \beta_4 x_4 + \omega_{43}(x_3 - x_4)
\end{aligned}$$

The parameters $\omega_{ij} \geq 0$ can be used to stabilize the dynamics even in situations where some of the β_j parameters are positive. Note that the dynamics can be written as $\dot{x} = Mx$, where M is a Metzler matrix. Hence, by Proposition 3.4, stability is equivalent to existence of numbers $x_1, \ldots, x_4 > 0$ such that

$$\begin{aligned}
\omega_{12}x_2 + (\beta_1 - \omega_{12})x_1 &< 0 \\
\omega_{21}x_1 + \omega_{23}x_3 + (\beta_2 - \omega_{21} - \omega_{23})x_2 &< 0 \\
\omega_{32}x_2 + \omega_{34}x_4 + (\beta_3 - \omega_{32} - \omega_{34})x_3 &< 0 \\
\omega_{43}x_3 + (\beta_4 - \omega_{43})x_4 &< 0
\end{aligned}$$

This can be implemented as a distributed test where the first node verifies the first inequality, the second node verifies the second inequality, and so on.

3.4.2 Distributed stabilization by linear programming

Given the distributed stability test set in the foregoing section, the next step is to search for stabilizing feedback laws using distributed optimization. This can be done using the following theorem:

Theorem 3.7: *Let $A \in \Re^{n\times n}, H \in \Re_+^{n\times m}, G \in \Re^{m\times n}$ and let D be the set of $m \times m$ diagonal matrices with entries in $[0, 1]$. Suppose for all $D \in \mathcal{D}$ that $A + HDG$ is nonnegative. Then the following statements are equivalent:*

1. *There is $D \in \mathcal{D}$ such that $A + HDG$ is Schur.*
2. *There exist $p \in \Re_+^n, q \in \Re_+^m$, with $q \leq H'p$ and $A'p + G'q < p$.*

Proof. Suppose that item 1 holds. Let $A + HDG$ be Schur and define $p \in \Re_+^n$ with $(A + HDG)'p < p$. Let $q = DH'p$. Then $q \leq H'p$ and $A'p + G'q = (A + HDG)'p < p$.

Conversely, suppose that item 2 holds. Choose $D \in D$ to get $q = DH'p$. Then

$$(A + HDG)'p = A'p + G'q < p$$

so $A + HDG$ is Schur. □

A continuous-time result can be stated as follows:

Theorem 3.8: *Let $A \in \Re^{n\times n}, H \in \Re_+^{n\times m}, G \in \Re^{m\times n}$ and let $\mathcal{D}$ be the set of $m \times m$ diagonal matrices with entries in $[0, 1]$. Suppose for all $D \in \mathcal{D}$ that $A + HDG$ is Metzler. Then the following statements are equivalent:*

1. *There exists $D \in \mathcal{D}$ such that $A + HDG$ is Hurwitz.*
2. *There exist $p \in \Re_+^n, q \in \Re_+^m$, with $q \leq H'p$ and $A'p + G'q < 0$.*

Remark 3.9: *It is natural to compare the expression $A + HDG$ with the 'state feedback' expression $A + BK$ of standard linear quadratic optimal control. A major difference is the matrix G, which makes the optimization of $A + HDG$ into a problem of 'static output feedback' rather than state feedback. Another difference is the diagonally structured D instead of the full matrix K. The diagonal structure gives a much higher degree of flexibility, particularly in the specification of distributed controllers. Moreover, if the diagonal elements of $\mathcal{D}$ are restricted to $\Re_+$ instead of $[0, 1]$, then the condition $q \leq H'p$ is replaced by $0 \leq H'p$. Additionally, each row of the inequalities $q \leq H'p$ and $A'p + G'q < 0$ can be verified separately to get a distributed test.*

3.4.3 Illustrative example 3.10

Consider the system in illustrative example 3.10 and given $\beta_1, \ldots, \beta_4$, the problem is to find feedback gains ω_{ij} [0, 1] that stabilize the system. This problem can be solved by applying a distributed linear programming method to condition item 2 in Theorem 3.8 with

$$A = diag\{\beta_1, \beta_2, \beta_3, \beta_4\}$$

$$H = \begin{bmatrix} 1 & 0 & 0 & 0 \\ 0 & 1 & 0 & 0 \\ 0 & 1 & 0 & 0 \\ 0 & 0 & 1 & 0 \\ 0 & 0 & 1 & 0 \\ 0 & 0 & 0 & 1 \end{bmatrix}$$

$$D = diag\{\omega_{12}, \omega_{21}, \omega_{23}, \omega_{32}, \omega_{34}, \omega_{43}\}$$

$$G = \begin{bmatrix} -1 & 1 & 0 & 0 & 0 & 0 \\ 1 & -1 & -1 & 1 & 0 & 0 \\ 0 & 0 & 1 & -1 & -1 & 1 \\ 0 & 0 & 0 & 0 & 1 & -1 \end{bmatrix}$$

It can be verified that $M = A + HDG$ is the state space matrix of the system in illustrative example A.

The next theorem adds further feature to the input–output performance.

Theorem 3.9: *Let $T(z) = C(zI - A)^{-1}B + N$ where $A \in \Re_+^{n\times n}, B \in \Re_+^{n\times 1}, C \in \Re_+^{1\times n}$ and $N \in \Re_+$. Then the following two conditions are equivalent:*

1. *The matrix A is Schur and $\|T(z)\|_\infty < \gamma$.*
2. *The matrix $\left[\begin{smallmatrix} A & B \\ \gamma^{-1}C & \gamma^{-1}N \end{smallmatrix}\right]$ is Schur.*

Moreover, $\|T(z)\|_\infty = C(I - A)^{-1}B + N$.

Proof. Note that the maximum $\max_\omega |T(e^{j\omega})|$ must be attained at $\omega = 0$ since

$$\begin{aligned} |T(e^{j\omega})| &= |N + \sum_{k=1}^{\infty} C(e^{-i\omega}A)^{k-1}B| \\ &\leq |N| + \sum_k |C(A)^{k-1}B| = C(I - A)^{-1}B + N = T(1) \end{aligned}$$

Hence, $\|T(z)\|_\infty < \gamma$ may equivalently be expressed as

$$C(I - A)^{-1}B + N < \gamma$$

Now, suppose that item 1 holds. By Proposition 3.3, there exists $\zeta \in \Re_+^{1\times n}$ such that $A\zeta < \zeta$. Define $x = \zeta + (I - A)^{-1}B$. Then $x > 0$ and

$$Ax + B - x = A\zeta - \zeta < 0$$

If ζ is sufficiently small, we also get $Cx + D < \gamma$ so

$$\begin{bmatrix} A & B \\ \gamma^{-1}C & \gamma^{-1}N \end{bmatrix} \begin{bmatrix} x \\ 1 \end{bmatrix} < \begin{bmatrix} x \\ 1 \end{bmatrix}$$

and item 2 of the theorem holds.

Conversely, item 2 of the theorem implies that $Ax < x$ and

$$(I - A)^{-1}B < x$$

$$0 \leq C(I - A)^{-1}B + N < Cx + D < \gamma$$

so item 1 of the theorem follows. □

A continuous-time result can be stated as follows:

Theorem 3.10: *Let $T(s) = C(sI - A)^{-1}B + N$ where $A \in \Re_+^{n\times n}$ is Metzler, while $B \in \Re_+^{n\times 1}, C \in \Re_+^{1\times n}$, and $N \in \Re_+$. Then the following two conditions are equivalent:*

1. *The matrix A is Hurwitz and $\|T(z)\|_\infty < \gamma$.*
2. *The matrix $\left[\begin{smallmatrix} A & B \\ C & N-\gamma I \end{smallmatrix}\right]$ is Hurwitz.*

Moreover, $\|T(z)\|_\infty = N - C(A)^{-1}B$.

It is interesting to record that by combining Theorems 3.8 and 3.10, we readily obtain a linear programming formulation of the problem to minimize input–output gain.

Theorem 3.11: *Let $T(s) = C(sI - A - HDG)^{-1}B + N$ where $B \in \Re_+^{n\times 1}, C \in \Re_+^{1\times n}$, $N \in \Re_+, H \in \Re_+^{n\times m}$, and $G \in \Re^{m\times n}$. Let $\mathcal{D}$ be the set of $m \times m$ diagonal matrices with entries in $[0, 1]$. Suppose that $A + HDG$ is Metzler for all $D \in \mathcal{D}$. Then the following two conditions are equivalent:*

1. *There exists $D \in \mathcal{D}$ such that $A + HDG$ is Hurwitz and $\|T(z)\|_\infty < \gamma$.*
2. *There exist $p \in \Re_+^n, q \in \Re_+^m$, with*

$$\begin{aligned} A^t p + G^t q + C^t &< 0 \\ B^t p + N - \gamma I &< 0 \\ q - H^t p &\leq 0 \end{aligned}$$

Moreover, if the conditions of item 2 are satisfied for some p, q, then the conditions of item 1 hold for every d such that $q = (HD)^t p$.

Proof. It follows from Theorem 3.10 that condition of item 1 holds if and only if the matrix

$$\begin{bmatrix} A + HDG & B \\ C & N - \gamma I \end{bmatrix}$$

is Hurwitz, or equivalently, there exists $p \in \Re_+^n$ with

$$\begin{bmatrix} A + HDG & B \\ C & N - \gamma I \end{bmatrix}^t \begin{bmatrix} p \\ 1 \end{bmatrix} < 0 \tag{3.135}$$

Given inequality (3.135), the inequalities of item 2 hold with $q = H^t p$.

Conversely, given item 2, inequality (3.135) follows provided that $q = (HD)^t p$. This proves the desired equivalence between items 1 and 2 of Theorem 3.11. □

3.4.4 Illustrative example 3.11

Consider the problem of finding $\omega_{ij} \in [0, 1]$ such that the system

$$\begin{aligned} \dot{x}_1 &= \beta_1 x_1 + \omega_{12}(x_2 - x_1) + v \\ \dot{x}_2 &= \beta_2 x_2 + \omega_{21}(x_1 - x_2) + \omega_{23}(x_3 - x_2) \\ \dot{x}_3 &= \beta_3 x_3 + \omega_{32}(x_2 - x_3) + \omega_{34}(x_4 - x_3) \\ \dot{x}_4 &= \beta_4 x_4 + \omega_{43}(x_3 - x_4) \end{aligned}$$

is stabilized, while the gain from v to x_1 is minimized. The solution is obtained by distributed linear programming solving the inequalities of item 2 in Theorem 3.11 with matrices A, H, G, and D as specified in Illustrative example 3.11, and

$$B^t = \begin{bmatrix} 1 & 0 & 0 & 0 \end{bmatrix}, \quad C = \begin{bmatrix} 1 & 0 & 0 & 0 \end{bmatrix}, \quad N = 0$$

It is obvious that once p and q are determined, D can be obtained from $q = (HD)^t p$.

3.5 Distributed control of power load frequency

This section discusses frequency control of large electrical power networks composed of more than one generation stations and each station including multiple power generation sources such as wind power and gas turbines. The control methodology is based on DC framework whose control gains are computed by iterative gradient method. The proposed scheme is compared with conventional proportional and integral (PI) and centralized control and the results show that it improves the performance by decreasing frequency fluctuation.

3.5.1 Introduction

Large power networks are designed such that multiple generation units contribute in powering up single electric supply line. Each of these units generate power using different sources such as biomass, solar energy, and wind power. The need for these unconventional sources of electrical power has peaked nowadays due to the adverse effects of fuel ignition on the environment. This not only ensures environmental safety but also cuts down the cost of generation. However, the price to pay is large frequency and voltage fluctuation, which makes it inevitable to control every generation unit explicitly in an optimal way.

Due to the nature of design, these huge networks are coupled with each other in a sense that the frequency fluctuation in one unit affects the performance of the neighboring units. Therefore, by proposing a decentralized control strategy one needs to keep in mind that ignoring this coupling effect has adverse results on the overall performance of the system. On the other hand, considering centralized control strategy can jeopardize the whole system's power output if the controller experiences malfunction. Hence, the control framework proposed for power networks is distributed in which the controllers can exchange the information of interconnected units in order to take into account the effects of coupled parameters or states.

System frequency control is done by using DC methodology and the distributed gains are continuously tuned and optimized by iterative gradient algorithm. The advantage of this scheme is that the controller gains of a subsystem and its neighboring units are changed as the system configuration changes. This also takes into account malfunctions such as load fluctuations and generator breakdown as demonstrated in the simulation results presented later in the report.

3.5.2 Distributed systems

The framework of distributed systems is described here in the context of large power system consisting of more than two subsystems or generation units. If $N(\geq 2)$ is the number of subsystems then ith subsystem ($i = 1, \ldots, N$) is given by

$$x_i(t+1) = \sum_{j=1}^{N} A_{ij}x_j(t) + B_iu_i(t) + w_i(t) \tag{3.136}$$

where $t \in \mathbb{Z}_+$, $x_i(t) \in \Re^{n_{xi}}$ is state vector of ith subsystem and $A_{ii} \in \Re^{n_{xi} \times n_{xi}}$. If j represents a subsystem interconnected with ith subsystem and if $j \neq i$ then $A_{ij} \in \Re^{n_{xj} \times n_{xj}}$, $u_i(t) \in \Re^{n_{ui}}$ is control signal of ith subsystem, $w_i(t) \in \Re^{n_{wi}}$ is white noise with variance W, and it is independent of $x_i(s)$ for $s \leq t$.

The interconnection between subsystems i and j is represented as $(i,j) \in E$, where the two subsystems are said to be neighbors. If there is no interconnection between the subsystems, that is, $(i,j) \notin E$ then

$$A_{ij} = 0 \tag{3.137}$$

Moreover, A_{ij} shows how the subsystem i has an influence on j. In distributed systems framework, signal flow diagrams are used to show how different subsystems are interconnected. Figure 3.6 shows how four subsystems of an electric power network are interconnected. The system matrix A and input matrix B for the overall system shown in Figure 3.6 are given by

$$A = \begin{bmatrix} A_{11} & A_{12} & \mathbf{0} & \mathbf{0} \\ A_{21} & A_{22} & A_{23} & \mathbf{0} \\ \mathbf{0} & A_{32} & A_{33} & A_{34} \\ \mathbf{0} & \mathbf{0} & A_{43} & A_{44} \end{bmatrix} \tag{3.138}$$
$$B = diag[B_{11}, \ldots, B_{44}]$$

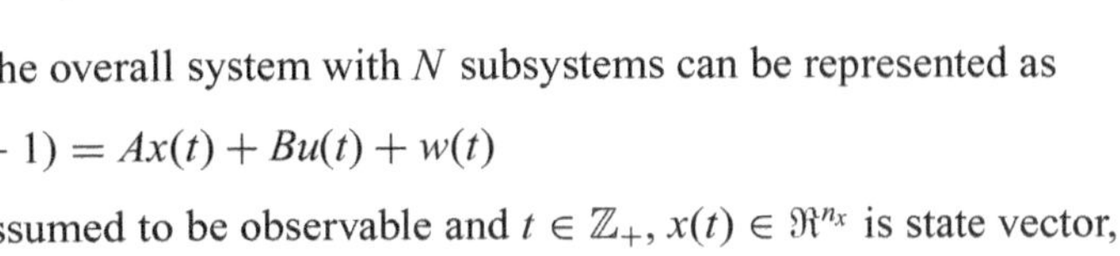

Figure 3.6 Representation of a distributed system with four subsystems

Now the overall system with N subsystems can be represented as

$$x(t+1) = Ax(t) + Bu(t) + w(t) \tag{3.139}$$

which is assumed to be observable and $t \in \mathbb{Z}_+$, $x(t) \in \Re^{n_x}$ is state vector, $u(t) \in \Re^{n_u}$ is control signal, $w(t) \in \Re^{n_w}$ is white noise and

$$\begin{aligned} \mathbf{E}w(t)w(s)^T &= W\delta_{ts} \\ \mathbf{E}w(t)x(s)^T &= 0, \quad \text{if } t \geq s \end{aligned}$$

$$\begin{aligned} \delta_{ts} &= 1 \quad \text{if } t = s \\ \delta_{ts} &= 0 \quad \text{if } t \neq s \end{aligned} \tag{3.140}$$

The system is controlled by state feedback and the input is of following form

$$u = -Lx(t) \tag{3.141}$$

An important point to note here is that the structure of control gain matrix L is exactly the same as that of system matrix due to distributed framework. This implies that the control gain matrix will not introduce any interconnection term and only the measurement value from the neighboring subsystem will be required to calculate the control input of the ith subsystem. Hence, the structure of gain matrix for the distributed system composed of four subsystems described by (3.138) can be given as

$$L = \begin{bmatrix} L_{11} & L_{12} & \mathbf{0} & \mathbf{0} \\ L_{21} & L_{22} & L_{23} & \mathbf{0} \\ \mathbf{0} & L_{32} & L_{33} & L_{34} \\ \mathbf{0} & \mathbf{0} & L_{43} & L_{44} \end{bmatrix} \tag{3.142}$$

It can be seen that the dynamics of the closed-loop system $A - BL$ has the same structure as A.

3.5.3 Iterative gradient algorithm

An iterative method to update the control gains of a distributed linear system driven by random disturbances is provided in [259]. In this methodology, the control objective aims to minimize the sum of variances of states and inputs in the constituent subsystems. The control gains are continuously computed using gradient of the cost function that in turn is calculated by a forward simulation of the states and a backward simulation of the adjoint states. This framework converges the DC gains to optimal values.

The cost function is defined as follows

$$J(L) = \mathbb{E}(|x|_Q^2 + |u|_R^2) \tag{3.143}$$

where $|x|_Q^2 = x^T Qx$ and $|u|_R^2 = u^T Ru$ and Q and R are positive definite. The control gain matrix L is updated as

$$L_{k+1} = L_k - \gamma \nabla_L J \tag{3.144}$$

where γ is the step size that is small enough. The gradient of J is given by the following equation

$$\nabla_L J = 2[RL - B^T P(A - BL)]X \tag{3.145}$$

where X and P satisfy the following Lyapunov equations:

$$\begin{aligned} X &= (A - BL)X(A - BL)^T + W \\ P &= (A - BL)^T P(A - BL) + Q + L^T RL \end{aligned} \tag{3.146}$$

In order to formulate a distributed method of calculation of gradient, adjoint system is introduced, which is computed by backward iteration and is given as follows:

$$\lambda(t-1) = (A - BL)^T \lambda(t) - (Q + L^T RL)x(t) \tag{3.147}$$

With all this information, the gradient is given by

$$\nabla_L J = 2(RL\mathbf{E}xx^T + B^T \mathbf{E}\lambda x^T) \tag{3.148}$$

However, this gradient cannot be computed in a distributed way because covariance between all the states and between states and adjoint states needs to be determined. In order to solve this problem, first the matrices Q and R are restricted to be block diagonal and dimensionally consistent with the system description. Second, subsystem i must compute the adjoint state vector using only local information and information of the neighboring subsystems. With these modifications (3.147) can be represented as

$$\lambda_i(t-1) = \sum_{j \in E_i} (A - BL)_{ji}^T \lambda_j(t) - \left(Q_i x_i(t) - \sum_{j \in E_i} L_{ji}^T R_j u_j(t) \right) \tag{3.149}$$

Using this information the gradient of cost can be calculated. As the feedback gain matrix has the structure given by (3.142), the gradient matrix will have same structure and is represented as $G_{ij}(= \{\nabla_L J\}_{ij})$. From (3.141), replacing Lx by $-u$ in the covariance terms of (3.148), we get

$$\begin{aligned} (RL\mathbf{E}xx^T)_{ij} &= -R_i \mathbf{E} u_i x_j^T \\ (B^T \mathbf{E} \lambda x^T)_{ij} &= B_i^T \mathbf{E} \lambda_i x_j^T \end{aligned} \tag{3.150}$$

The algorithm for computing the feedback gain update using iterative gradient scheme is based on the idea that the iterations are done M times for computing single update of the control gain matrix. This implies that the covariance is calculated for a window with length of M samples of system and adjoint states, and increasing M will translate into more accurate values of covariance terms. Every time the update is computed, M number of values are taken. The algorithm is outlined as follows for $M = 5$:

Algorithm 3.1: *The initial value of L matrix is taken such that the eigenvalues of the closed-loop $(A - BL)$ lie inside the unit circle. Initial values of states and adjoint state vectors are taken as zero. If initial time is denoted by t_k then the initial value gain matrix will be $L^{(k)}$. For each update the time will be incremented by a value equal to the window length M. To update the feedback gain matrix of ith subsystem*

1. *simulate the state $x_i(t)$ for $t = t_k, \ldots t_k + M$ using following equation*

$$x_i(t+1) = \sum_{j \in E_i} (A - BL)_{ij} x_j(t) + w_j(t); \tag{3.151}$$

2. *simulate the adjoint state vector for $t = t_k, \ldots t_k + M$ in backward direction using (3.149);*
3. *calculate the covariances as follows:*

$$\begin{aligned} (\mathbf{E} u_i x_j^T)_{est} &= \tfrac{1}{M+1} \textstyle\sum_{t=t_k}^{t_k+M} u_i(t) x_j(t)^T \\ (\mathbf{E} \lambda_i x_j^T)_{est} &= \tfrac{1}{M+1} \textstyle\sum_{t=t_k}^{t_k+M} \lambda_i(t) x_j(t)^T \end{aligned} \tag{3.152}$$

   Note that only estimates of covariances are computed rather than the exact values because of limited number of samples taken.
4. *The estimated value of gradient for the feedback matrix of ith subsystem will then be given by*

$$G_{ij} = -2[R_i(\mathbf{E} u_i x_j^T)_{est} + B_j^T (\mathbf{E} \lambda_i x_j^T)_{est}] \tag{3.153}$$

5. *Then the update for gain matrix will be given as*

$$L^{(k+1)} = L^{(k)} - \gamma G_{ij} \tag{3.154}$$

6. *For next window of M samples, add $M + 1$ samples to t_k and go to step 1.*

3.5.4 Power system model

As mentioned earlier, the power system is composed of subsystems and each of these units uses multiple power generation plants for power generation. A general framework is shown in Figure 3.7. As described in [260], the power system taken for analysis is composed of identical subsystems such as turbine and wind power generators. The control objective is to bring the frequency deviation Δf, which is introduced due to hybrid power generation, close to zero. The frequency analysis model is shown in Figure 3.8 and each system is assumed to deliver 40 MW of electric power. Also, the generation systems are simplified to give only first-order equivalent models.

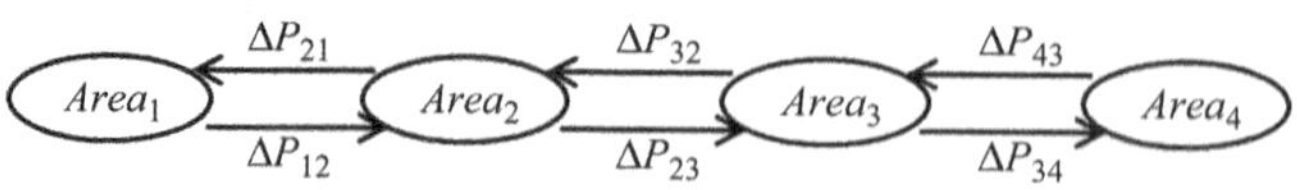

Figure 3.7 Distributed power network

Referring to the frequency analysis model, the battery electric storage system and water heaters are at 5% and 15% of the system capacity, respectively. This implies that the system capacity constants are $a_E = 0.05$ and $a_H = 0.15$, and from the assumption that $a_E + a_H + a_g = 1$, it follows that $a_g = 0.8$. For any ith subsystem, ΔP_{gi} is the output of gas turbine generator, Δx_{gi} is the governor input, ΔP_{wi} is the wind power generator output, ΔP_{Li} represents all the load fluctuation except due to controllable load, ΔP_{Ei} is the output of the electric storage system, ΔP_{Hi} is the output of electric water heater. All the generation units power up single line; therefore, frequency fluctuation in one unit will affect all other units. The tie-line power flow deviation ΔP_{tie_i} takes into account the effects of such disturbances. Following equation gives the power generated by the ith unit:

$$\Delta P_i = \Delta P_{gi} + \Delta P_{Wi} - \Delta P_{Li} + \Delta P_{tie_i} + \Delta P_{Ei} - \Delta P_{Hi} \tag{3.155}$$

If the synchronizing coefficient for each generation unit i with its neighboring unit j is represented by T_{ij} then the tie-line power flow is given by $\Delta P_{tie_i} = \int T_{ij}(\Delta f_j - \Delta f_i)dt$. The output of the PI controller for load frequency control (LFC) is given as $U_{AR_i} = \int AR_i dt$ where AR_i represents the regional demand, given by $\Delta P_{tie_i} - B_i \Delta f_i$, and B_i is the frequency bias. Also, regulation constant R_{gi} contributes to the LFC signal.

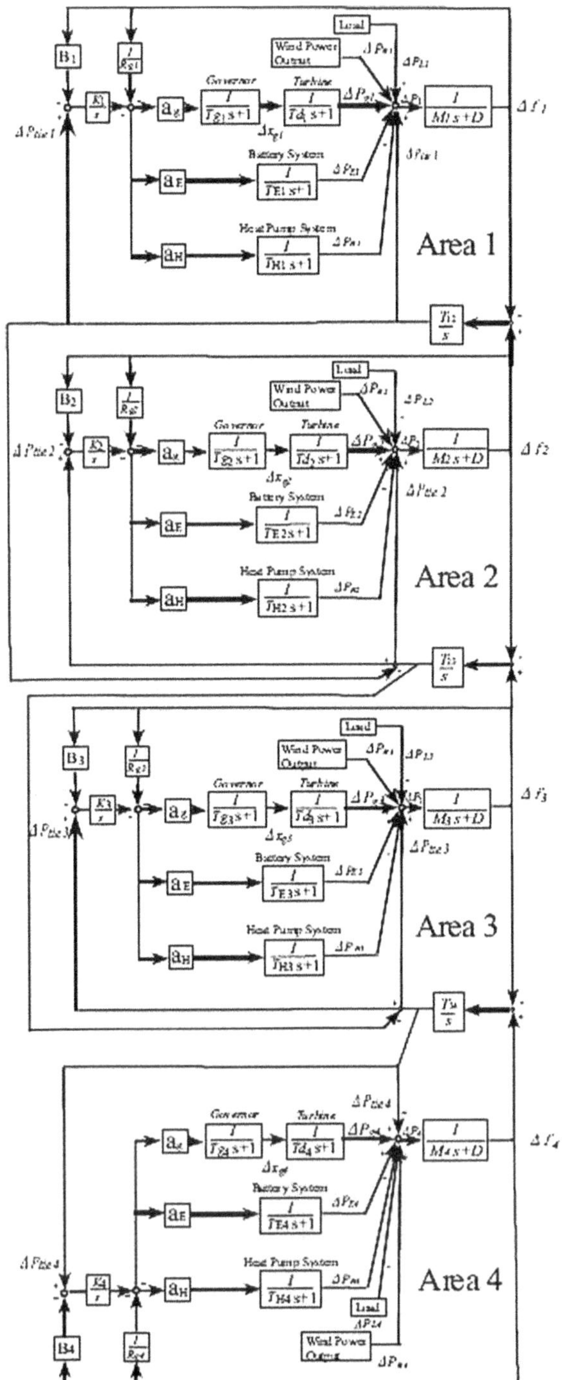

Figure 3.8 Frequency analysis model of power network

3.5.5 State-space representation

The continuous time state-space representation of the power system consisting of N subsystems is given as

$$\dot{x} = A_c x(t) + B_c u(t) + w(t) \tag{3.156}$$

where $x(t) = [x_1^T(t), \ldots, x_N^T(t)]^T \in \Re^{7N}$, $u(t) = [u_1^T(t), \ldots, u_N^T(t)]^T \in \Re^{7N}$, A_c has same structure as defined by (3.138) and $B_c = diag[B_{11}, \ldots, B_{NN}]$.

The structure of system and state matrices is given in (3.158) to (3.160) and the values of constants are given in Table 3.2. In order to compute control gains for centralized and distributed frameworks, the system is converted from continuous to discrete time using sampling time $T_s = 1$ [s]. Whole system is first discretized using c2d in MATLAB to get discrete equivalent as follows:

$$x(t+1) = \Phi x(t) + \Gamma u(t) + w(t) \tag{3.157}$$

where $w(t)$ is white noise given by $10^{-4} \times I^{28\times 28}$ and represents state turbulence. The wind power generation and load fluctuation are also taken to be white noise of small variance. The centralized controller is calculated using LQR for $Q = 0.1 \times I^{28\times 28}$ and $R = I^{28\times 28}$.

For distributed framework, each subsystem is discretized separately. The weight matrices are taken similar to that for centralized controller, i.e., $Q = 0.1 \times I^{7\times 7}$ and $R = I^{7\times 7}$

$$x_i = \begin{bmatrix} \Delta P_{tie_i} \\ \Delta f_i \\ \Delta P_{gi} \\ \Delta x_{vgi} \\ \Delta P_{Ei} \\ \Delta P_{Hi} \\ U_{AR_i} \end{bmatrix}, \quad B_{ii} = \begin{bmatrix} 0 & 0 & 0 & 0 & 0 & 0 & 0 \\ 0 & 0 & 0 & 0 & 0 & 0 & 0 \\ 0 & 0 & 0 & 0 & 0 & 0 & 0 \\ 0 & 0 & 0 & \frac{1}{T_{gi}} & 0 & 0 & 0 \\ 0 & 0 & 0 & 0 & \frac{1}{T_{Ei}} & 0 & 0 \\ 0 & 0 & 0 & 0 & 0 & \frac{1}{T_{Hi}} & 0 \\ 0 & 0 & 0 & 0 & 0 & 0 & 0 \end{bmatrix} \tag{3.158}$$

Table 3.2 Parameter values of power network

Parameters	Symbol	Value	Unit
Inertia constant	M	0.2	pu MWEs/Hz
Damping constant	D	0.26	pu MW/Hz
Governor time constant	T_g	0.2	s
Gas turbine constant	T_d	5.0	s
BESS time constant	T_E	0.2	s
HP time constant	T_H	4.5	s
Regulation constant	R_g	2.5	Hz/pu MW
Synchronizing coefficient	T_{ij}	0.5	pu MW

$$A_{ii} = \begin{bmatrix} 0 & -\sum_{j\in E_i} T_{ij} & 0 & 0 & 0 & 0 & 0 \\ \frac{1}{M_i} & -\frac{D}{M_i} & \frac{1}{M_i} & 0 & \frac{1}{M_i} & -\frac{1}{M_i} & 0 \\ 0 & 0 & -\frac{1}{T_{di}} & \frac{1}{T_d i} & 0 & 0 & 0 \\ 0 & -\frac{1}{T_{gi}R_{gi}} & 0 & -\frac{1}{T_{gi}} & 0 & 0 & \frac{a_g K_i}{T_{gi}} \\ 0 & -\frac{1}{T_{Ei}R_{gi}} & 0 & 0 & -\frac{1}{T_{Ei}} & 0 & \frac{a_E K_i}{T_{Ei}} \\ 0 & -\frac{1}{T_{Hi}R_{gi}} & 0 & 0 & 0 & \frac{1}{T_{Hi}} & \frac{a_H K_i}{T_{Hi}} \\ 1 & -B_i & 0 & 0 & 0 & 0 & 0 \end{bmatrix} \tag{3.159}$$

$$A_{ij} = A_{ji} = \begin{bmatrix} 0 & T_{ij} & 0 & 0 & 0 & 0 & 0 \\ 0 & 0 & 0 & 0 & 0 & 0 & 0 \\ 0 & 0 & 0 & 0 & 0 & 0 & 0 \\ 0 & 0 & 0 & 0 & 0 & 0 & 0 \\ 0 & 0 & 0 & 0 & 0 & 0 & 0 \\ 0 & 0 & 0 & 0 & 0 & 0 & 0 \\ 0 & 0 & 0 & 0 & 0 & 0 & 0 \end{bmatrix} \quad \text{if } (i,j) \in E$$

$$= 0 \quad \text{if } (i,j) \notin E \tag{3.160}$$

3.5.6 Simulation study

Three control techniques are compared:

1. DC using iterative gradient methods.
2. Centralized control using LQR.
3. Conventional PI control.

The effectiveness of these techniques is demonstrated using the cases in which load fluctuation occurs as a disturbance and a generator breaks down. These cases are discussed separately as follows:

Case 1: Load fluctuation

The first case studies the effect of disturbance, which is taken as load fluctuation. A disturbance of 0.2 is introduced at 450th second because at this point in time the DC gains are said to be converged. This can be seen in Figure 3.9. Performance of conventional PI control and proposed DC frameworks are compared by observing the frequency fluctuation of subsystem 1 (area 1). The results are shown in Figures 3.10 and 3.11. Former shows the result with noise and the latter without noise. It is evident that the proposed methodology results in significantly less frequency fluctuation than the conventional PI method.

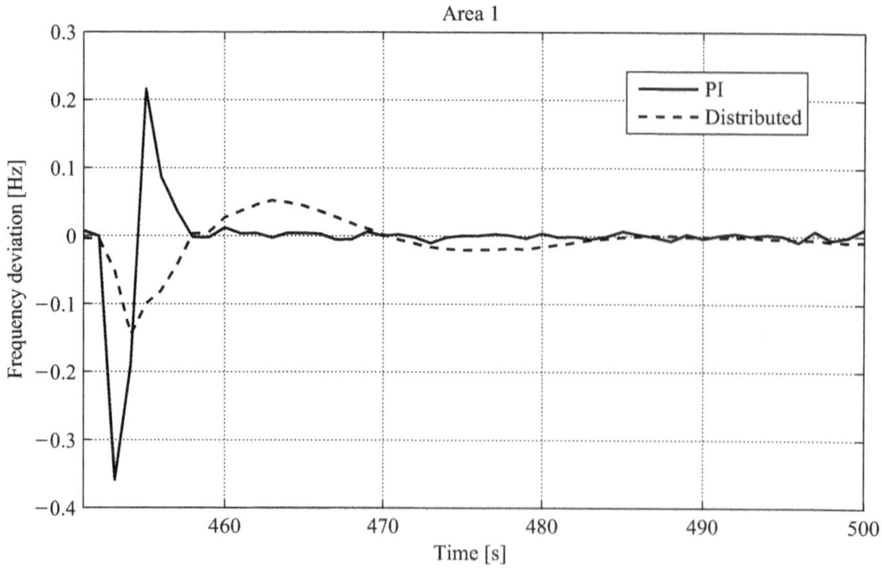

Figure 3.9 Comparison between PI and DC in case 1 with noise

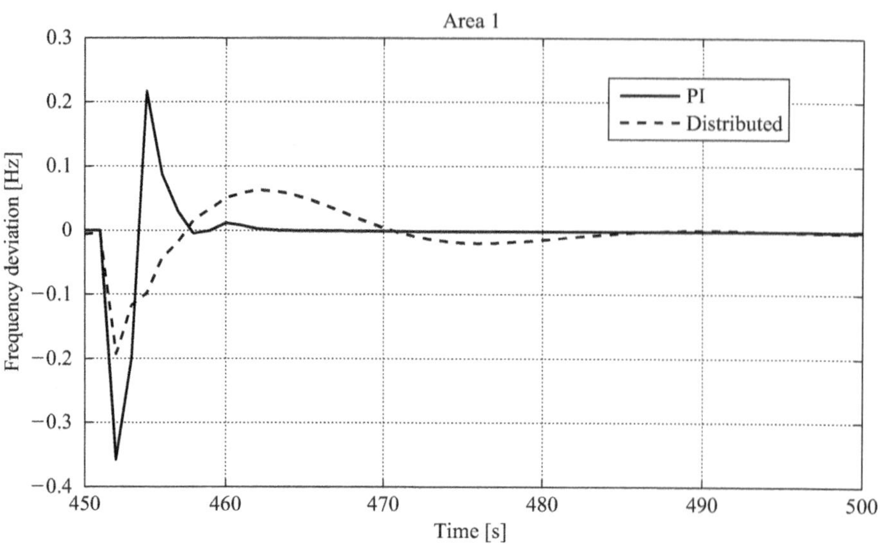

Figure 3.10 Comparison between PI and DC in case 1 without noise

Case 2: Generator breakdown

The second case takes into account the frequency fluctuation of area 1 when a generator breaks down. This condition translates into unstable system matrix. For simulation purpose, turbine generator of area 1 is made unstable at 0 [s] and its effect on frequency

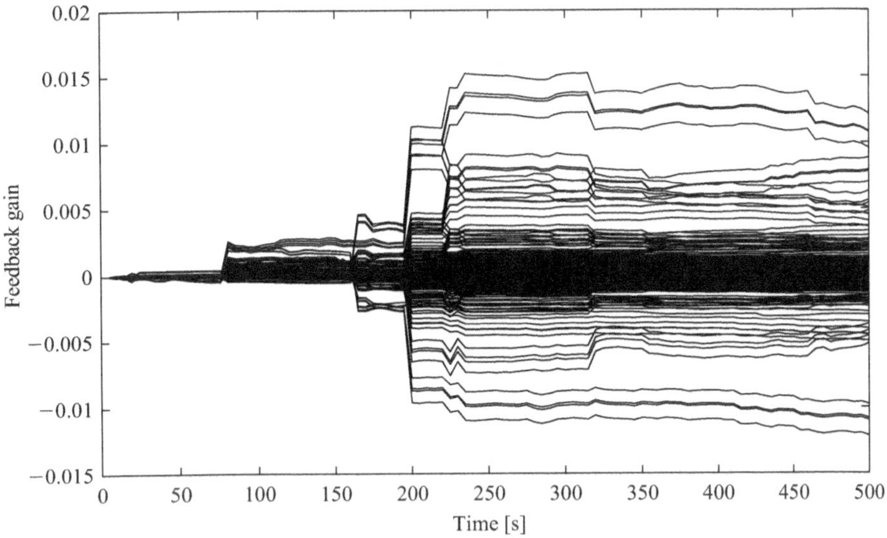

Figure 3.11 Feedback gain of DC

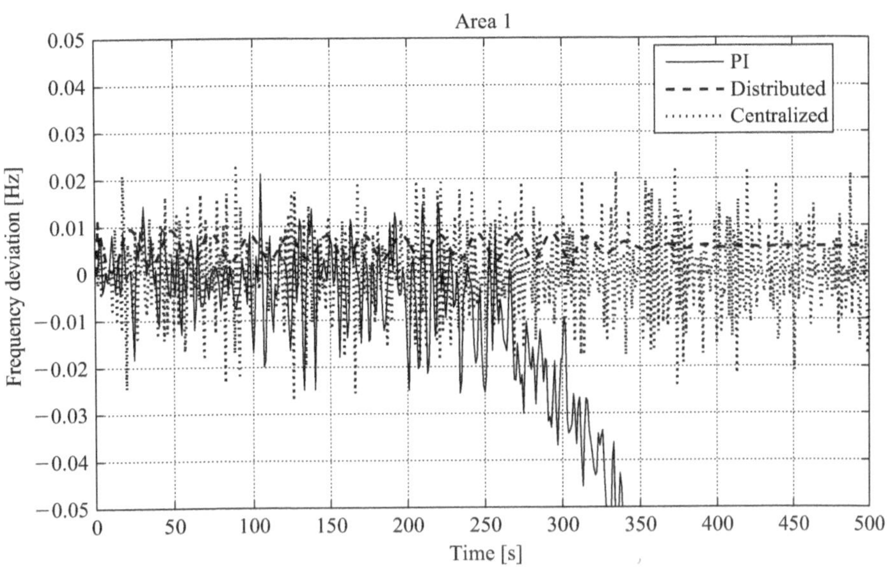

Figure 3.12 Comparison between PI, distributed, and centralized control for case 2

deviation of area 1 is observed for PI distributed and centralized control settings. Figure 3.12 shows that PI control cannot stabilize the power system and a huge blackout occurs. However, distributed and centralized control keep the system stable. Also, cost of system is shown in Figure 3.13 for centralized control.

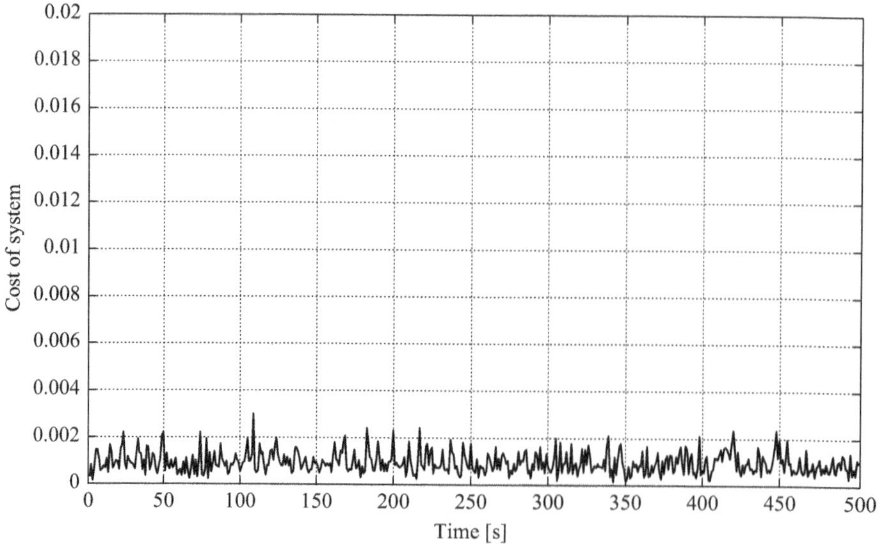

Figure 3.13 Cost of centralized control

3.6 Notes

In this chapter, we have introduced a stabilizing DC design method for interconnections of identical, dynamically decoupled, and independently actuated linear systems. As demonstrated, the method requires the solution of a single local LQR problem, whose size depends only on the maximum vertex degree of the interconnection graph. Salient properties of the local LQR problem have been derived, which enabled the construction of stabilizing distributed controllers independently of the choice of weighting matrices. Then the relationship between stability of the overall system, the robustness of local controllers, and the spectrum of a sparsity pattern matrix has been highlighted. It is interesting to observe that the distributed controller will be stabilizing even if subsystems and communication links are added to or removed from the overall system, as long as this does not change the maximum vertex degree or violate some prescribed conditions. This feature provides a very powerful modularity to our approach.

Next, the classical LQR theory was extended to include distributed semistable discrete-time systems. Using constrained optimization, a distributed semistable LQR control framework was developed for discrete-time linear dynamical systems. The attractive topic of nonnegative systems was discussed and a distributed stabilization scheme based on linear programming was presented. Finally, a gradient method was provided to iteratively compute feedback gains for DC framework as implemented on power generation network.

There is ample of extensions to the problems addressed above. These include, but not limited to, extending LQG theory to distributed semistable discrete-time systems and improving the performance of DC methods through parallel processing aspects.

Chapter 4
Distributed observer-based control

In this chapter, we investigate a distributed control approach to dynamical systems where the control loops are spread over a network. In this approach, the subsystems communicate to each other via a shared communication network. The properties of the network such as delay, packet dropout, varying sample interval, and induced errors must be studied when designing the control system. These properties add restrictions and difficulties in the control loop that are not present in traditional control loops. It is important in the design of the control strategy to identify how the control loop is structured and where the network comes into the picture. Also, the general structure of a decentralized networked control system is described, the main characteristics, major problems, network communication parameters, and the techniques of handling lost control data.

4.1 Observer-based control

A decentralized networked control system (DecNCS) is a class of decentralized control systems in which the different components (units) are connected over real-time communication channels or a data network. It is envisioned that there is a data link between the sensors (which collect information), controllers (which make decisions), and actuators (which apply the controller commands) [43,44]. In general, a system is said to be decentralized if there are multiple decision makers in the system (e.g., sensors, controllers, encoders) and these decision makers have access to different and imperfect information with regard to the system they operate in, and they need to either cooperate or compete with each other. Such systems are becoming ubiquitous, with applications ranging from automobile and inter-vehicle communications design, control of surveillance and rescue robot teams for access to hazardous environments, space exploration, and aircraft design, among many other fields of applications [45,50,51]. In such control applications, one major concern is the characterization of a sufficient amount of information transfer needed for a satisfactory performance. This information transfer can be between/various components of DecNCS (see Figure 4.1). Several necessities for satisfactory control performance are the ability for the controllers to track the plant state under communication constraints, the determination of the data rate required for the transmission of control

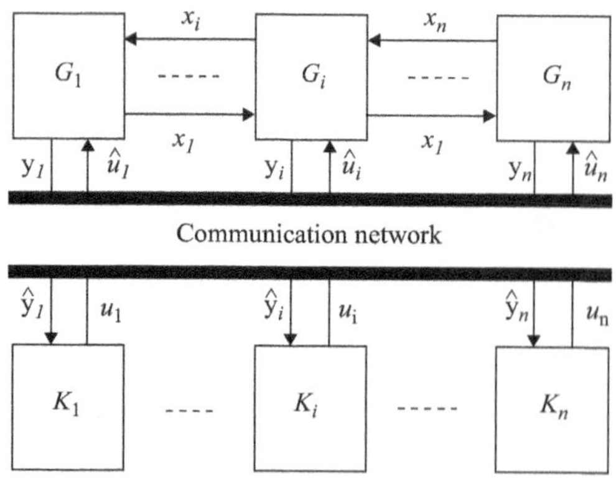

Figure 4.1 A decentralized networked control system

signals, and the construction of dynamic encoding, decoding, and control policies meeting some criteria [46,47]. Another important problem is the coordination among multiple sensors or multiple controllers/decision makers with the lowest information exchange possible. Even in cases when communication resources are not scarce, a strong understanding of the fundamentals can be useful in the system architecture, and finally, such an insight can help reduce the computation requirements and complexity.

The insertion of communication network can substantially improve the flexibility and fault-tolerance of an industrial control system, in addition to reducing the installation, material/labor cost, reconfiguration, maintenance time, and costs [8,261]. Currently, process control systems utilize dedicated, wired control networks to achieve key closed-loop properties like stability, set point tracking, and robustness to disturbances. Robustness and reliability are major concerns because the interference in the process control field and the consequence of a failure can be severe. Interference caused by environmental events and other signals impacts timely data transmits on which directly challenges the objective of real-time process control [262,263]. Timing scheme uses clock-driven sensing and actuation with event-driven control and the control loop can adapt to varying network conditions [20]. Moreover, the network-induced delay considered in [264] is composed of the sensor-to-controller delay and the controller-to-actuator delay as well as the computation delay and can be slowly or quickly time varying. Then, a controller design method is proposed based on a delay-dependent approach. The effect of unreliable channels have on overall system performance assuming that there are no quantization errors been shown in [43]. The optimal H_2 design [63,265] of semi-decentralized controllers is considered for a special class of spatially distributed systems. This class includes spatially invariant and distributed discrete-time systems with an inherent temporal delay in the interaction of neighboring sites [266]. Another approach introduces robust stabilization of discrete-time delay systems under non-linear perturbations and this is transformed to a constrained convex optimization. Sufficient conditions on the existence of state-feedback controllers are established in terms of linear matrix inequality (LMI), which guarantee stability of the closed-loop system and at the same time, maximize the non-linearity bound [34].

In this section, we consider several network side effects in our models because the controllers are communicating to the plants over a shared communication network. The basic features of the controller are discrete, decentralized, and output based. The controller will do a control over network not through network and design shall be robust to the following:

- Uncertain time delays due to communication, processing, and queuing.
- Transmission constraints where not all outputs and inputs can be transmitted at same time.
- Quantization error.
- Fixed and varying sampling intervals.
- Sampling interval selection.
- Unreliable transmission and packets dropout.
- Network-induced errors.
- Interconnected communication.

The main aim is to design a controller that can adapt to combined effects from these points mentioned above all together which is not done in the literature for DecNCS. Each point separately will be explained as will be shown in the next sections.

4.1.1 Problem definition and system modeling

We consider a class of linear continuous-time systems represented by

$$\begin{aligned} \dot{x}(t) &= Ax(t) + Bu(t) \\ y(t) &= Cx(t) \end{aligned} \tag{4.1}$$

where $x(t) \in \Re^n$ is the state vector, $u(t) \in \Re^m$ is the control input, and $y(t) \in \Re^q$ is the output vector, which represents the real measured output taken directly from the plant G. This is an ideal presentation of system model without the presence of the network that is our focus in this study. The same model in (4.1) will have some changes in the presentation due to the network effects. For example, the actual control input will be $\hat{u}$ that controls the plant after considering the network effects on it or we can call it the network version of original control signal u. Similarly, the output measurement that are sent from the plant back to the controller via the network will be $\hat{y}$, that is, the network version of y.

$$\begin{aligned} \dot{x}(t) &= Ax(t) + B\hat{u}(t) \\ y(t) &= Cx(t) \end{aligned} \tag{4.2}$$

All decentralized controllers $K_i, i = 1, \ldots, n$, are communicating with sensors and actuators via a shared network. Each plant G_i is controlled by discrete-time observer-based controller. The plant model described in (4.2) will be discretized with a zero-order hold to

$$\begin{aligned} x_{k+1} &= A_d x_k + B_d \hat{u}_k \\ y_k &= C x_k \\ A_d &= e^{Ah}, \quad B_d = \int_0^h e^{As} ds\, B \end{aligned} \tag{4.3}$$

where $x_k = x(t_k) = x(kh), y_k = y(t_k) = y(kh)$ and t_k are discrete points of time and k is an integer time index. The distance in time between each point of time is the time-step, denoted here by h, which is the sampling interval $h = t_k - t_{k-1}$. The discrete-time control input $\hat{u}_k$ is available at the plant at time $t = t_k$. Basically, what is modeled up to now is for one system and since we have a number of systems connected in decentralized structure then we need to consider the interconnection term as the following with the superscript i used to indicate association of the term to subsystem i:

$$\begin{aligned} x^i_{k+1} &= A^i_d x^i_k + B^i_d \hat{u}^i_k + \sum_{j=1, j\neq i}^{N} A^{i,j}_d \hat{x}^j_k \\ y^i_k &= C^i_d x^i_k + \sum_{j=1, j\neq i}^{N} C^{i,j}_d \hat{x}^j_k \\ \hat{u}^i_k &= K^i \hat{x}^i_k + \sum_{j=1, j\neq i}^{N} K^{ij} \hat{x}^j_k \end{aligned} \tag{4.4}$$

The interconnection term $A^{i,j}_d \hat{x}^j_k$, $j \neq i$ describes how the dynamics of the ith unit are influenced by the jth unit in the plant. Note from the summation notation that each processing unit can in general be connected to all other units in the plant. Also, the system is decentralized that means the interactions between subsystems shall be as minimum as possible to make the decentralized design successful. The term $K^i \hat{x}^i_k$ in the control law equation represents the local feedback component responsible for stabilizing the ith subsystem in the absence of interconnections, and $\sum_{j=1, j\neq i}^{N} K^{ij} \hat{x}^j_k$ is a 'feedforward' component that compensates for the effect of the jth neighboring subsystem on the dynamics of the ith unit. Note in this regard that the implementation of the control law requires the availability of state measurements from both the local subsystem that is being controlled as well as the other connected units. It is significant to observe that a choice of $K^{ij} = 0$ reduces the control strategy to a fully decentralized one where only measurements of the process variables of the ith unit are collected and processed with no signal transfer taking place across the network from other systems.

All communication between sensors, actuators, and controllers are going via a shared communication network. The sensors and controller nodes are clock-driven while actuator is event-driven and it can also be clock-driven. This means that the controller will not compute new control command until it receives the sensors measurements and the actuator continues using the old command until new one is arrived. By event-driven we mean that the node starts its activity when an event occurs, for instance, when it receives information from another node over the data network. Clock-driven means that the node starts its activity at a pre-specified time, for instance, the node can run periodically. In the following sections we will highlight the impacts of having the communication between the system components over share network and try to explain each effect separately with details.

4.1.2 Induced delays

Communication over shared network causes time delays in various sections of DecNCS. These time delays cannot be neglected, especially when the time constant of the controlled plant is short and the order of the plant model is high because they may affect network QoS and degrade control performance. The delays can be classified into different types using different classification criteria. It can be categorized from the direction of data transfers as the sensor-to-controller delay, computation delay in the controller and the controller-to-actuator delay. Also, it can be separated as device delay and network delay where the device delay is divided into several subtypes of delays. There are several factors that may affect the network time delays like the network load, network schedule, network bandwidth, size of the messages, and message priority.

Back to delays classifications, in the first classification, there are essentially three kinds of delays that will affect the system:

1. Communication delay between the sensor and the controller, τ_{sc}.
2. Computational delay in the controller, τ_c.
3. Communication delay between the controller and the actuator, τ_{ca}.

For a discrete-time system representation we can use the following indications, $\tau_{sc}^k, \tau_c^k, \tau_{ca}^k$ to show the delay at certain time instant k. The computational delay of the controller is very small and it can be neglected or added as a part of τ_{ca} delay. As a result, the round trip time for one packet can be shown as sum of τ_{sc} and τ_{ca}, that is

$$\tau_{rt}^k = \tau_{sc}^k + \tau_{ca}^k \tag{4.5}$$

There are different ways to calculate τ_{ca} and τ_{sc}, however, in the present work we determine it from the following:

$$\tau_{sc} = \tau_{cs} - \tau_{se}, \quad \tau_{sc} = \tau_{as} - \tau_{ce} \tag{4.6}$$

where

- t_{se}: time instant when remote sensor encapsulates the measurement into sent packet,
- t_{cs}: time instant when controller starts processing the delivered measurement packet,

- t_{ce}: time instant when main controller encapsulates the control signal into sent packet,
- t_{as}: time instant when remote system starts processing the control signal.

Moreover, the delays τ_{sc} and τ_{ca} are composed of at least the following parts:

- **Waiting time delay τ_w:** is the time for which a source (the main controller or the remote system) has to wait for queuing and network availability before actually sending a frame or a packet out. Also, it can be called queuing delay τ_q.
- **Frame time delay τ_F:** is the time during the moment the source is placing a frame or a packet on the network.
- **Propagation delay τ_P:** is the delay for a frame or a packet traveling through a physical media. It depends on the speed of signal transmission and the distance between the source and destination.

These delay parts are fundamental delays that occur on a communication network. When the control or sensory data travel across networks, there can be additional delays such as the queuing delay at a switch or a router, and the propagation delay between network hops. The delays τ_{ca} and τ_{sc} also depend on other factors such as maximal bandwidths from protocol specifications, and frame or packet sizes.

Figures 4.2 and 4.3 show system with short or constant time delays, which means the delay is less than one sampling period and this presents (4.3) in the following form:

$$\begin{aligned} x_{k+1} &= A_d x_k + B_{d0}(\tau_{rt}^k)\hat{u}_k \\ y_k &= C x_k \\ A_d &= e^{Ah}, \quad B_d = \int_0^h e^{As}\, ds\, B \end{aligned} \tag{4.7}$$

Two control samples u_k and u_{k-1} were applied during kth sampling period and it can be explained clearly from Figure 4.2 and it is shown as a network version since it is being sent over a network. On considering the N interconnected term, then (4.7) will be

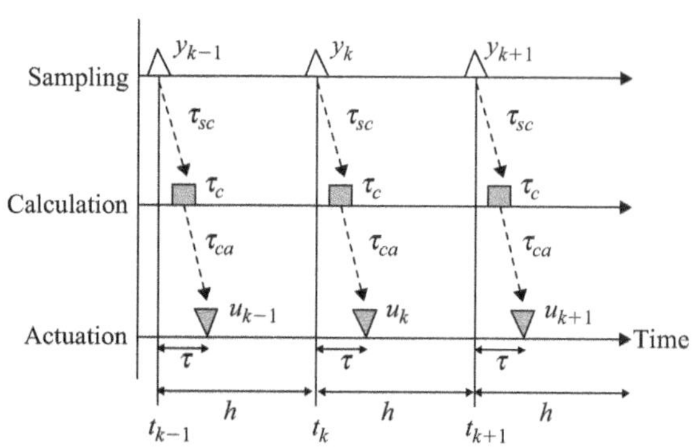

Figure 4.2 Control system with instantaneous input-to-output latency

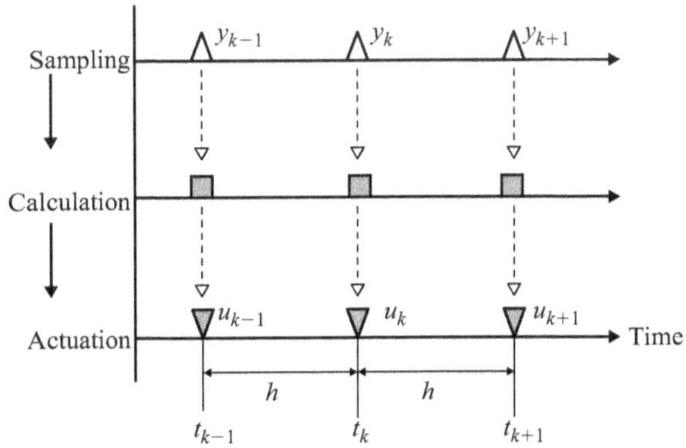

Figure 4.3 Network-induced delays that make input-to-output latency > 0

$$
\begin{aligned}
x_{k+1}^i &= A_i x_k^i + \Phi_i \hat{u}_k^i + \Theta_i \hat{u}_{k-1}^i + \sum_{j=1, j\neq i}^{N} A_{i,j} \hat{x}_k^j \\
\hat{y}_k^i &= C_i x_k^i + \sum_{j=1, j\neq i}^{N} C_{i,j} \hat{x}_k^j \\
\hat{u}_k^i &= K^i x_k^i + \sum_{j=1, j\neq i}^{N} K^{ij} \hat{x}_k^j
\end{aligned}
\tag{4.8}
$$

where $\Phi_i = B_{d0}(\tau_{rt}^k), \Theta_i = B_{d1}(\tau_{rt}^k), A_i = A_d^i$ are used for simplicity in exposition. It must be observed that we have not considered the delay explicitly in the interconnection term due to the fact that if it is was not occurring within the same sampling period, it will be eventually ignored. In general, if the delay is less than one sampling interval h for the system i, we can estimate the maximum delay as h. However, when the delays are longer than one sampling interval h, in this case the delayed packets will be dropped (Figure 4.4).

4.1.3 Packet dropout

From (4.4), if delay is $\tau_{rt}^k > h$ for a packet, then it will be lost. This means that it is not arrived before the end of the sample period. Let us introduce two parameters to be used for the packet drop formulation namely α_k and β_k, where $\alpha_k, \beta_k \in \{0, 1\}$. At time instant k the following could happen:

- Sensor packet is lost, which implies that $\beta_k = 0$.
- Control command packet is lost, which implies that $\alpha_k = 0$.

The assumption used herein is that the actuator uses the previous control command $\hat{u}_{k-1}$ if the current control command u_k is not reached. It uses the most recent control input that remains active [267], if an actuation packet lost and the controller is passive. This implies that it will provide actuation when a packet is received from sensor.

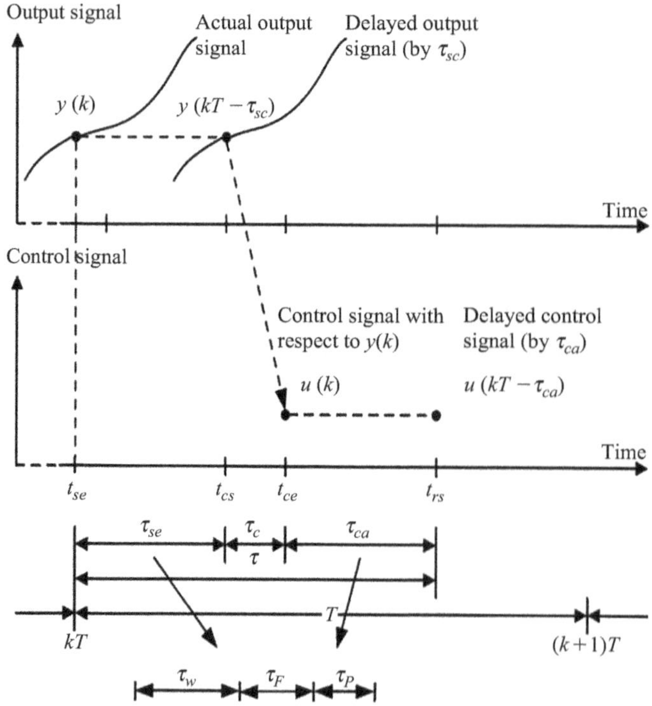

Figure 4.4 Timing diagram of network delay propagations

The same took place for the measurement data y_k, if the same is not received then the previous data will be used $\hat{y}_{k-1}$. These can be represented by the following:

$$\begin{aligned} y_k &= \beta_k \hat{y}_k + (1-\beta_k) y_{k-1} \\ \hat{u}_k &= \alpha_k u_k + (1-\alpha_k)\hat{u}_{k-1} \end{aligned} \tag{4.9}$$

When the control packet is lost, the DecNCS model described in (4.8) will be

$$\begin{aligned} x^i_{k+1} &= A_i x^i_k + (B_{d,prev})_i \hat{u}^i_{k-1} + \sum_{j=1, j\neq i}^{N} A_{i,j}\hat{x}^j_k \\ (B_{d,prev})_i &= \left[\int_0^h e^{As} ds\right] B \end{aligned} \tag{4.10}$$

Furthermore, the number of subsequent packet dropouts is upper bounded by ε and guarantees that from the sequence of previous control inputs $\{u_{k-\varepsilon}, u_{k-\varepsilon+1}, \ldots\}$ at least one is implemented. This means that

$$\sum_{i=k-\varepsilon}^{k} \alpha_k \leq \varepsilon \tag{4.11}$$

4.1.4 Variable sampling interval

Due to the nature of the network, the actual sampling times are not necessarily equidistant in time. In (4.2), we have used a constant sampling interval h but this actually will vary for each instant k to be h_k, see Figure 4.2, and hence (4.2) becomes

$$\begin{aligned} x_{k+1}^i &= A_p^i x_k^i + B_p^i \hat{u}_k^i + \sum_{j=1, j\neq i}^{N} A_{i,j} x_k^j \\ y_k^i &= C_p^i x_k^i + \sum_{j=1, j\neq i}^{N} C_p^{i,j} x_k^j \\ A_p^i &= e^{Ah_k}, \quad B_p = \left[\int_0^{h_k} e^{As} ds \right] B \end{aligned} \tag{4.12}$$

where A_p^i and B_p^i are basically the discretized versions of A_d and B_d that were mentioned earlier for constant sampling interval. Here, however, they come with varying sample interval. The state measurements are sampled at the sampling times t_k

$$t_k = \sum_{i=0}^{k-1} h_i, \quad \forall k > 0 \tag{4.13}$$

which are nonequidistantly spaced in time due to the time varying sampling intervals $h_k > 0$. The sequence of sampling instants $\{t_0, t_1, \ldots, t_k\}$ is strictly increasing in the sense that $t_{k+1} > t_k$. It is obvious that the sampling intervals are bounded and lie in the set

$$[h_{\min}, h_{\max}], \quad h_{\min} < h_{\max}$$

4.1.5 Sampling interval selection

Sampling is a major process in a control system [268] where the sampling interval determines how frequently the sensors, controllers, and actuator exchange their data through the shared communication networks. The sampling rate must be chosen properly to satisfy the Shannon's sampling theorem and also to minimize the data transmission. A higher sampling rate can improve performance and achieve higher disturbance rejection. However, this puts more data into the communication links, causes longer time delay and packet loss, and degrades system performance. Therefore, there is an upper bound for the sampling period where the network traffic is saturated in this sampling period. Any sampling period smaller than that would cause longer time delay and packet losses, and degrades the performance in DecNCS system. The lower sampling interval bound can be estimated using schedulability conditions. Similar to NCS period tasks, in DecNCS system periodical tasks might be scheduled using rate-monotonic (RM) and Earliest Deadline First (EDF) algorithms. RM is a scheduling algorithm [34,269] used in real-time operating systems with a static-priority scheduling class. The static priorities are assigned on the basis of the cycle duration of the job: the shorter the cycle duration means a higher job's priority. EDF or least time to go is a dynamic scheduling algorithm used in real-time operating systems. It places processes in a priority queue. Whenever a scheduling event occurs (task finishes, new task released, etc.) the queue will be searched for the process closest to its deadline. This process is the next to be scheduled for execution. EDF can guarantee that all deadlines are met provided that the total utilization is not more than 100%. So, compared to fixed priority scheduling techniques like RM scheduling, EDF can guarantee all the deadlines in the system at higher loading. Back to our sampling interval selection based on comparison between EDF and RM is as follows:

$$\min(h) = \max\left(\frac{h+\tau_{rt}}{1}, \frac{h+\tau_{rt}}{n(2^{1/n}-1)}\right) = \frac{h+\tau_{rt}}{n(2^{1/n}-1)} \cong \frac{\tau_{rt}}{0.69} \tag{4.14}$$

where n is an integer number that represents the scheduled tasks that will be executed for control system over the network, for example. The denominator, that represents the maximum ratio of utilization to meet the sufficient schedulability, can be estimated by iterative approach and it was found to be 0.693. In general, the sampling interval h is too small when compared to network delay and other delays like device processing time will be added if it is considered in the delays calculations (Figure 4.5).

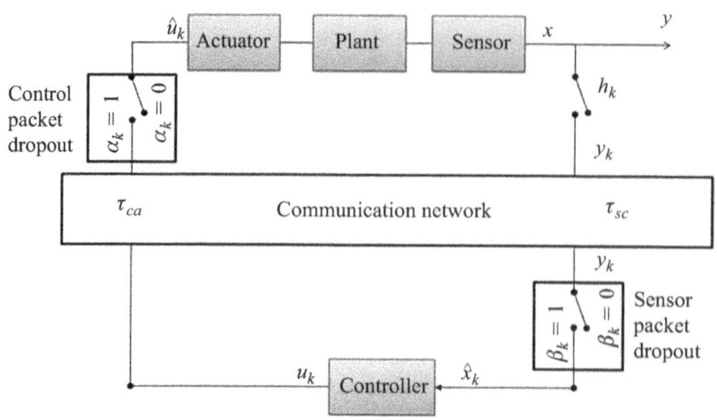

Figure 4.5 DecNCS with packet dropout and varying sample interval

4.1.6 Transmission constraints

Since the plant and controller are communicating through a network, it is possible to have a type of network that allows one node to access the network and transmit its corresponding values at each sampling time. This will add constraints [10,270] on the transmission, and we know that the actual input of the plant $\hat{u}_k$ is not equal to the controller output u_k and the actual input of the controller $\hat{y}_k$ is not equal to the plant output y_k due to network effects. In other words, we can say that $\hat{u}_k$ and $\hat{y}_k$ are networked versions of u_k and y_k, respectively, or the noise corrupted signals. To explain the effect of transmission constraints, assume that the plant has N sensors and M actuators where only one node can be sent at a time, and then only the transmitted values will be updated, while other values remain unchanged. This means that the constrained data exchange can be expressed as the following:

$$\begin{aligned} \hat{u}^i_k &= \Gamma^u_{\sigma_k} u_k + (1 - \Gamma^u_{\sigma_k})\hat{u}_{k-1} \\ \hat{y}^i_k &= \Gamma^y_{\sigma_k} y_k + (1 - \Gamma^y_{\sigma_k})\hat{y}_{k-1} \end{aligned} \tag{4.15}$$

where $\sigma_k = 1, \ldots, M + N$ is used as switched function to determine which node will have the access to transmit, Γ^u_ℓ and Γ^y_ℓ are diagonal matrices where the jth diagonal value is 1 if input/output belongs to node ℓ and zero otherwise.

4.1.7 Quantization effects

In modeling communication channel's capacity and buffers for control loops, the channel can be digital and due to the finite word length effects only a finite number of bits can be transmitted over the channel at any transmission instant. The main issue in control systems with such channels is that of quantization because the use of quantizer will add a quantization error and this error will vary based on different parameters. The resulting quantization error is often modeled as uniform and white noise [271]. Also, when we have to convert A/D or D/A during the communication between sensors/controller and controller/actuator we need to use the quantizer [272]. Designing the quantizer can impact the system performance in different manner. If the quantization regions is defined as the quantizer that do not change with time then this is a static quantizer. This type is simple to implement in both hardware and software and not computationally expensive. It has two types, namely uniform quantizers and logarithmic quantizers. The dynamic quantization has a varying quantized region and quantization error at each transmission time that will make the quantizer more complex and needs to compute new quantization regions and detect the plant state presence within this region each time.

From Figure 4.6 we have the networked and quantized version of actual input u that is $\hat{u}_q$ and the zero-order hold (ZOH) function is applied to transform the discrete-time control input to a continuous-time control input being the actual actuation signal of the plant.

$$u = u_k = K\hat{x}_k, \quad (\hat{u}_k)_q = f(\hat{u}_k) \tag{4.16}$$

The quantization error can be defined as the difference between u_q and u:

$$\begin{aligned} (e_k)_q &= (\hat{u}_k)_q - u_k = f(\hat{u}_k) - K\hat{x}_k \\ &= \Delta_k f(\hat{u}_k) \end{aligned} \tag{4.17}$$

where Δ_k is a quantized error parameter that is bounded by a certain threshold value.

4.1.8 Network-induced errors

The network-induced error can simply be shown as discrepancies between current and the most recently transmitted input/output values of nodes' signals and it can be used as shown in [237] to design dynamic output feedback and communication protocol without the need for knowledge about the controller and plant states.

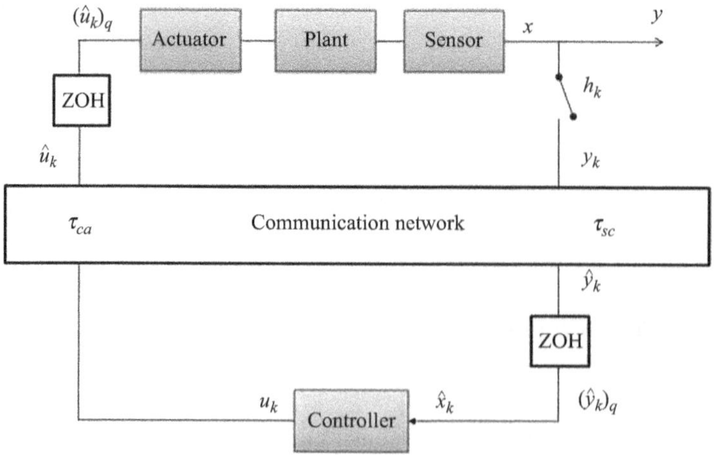

Figure 4.6 Control system with delays, quantizer, and varying sample interval

Also, the network-induced error can be used for transmission scheduling where the node with highest error will have the higher chance to obtain the network access for transmission.

$$e_k^u = \hat{u}_{k-1} - u_k, \quad e_k^y = \hat{y}_{k-1} - y_k \tag{4.18}$$

We can define threshold levels γ_i^u, γ_i^y for the induced error based on (4.18) where $e_k^u < \gamma_i^u$ and $e_k^y < \gamma_i^y$ for each subsystem i.

4.1.9 Control design

In DecNCS we may not have all state space variables available for measurements and it could be not practical to measure all of them or it is too expensive to measure all state space variables. To be able to apply the state-feedback control to a system, all of its state space variables must be available at all times. One of the solutions can be achieved by estimating system state space variables. This can be done by constructing another dynamical system called the observer or estimator, connected to the system under consideration, whose role is to produce good estimates of the state space variables of the original system. Since the whole state vector is not available for feedback, then we can measure only $y = Cx$ from linear systems, a simple state-feedback control law would be $u = Kx$, where the gain K is chosen so that the closed-loop matrix $A + BK$ is stable. In practice, we cannot measure x, so that we have to use an output feedback design. Static output feedback design, that is $u = Ky$ turns out to be relatively hard to solve and does not guarantee closed-loop stability. The most common and systematic approach is to use a dynamic output feedback, where the controller (or compensator) has its own dynamics. The simplest form is an observer structure:

$$\begin{aligned} \dot{\tilde{x}}_i &= A_i\tilde{x}_i + B_iu_i + L_i(y_i - C_i\tilde{x}_i) \\ u_i &= -K_i\tilde{x}_i, \quad i = 1, \ldots, n \end{aligned} \tag{4.19}$$

In this simple approach, $\tilde{x}_i$ is an estimate for the actual x for each subsystem i and we need to pick a good observation gain L_i such that $\tilde{x}_i \to x$ as fast as possible. In this work, we will use decentralized observer-based controllers in the sense that for each subsystem of the plant we have one observer-based controller and the controller exchange information. Therefore, the individual observers have information about external states. Furthermore, the model-based controllers will adopt switching gains to deal with the communication constraints effectively. The ith networked observer-based controller is given by considering all network side effects we have discussed in this work:

$$C^i_{\sigma_k} = \begin{cases} \tilde{x}^i_{k+1} = A^i_p \tilde{x}^i_k + B^i_{p1}\hat{u}^i_k + B^i_{p2}\hat{u}^i_{k-1} + L^i_{\sigma_k}\Gamma^y_{i,\sigma_k}(\hat{y}^i_k - C^i_d \tilde{x}^i_k) \\ \qquad + \sum_{j=1, j\neq i}^{N} A^{i,j}_p \hat{x}^j_k \\ \hat{u}^i_k = -K^i_{\sigma_k}\tilde{x}^i_k \end{cases} \tag{4.20}$$

where $\tilde{x}^i_{k+1}$ represents the state estimate at time $k+1$ for the plant state x^i_{k+1}, $B^i_{p1} = [\int_0^{h-\tau^k_{rt}} e^{As}ds]B$ and $B^i_{p2} = [\int_{h-\tau^k_{rt}}^{h} e^{As}ds]B$ when $\tau^k_{rt} \leq h$. The output-related matrices $L^i_{\sigma_k}, K^i_{\sigma_k}, i = 1, \ldots, N$ are the subsystem gain matrices. The state estimation error is

$$\psi^i_k = \tilde{x}^i_k - x^i_k$$

To deal with the communication constraints, the switched observer structure is used where the standard output is applied only when a new measurement is received. The dynamic of all controllers can be shown in discrete model that comprises block-diagonal matrices due the decentralized nature of the controllers and the same (4.20) will be rewritten without the superscript i.

4.1.10 Closed-loop system

Since the quantization error is usually too small compared to all other network side effects, it will henceforth be dropped out. For simplicity in exposition, we introduce

$$\begin{aligned} d^{i,j}_k &:= \sum_{j=1, j\neq i}^{N} A^{i,j}_p \hat{x}^j_k, \quad w^{i,j}_k := \sum_{j=1, j\neq i}^{N} C_{i,j}\hat{x}^j_k \\ \xi_k &:= [x_k \quad \psi_k \quad e^u_k \quad e^y_k \quad d_k \quad w_k] \end{aligned} \tag{4.21}$$

and combining the foregoing relations, the overall closed-loop dynamics can be expressed as

$$\xi_{k+1} = A_{cl}\xi_k$$

$$A_{cl} = \begin{bmatrix} a_{11} & a_{12} & a_{13} & 0 & I & 0 \\ a_{21} & 0 & 0 & 0 & 0 & 0 \\ a_{31} & a_{32} & a_{33} & 0 & 0 & 0 \\ a_{41} & a_{42} & a_{43} & a_{44} & 0 & I \\ 0 & 0 & 0 & 0 & I & 0 \\ 0 & 0 & 0 & 0 & 0 & I \end{bmatrix} \tag{4.22}$$

$$
\begin{aligned}
a_{11} &= A_p^i + B_p^i K_{\sigma_k}^i, \quad a_{12} = -\alpha_k \Gamma_{\sigma_k}^u K_{\sigma_k}^i B_p^i \\
a_{13} &= -B_p^i (I - \alpha_k \Gamma_{\sigma_k}^u), \quad a_{21} = A_p^i - L_{\sigma_k}^i \Gamma_{i,\sigma_k}^y C_p^i \\
a_{31} &= K_{\sigma_k}^i (A_p^i - B_p^i K_{\sigma_k}^i - L_{\sigma_k}^i \Gamma_{i,\sigma_k}^y C_p^i + I) \\
a_{32} &= K_{\sigma_k}^i (A_p^i - K_{\sigma_k}^i B_p^i - I) \\
a_{33} &= (K_{\sigma_k}^i B_p^i + I)(I - \alpha_k \Gamma_{\sigma_k}^u), \quad a_{41} = \alpha_k K_{\sigma_k}^i B_p^i C_p^i \\
a_{42} &= C_p^i (-A_p^i + \alpha_k K_{\sigma_k}^i B_p^i + I), \quad a_{43} = C_p^i B_p^i (I - \alpha_k \Gamma_{i,\sigma_k}^u) \\
a_{44} &= (I - \beta_k \Gamma_{i,\sigma_k}^y)
\end{aligned} \tag{4.23}
$$

However, if we seek fully decentralized structure in which case the exchange of state information among subsystems is not allowed, then (4.21)–(4.23) will be reduced to

$$
\begin{aligned}
\zeta_k &:= \begin{bmatrix} x_k & \psi_k & e_k^u & e_k^y \end{bmatrix} \\
\zeta_{k+1} &= A_{de} \zeta_k
\end{aligned} \tag{4.24}
$$

$$
A_{de} = \begin{bmatrix} a_{11} & a_{12} & a_{13} & 0 \\ a_{21} & 0 & 0 & 0 \\ a_{31} & a_{32} & a_{33} & 0 \\ a_{41} & a_{42} & a_{43} & a_{44} \end{bmatrix} \tag{4.25}
$$

where $a_{11}, \ldots, a_{44}$ are given by (4.23). The design complexity can be further reduced by assuming that each controller has the accessibility to send the control signals to actuators at any time instant, hence there is no constraint on sending. This implies that $e_k^u = 0, \Gamma_{i,\sigma_k}^u = I$. As a result, the closed-loop system matrix in (4.25) will be reduced to

$$
A_{dd} = \begin{bmatrix} a_{11} & a_{12} & a_{13} \\ a_{21} & 0 & 0 \\ a_{31} & a_{32} & a_{33} \end{bmatrix} \tag{4.26}
$$

Same will take place to (4.23).

4.1.11 Stability analysis

We have previously introduced the decentralized networked observer-based controller. In what follows, we shed light on this type of controller.

Remark 4.1: *Note that the observer has the same structure as the system plus the driving feedback term, which is send via communication network, that contain information about the observation error. The role of the feedback term is to reduce the observation error to zero (at steady state). Figure 4.7 shows the ideal observer before introducing the network and Figure 4.8 gives networked controller where the observer-based elements are placed in the feedback loop.*

In view of the block-diagonal structure, let $A_{cl} = blockdiag\{A_{1,cl}^t, \ldots, A_{N,cl}^t\}$. The following is a preliminary result.

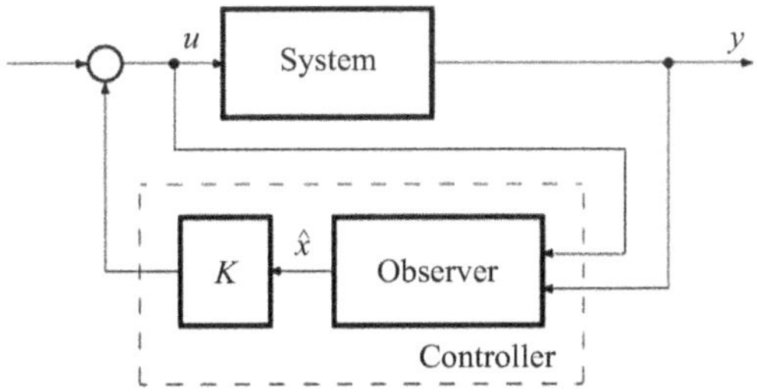

Figure 4.7 Closed-loop observer-based control system

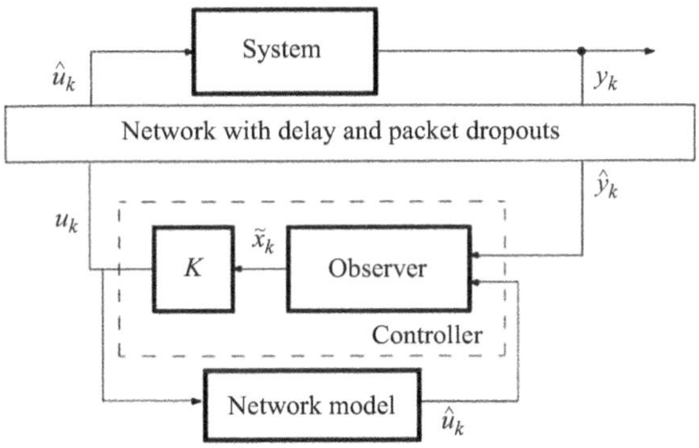

Figure 4.8 Closed-loop observer-based networked control system

Lemma 4.1: *Given the gains Kandl system (4.22) is said to be asymptotically stable if there exists matrices $0 < P_i = P_i^t \in \Re^{n_i \times n_i}, i = 1, \ldots, N$ such that the following LMIs*

$$\begin{bmatrix} -P_i & A_{i,cl}^t P_i \\ \bullet & -P_i \end{bmatrix} < 0, \quad i = 1, \ldots, N \tag{4.27}$$

have a feasible solution.

Proof. Define $V = \xi_k^t Q \xi_k, Q = blockdiag\{Q_1, \ldots, Q_N\}$. Evaluating the first difference ΔV along the solutions of (4.22) yields $\Delta V = A_{cl}^t Q A_{cl} - Q$. Using $P = Q_{cl}^{-1}$ and Schur complements, we readily obtain (4.27). The main design result is established by the following theorem. □

Theorem 4.1: *System (4.22) is said to be asymptotically stable if there exists matrices $0 < P_i = P_i^t \in \Re^{n_i \times n_i}$, $i = 1, \ldots, N$ such that the following LMIs have a feasible solution.*

4.1.12 Illustrative example 4.1

To illustrate the theoretical developments, we consider a plant comprised of three chemical reactors as shown in Figure 4.9.

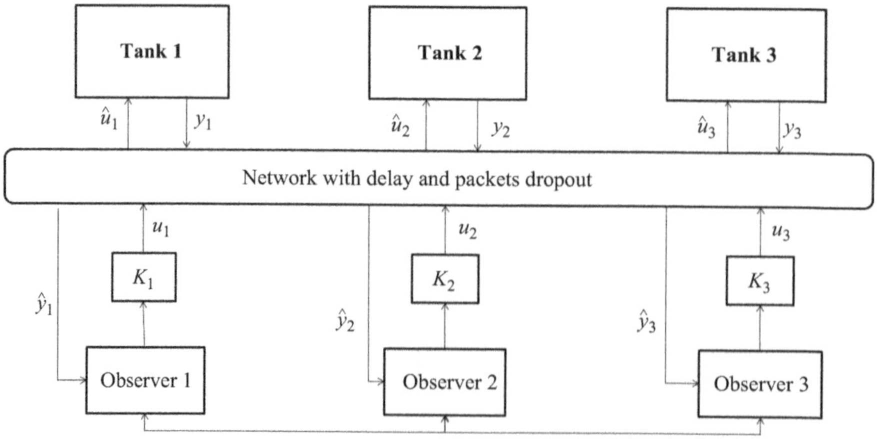

Figure 4.9 Closed-loop three-tanks observer-based control system

The linearized model can be described by the following matrices:

$$A_j = \begin{bmatrix} -a_{1j} & -1.01 & 0 & 0 \\ -3.2 & -a_{2j} & -12.8 & 0 \\ 6.4 & 0.347 & -a_{3j} & -1.04 \\ 0 & 0.833 & 11.0 & -a_{4j} \end{bmatrix}$$

$$B_j^t = \begin{bmatrix} 1 & 0 & 0 & 0 \\ 0 & 1 & 0 & 0 \end{bmatrix} \tag{4.28}$$

$$A_{dj} = \begin{bmatrix} b_{1j} & 0 & 0 & 0 \\ 0 & b_{2j} & 0 & 0 \\ 0 & 0 & b_{3j} & 0 \\ 0 & 0 & 0 & b_{4j} \end{bmatrix}$$

$$C_j = \begin{bmatrix} 10 & 0 & 0 & 0 \end{bmatrix}$$

where

$$\begin{aligned} a_{11} &= 4.931, & a_{12} &= 4.886, & a_{13} &= 4.902 \\ a_{21} &= 5.301, & a_{22} &= 5.174, & a_{23} &= 5.464 \\ a_{31} &= 35.511, & a_{32} &= 30.645, & a_{33} &= 31.773 \\ a_{41} &= 3.961, & a_{42} &= 3.878, & a_{43} &= 3.932 \\ b_{11} &= 1.921, & b_{12} &= 1.915, & b_{13} &= 1.908 \\ b_{21} &= 1.921, & b_{22} &= 1.914, & b_{23} &= 1.907 \\ b_{31} &= 1.878, & b_{32} &= 1.866, & b_{33} &= 1.869 \\ b_{41} &= 0.724, & b_{42} &= 0.715, & b_{43} &= 0.706 \end{aligned} \tag{4.29}$$

Using the LMI toolbox in MATLAB, the ensuing results are summarized as follows.

LMI matrices:

$$P_1 = \begin{bmatrix} 0.1448 & -0.0020 & 0.0005 & 0.0002 \\ -0.0020 & 0.1442 & -0.0005 & 0.0009 \\ 0.0005 & -0.0005 & 0.1420 & 0.0001 \\ 0.0002 & 0.0009 & 0.0001 & 0.1463 \end{bmatrix}$$

$$P_2 = \begin{bmatrix} 0.1022 & -0.0021 & 0.0006 & 0.0002 \\ -0.0021 & 0.1025 & -0.0005 & 0.0020 \\ 0.0006 & -0.0005 & 0.0993 & 0.0001 \\ 0.0002 & 0.0020 & 0.0001 & 0.1032 \end{bmatrix}$$

$$P_3 = \begin{bmatrix} 0.6502 & -0.0131 & 0.0042 & 0.0012 \\ -0.0131 & 0.6406 & -0.0032 & 0.0069 \\ 0.0042 & -0.0032 & 0.6280 & 0.0006 \\ 0.0012 & 0.0069 & 0.0006 & 0.6554 \end{bmatrix}$$

$$X_1 = \begin{bmatrix} 0.2142 & -0.0560 & 0.0421 & -0.0153 \\ -0.0560 & 0.0383 & -0.0138 & 0.0515 \\ 0.0421 & -0.0138 & 0.0292 & 0.0435 \\ -0.0153 & 0.0515 & 0.0435 & 0.2597 \end{bmatrix}$$

$$X_2 = \begin{bmatrix} 0.1625 & -0.0200 & 0.0314 & -0.0291 \\ -0.0200 & 0.0275 & -0.0033 & 0.0672 \\ 0.0314 & -0.0033 & 0.0222 & 0.0341 \\ -0.0291 & 0.0672 & 0.0341 & 0.2809 \end{bmatrix}$$

$$X_3 = \begin{bmatrix} 0.1106 & -0.0172 & 0.0222 & -0.0131 \\ -0.0172 & 0.0144 & -0.0044 & 0.0257 \\ 0.0222 & -0.0044 & 0.0147 & 0.0207 \\ -0.0131 & 0.0257 & 0.0207 & 0.1264 \end{bmatrix}$$

$$Z_1 = \begin{bmatrix} 0.1595 & -0.0419 & 0.0314 & -0.0118 \\ -0.0419 & 0.0284 & -0.0101 & 0.0387 \\ 0.0314 & -0.0101 & 0.0216 & 0.0325 \\ -0.0118 & 0.0387 & 0.0325 & 0.1943 \end{bmatrix}$$

$$Z_2 = \begin{bmatrix} 0.1115 & -0.0141 & 0.0214 & -0.0212 \\ -0.0141 & 0.0190 & -0.0020 & 0.0475 \\ 0.0214 & -0.0020 & 0.0151 & 0.0239 \\ -0.0212 & 0.0475 & 0.0239 & 0.1982 \end{bmatrix}$$

$$Z_3 = \begin{bmatrix} 0.7193 & -0.1135 & 0.1443 & -0.0893 \\ -0.1135 & 0.0925 & -0.0271 & 0.1686 \\ 0.1443 & -0.0271 & 0.0945 & 0.1350 \\ -0.0893 & 0.1686 & 0.1350 & 0.8279 \end{bmatrix}$$

Observe that $P_j, Z_j > 0, j = 1, \ldots, 3$.

Observer–controller gains:

$$L_1^t = \begin{bmatrix} 0.0130 & 0.0681 & 0.0164 & -0.0385 \end{bmatrix}$$

$$L_2^t = \begin{bmatrix} -0.0158 & 0.1235 & 0.0162 & -0.0654 \end{bmatrix}$$

$$L_3^t = \begin{bmatrix} -0.0456 & 0.0701 & 0.0706 & -0.0201 \end{bmatrix}$$

$$K_1 = \begin{bmatrix} 0.7573 & 0.7142 & 0.3973 & 0.8391 \\ 0.2138 & 8.2185 & 13.8882 & -3.4177 \end{bmatrix}$$

$$K_2 = \begin{bmatrix} 0.3144 & -0.7983 & -3.8703 & 1.7806 \\ -0.6559 & 7.2776 & 14.8651 & -6.8895 \end{bmatrix}$$

$$K_3 = \begin{bmatrix} 0.2634 & -0.1587 & -3.1912 & 1.5713 \\ -1.0803 & 8.5794 & 12.2875 & -3.9658 \end{bmatrix}$$

Simulation of the closed-loop system is performed and the ensuing state trajectories are presented in Figures 4.10–4.12. It is clearly evident that the interconnected chemical reactors are decentralized, asymptotically stabilizable with guaranteed performance.

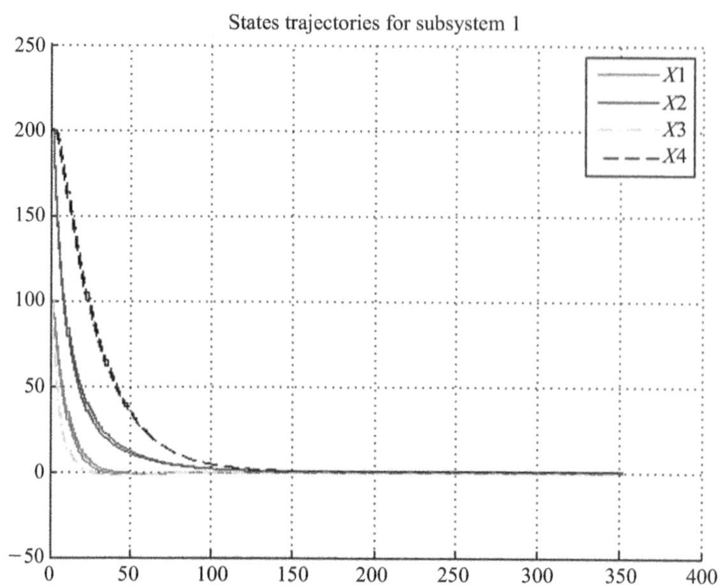

Figure 4.10 Closed-loop state trajectories: subsystem 1

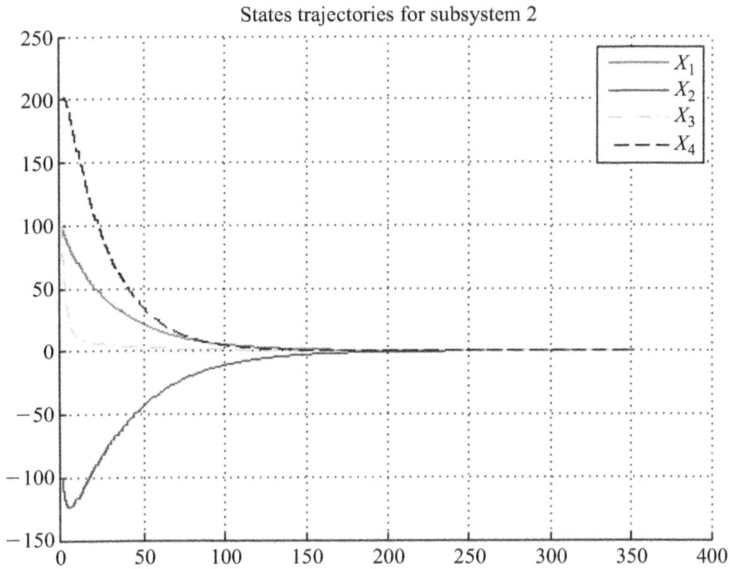

Figure 4.11 Closed-loop state trajectories: subsystem 2

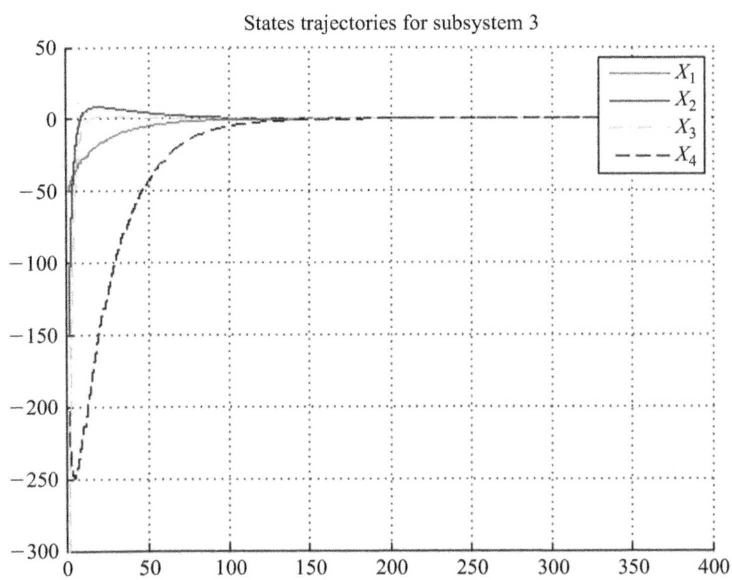

Figure 4.12 Closed-loop state trajectories: subsystem 3

4.2 Approximate distributed feedback control

This section presents an approach for feedback control of interconnected systems that lies between decentralized (no communications between local controllers) and distributed (prescribed communication between controllers) approaches. It is termed 'approximate distributed' since it relies on partial communications among controllers. The information is exchanged over a wireless sensor network (WSN) with limited resources. Due to limited resources and availability of states, the communication rate over the wireless network is minimized and an observer-based output feedback controller is designed to ensure closed-loop stability. Each subsystem has models for other units in the plant that generate necessary control action until the communication between the subsystems is re-established. This communication suspension is done periodically to conserve the battery power of the wireless sensor suits. When the subsystems transmit their state estimates at discrete-time instants according to certain schedule, the states of the models embedded in the target units are updated. Analysis of this discrete-continuous behavior of the whole system yields explicit characterization of the stability of the closed-loop system in terms of the communication time, transmission schedule, accuracy of the embedded models, and controller and observer gains. The results are used for a chemical plant example where it is demonstrated using simulations, the role of transmission periods and scheduling, model estimates, and closed-loop parameters in the stability of the overall system.

4.2.1 Introduction

Industrial systems, such as chemical plants and manufacturing processes, are typically interconnected dynamical systems with complex, distributed arrangements of interconnected subsystems that are heavily integrated through mass, energy, and information flows. The control of geographically distributed large-scale systems with coupled subsystems has been studied within centralized or decentralized control frameworks. The centralized scheme issues control commands to the whole plant using the collected measurements. This gives best performance due to least constraints imposed on the plant. However, due to high computational and organizational complexity their implementation is limited. Moreover, failure in the centralized controller may become detrimental to the entire plant [507–510].

These considerations resulted in significant work on decentralized/distributed control. In this setting, the plant is decomposed into a number of simpler subsystems with interconnections, and local controllers are connected to each unit without communication between different local controllers. This scheme reduces the complexity in the controller design and implementation and can also provide flexibility in dealing with local controller failures. However, since in this structure the interconnections between the constituent subsystems are completely ignored, the closed-loop performance of the plant may deteriorate, and in some cases stability may be lost. Significant research work has explored in depth, the benefits and limitations of decentralized controllers as well as possible ways of overcoming some of their limitations [273].

An approach that provides a compromise between the complexity of traditional centralized control schemes on one hand and the performance limitations of decentralized control approaches on the other, is quasidecentralized control, which refers to a control strategy in which most signals used for control are collected and processed locally, while some signals are transferred between the local units and controllers to adequately account for the interactions and minimize the propagation of process upsets from one unit to another. One of the key problems that has been addressed in this design is how to coordinate the control and communication functions and how to account for possible limitations of the communication medium in the formulation and solution of the control problem.

An important consideration in the design and implementation of quasidecentralized control systems is the selection of the communication medium over which the local control systems must communicate. The traditional solution for communication in an industrial plant is point-to-point wired communication connecting the central controller with each sensor or actuator point, and has been successfully implemented for years. Due to growth in size and complexity of industrial systems, the complexity and cost of installing and maintaining wired control systems become significant. These issues, coupled with the significant growth in computing and networking abilities, as well as the rapid advances in actuator and sensor technologies, have led to an increased reliance on distributed computing and process operations across computer networks.

A better solution is the use of wireless communication networks due to the fact that the recent advances in sensor and actuator technologies embedded with communication feature have produced low-cost wireless sensors that can be installed for a fraction of the cost of wired devices. WSNs have unmatched advantages such as high-density sensing capabilities and ability to be deployed in areas where wired devices may be difficult or impossible to install. Augmenting existing process control and monitoring systems with WSNs has the potential to expand the capabilities of the existing control technology beyond what is feasible with wired architectures alone. These are appealing goals that coincide with the recent calls for expanding the traditional process control and operations paradigm in the direction of smart plant operations [42,506].

One of the main challenges to be addressed when deploying a low-cost WSN for control is that of handling the inherent constraints on network resources, including the limitations on the computation, processing, and communication capabilities. This problem has been addressed in [273] where a quasidecentralized model-based networked control architecture that enforces closed-loop stability with minimal cross-communication between the constituent subsystems, was developed. The networked control structure was subsequently generalized in [274] to address the problem of limited state measurements.

An approximate distributed approach for model-based control, state estimation, and sensor scheduling in plants with limited state measurements and interconnected processing units that communicate over a resource-constrained WSN, is presented in the sequel. The plant is distributed in different subsystems and following the framework of quasi-decentralized control scheme, each of these subsystems is equipped with hard-wired local controller that is responsible for the stability of respective subsystem. Also, an additional module of wireless sensors is installed for each unit that communicates the states over WSN, both to the plant supervisor and to other subsystems. All these sensor suits form a plant-wide WSN. The objective is to find a strategy for establishing and terminating communication between the sensor suites of the WSN and the local control systems in a way that minimizes the rate at which each node in the WSN broadcasts data to the rest of the plant without jeopardizing closed-loop stability.

4.2.2 Plant description

The state space description of a large-scale system composed of n interconnected subsystems is given by (4.30) where $x_i := [x_i^{(1)} x_i^{(2)} \dots x_i^{(p_i)}]^T \in R^{p_i}$ denotes the vector of process state variables associated with the ith processing unit, p_i is the number of state variables in the ith unit, $y_i := [y_i^{(1)} y_i^{(2)} \dots y_i^{(q_i)}]^T \in R^{q_i}$ and $u_i := [u_i^{(1)} u_i^{(2)} \dots u_i^{(r_i)}]^T \in R^{r_i}$ denote the vector of measured outputs and manipulated inputs associated with the ith processing unit, respectively, x^T denotes the transpose of a column vector x, A_i, B_i, A_{ij}, and C_i are constant matrices. The interconnection term $A_{ij}x_j$, where $i \neq j$ describes how the dynamics of the ith unit are influenced by the jth unit in the plant. Note from (4.30) that each processing unit can in general be connected to all the other units in the plant.

$$
\begin{aligned}
\dot{x}_1 &= A_1x_1 + B_1u_1 + \sum_{j=2}^{n} A_{1j}x_j, & y_1 &= C_1x_1 \\
\dot{x}_2 &= A_2x_2 + B_2u_2 + \sum_{j=1, j\neq 2}^{n} A_{2j}x_j, & y_2 &= C_2x_2 \\
&\vdots \quad \vdots & &\vdots \\
\dot{x}_n &= A_nx_n + B_nu_n + \sum_{j=1}^{n-1} A_{nj}x_j, & y_n &= C_nx_n
\end{aligned}
\tag{4.30}
$$

4.2.3 Problem formulation

The control objective is to stabilize all the plant units at zero steady state while simultaneously: (a) keeping the data dissemination and exchange over the WSN to a minimum and (b) accounting for the lack of full-state measurements within each unit. To address the resource constraints problem, an integrated model-based quasi-decentralized output feedback control and scheduling strategy is given in section 4.2.4 that reduces the exchange of information between the plant units without loss of stability. This is accomplished by

(a) designing for each local control system, an appropriate state observer to generate estimates of the local state variables from the measured outputs,
(b) including models within each control system to estimate the interaction terms when measurements are not available through the WSN, and
(c) limiting the number of WSN nodes that, at any given time, transmit their data to update the corresponding target models.

The problem is to find an optimal scheduling strategy for establishing and terminating communication between the sensor suites and the target controllers. To illustrate the main ideas, as an example, the configuration where only one wireless sensor suite is allowed to transmit its data to the appropriate units at any given time, while the other nodes remain dormant until the next suite is allowed to transmit its data, is considered.

4.2.4 Approximate distributed controllers

Referring to (4.30), the output feedback controller is given by

$$\begin{aligned} u_i &= K_i\bar{x}_i + \sum_{j=1,j\neq i}^{n} K_{ij}\bar{x}_j \\ \bar{x}_i &= (A_i - L_iC_i)\bar{x}_i + \sum_{j=1,j\neq i}^{n} A_{ij}\bar{x}_j + B_iu_i + L_iy_i \end{aligned} \tag{4.31}$$

where $\bar{x}_i$ is an estimate of the state of the ith unit generated by an observer embedded within the local control system of the ith unit, K_i is the local feedback gain responsible for stabilizing the ith subsystem in the absence of interconnections. K_{ij} is a gain that compensates for the effect of the jth neighboring subsystem on the dynamics of the ith unit, and L_i is the observer gain. Both controller and observer gains are calculated using pole-placement method. The implementation of the controller of (4.31) requires the availability of observer-generated state estimates from the other units in the plant, which can be transmitted only through the WSN. A copy of the local observer must therefore be included within the wireless sensor suite of each unit in order to generate the state estimates that are then broadcasted to the rest of the plant. This setup is possible given the computational capabilities of wireless sensors.

4.2.5 Model-based networked control structure

The motivation behind sensor scheduling is conservation of sensor suit battery power. Therefore, initially the rate at which the information (i.e., $\bar{x}_j$) is transferred from the wireless sensor suite of each unit to the target control system in the neighboring units is reduced as much as possible without loosing closed-loop stability. Each unit is embedded with a set of estimated dynamic models that provide estimates of the evolution of the states of the neighboring units when communication is suspended. The model estimates are used to generate both the local state estimates and the local control action. The state of each model is then reset using the state estimate generated by the observer of the corresponding unit when the wireless sensor suite of the latter is allowed to transmit its data at discrete-time instances. Mathematically, the local control and update laws for unit i are implemented as (4.32), where $\hat{x}_j^i$ is the estimate of x_j provided by a model of unit j embedded in unit $i, \hat{A}_j, \hat{B}_j$, and $\hat{A}_{jl}$ are constant matrices, t_k^j indicates the kth transmission time for the jth sensor suite in the WSN. The models of neighboring units embedded in the local control systems might not be accurate, that is, $\hat{A}_j \neq A_j, \hat{B}_j \neq B_j, \hat{A}_{jl} \neq A_{jl}$. This translates into the parametric uncertainty in the embedded plant models for the chemical plant example discussed in this report.

4.2.6 WSN transmission scheduling and model updates

The frequency at which jth subsystem sends its state update to other units determines the update period, $h^j := t^j_{k+1} - t^j_k$, which also reveals the extent of network utilization. Referring to Figure 4.13, it can be seen that larger h translates into larger savings in WSN resource utilization. In this work, only the case when the update period is constant and same for all the units is considered. Moreover, only one sensor suit is allowed to transmit its states any one time while other units remain passive and wait for their turn in the schedule. The transmission schedule is defined by

- the sequence (or order) of transmitting nodes: $s_j,\ j = 1, 2, \ldots, n,\ s_j \in N := 1, 2, \ldots, n$, where s_j is a discrete variable that denotes the jth transmitting entity in the sequence, and
- the time at which each node in the sequence transmits observer estimates. Also, $\Delta t_j := t_k^{s_{j+1}} - t_k^{s_j},\ j = 1, 2, \ldots, n$, which is the time interval between the transmissions of two consecutive nodes in the sequence.

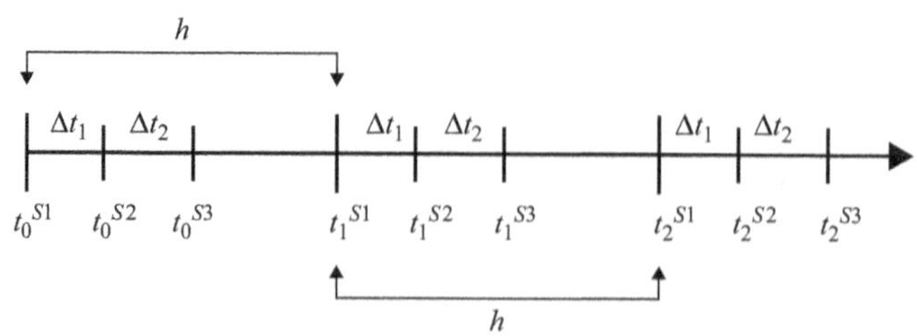

Figure 4.13 Time line for transmission of each sensor suit in an h-periodic schedule for three subsystems

It must be noted that $\sum_{j=1}^{n-1} \Delta t_j < h$. By manipulating the time intervals Δt_j and the schedule or sequence of transmission, we can search for optimal condition in which the communication rate is as minimum as possible, that is, lowest power consumption for WSN suites.

$$
\begin{aligned}
u_i(t) &= K_i\bar{x}_i(t) + \sum_{j=1, j\neq i}^{n} K_{ij}\hat{x}^i_j(t), \quad t \neq t^j_k, \quad i = 1, 2, \ldots, n \\
\dot{\bar{x}}_i(t) &= (A_i - L_iC_i)\bar{x}_i(t) + \sum_{j=1, j\neq i}^{n} A_{ij}\hat{x}^i_j(t) + B_iu_i(t) + L_iy_i(t) \\
\dot{\hat{x}}_i(t) &= \hat{A}_j\hat{x}^i_j(t) + \hat{B}_j\hat{u}^i_j(t) + \hat{A}_{ji}\bar{x}_i(t) + \sum_{l=1, l\neq i, l\neq j}^{n} \hat{A}_{jl}\hat{x}^i_l(t), \quad t \neq t^j_k \\
\hat{u}^i_j(t) &= K_j\hat{x}^i_j(t) + K_{ji}\bar{x}_i(t) + \sum_{l=1, l\neq i, l\neq j}^{n} K_{jl}\hat{x}^i_l(t), \quad t \neq t^j_k \\
\hat{x}^i_j(t^j_k) &= \bar{x}_j(t^j_k), \quad j = 1, \ldots, n, \quad j \neq i, \quad k = 0, 1, 2, \ldots
\end{aligned}
\tag{4.32}
$$

4.2.7 Networked closed-loop stability

To form a closed-loop system and characterize its stability, (4.30) and (4.32) along with model estimation errors are used to form a combined discrete-continuous system. The difference between the actual states of the jth subsystem and the estimated one by ith subsystem, that is, the model estimation errors, are defined as $e_j^i = \bar{x}_j - \hat{x}_j^i$, for $j \neq i$. To account for state evolution of the system along with its estimation and model estimation errors, the state and error vectors are augmented as, $x := [x_1^T x_2^T \dots x_n^T]^T, \bar{x} := [\bar{x}_1^T \bar{x}_2^T \dots \bar{x}_n^T]^T$ and $e_j := [(e_j^1)^T (e_j^2)^T \dots (e_j^n)^T]^T, e := [e_1^T e_2^T \dots e_n^T]^T$.

The overall closed-loop system thus obtained is of the form

$$\begin{aligned} \dot{x}(t) &= \Lambda_{11}x(t) + \Lambda_{12}\bar{x}(t) + \Lambda_{13}e(t) \\ \dot{\bar{x}}(t) &= \Lambda_{21}x(t) + \Lambda_{22}\bar{x}(t) + \Lambda_{23}e(t) \\ \dot{e}(t) &= \Lambda_{31}x(t) + \Lambda_{32}\bar{x}(t) + \Lambda_{33}e(t), \quad t \neq t_k^j \\ e_j(t_k^j) &= 0, j = 1, 2, \dots, n, k = 0, 1, 2, \dots \end{aligned} \tag{4.33}$$

where Λ_{ij}'s are constant matrices that can be obtained by substituting (4.32) into (4.30) and noting that e_j^i. These matrices constitute individual system matrices, the closed-loop system terms, observer terms, and the estimated model terms. They are given later in the illustration of the chemical plant example of section 4.3.

Defining the augmented state vector as $\xi(t) := [x^T(t) \quad \bar{x}(t) \quad e^T(t)]$, the dynamics of the overall closed-loop system can be represented as

$$\begin{aligned} \dot{\xi}(t) &= \Lambda_o \xi(t), \quad t \neq t_k^j \\ \xi(t_k^j) &= [x^T(t_k^j) \quad \bar{x}(t_k^j) \quad e^T(t_k^j)], \quad k = 0, 1, 2, \dots \\ e^T(t_k^j) &= [e_1^T(t_k^j) \dots e_{j-1}^T(t_k^j) \quad 0 \quad e_{j+1}^T(t_k^j) \dots e_n^T(t_k^j)]^T \\ \Lambda_o &= \begin{bmatrix} \Lambda_{11} & \Lambda_{12} & \Lambda_{13} \\ \Lambda_{21} & \Lambda_{22} & \Lambda_{23} \\ \Lambda_{31} & \Lambda_{32} & \Lambda_{33} \end{bmatrix} \end{aligned} \tag{4.34}$$

The closed-loop response in terms of update period and transmission schedule can be characterized for transmission sequence $s_1, s_2, \dots, s_n$ and the initial condition $\xi(t_o^{s1}) = [x^T(t_o^{s1}) \quad \bar{x}(t_o^{s1}) \quad e^T(t_o^{s1})]^T = \xi_o$, with $e_{s_1}(t_o^{s1}) = 0$. Then for $k = 0, 1, 2, \dots,$

$$\xi(t) = \begin{cases} \exp^{\Lambda_o(t - t_k^{s_j})} \Gamma_j(\Delta t_j, I_o^{s_j}) M_o^k \xi_0, & t \in [t_k^{s_j}, t_k^{s_j+1}) \\ \exp^{\Lambda_o(t - t_k^{s_n})} \Gamma_n M_o^k \xi_0, & t \in [t_k^{s_n}, t_{k+1}^{s1}) \end{cases}$$

where $j = 1, 2, \dots, n-1$ and

$$\Gamma_j = \prod_{\mu = j-1}^{j-2} I_o^{s_\mu + 1} \exp^{\Lambda_o \Delta t_\mu}, \text{ for } j \geq 2, \quad \text{and} \quad \Gamma_j = I, \text{ for } j = 1 \tag{4.35}$$

$$M_o = I_o^{s_1} \exp^{\Lambda_o\left(h - \sum_{j=1}^{n-1} \Delta t_j\right)} \Gamma_n \tag{4.36}$$

$$I_o^{s_j} = \begin{bmatrix} I & O & \dots & O \\ O & H_1 & \dots & O \\ \vdots & \vdots & & \vdots \\ O & O & \dots & H_n \end{bmatrix} \tag{4.37}$$

where $H_i = I$ for $i \neq s_j$ and $H_i = O$ for $i = s_j$, for $j = 1, 2, \dots, n, t_{k+1}^{s_j} - t_k^{s_j} = h$, and $\Delta t_j = t_k^{s_{j+1}} - t_k^{s_j}, j = 1, 2, \dots, n-1$. Note that the H_i matrices in (4.37) are square matrices with dimensions '$n + n$', where n is the number of subsystems and the square matrix $I_o^{s_j}$ has the dimension equal to the Λ_o matrix in (4.34).

The necessary and sufficient condition for stability of the scheduled closed-loop plant is

The zero solution, $\xi = [x^T \quad \bar{x}^T \quad e^T]^T = [0 \quad 0 \quad 0]^T$, *is globally exponentially stable if and only if the eigen values of the matrix in (4.36) are strictly inside the unit circle.*

As can be seen from (4.36), that the test matrix M_o depends on the transmission schedule, update period, and transmission time. This type of structure and the stability condition stated above provides us with the flexibility to search for optimal transmission periods and schedules along with ensuring the closed-loop stability. Hence, achieving the goal of optimal WSN power consumption.

4.3 Application to chemical reactors

In the sequel, an example of a chemical reactor plant is considered with three non-isothermal Continuously Stirred-Tank Reactors (CSTRs) in a cascade as shown in Figure 4.14. Three parallel irreversible exothermic reactions of the form $A \rightarrow^{k_1} B, A \rightarrow^{k_2} U$ and $A \rightarrow^{k_3} R$ take place, where A is the reactant, B is the desired product,

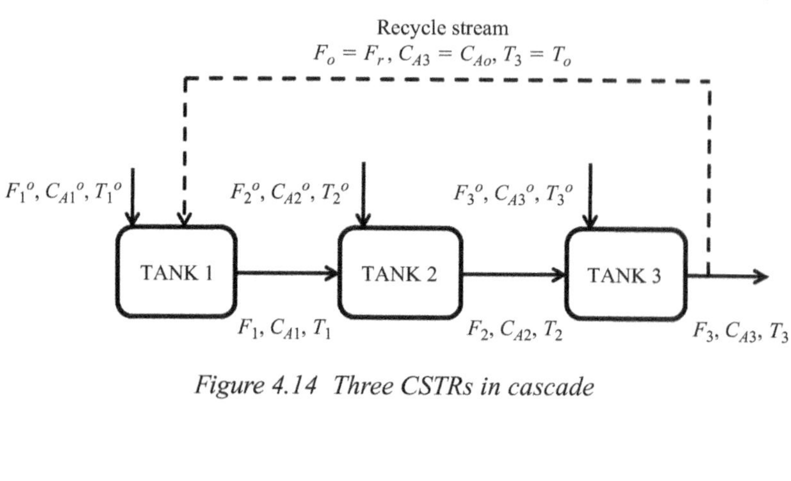

Figure 4.14 Three CSTRs in cascade

and U and R are the undesired by-products. As can be seen from Figure 4.14, the feed to all the tanks consists of two streams, one containing fresh A at flow rate F_j^o, molar concentration C_{Aj}^o, and temperature T_j^o, where $j = 1, 2, 3$. For the second and third tank, the second stream contains the products from the first and second tank, respectively. For the first tank, the output of the third tank is passed through a separator that removes the products and recycles unreacted A. F_j, T_j, and C_{A_j} are the flow rates, temperature values, and molar concentrations of the second stream for the respective tanks.

The plant model can be given by following non-linear set of equations:

$$\begin{aligned}
\frac{dT_j}{dt} &= \frac{F_j^o}{V_j}(T_j^o - T_j) + \frac{F_{j-1}}{V_j}(T_{j-1} - T_j) \\
&\quad + \sum_{i=1}^{3} \frac{-\Delta H_i}{\rho c_p} R_i(C_{A_j}, T_j) + \frac{Q_j}{\rho c_p V_j} \\
\frac{dC_{Aj}}{dt} &= \frac{F_j^o}{V_j}(C_{Aj}^o - C_{Aj}) + \frac{F_{j-1}}{V_j}(C_{A(j-1)} - C_{Aj}) \\
&\quad - \sum_{i=1}^{3} R_i(C_{A_j}, T_j), \quad j = 1, 2, 3
\end{aligned} \tag{4.38}$$

where Q_j is the rate of heat input for the jth tank, $R_i(C_{A_i}, T_j) = k_{i0} \exp\left(\frac{-E_i}{RT_j}\right) C_{A_j}$ is the rate of ith reaction, where $i = 1, 2, 3$. See the table in appendix for parameter values and their description. The plant with $Q_j = 0, C_{A_j}^o = C_{Aj}^{os}$ and a recycle ratio of $r = 0.5$, has three steady states (two locally asymptotically stable and one unstable) at $T_1^s, C_{A1}^s, T_2^s, C_{A2}^s, T_3^s, C_{A3}^s$. The steady-state values are also given in the table. Only the temperature of three reactions is assumed to be available as measurements. The plant-wide WSN composed of three wireless sensor suits is deployed and each of them broadcasts the state estimates of its respective tank, provided by the state observer, to other units.

The plant given in (4.38) is linearized around the steady states using Taylor series and the model thus obtained is given as

$$\begin{aligned}
\frac{\dot{T}_1}{T_1^s} &= 25.2075\left(\frac{T_1 - T_1^s}{T_1^s}\right) + 31.7573\left(\frac{T_3 - T_3^s}{T_3^s}\right) + 4.9639\left(\frac{C_{A1} - C_{A1}^s}{C_{A1}^s}\right) \\
&\quad + 9.454 \times 10^{-6} Q_1 \\
\frac{\dot{C}_{A1}}{C_{A1}^s} &= -78\left(\frac{T_1 - T_1^s}{T_1^s}\right) - 45.9295\left(\frac{C_{A1} - C_{A1}^s}{C_{A1}^s}\right) + 2.8237\left(C_{A1}^o - C_{A1}^{os}\right) \\
&\quad + 34.6025\left(\frac{C_{A3} - C_{A3}^s}{C_{A3}^s}\right)
\end{aligned}$$

$$\frac{\dot{T}_2}{T_2^s} = -2.8393\left(\frac{T_2 - T_2^s}{T_2^s}\right) + 14.6925\left(\frac{T_1 - T_1^s}{T_1^s}\right) + 1.4157\left(\frac{C_{A2} - C_{A2}^s}{C_{A2}^s}\right) + 3.47 \times 10^{-6} Q_2$$

$$\frac{\dot{C}_{A2}}{C_{A2}^s} = -22.4845\left(\frac{T_2 - T_2^s}{T_2^s}\right) - 24.8849\left(\frac{C_{A2} - C_{A2}^s}{C_{A2}^s}\right) + 5.7143(C_{A2}^o - C_{A2}^{os}) + 13.4844\left(\frac{C_{A1} - C_{A1}^s}{C_{A1}^s}\right)$$

$$\frac{\dot{T}_3}{T_3^s} = -2.8393\left(\frac{T_3 - T_3^s}{T_3^s}\right) + 13.332\left(\frac{T_2 - T_2^s}{T_2^s}\right) + 1.4157\left(\frac{C_{A3} - C_{A3}^s}{C_{A3}^s}\right) + 3.47 \times 10^{-6} Q_2$$

$$\frac{\dot{C}_{A3}}{C_{A3}^s} = -22.4845\left(\frac{T_3 - T_3^s}{T_3^s}\right) - 24.8849\left(\frac{C_{A3} - C_{A3}^s}{C_{A3}^s}\right) + 5.7143(C_{A3}^o - C_{A3}^{os}) + 13.332\left(\frac{C_{A2} - C_{A2}^s}{C_{A2}^s}\right)$$

Note that the states are normalized. The state and input vectors and the state matrices are thus given by

$$x_1 = \begin{bmatrix} \dfrac{T_1 - T_1^s}{T_1^s} \\ \dfrac{C_{A1} - C_{A1}^s}{C_{A1}^s} \end{bmatrix}, \quad x_2 = \begin{bmatrix} \dfrac{T_2 - T_2^s}{T_2^s} \\ \dfrac{C_{A2} - C_{A2}^s}{C_{A2}^s} \end{bmatrix}, \quad x_3 = \begin{bmatrix} \dfrac{T_3 - T_3^s}{T_3^s} \\ \dfrac{C_{A3} - C_{A3}^s}{C_{A3}^s} \end{bmatrix}$$

$$u_1 = \begin{bmatrix} Q_1 \\ C_{A1}^o - C_{A1}^{os} \end{bmatrix}, \quad u_2 = \begin{bmatrix} Q_2 \\ C_{A2}^o - C_{A2}^{os} \end{bmatrix}, \quad u_3 = \begin{bmatrix} Q_3 \\ C_{A3}^o - C_{A3}^{os} \end{bmatrix} \tag{4.39}$$

$$A_1 = \begin{bmatrix} 25.2075 & 4.9639 \\ -78 & -45.9295 \end{bmatrix} \quad B_1 = \begin{bmatrix} 9.454 \times 10^{-6} & 0 \\ 0 & 2.8237 \end{bmatrix}$$

$$A_{13} = \begin{bmatrix} 31.7573 & 0 \\ 0 & 34.6025 \end{bmatrix}$$

$$A_2 = \begin{bmatrix} -2.8393 & 1.4157 \\ -22.4845 & -24.8849 \end{bmatrix}, \quad B_2 = \begin{bmatrix} 3.47 \times 10^{-6} & 0 \\ 0 & 5.7143 \end{bmatrix} \tag{4.40}$$

$$A_{21} = \begin{bmatrix} 14.6925 & 0 \\ 0 & 13.4844 \end{bmatrix}$$

$$A_3 = \begin{bmatrix} -2.8393 & 1.4157 \\ -22.4845 & -24.8849 \end{bmatrix}, \quad B_3 = \begin{bmatrix} 3.47 \times 10^{-6} & 0 \\ 0 & 5.7143 \end{bmatrix} \tag{4.41}$$

$$A_{32} = \begin{bmatrix} 13.332 & 0 \\ 0 & 13.332 \end{bmatrix}$$

where $A_{ij}, i \neq j$ are the interconnection terms that define the coupling among the tanks. Note that A_{12}, A_{23}, and A_{31} are zero.

4.3.1 Controller and observer design

Referring to (4.31), two controllers are designed for each subsystem such that one of them is responsible for the control of subsystem itself and the second one is to suppress the interconnection with its neighboring unit. These controllers are designed using pole-placement method such that the eigen values for the closed-loop $A_j + B_jK_j$, where $j = 1, 2, 3$, are -5 and -1, and for the interconnection terms, that is, $A_{ij} + B_iK_{ij}$, where j is the dependent interconnected system, are zero. Similarly, Luenberger observer is designed for all three tanks because only temperatures are available for measurement. The observers L_1, L_2, and L_3 are designed to place the eigen values of $A_j - L_jC_j$ at -50 and -25 [274].

4.3.2 State space form of the networked closed-loop system

Now the whole system is formed as described in (4.33) and the augmented state matrices are obtained. Then these matrices are combined to form the matrix Λ_o as in (4.34).

The matrices for the estimated model, $\hat{A}$ and $\hat{B}$, are obtained by introducing 10% parametric uncertainty in the temperature of the reaction. The key matrix that defines scheduling is given in (4.37). The dimensions of the individual identity and zero matrices determine the subsystem that transmits at a particular time. As an example, the dimensions of these matrices are given for three-tank plant when sensor suit for first tank transmits its states.

$$I_o^1 = \begin{bmatrix} I^{(12)} & O^{(6)} & O^{(6)} & O^{(6)} \\ O^{(6\times12)} & O^{(6)} & O^{(6)} & O^{(6)} \\ O^{(6\times12)} & O^{(6)} & I^{(6)} & O^{(6)} \\ O^{(6\times12)} & O^{(6)} & O^{(6)} & I^{(6)} \end{bmatrix} \tag{4.42}$$

4.3.3 Simulation results

In what follows, the effects of transmission times and different schedules on the closed-loop stability of the networked system are studied. Six possible schedules are shown in Table 4.1. Since closed-loop stability requires all eigen values of the

Table 4.1 Six schedules

Schedule	$s_1, s_2, s_3, s_1, s_2, s_3, \ldots$
1	$1, 2, 3, 1, 2, 3, \ldots$
2	$1, 3, 2, 1, 3, 2, \ldots$
3	$2, 1, 3, 2, 1, 3, \ldots$
4	$2, 3, 1, 2, 3, 1, \ldots$
5	$3, 1, 2, 3, 1, 2, \ldots$
6	$3, 2, 1, 3, 2, 1, \ldots$

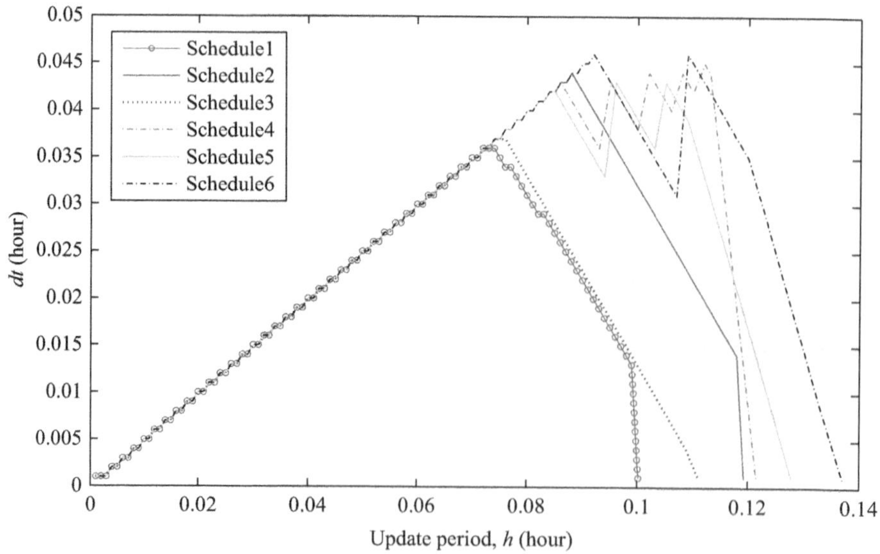

Figure 4.15 Dependence of $\lambda_{\max}(M_o)$ on Δt and h for different sensor transmission sequences

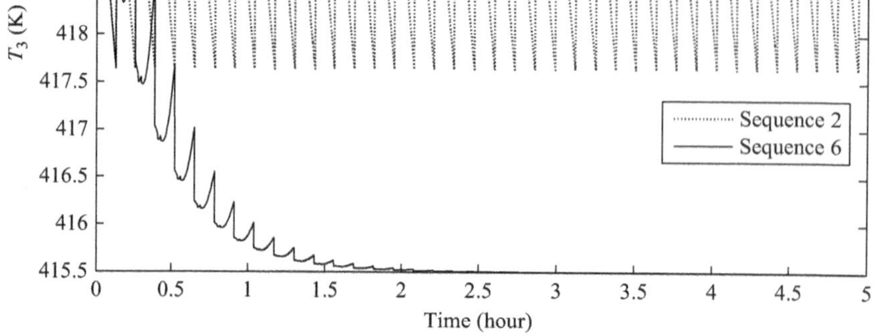

Figure 4.16 Closed-loop temperature profile for CSTR3 using two different sensor transmission schedules with same update period and transmission interval

test matrix M_o to lie inside unit disc, only the maximum eigen value is considered for stability analysis for simplicity, and is denoted by $\lambda_{\max}(M_o)$. Two cases are considered, first when the transmission intervals are equal, that is, $\Delta t_1 = \Delta t_2$ and second when they are unequal, that is, $\Delta t_1 \neq \Delta t_2$ (Figure 4.13).

$\Delta t_1 = \Delta t_2$

Figure 4.15 is a contour plot showing the dependence of $\lambda_{\max}(M_o)$ on transmission interval, Δt_1 and update period, h. The contour line shows the unit line and the area inside this contour corresponds to stable region, that is, when $\lambda_{\max}(M_o) < 1$. It can be seen that the largest update period, $h \approx 0.13$ h, is allowed under schedule 6. Comparing this with schedule 1, we can only get update period of 0.1 h maximum. That is, schedule 6 allows highest savings in WSN battery power while ensuring closed-loop stability. To further demonstrate this result, the temperature profile for third tank, T_3 is plotted for schedule 2 and 6 for $h = 0.13$ h and $\Delta t = 0.02$ h in Figure 4.16. From the contour plots, sequence 2 gives unstable closed loop under suggested transmission interval and update periods, hence we get unstable state profile. The same is not true for sequence 6.

$\Delta t_1 \neq \Delta t_2$

Next, the case is studied when transmission intervals are not equal. Figure 4.17 shows the contour plots for different values of Δt_2 under schedule 2. It can be seen that when $\Delta t_2 = 0.03$ h, the maximum allowable update period, h is the largest of all other cases. Furthermore, state profile is plotted in Figure 4.18 for the case when for sequence 2, $\Delta t_1 = 0.05$ h, $\Delta t_2 = 0.03$ h, and $h = 0.08$ h.

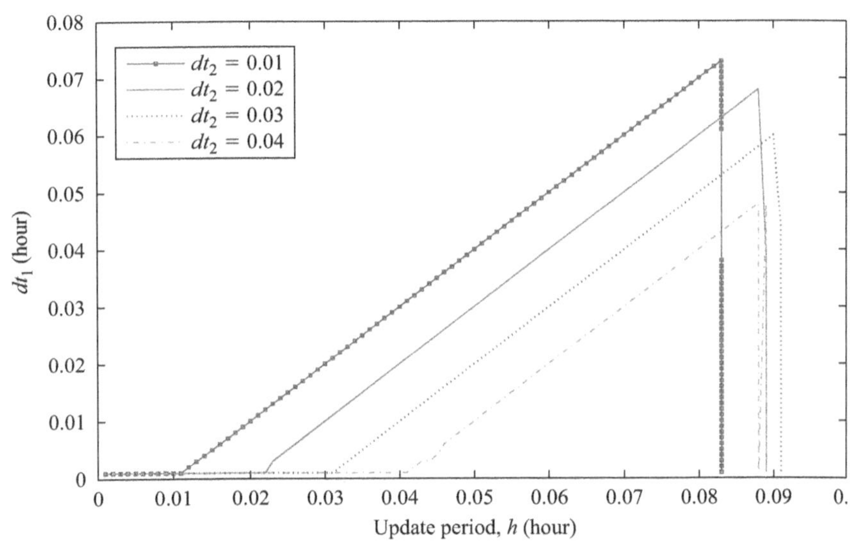

Figure 4.17 Dependence of $\lambda_{\max}(M_o)$ on Δt_1 and h for different values of Δt_2 under schedule 2 with a fixed model

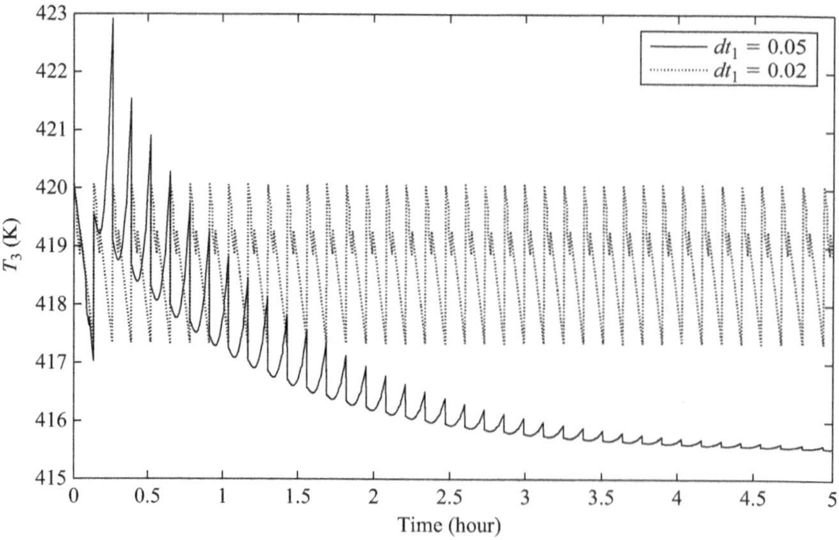

Figure 4.18 Closed-loop temperature profile for CSTR3 when $\Delta t_2 = 0.03$ h and $h = 0.08$ h for two different values of Δt_1

Appendix A: System matrices and data

$$\Lambda_{11} = \begin{bmatrix} A_1 & 0 & A_{13} \\ A_{21} & A_2 & 0 \\ 0 & A_{32} & A_3 \end{bmatrix},$$

$$\Lambda_{12} = \begin{bmatrix} B_1K_1 & 0 & B_1K_{13} \\ B_2K_{21} & B_2K_2 & 0 \\ 0 & B_3K_{32} & B_3K_3 \end{bmatrix},$$

$$\Lambda_{13} = \begin{bmatrix} 0 & 0 & 0\ 0\ 0 & 0 & -B_1K_{12} & 0\ 0 \\ 0 & -B_2K_{21} & 0\ 0\ 0 & 0 & 0 & 0\ 0 \\ 0 & 0 & 0\ 0\ 0 & -B_3K_{32} & 0 & 0\ 0 \end{bmatrix},$$

$$\Lambda_{21} = \begin{bmatrix} L_1C_1 & 0 & 0 \\ 0 & L_2C_2 & 0 \\ 0 & 0 & L_3C_3 \end{bmatrix},$$

$$\Lambda_{22} = \begin{bmatrix} A_1 - L_1C_1 + B_1K_1 & 0 & A_{13}B_1K_{13} \\ A_{21}B_2K_{21} & A_2 - L_2C_2 + B_2K_2 & 0 \\ 0 & A_{32}B_1K_{32} & A_3 - L_3C_3 + B_3K_3 \end{bmatrix},$$

$$\Lambda_{23} = \begin{bmatrix} 0 & 0 & 0\ 0\ 0 & 0 & -A_{13} - B_1K_{13} & 0\ 0 \\ 0 & -A_{21} - B_2K_{21} & 0\ 0\ 0 & 0 & 0 & 0\ 0 \\ 0 & 0 & 0\ 0\ 0 & -A_{32} - B_3K_{32} & 0 & 0\ 0 \end{bmatrix},$$

$$\Lambda_{31} = \begin{bmatrix} 0 & L_1C_1 & L_1C_1 & 0 & 0 & 0 & 0 & 0 & 0 \\ 0 & 0 & 0 & L_2C_2 & 0 & L_2C_2 & 0 & 0 & 0 \\ 0 & 0 & 0 & 0 & 0 & 0 & L_3C_3 & L_3C_3 & 0 \end{bmatrix}^T,$$

$$\Lambda_{32} = \begin{bmatrix} 0 & 0 & 0 \\ A_1 - L_1C_1 + B_1K_1 - \hat{A}_1 - \hat{B}_1K_1 & 0 & A_{13} + B_1K_{13} - \hat{A}_{13} - \hat{B}_1K_{13} \\ A_1 - L_1C_1 + B_1K_1 - \hat{A}_1 - \hat{B}_1K_1 & 0 & A_{13} + B_1K_{13} - \hat{A}_{13} - \hat{B}_1K_{13} \\ A_{21} + B_2K_{21} - \hat{A}_{21} - \hat{B}_2K_{21} & A_2 - L_2C_2 + B_2K_2 - \hat{A}_2 - \hat{B}_2K_2 & 0 \\ 0 & 0 & 0 \\ A_{21} + B_2K_{21} - \hat{A}_{21} - \hat{B}_2K_{21} & A_2 - L_2C_2 + B_2K_2 - \hat{A}_2 - \hat{B}_2K_2 & 0 \\ 0 & A_{32} + B_3K_{32} - \hat{A}_{32} - \hat{B}_3K_{32} & A_3 - L_3C_3 + B_3K_3 - \hat{A}_3 - \hat{B}_3K_3 \\ 0 & A_{32} + B_3K_{32} - \hat{A}_{32} - \hat{B}_3K_{32} & A_3 - L_3C_3 + B_3K_3 - \hat{A}_3 - \hat{B}_3K_3 \\ 0 & 0 & 0 \end{bmatrix},$$

$$\Lambda_{33} = \begin{bmatrix} 0 & 0 & 0 & 0 & 0 & 0 & 0 & 0 & 0 \\ 0 & \hat{A}_1 + \hat{B}_1K_1 & 0 & 0 & 0 & 0 & -A_{13} - B_1K_{13} & -\hat{A}_{13} - \hat{B}_1K_{13} & 0 \\ 0 & 0 & -\hat{A}_1 - \hat{B}_1K_1 & 0 & 0 & 0 & -A_{13} - B_1K_{13} & 0 & 0 \\ 0 & -A_{21} - B_2K_{21} & 0 & \hat{A}_2 + \hat{B}_2K_2 & 0 & 0 & 0 & 0 & 0 \\ 0 & 0 & 0 & 0 & 0 & 0 & 0 & 0 & 0 \\ 0 & -A_{21} - B_2K_{21} & \hat{A}_2 + \hat{B}_2K_{21} & 0 & 0 & \hat{A}_2 + \hat{B}_2K_2 0 & 0 & 0 & \\ 0 & 0 & 0 & \hat{A}_{32} + \hat{B}_3K_{32} & 0 & -A_{32} - B_3K_{32} & \hat{A}_3 + \hat{B}_3K_3 & 0 & 0 \\ 0 & 0 & 0 & 0 & 0 & -\hat{A}_{32} - \hat{B}_3K_{32} & 0 & \hat{A}_3 + \hat{B}_3K_3 & \\ 0 & 0 & 0 & 0 & 0 & 0 & 0 & 0 & 0 \end{bmatrix}$$

Table A.1 Parameter values and description for three CSTRs.

Parameter	Description	Value
ρ	Density of fluid in reactor	1000 kg/m^3
c_p	Heat capacity	0.231 kJ/kg K
E_1	Activation energy of first reaction	5×10^4 kJ/kmol
E_2	Activation energy of second reaction	7.53×10^4 kJ/kmol
E_3	Activation energy of third reaction	7.53×10^4 kJ/kmol
k_{10}	Pre-exponential constant of first reaction	3×10^6 h^{-1}
k_{20}	Pre-exponential constant of second reaction	3×10^5 h^{-1}
k_{30}	Pre-exponential constant of third reaction	3×10^5 h^{-1}
ΔH_1	Enthalpy of first reaction	-5×10^4 kJ/kmol
ΔH_2	Enthalpy of second reaction	-5.2×10^4 kJ/kmol
ΔH_3	Enthalpy of third reaction	-5.4×10^4 kJ/kmol
R	Constant	8.314 kJ/kmol K
V_1	Volume of first tank	1 m^3
V_2	Volume of second tank	3 m^3
V_3	Volume of third tank	3 m^3
F_1^o	Flow rate of fresh feed stream in tank 1	4.998 m^3/h
F_2^o	Flow rate of fresh feed stream in tank 2	39.996 m^3/h
F_3^o	Flow rate of fresh feed stream in tank 3	39.996 m^3/h
C_{A1}^{os}	Steady-state value of reactant concentration fresh feed stream in tank 1	4 kmol/m^3

(*Continues*)

Table A.1 (Continued)

Parameter	Description	Value
C_{A2}^{os}	Steady-state value of reactant concentration fresh feed stream in tank 2	2 kmol/m^3
C_{A3}^{os}	Steady-state value of reactant concentration fresh feed stream in tank 3	2 kmol/m^3
T_1^o	Temperature of fresh feed stream in tank 1	300 K
T_2^o	Temperature of fresh feed stream in tank 2	300 K
T_3^o	Temperature of fresh feed stream in tank 3	300 K
F_1	Flow rate of feed stream from tank 1 to tank 2	39.996 m^3/h
F_2	Flow rate of feed stream from tank 2 to tank 3	39.996 m^3/h
$F_r = F_o$	Flow rate of recycle stream	34.998 m^3/h
C_{A1}^s	Steady-state value of reactant concentration feed stream from tank 1 to tank 2	1.77 kmol/m^3
C_{A2}^s	Steady-state value of reactant concentration feed stream from tank 2 to tank 3	1.75 kmol/m^3
C_{A3}^s	Steady-state value of reactant concentration of recycle stream	1.75 kmol/m^3
T_1^s	Steady-state value of temperature of feed stream from tank 1 to tank 2	457.9 K
T_2^s	Steady-state value of temperature of feed stream from tank 2 to tank 3	415.5 K
T_3^s	Steady-state value of temperature of recycle stream	415.5 K

4.4 Notes

In this chapter, architectures for the decentralized and distributed networked control for interconnected dynamical systems have been carefully examined. The work discussed the underlying rationale for the individual architectures and illustrated the fields of application and the merits/demerits as reported in the literature. A single discretized system without any extra parameters is initiated and then added to it the inter-communication between systems in the decentralized/distributed design. The networked induced delays are then considered. Packet dropout and varying sample interval were also included. A criterion to select the minimum sample interval was set and we defined a switched function for the transmission constraints. More work is expected to enhance the computational efficiency of the developed approach, see [515–517] for further discussions.

Chapter 5

Distributed consensus control

This chapter is concerned with the consensus of a network of agents with general linear or linearized dynamics, whose communication topology contains a directed spanning tree. An observer-type consensus protocol based on the relative outputs of the neighboring agents is adopted. The notion of consensus region is introduced, as a measure of the robustness of the protocol and as a basis for the protocol design. For neutrally stable agents, it is shown that there exists a protocol achieving consensus together with a consensus region that is the entire open right-half plane if and only if each agent is stabilizable and detectable. An algorithm is further presented for constructing such a protocol. For consensus with a prescribed convergence speed, a multistep protocol design procedure is given, which yields an unbounded consensus region and at the same time maintains a favorable decoupling property. Finally, the consensus algorithms are extended to solve the formation control problems.

5.1 Consensus of multiagent systems

In recent years, the consensus problem of multiagent systems has received increasing attention from scientific communities, for its broad applications in such areas as satellite formation flying, cooperative unmanned air vehicles, scheduling of automated highway systems, and air traffic control.

A large volume of research work on consensus has emerged in the past few years. In [275], a simple model is proposed for phase transition of a group of self-driven particles with numerical analysis on its complexity. In [276], a theoretical explanation is provided for the behavior observed in [275] based on graph theory. In [247], a general framework of consensus problem for networks of dynamic agents with fixed or switching topologies is established. The conditions given by Olfati-Saber and Murray [247] are further relaxed in [277]. In [278], tracking control for multiagent consensus with an active leader is studied and a local controller is designed together with a neighbor-based state-estimation rule. In [248], the consensus problem is addressed for directed networks of agents with external disturbances and model uncertainties. The consensus problem of networks with double-integrator agents is studied in [279] and [280]. In [281], a distributed algorithm is developed, which asymptotically achieves consensus in finite time. Average consensus of multiagent systems under communication constraints and control of networked systems with limited channels are studied in [282–285]. The controlled agreement problem for multiagent networks is considered from a graph-theoretic perspective in [286]. In most existing works on consensus, the agent dynamics are restricted to be first-, second-, and, sometimes, high-order integrators, and the proposed consensus protocols are based on information of relative states among neighboring agents, which in many cases are not available.

The consensus of networks of identical agents with linear or linearized dynamics whose communication topology contains a directed spanning tree is of particular interest, see [287–290]. A unified framework was introduced in [287] to address the consensus of multiagent systems and the synchronization of complex networks studied in the same line as [291–293]. In [288–290], static consensus protocols based on relative states of the neighboring agents were used. The dynamic consensus protocol in [294] requires the absolute output measurement of each agent to be available, which is impractical in many cases, for example, in deep-space formation flying [295]. In contrast, an observer-type protocol proposed in [287] using only the relative outputs of the neighboring agents is adopted here. A decomposition approach is utilized to cast the consensus problem of the multiagent system into the stability of a set of matrices that have the same dimension as that of a single agent.

Suppose that there are m nodes in the graph. The adjacency matrix $A \in \Re^{m\times m}$ is defined by $a_{ii} = 0$, and $a_{ij} > 0$ if $(j,i) \in E$ but 0 otherwise. The Laplacian matrix $L \in \Re^{m\times m}$ is defined as $L_{ii} = \sum_{j\neq i} a_{ij}, L_{ij} = -a_{ij}$ for $i \neq j$. It follows immediately that 0 is an eigenvalues of L with 1 as the corresponding right eigenvector and all the non-zero eigenvalues have positive real parts. For a directed graph, 0 is a simple eigenvalue of L if and only if the graph has a directed spanning tree [277,514].

5.1.1 Dynamic consensus with observer-type protocol

Consider a network of N identical agents with linear or linearized dynamics, where the dynamics of the ith agent are described by

$$\begin{aligned} \dot{x}_i &= Ax_i + Bu_i \\ y_i &= Cx_i \end{aligned} \tag{5.1}$$

where $x_i \in \Re^n$ is the state, $u_i \in \Re^p$ is the control input, $y_i \in \Re^q$ is the measured output and A, B, and C are constant matrices with compatible dimensions.

The communication topology among agents is represented by a directed graph $G = (V,E)$, where $V = \{1,\ldots,N\}$ is the set of nodes (i.e., agents), and $E \subset V \times V$ is the set of edges. An edge (i,j) in graph G means that agent j can obtain information from agent i, but not conversely.

At each time instant, the information available to agent i is the relative measurements of other agents with respect to i itself, given by

$$\zeta_i = \sum_{j=1}^{N} a_{ij}(y_i - y_j) \tag{5.2}$$

where $\{a_{ij}\}$ is the $N \times N$ adjacency matrix of graph G. The consensus protocol takes the following observer-type form:

$$\dot{v}_i = (A + BK)v_i + cL\left(\sum_{j=1}^{N} a_{ij}C(v_i - v_j) - \zeta_i\right)$$
$$u_i = Kv_i \tag{5.3}$$

where $v_i \in \Re^n$ is the protocol state, $i = 1, 2, \ldots, N$, $c > 0$ denotes the coupling strength, and $L \in \Re^{q\times n}$ and $K \in \Re^{p\times n}$ are feedback gain matrices to be determined. In (5.3), the term $\sum_{j=1}^{N} a_{ij}C(v_i - v_j)$ denotes the information exchanges between the protocol of agent i and those of its neighboring agents. It is observed that the protocol in (5.3) maintains the same communication topology of the agents in (5.1).

Let $z_i = [x_i^T, v_i^T]^T$. Then, systems (5.1) and (5.3) together can be written as

$$\dot{z}_i = \mathcal{A}z_i + c\sum_{j=1}^{N} L_{ij}\mathcal{H}z_j, \quad i = 1, 2, \ldots, N \tag{5.4}$$

where $L = (L_{ij})N \times N$ is the Laplacian matrix of graph G, and

$$\mathcal{A} = \begin{bmatrix} A & BK \\ 0 & A + BK \end{bmatrix}, \quad \mathcal{H} = \begin{bmatrix} 0 & 0 \\ -LC & LC \end{bmatrix}$$

5.1.2 Dynamic consensus

We start by introducing the concept of dynamic consensus.

Definition 5.1: Given agents (5.1), the protocol (5.3) is said to solve the dynamic consensus problem if the states of system (5.4) satisfy

$$\lim_{t\to\infty} \|z_i(t) - z_j(t)\| = 0, \quad \forall i, j = 1, 2, \ldots, N \tag{5.5}$$

Let $r \in \Re^N$ be such that $r^T L = 0$ and $r^T 1 = 1$.

Remark 5.1: *Introduce a new variable*

$$\begin{aligned} \delta(t) &= z(t) - ((1r^T) \otimes I_{2n})z(t) \\ &= ((I_N - 1r^T) - I_{2n})z(t) \end{aligned} \tag{5.6}$$

where $z = [z_1^T, \ldots, z_N^T]^T$ *and* $\delta \in \Re^{2Nn\times 2Nn}$ *satisfies*

$$(r^T \otimes I_{2n})\delta = 0$$

By similarity to [247], δ *is referred to as the disagreement vector. It is easy to see that 0 is a simple eigenvalue of* $I_N - 1r^T$ *with **1** as the right eigenvector, and 1 is another eigenvalue with multiplicity* $N - 1$.

It follows from (5.6) that $\delta = 0$ if and only if $z_1 = \cdots = z_N$, that is, the dynamic consensus problem can be recast into the asymptotic stability of vector δ, which evolves according to the following dynamics:

$$\dot{\delta} = (I_N \otimes A + cL \otimes H)\delta \tag{5.7}$$

In the sequel, we provide a decomposition approach to the dynamic consensus problem, based on [287].

Theorem 5.1: *For the communication topology G containing a directed spanning tree, protocol (5.3) solves the dynamic consensus problem if and only if all the matrices*

$$A + BK, \quad A + c\lambda_i LC, \quad i = 2, 3, \ldots, N$$

are Hurwitz, where λ_i, $i = 2, 3, \ldots, N$, are the non-zero eigenvalues of the Laplacian matrix L.

Proof. For a complete proof, the reader is referred to [287]. □

In the sequel, we let $Y_1 \in \Re^{N\times(N-1)}$, $Y_2 \in \Re^{(N-1)\times N}$, $T \in \Re^{N\times N}$, and an upper-triangular matrix $\Delta \in \Re^{(N-1)\times(N-1)}$ be such that

$$\begin{aligned} T &= [1 \quad Y_1], \quad T_{-1} = \begin{bmatrix} r^T \\ Y_2 \end{bmatrix} \\ T^{-1}LT &= \Lambda = \begin{bmatrix} 0 & 0 \\ 0 & \Delta \end{bmatrix} \end{aligned} \tag{5.8}$$

where the diagonal entries of Δ are the non-zero eigenvalues of L. Then, (5.7) can be rewritten in terms of ξ, where $\xi = (T^{-1} \otimes I_{2n})\delta$ with $\xi = [\xi_1^T, \ldots, \xi_N^T]^T$, as follows:

$$\dot{\xi} = (I_N \otimes A + c\Lambda \otimes H)\xi \tag{5.9}$$

As to ξ_1, it can be seen from (5.6) that

$$\xi_1 = (r^T \otimes I_{2n})\delta = 0 \tag{5.10}$$

Note that the elements of the state matrix of (5.9) are either block diagonal or block upper triangular. Hence, $\xi_i, i = 2, \ldots, N$, converge asymptotically to zero, if and only if the $N - 1$ subsystems

$$\dot{\xi}_i = (A + c\lambda_i H)\xi_i, \quad i = 2, 3, \ldots, N \tag{5.11}$$

are asymptotically stable, which leads to the assertion by noting that matrices $A + \lambda_i H$ are similar to $\left[\begin{smallmatrix} A+\lambda_i LC & 0 \\ -\lambda_i LC & A+BK \end{smallmatrix}\right]$, $i = 2, \ldots, N$.

Remark 5.2: *The significance of Theorem 5.1 lies in the fact that it converts the consensus of the multiagent system under the dynamic protocol (5.3) to the stability of a set of matrices with the same low dimensions as a single agent. The effects of the communication topology on the consensus are characterized by the eigenvalues* *of the corresponding Laplacian matrix L. Moreover, protocol (5.3) is based only on relative output measurements between neighboring systems, in contrast to [288–290] where the relative-state coupling laws are required.*

Lemma 5.1 [287]: *Consider network (5.4) whose communication topology G has a directed spanning tree. If protocol (5.3) satisfies Theorem 5.1, then*

$$\begin{aligned} x_i(t) &\to \varpi(t) \triangleq (r^T \otimes e^{At}) \begin{bmatrix} x_1(0) \\ \vdots \\ x_N(0) \end{bmatrix} \\ v_i(t) &\to i = 1, 2, \ldots, N, \ as\ t \to \infty \end{aligned} \tag{5.12}$$

where $r \in \Re^N$ is such that $r^T L = 0$ and $r^T 1 = 1$.

From (5.12), it follows that agents (5.1) are excluded from having poles in the open right-half plane; otherwise, the consensus value reached by the states of (5.1) will tend to infinity exponentially. On the other hand, if matrix A is Hurwitz, then the agents will reach consensus onto 0. Therefore, it is critical for matrix A in (5.1) to have eigenvalues along the imaginary axis, so that the systems can reach consensus on a non-zero value, a special case of which is that matrix A is neutrally stable.

5.1.3 Consensus region

Given a protocol of the form (5.3), the dynamic consensus problem can be cast into analyzing the system

$$\dot{\varsigma} = (A + \sigma H)\varsigma \tag{5.13}$$

where $\varsigma \in \Re^{2n}, \sigma \in C$. The stability of system (5.13) depends on the parameter σ, based on which the notion of consensus region is introduced following [287].

Definition 5.2: The region S of the complex parameter σ, such that (5.13) is asymptotically stable, is called the consensus region of network (5.4).

It follows from Theorem 5.1 that consensus is reached if and only if

$$c(\alpha_k + i\beta_k) \in S, \quad k = 2, 3, \ldots, N$$

where $i = \sqrt{-1}, \alpha_k = \mathrm{Re}(\lambda_k)$ and $\beta_k = Im(\lambda_k)$. For an undirected communication graph, its consensus region S is an interval or a union of several intervals on the real axis. However, for a directed graph, where the eigenvalues of L are generally complex numbers, its consensus region S is a region or a union of several regions on the complex plane, which can be bounded or unbounded. For a detailed discussion on consensus regions, the reader is referred to [287].

Remark 5.3: *It should be noted that the consensus region serves in certain sense as a measure for the robustness of the protocol (5.3) to parametric uncertainties of its feedback gain matrix L and the communication topology. Given a consensus protocol,* *the consensus region should be large enough for the protocol to maintain a desirable robustness margin.*

5.1.4 Consensus with neutrally stable matrix

In this subsection, for the case when matrix A is neutrally stable, it is shown that an unbounded consensus region in the form of the open right-half plane can be achieved. A constructive design algorithm for protocol (5.3) is then presented [296].

Lemma 5.2: *A complex matrix $A \in C^{n\times n}$ is Hurwitz if and only if there exists a positive-definite matrix $Q = Q^H$ and a matrix $C \in C^{m\times n}$ such that (A, C) is observable and $A^H Q + QA = -C^H C$.*

Lemma 5.3: *For matrices $S \in \Re^{n\times n}, H \in \Re^{m\times n}$, where S is skew-symmetric and (S, H) is observable, the matrix $S - (x + iy)H^T H$ is Hurwitz for any $x > 0, y \in R$.*

Proof. Let $\tilde{S} = S - (x + iy)H^T H$. Then

$$\begin{aligned}\tilde{S} + \tilde{S}^H &= S - (x + iy)H^T H + S^T - (x - iy)H^T H \\ &= -2xH^T H \leq 0, \quad \forall x > 0\end{aligned} \tag{5.14}$$

Obviously, $(\tilde{S}, H)$ is observable, for (S, H) is observable. By Lemma 5.2, (5.14) directly leads to the assertion. □

Next, a constructive algorithm for protocol (5.3) is presented, which will be used later.

Algorithm 5.1: *Given that $A \in \Re^{n\times n}$ is neutrally stable and that the pair (A, B, C) is stabilizable and detectable, the dynamic protocol (5.3) can be constructed as follows:*

- *Let K be such that $A + BK$ is Hurwitz.*
- *Choose $U \in \Re^{n\times n_1}$ and $W \in \Re^{n\times(n-n_1)}$ such that*

$$[U \quad W]^{-1} A [U \quad W] = \begin{bmatrix} S & 0 \\ 0 & X \end{bmatrix} \tag{5.15}$$

where $S \in \Re^{n_1\times n_1}$ is skew-symmetric and $X \in R^{(n-n_1)\times(n-n_1)}$ is Hurwitz.
- *Let $L = -UU^T C^T$.*
- *Select the coupling strength $c > 0$.*

In the above algorithm, note that matrices U and W can be derived by rendering matrix A into the real Jordan canonical form [297].

Theorem 5.2: *Given that $A \in \Re^{n\times n}$ is neutrally stable and that G has a directed spanning tree, there exists a distributed protocol in the form of (5.3) that solves the dynamic consensus problem and meanwhile yields an unbounded consensus region $(0, \infty) \times (-\infty, \infty)$, if and only if (A, B, C) is stabilizable and detectable.*

Proof. $\Rightarrow$ Follows Theorem 5.1.

$\Leftarrow$ Let the related variables be defined as in Algorithm 5.1. Construct the protocol (5.3) by Algorithm 5.1, and let $H = CU$. Then, (S, H) observed for (A, C) is detectable. Let $U^\dagger \in \Re^{n_1 \times n}$ and $W^\dagger \in \Re^{(n-n_1)\times n}$ be such that $\begin{bmatrix} U^\dagger \\ W^\dagger \end{bmatrix} = [U \quad W]^{-1}$, where $U^\dagger U = I$, $W^\dagger W = I$, $U^\dagger W = 0$, and $W^\dagger U = 0$. It can be verified by some algebraic manipulations that

$$
\begin{aligned}
&[U \quad W]^{-1}(A + (x+iy)LC)[U \quad W] \\
&\quad = \begin{bmatrix} S + (x+iy)U^\dagger LCU & (x+iy)U^\dagger LCW \\ (x+iy)W^\dagger LCU & X + (x+iy)W^\dagger LCW \end{bmatrix} \\
&\quad = \begin{bmatrix} S - (x+iy)H^T H & -(x+iy)H^T SW \\ 0 & X \end{bmatrix}
\end{aligned} \tag{5.16}
$$

which implies that matrix $A + (x+iy)LC$ is Hurwitz for all $x > 0$ and $y \in R$, because by Lemma 5.1 matrix $S - (x+iy)H^T H$ is Hurwitz for any $x > 0$ and $y \in R$. Hence, by Theorem 5.1, the protocol given by Algorithm 5.1 solves the dynamic consensus problem with an unbounded consensus region $(0, \infty) \times (-\infty, \infty)$. □

Remark 5.4: *The consensus region $(0, \infty) \times (-\infty, \infty)$ achieved by the protocol constructed by Algorithm 5.1 means that such a protocol can reach consensus for any communication topology containing a directed spanning tree and for any positive coupling strength. However, the communication topology and the coupling strength do affect the performances of consensus, for example, the convergence speed, which is the topic of the next subsection. It should be pointed out that similar results were reported in [289]. Nevertheless, a general case of dynamic consensus protocol based on relative measurements of the neighboring agents is investigated here, containing the static protocol study of [289] as a special case. Moreover, the method leading to Theorem 5.2 here is quite different from that in [289], and indeed is comparatively much simpler.*

5.1.5 Consensus with prescribed convergence speed

For the general case where matrix A has no eigenvalues in the open right-half plane, the protocol (5.3) is designed in this subsection to achieve consensus with a prescribed convergence speed. The results given in [287] can be regarded as a special case without specification on the convergence speed. Previous works along this line include [247,298,299], where the convergence speed of consensus for networks of integrators was analyzed.

From the proof of Theorem 5.1, it is easy to see that the convergence speed of N agents in (5.1) reaching consensus under protocol (5.3) is equal to the minimal decay rate of the $N-1$ systems in (5.11). The decay rate of system $\dot{x} = Ax$ is defined as the maximum of negative real parts of the eigenvalues of matrix A [300].

Thus, the convergence speed of agents (5.1) reaching consensus can be manipulated by properly assigning the eigenvalues of matrices

$$A + BK, \quad A + c\lambda_i LC, \quad i = 2, 3, \ldots, N$$

We know from [300] that the decay rate of system $\dot{x} = Ax$ is larger than $\alpha > 0$, if and only if there exists a matrix $Q > 0$ such that

$$A^T Q + QA + 2\alpha Q < 0$$

Hence, we provide the following result.

Proposition 5.1: *Given the agents (5.1), there exists a matrix L such that $A + (x + iy)LC$ is Hurwitz with a decay rate larger than α for all $x \in [1, \infty)$, $y \in (-\infty, \infty)$, if and only if there exists a matrix $Q = Q^T > 0$ such that*

$$A^T Q + QA - 2C^T C + 2\alpha Q < 0 \tag{5.17}$$

Proof. $\Leftarrow$ By Lemma 5.3, there exists a matrix L such that $A + LC$ is Hurwitz with a decay rate larger than a if and only if there exists a matrix $Q = Q^T > 0$ such that

$$(A + LC)^T Q + Q(A + LC) + 2\alpha Q < 0$$

Let $QL = V$. Then, the above inequality becomes

$$A^T Q + QA + VC + C^T V^T + 2\alpha Q < 0$$

By Finsler's Lemma [301], there exists a matrix V satisfying the above inequality if and only if there exists a scalar $\tau > 0$ such that

$$A^T Q + QA - \tau C^T C + 2\alpha Q < 0 \tag{5.18}$$

Without loss of generality, letting $\tau = 2$ in (5.18) leads to (5.17). Take $V = -C^T$, that is, $L = -Q^{-1}C^T$. By the above inequalities, one has

$$\begin{aligned}
&(A + (x + iy)LC)^H Q + Q(A + (x + iy)LC) + 2\alpha Q \\
&\quad = (A + (x - iy)LC)^T Q + Q(A + (x + iy)LC) + 2\alpha Q \\
&\quad = AQ + QA^T - 2xC^T C + 2\alpha Q < 0
\end{aligned}$$

for all $x \geq 1$, that is, $A + (x + iy)LC$ is Hurwitz with a decay rate larger than a for all $x \in [1, \infty), y \in (-\infty, \infty)$.

$\Rightarrow$ Follows by letting $x = 1, y = 0$. □

Combining Proposition 5.1 and Theorem 5.1 led to the following result.

Theorem 5.3: *For network (5.4) with G containing a directed spanning tree, there exists a protocol (5.3) that solves the consensus problem with a convergence rate larger than a and yields an unbounded consensus region $[1, \infty) \times (-\infty, \infty)$, if and* *only if there exist matrices K and L such that both $A + BK$ and $A + LC$ are Hurwitz with a decay rate larger than α.*

Algorithm 5.2: *For graph G containing a directed spanning tree, a protocol (5.3) solving the dynamic consensus problem with a convergence speed larger than α can be constructed as follows:*

- *Obtain the feedback gain matrix K, for example, by using the Ackermann's formula, such that the poles of matrix $A + BK$ lie in the left-half plane of $x = -\alpha$.*
- *Choose the feedback gain matrix $L = -Q^{-1}C^T$, where $Q > 0$ is one solution to (5.17).*
- *Select the coupling strength c larger than the threshold value c_{th} given by*

$$c_{th} = \frac{1}{\min_{i=2,\ldots,N} \mathrm{Re}(\lambda_i)} \tag{5.19}$$

where λ_i, $i = 2, \ldots, N$, are the non-zero eigenvalues of matrix L.

The following remarks stand out:

Remark 5.5: *Algorithm 5.2 has a favorable decoupling feature. To be specific, steps 1 and 2 deal with only the agent dynamics and the feedback gain matrices of the consensus protocol, leaving the communication topology of the multiagent network to be handled in step 3 by manipulating the coupling strength. The protocol designed by Algorithm 5.2 for one communication graph is applicable to any other graph with larger minimum real parts of eigenvalues, thereby, is robust in this sense to the communication topology. For the case where the agent number N is large, for which the eigenvalues of the corresponding Laplacian matrix are hard to determine or even troublesome to estimate, one only needs to choose the coupling strength to be large enough.*

Remark 5.6: *Compared to the consensus when A is neutrally stable, where the coupling strength can be chosen as any positive scalar, for the case where A is critically unstable or a prescribed convergence speed is desired, the coupling strength generally has to be larger than a threshold value, which is related to the specific communication topology. This is consistent with the intuition that unstable behaviors are more difficult to synchronize than stable behaviors.*

Remark 5.7: *One sufficient condition satisfying Theorem 5.3 is that (A, B, C) is controllable and observable. Under such a condition, the protocol achieving consensus with a convergence speed larger than an arbitrary given positive value can be constructed by Algorithm 5.2. However, larger a implies higher feedback gains in protocol (5.3). Thus, a trade-off has to be made between the convergence speed and the cost of the consensus protocol.*

5.1.6 Illustrative example 5.1

The agent dynamics are given by (5.1), with

$$A = \begin{bmatrix} 0 & -1 & 0 \\ 1 & 0 & 0 \\ 0 & 1 & 0 \end{bmatrix}, \quad B = \begin{bmatrix} 1 \\ 0 \\ 0 \end{bmatrix}, \quad C = [0 \quad 0 \quad 1]$$

Obviously, matrix A is neutrally stable, and (A, B, C) is controllable and observable. A third-order consensus protocol is in the form of (5.3).

Using MATLAB, the feedback gain matrix K of (5.3) is given as $K = [\% \quad -4.5 \quad -5.5 \quad -3]$ such that the poles of $A + BK$ are $-1, -1.5, -2$. The matrix U such that $U^{-1}AU = J$ is of the real Jordan canonical form

$$U = \begin{bmatrix} 0 & 0.5774 & 0 \\ 0 & 0 & -0.5774 \\ 1 & -0.5774 & 0 \end{bmatrix} \quad J = \begin{bmatrix} 0 & 0 & 0 \\ 0 & 0 & 1 \\ 0 & -1 & 0 \end{bmatrix}$$

By Algorithm 5.1, the feedback gain L of (5.3) is obtained as $L = [0.3333 \quad 0 \quad -1.3333]^T$. By Theorem 5.2, the agents under this protocol can reach consensus with respect to any communication graph containing a spanning tree and for any positive coupling strength. One such graph with six nodes is shown in Figure 5.1, whose non-zero eigenvalues are 3 and 1 with multiplicity 4. Select the coupling strength $c = 1$ for simplicity. It can be verified that the convergence speed in this case equals 0.0303.

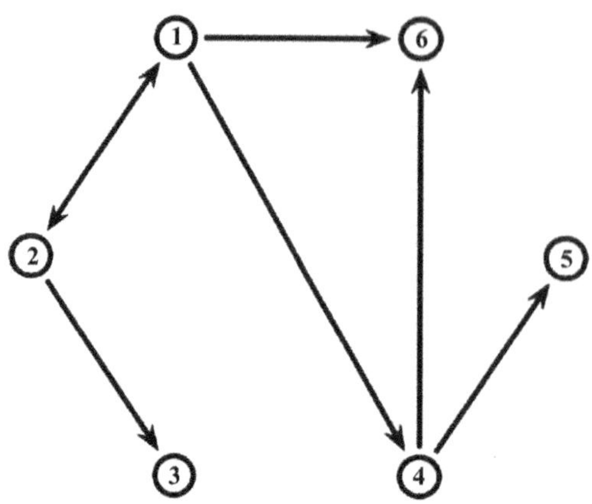

Figure 5.1 Communication graph

Next, protocol (5.3) is redesigned to achieve consensus with a specified convergence speed larger than 1. The feedback gain K is the same as above. Solving linear matrix inequality (LMI) (5.17) with $a = 1$ by using Sedumi toolbox [94] gives $L = [\% \quad -8.5763 \quad -15.2128 \quad -5.6107]^T$. For the graph in Figure 5.1, the threshold value for the coupling strength is $c_{th} = 1$ by (5.19). Select $c = 1$, the same as before. The consensus errors $x_i - x_1, i = 2, \ldots, 6$, for the graph in Figure 5.1 under

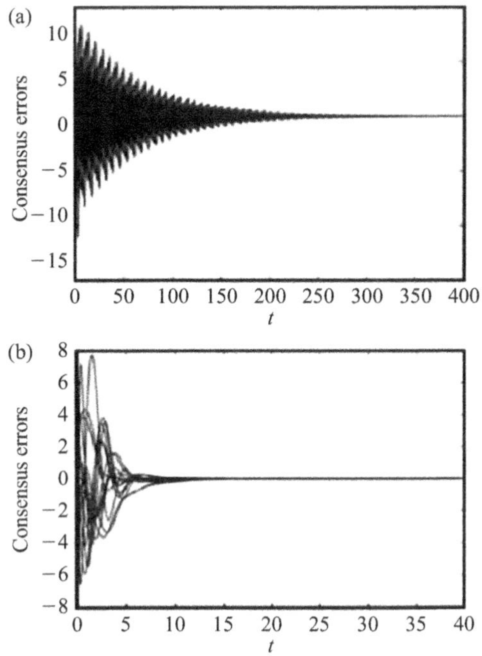

Figure 5.2 Consensus errors. (a) Algorithm 5.1 (b) Algorithm 5.2 with $\alpha = 1$

the protocols generated by Algorithms 5.1 and by Algorithm 5.2 with $\alpha = 1$, are depicted in Figure 5.2(a and b), respectively. It can be observed that the consensus process of the former case is indeed much slower than the latter.

5.1.7 Consensus with static protocols

In this section, a special case where the relative states between neighboring agents are available is considered. For this case, a distributed static protocol is proposed as

$$u_i = cF \sum_{j=1}^{N} a_{ij}(x_i - x_j) \tag{5.20}$$

where $c > 0$ and a_{ij} are the same as those defined in (5.3), and $F \in \Re^{p \times n}$ is the feedback gain matrix to be determined. For protocol (5.20), the dynamic consensus problem studied in Section 5.1.1 reduces to the following static consensus problem.

Definition 5.3: Protocol (5.20) is said to solve the (static) consensus problem if the states of agents (5.1) with (5.20) satisfy

$$\lim_{t\to\infty} \|x_i(t) - x_j(t)\| = 0, \quad \forall i,j = 1, 2, \ldots, N \tag{5.21}$$

The following result is a direct consequence of Theorem 5.1 in the static case.

Corollary 5.1: *For graph G containing a directed spanning tree, there exist a protocol (5.20) solving the consensus problem if and only if all the matrices* $A + c\lambda_i BF$, $i = 2, 3, \ldots, N$, *are Hurwitz, where* $\lambda_i, i = 2, 3, \ldots, N$, *is the same as in Theorem 5.1. For the case where matrix A is neutrally stable, one has the following.*

Corollary 5.2: *Given that* $A \in \Re^{n\times n}$ *is neutrally stable and that G has a directed spanning tree, there exists a protocol (5.20) solving the consensus problem and yielding a consensus region* $(0, \infty) \times (-\infty, \infty)$, *if and only if* (A, B) *is stabilizable.*

Contrary to the sufficient condition given in [289], a necessary and sufficient condition is obtained here as a consequence of Theorem 5.2. The method for constructing the protocol (5.20) is similar to Algorithm 5.1, therefore is omitted for brevity.

The design procedure for the protocol (5.20) solving consensus with a specification on the convergence speed is now presented.

Algorithm 5.3: *For a controllable pair* (A, B), *a protocol (5.20) solving the consensus problem with a convergence speed larger than* α *can be constructed as follows:*

- *Choose the feedback gain matrix* $F = -B^T P^{-1}$, *where* $P > 0$ *is a solution to*

  $$AP + PA^T - 2BB^T + 2\alpha P < 0$$

- *Select the coupling strength* $c \geq c_{th}$, *with* c_{th} *given by (5.19).*

5.1.8 Formation control

In this section, the consensus algorithms are modified to solve formation control problems of multiagent systems.

Let $\tilde{H} = (h_1, h_2, \ldots, h_N) \in \Re^{n\times N}$ describe a constant formation structure of the agent network in a reference coordinate frame, where $h_i \in R_n$, is the formation variable corresponding to agent i. Then, variable $h_i - h_j$ can be used to denote the relative formation vector between agents i and j, which is independent of the reference coordinate. For the agents (5.1), a distributed formation protocol is proposed as

$$\begin{aligned} v_i^+ &= (A + BK)v_i \\ &\quad + cL\left(\sum_{j=1}^{N} d_{ij}C(v_i - -v_j) - \sum_{j=1}^{N} d_{ij}(y_i - y_j - C(h_i - h_j))\right) \\ u_i &= Kv_i \end{aligned} \tag{5.22}$$

where the variables are the same as in (5.3). It should be noted that (5.22) reduces to the consensus protocol (5.3), when $h_i - h_j = 0, \forall i, j = 1, 2, \ldots, N$.

Definition 5.4: The agents (5.1) under protocol (5.22) achieve a given formation $\tilde{H} = (h_1, h_2, \ldots, h_N)$, if

$$\begin{aligned} \lim_{t\to\infty} \ & \|(x_i(t) - h_i) - (x_j(t) - h_j)\| \to 0 \\ & \forall i,j = 1, 2, \ldots, N \end{aligned} \tag{5.23}$$

Theorem 5.4: *For graph G containing a directed spanning tree, the agents (5.1) reach the formation $\tilde{H}$ under protocol (5.22) if all the matrices*

$$A + BK, \quad A + c\lambda_i LC, \quad i = 2, \ldots, N$$

are Hurwitz, and $Ah_i = 0, \forall i = 1, 2, \ldots, N$, where λ_i, $i = 2, 3, \ldots, N$, is the non-zero eigenvalue of matrix L.

Proof. Let $\tilde{x}_i = x_i - h_i$ and $\tilde{z}_i = [\tilde{x}_i^T, v_i^T]^T$, $i = 1, 2, \ldots, N$. Then, systems (5.1) and (5.22) together can be written as

$$\dot{\tilde{z}}_i = A\tilde{z}_i + c\sum_{j=1}^{N} L_{ij} H\tilde{z}_j + \begin{bmatrix} Ah_i \\ 0 \end{bmatrix}, \quad i = 1, 2, \ldots, N \tag{5.24}$$

where matrices A and H are defined in (5.4). Note that the formation $\tilde{H}$ is achieved if system (5.24) reaches consensus, which by (5.6) implies that $Ah_i = 0$, $i = 1, 2, \ldots, N$. The rest is similar to the proof of Theorem 5.1. □

Remark 5.8: *Note that not all kinds of formation structure can be achieved for the agents (5.1) by using protocol (5.22). The achievable formation structures have to satisfy the constraints $Ah_i = 0, \forall i = 1, 2, \ldots, N$. Note that h_i can be replaced by $h_i - h_1, i = 2, \ldots, N$, in order to be independent of the reference coordinate, by simply choosing h_1 corresponding to agent 1 as the origin. The formation protocol (5.22) satisfying Theorem 5.4 can be constructed by using Algorithm 5.1 or 5.2. Theorem 5.4 generalizes the results given in [279,302,303], where the agent dynamics in [279,302,303] are (generalized) double integrators.*

5.1.9 Illustrative example 5.2

Consider a network of six double integrators, described by

$$\begin{aligned} \dot{x}_i &= v_i \\ \dot{\tilde{v}}_i &= u_i \\ y_i &= x_i, \quad i = 1, 2, \ldots, 6 \end{aligned}$$

where $x_i \in \Re^2, \tilde{v}_i \in \Re^2, y_i \in \Re^2$, and $u_i \in R^2$ are the position, velocity, measured output, and acceleration input of agent i, respectively.

The objective is to design a dynamic protocol (5.22) such that the agents will evolve to form a regular hexagon with edge length 4. In [279], a different dynamic protocol is proposed to study formation control of double-integrator agents, which however, is applicable only to undirected communication topologies. As pointed out in [295], the spacecraft's dynamics in deep space can be modeled as a double integrator. Therefore, this example may have possible applications in deep-space formation flying mission.

In this case, choose

$$h_1 = [0 \quad 0 \quad 0 \quad 0]^T, \qquad h_2 = [4 \quad 0 \quad 0 \quad 0]^T$$
$$h_3 = [6 \quad 2\sqrt{3} \quad 0 \quad 0]^T, \qquad h_4 = [4 \quad 4\sqrt{3} \quad 0 \quad 0]^T$$
$$h_5 = [0 \quad 4\sqrt{3} \quad 0 \quad 0]^T, \qquad h_6 = [-2 \quad 2\sqrt{3} \quad 0 \quad 0]^T$$

We take

$$K = [-1.5 \quad -2.5] \otimes I_2$$

in (5.22) such that matrix

$$\lambda(A + BK) := \{-1, -1.5\}$$

By solving LMI (5.17) with $\alpha = 1$, one obtains

$$L = [-3.6606 \quad -4.8221]^T \otimes I_2$$

By Theorem 5.4 and Algorithm 5.2, the six agents under protocol (5.22) with K, L given as above, and $c = 1$ will form a regular hexagon with a convergence rate larger than 1 for the communication topology given in Figure 5.1. The state trajectories of the six agents are depicted in Figure 5.3.

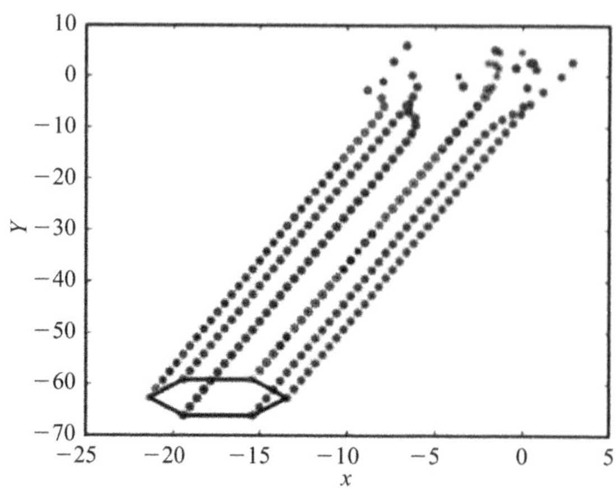

Figure 5.3 Six agents form a hexagon (a) G_1 for four agents (b) G_2 for five agents

5.2 Consensus control for time-delay systems

Mathematical modeling and analysis for consensus problems will be conducted for a group of autonomous mobile agents with double-integrator dynamics and time-varying interconnection delays. To solve the problems, distributed control scheme for each agent will be proposed first. Then the consensus stability of the multiagent system is obtained for the problem, where the dynamics of each agent is second order with time-varying interconnection delays. In the convergence analysis, both fixed and switched interconnection topologies of the considered multiagent system are investigated.

5.2.1 Introduction

In recent years, many research communities from mathematics, physics, computer, biology, and engineering are throwing themselves into investigating the collective behaviors of multiple autonomous mobile agents. The studies focus on understanding the complex mechanisms and interconnection rules in cooperative phenomena and solving coordination problems in engineering applications.

The model in the present section is closely related to some models reported in the literature. A prominent and well-studied example concerns synchronization of coupled oscillators modeled by Kuramoto equation in [304] and [305]. Another interesting example concerns swarming behaviors of mobile agents (such as birds, fish, robots, and vehicles) [276,306,307]. In the swarm model, the interconnection rule is based on nearest neighborhood, which has been applied in formation control for engineers. A third and last example of multiagent systems that fall within the scope of the present paper are consensus algorithms studied in [247,308] and [309]. An important feature of the model is to enable networked dynamic agents to agree upon certain quantities of interest through a distributed decision process.

Time delays resulting from interconnection links have been paid much attention to multiagent systems because of the practical background [310]. A necessary and sufficient condition for a time-delay consensus problem was presented for the agents with first-order dynamics and undirected interconnection graph in [247]. In [308], the authors generalized the results of [247] in considering uniform or nonuniform time-varying delays and also providing the bounds of maximal delays.

Consensus problems for multiple agents with interconnection time delays are discussed. Local controller for each agent is neighbor-based as was done in many references related to agent-based control systems. In [311], the authors considered delayed consensus problems for multiple agent with a leader–follower architecture, which has poor disturbance rejection properties (see [312]). In many applications, there may not exist any leader agents in the multiagent systems [313,314].

5.2.2 Problem formulation

In this section, a consensus problem will be formulated mathematically for a leaderless multiagent system with double-integrator dynamics and time-varying interconnection delays. Then we introduce some basic concepts and notations in algebraic graph theory and functional differential equations that will be used [315,316].

Consider a group of n identical agents move in an m-dimensional space and the agents are indexed by $1, \ldots, n$. A continuous-time model of the ith agent is described by a second-order differential equation as follows:

$$\begin{cases} \dot{x}_i = v_i \\ \dot{v}_i = u_i \end{cases} \tag{5.25}$$

where $x_i \in \Re^m$ can be the position (or angle) of agent $i, v_i \in \Re^m$ its velocity (or angular velocity) and $u_i \in \Re^m$ its interconnection control inputs for $i = 1, \ldots, n$.

In this paper, a free consensus problem of system (5.25) is solved if $x_i = x_j$, $v_i = v_j$ for all $i,j \in I = \{1, \ldots, n\}$. In order to enable the agents to achieve consensus along with interconnection delays, consensus control schemes have to be constructed for all mobile agents. The design of neighbor-based feedback $u_i(t)$ usually depends on $x_j(t)$, $v_j(t)$ for some $j \in N_i$, which denotes the index set of the neighbors of agent i. However, in practice, there may be interconnection delays, and each agent cannot instantly obtain the information from others. Thus, the feedback $u_i(t)$ should be constructed based on the information of $x_j(t - r(t))$ and $v_j(t - r(t))$ for some $j \in N_i$ and time-varying delay $r(t) > 0$, a continuously differentiable function satisfying

$$0 < r(t) < \tau \tag{5.26}$$

To deal with this consensus problem for system (5.25), we propose the following local control scheme

$$u_i(t) = k^2 \left(\sum_{j \in N_i(\sigma)} a_{ij}(x_j(t-r) - x_i(t-r)) \right.$$
$$\left. + k \sum_{j \in N_i(\sigma)} a_{ij}(v_j(t-r) - v_i(t-r)) \right), \quad k > 0 \tag{5.27}$$

where $\sigma : [0, \infty) \to I_\Gamma = \{1, \ldots, N\}$ (N denotes the total number of all possible interconnection topologies). If the system (5.25) is free of time delay (i.e., $r \equiv 0$), then the consensus control (5.27) becomes

$$u_i(t) = k^2 \left(\sum_{j \in N_i(\sigma)} a_{ij}(x_j(t) - x_i(t)) + k \sum_{j \in N_i(\sigma)} a_{ij}(v_j(t) - v_i(t)) \right)$$

which is consistent with the consensus rules in [317].

If each agent is regarded as a node, then the interconnection topology associated with the agents is conveniently described by a simple graph, see [247,316,318] and the appendix. Let $G = (V, E, A)$ be a weighted digraph of order n with the set of nodes $V = \{1, 2, \ldots, n\}$, set of arcs $E \subseteq V \times V$, and a weighted adjacency matrix $A = [a_{ij}] \in \Re^{n \times n}$ with nonnegative elements. The node indexes belong to a finite index set $I = \{1, 2, \ldots, n\}$. An arc of G is denoted by (i,j), which starts from i and ends on j in case agent j is a neighbor of agent i. The element a_{ij} associated with the arc of the digraph is positive, that is $a_{ij} > 0 \Leftrightarrow (i,j) \in E$. Moreover, we assume $a_{ii} = 0$ for all $i \in I$. The index set of neighbors of node i is denoted by

$N_i = \{j \in V : (i,j) \in E\}$. If $\sum_{j \in N_i(\sigma)} a_{ij} = \sum_{j \in N_i(\sigma)} a_{ji}$ for all $i = 1, \ldots, n$, the digraph G is called balanced. A cluster is any subset J, V of the nodes of the digraph. The set of neighbors of a cluster J is defined by $N_J = \cup_{i \in J} N_i = \{j \in V : i \in J, (i,j) \in E\}$. A path in a digraph is a sequence $i_0, i_1, \ldots, i_f$ of distinct nodes such that (i_{j-1}, i_j) is an arc for $j = 1, 2, \ldots, f, f \in Z^+$. If there exists a path from node i to node j, we say that j is reachable from i. A digraph G is strongly connected if any two distinct nodes are reachable from each other. A strong component of a digraph is an induced subgraph that is maximal, subject to being strongly connected. If a node i is reachable from every other node in G, then we say it is globally reachable, which is much weaker than strong connectedness.

A diagonal matrix $D = diag\{d_1, \ldots, d_n\} \in R^{n \times n}$ is a degree matrix of G, whose diagonal elements $d_i = \sum_{j \in N_i} a_{ij}$ for $i = 1, \ldots, n$. Then the Laplacian of the weighted digraph is defined as

$$L = D - A$$

The next lemma shows an important property of Laplacian L associated with G [247].

Lemma 5.4: *L has at least one zero eigenvalue with $1 \in \Re^n$ as its eigenvector, and all the non-zero eigenvalues of L have positive real parts. Laplacian L has a simple zero eigenvalue if and only if G has a globally reachable node.*

Let $S_1, S_2, \ldots, S - p$ be the strong components of $G = (V, \varepsilon, A)$ and N_{S_i} be the neighbor sets for $S_i, i = 1, \ldots, p,\ p > 1$. The following lemma is a revised version of a result reported in [314].

Lemma 5.5: *A digraph $G = (V, E, A)$ has a globally reachable node if and only if every pair of S_i, S_j satisfies $N_{S_i} \cup N_{S_j} \neq \varnothing$. Moreover, if the graph is strongly connected, then each node is globally reachable from every other node.*

In the sequel, the interconnection topologies will be discussed in two cases. A fixed topology, described by a digraph, is considered first, and then variable topologies described by balanced digraphs are analyzed. To model varying interconnection topology, we introduce a function $\sigma : [0, \infty) \rightarrow I_\Gamma = \{1, \ldots, N\}$, which is a switching signal to show the sequence of the switched interconnection topologies over time. The set $\Gamma = \{G_1, \ldots, G_N\}$ is a finite collection of graphs with a common node set V. If σ is a constant function, then the corresponding interconnection topology is fixed. In addition, $N_i(\sigma)$ is the index set of neighbors of agent i in the digraph g_σ whereas $a_{ij}(i,j = 1, \ldots, n)$ are elements of the adjacency matrix of g_σ.

Introducing

$$x = \begin{bmatrix} x_1 \\ x_2 \\ \vdots \\ x_n \end{bmatrix} \in \Re^{mn}, \quad v = \begin{bmatrix} v_1 \\ v_2 \\ \vdots \\ v_n \end{bmatrix} \in \Re^{mn}, \quad u = \begin{bmatrix} u_1 \\ u_2 \\ \vdots \\ u_n \end{bmatrix} \in \Re^{mn}$$

Then, with the control scheme (5.27), the closed-loop system (5.25) can be rewritten in the following form:

$$\begin{cases} \dot{x} = v \\ \dot{v} = u = -k^2(L_\sigma \otimes I_m)(x(t-r) + kv(t-r)) \end{cases} \tag{5.28}$$

where $\otimes$ denotes the Kronecker product [297].

Before the consensus stability of system (5.28) is analyzed, we refer the reader to Section 9.1 where the stability notions of time-delay systems were treated.

In the sequel, we will analyze the consensus stability of multiagent systems with time delays under controller (5.27).

5.2.3 Fixed interconnection topology

In this section, we will focus on the convergence analysis of the system (5.28) when the switching signal is constant (or equivalently, the interconnection topology is fixed). Then the subscript σ is dropped for simplicity and the system (5.28) can be expressed with the following linear delayed differential equations:

$$\begin{cases} \dot{x} = v \\ \dot{v} = -k^2(L \otimes I_m)(x(t-r) + kv(t-r)) \end{cases} \tag{5.29}$$

To solve the consensus problem of the system (5.29), we recall the following result.

Lemma 5.6: *For Laplacian L associated with digraph G, then there exists a nonsingular matrix*

$$U = \begin{pmatrix} 1 & * & \dots & * \\ 1 & * & \dots & * \\ \vdots & \vdots & & \vdots \\ 1 & * & \dots & * \end{pmatrix} \in \Re^{n\times n} \tag{5.30}$$

*where * represents arbitrary value such that*

$$U^{-1}LU = \begin{pmatrix} 0 & \alpha^T \\ 0_{n-1} & H \end{pmatrix} = \Lambda \in \Re^{n\times n}, \quad \alpha \in \Re^{n-1}, \quad H \in \Re^{(n-1)\times(n-1)} \tag{5.31}$$

According to Lemma 5.6, with a coordinate transformation

$$\bar{x} = (U^{-1} \otimes I_m)x, \quad \bar{v} = (U^{-1} \otimes I_m)v \tag{5.32}$$

the system (5.29) becomes

$$\begin{cases} \dot{\bar{x}} = \bar{v} \\ \dot{\bar{v}} = -k^3(\Lambda \otimes I_m) - k^2(\Lambda \otimes I_m)\bar{x}(t-r) \end{cases}$$

or equivalently

$$\begin{cases} \dot{\bar{x}}_1 = \bar{v}_1 \\ \dot{\bar{v}}_1 = -k^3(\alpha^T \otimes I_m)\bar{v}_2(t-r) - k^2(\alpha^T \otimes I_m)\bar{x}_2(t-r) \end{cases} \tag{5.33}$$

and

$$\begin{cases} \dot{\bar{x}}_2 = \bar{v}_2 \\ \dot{\bar{v}}_2 = -k^3(H \otimes I_m)\bar{v}_2(t-r) - k^2(H \otimes I_m)\bar{x}_2(t-r) \end{cases} \tag{5.34}$$

where

$$\bar{x} = \begin{pmatrix} \bar{x}_1 \\ \bar{x}_2 \end{pmatrix}, \quad \bar{v} = \begin{pmatrix} \bar{v}_1 \\ \bar{v}_2 \end{pmatrix}, \quad \bar{x}_1, \bar{v}_1 \in \Re^m, \quad \bar{x}_2, \bar{v}_2 \in \Re^{m(n-1)}$$

for the subsystem (5.34), let $\varepsilon = (\bar{x}_2^T, \bar{v}_2^T)^T \in \Re^{2m(n-1)}$. Then we have a compact form

$$\dot{\varepsilon} = B\varepsilon(t) + E\varepsilon(t-r) \tag{5.35}$$

where

$$B = \begin{pmatrix} O_{(n-1)\times(n-1)} & I_{n-1} \\ 0_{(n-1)\times(n-1)} & 0_{(n-1)\times(n-1)} \end{pmatrix} \otimes I_m$$

$$E = \begin{pmatrix} O_{(n-1)\times(n-1)} & 0_{(n-1)\times(n-1)} \\ -k^2H & -k^3H \end{pmatrix} \otimes I_m$$

Remark 5.9: *From Lemmas 5.4 and 5.6, if graph G has a globally reachable node, the real parts of all the eigenvalues of $H \in \Re^{(n-1)\times(n-1)}$ are positive, or equivalently, $-H$ is Hurwitz stable. Therefore, there exist a positive-definite matrix $\bar{P} \in \Re^{(n-1)\times(n-1)}$ such that*

$$\bar{P}H + H^T\bar{P} = I_{n-1} \tag{5.36}$$

Let $\bar{\lambda}_l$ (or λ) denote the minimum (or maximum) eigenvalue of $-P$ and μ the maximum eigenvalue of $H^T\bar{P}^T\bar{P}H$. Let $\lambda_{\min}$ be the minimum eigenvalue of $-(PF + F^TP)$ with $F = B + E$. Then a result can be obtained for system (5.29).

Theorem 5.5: *For system (5.29), take*

$$k > k_1^* = \max\left\{\sqrt{\frac{1}{2}\lambda + 1}, \frac{\mu}{\bar{\lambda}} + 1\right\} \tag{5.37}$$

and assume that

$$\tau \leq \tau_1^* = \frac{\lambda_{\min}}{k^4 + k^6 + \|PE^2P^{-1}E^{2^T}P\| + 2q(k+1)\lambda} \tag{5.38}$$

where $q > 1$. Then the free consensus problem of the system (5.29) is solved if and only if G has a globally reachable node.

Proof. $\Leftarrow$ Since G has a globally reachable node, zero is a simple eigenvalue of Laplacian L whereas other eigenvalues have positive real parts, see Lemma 5.4. By Lemma 5.6, there exists a nonsingular matrix U given in (5.30) such that L can be transformed to (5.31), where H has eigenvalues with positive real parts. Then there is a positive-definite matrix $\bar{P}$ satisfying (5.36).

For system (5.35), take a Lyapunov–Razumikhin function

$$V(\varepsilon) = \varepsilon^T P\varepsilon \tag{5.39}$$

where

$$P = \begin{pmatrix} k\bar{P} & \bar{P} \\ \bar{P} & k\bar{P} \end{pmatrix} \otimes I_m$$

is positive definite since $k > 1$.

Furthermore, by Leibniz–Newton formula

$$\begin{aligned} \varepsilon(t-r) &= \varepsilon(t) - \int_{-r}^{0} \dot{\varepsilon}(t+s)ds \\ &= \varepsilon(t) - B\int_{-r}^{0} \varepsilon(t+s)ds - E\int_{-r}^{0} \varepsilon(t-r+s)ds \end{aligned}$$

Therefore, (5.35) can be rewritten as

$$\dot{\varepsilon} = F\varepsilon - EB\int_{-r}^{0} \varepsilon(t+s)ds - E^2\int_{-r}^{0} \varepsilon(t-r+s)ds \tag{5.40}$$

for arbitrary initial function on $[-2\tau, 0]$. If the zero solution of (5.40) is asymptotically stable, then the zero solution of (5.35) is asymptotically stable since (5.35) is a special case of (5.40) with continuous initial function $\tilde{\varphi}(s)$ given by $\tilde{\varphi}(s)$ arbitrary for $s \in [-2\tau, -\tau - r(0)]$, $\tilde{\varphi}(s) = \varphi(s + r(0))$, $-\tau - r(0) \le s \le -r(0)$, and $\tilde{\varphi}(s) = \varepsilon(t+s)$, $-r(0) \le s \le 0$, where $\varepsilon(t)$ is the solution of (5.35) with initial function φ on $[-\tau, 0]$.

Setting $\eta_1 = \varepsilon(t), \eta_2 = \varepsilon(t+s), \eta_3 = E^{2^T}P\varepsilon(t-r+s), \eta_4 = \varepsilon(t-r+s)$, $M_1 = PEB = Y_1, M_2 = I_{2m(n-1)} = Y_2, X_1 = k^4(1+k^2)I_{2m(n-1)}, X_2 = P^{-1}$, and $Z_1 = Z_2 = P$. Invoking a standard bounding inequality [30] gives (5.37), and leads to

$$\begin{aligned} \dot{V} &= \varepsilon^T(F^TP + PF)\varepsilon - 2\int_{-r}^{0} \varepsilon^T PEB\varepsilon(t+s)ds \\ &\quad - 2\int_{-r}^{0} \varepsilon^T PE^2\varepsilon(t-r+s)ds \\ &\le \varepsilon^T(FTP + PF)\varepsilon + rk^4(1+k^2)\varepsilon^T\varepsilon \\ &\quad + \int_{-r}^{0} \varepsilon^T(t+s)P\varepsilon(t+s)ds + r\varepsilon^T PE^2P^{-1}E^{2^T}P\varepsilon \\ &\quad + \int_{-r}^{0} \varepsilon^T(t-r+s)P\varepsilon(t-r+s)ds \end{aligned}$$

Take $\phi(s) = qs$ for some constant $q > 1$. In the case of

$$V(\varepsilon(t+\theta)) \le qV(\varepsilon(t)), \quad -2\tau \le \theta \le 0 \tag{5.41}$$

we have, with Remark 5.9

$$\dot{V} \le -\varepsilon^T Q\varepsilon + r\varepsilon^T((k^4 + k^6)I_{2m(n-1)} + PE^2P^{-1}E^{2^T} + 2qP)\varepsilon$$

where

$$Q = -(F^T P + PF) = \begin{pmatrix} k^2 I_{n-1} & k^3 I_{n-1} - k\bar{P} \\ k^3 I_{n-1} - k\bar{P} & k^4 I_{n-1} - 2\bar{P} \end{pmatrix} \otimes I_m$$

Here Q is positive definite if k satisfies (5.37), according to Schur complements [319]. Let $\lambda_{\min}$ denote the minimum eigenvalues of Q. If we take the upper bound τ_1^* in (5.38), then $\dot{V}(\varepsilon) \leq -\eta\varepsilon^T\varepsilon$ for some $\eta > 0$. By the Lyapunov–Razumikhin theorem from the appendix, we conclude that $\bar{x}_2 \to 0_{m(n-1)}, \bar{v}_2 \to 0_{m(n-1)}$ as $t \to \infty$.

On the other hand, for the system (5.33), let $\bar{x}_1(0), \bar{v}_1(0)$ be the initial values of $\bar{x}_1(t), \bar{v}_1(t)$ and take a variable of change $\tilde{x}_1 = \bar{x}_1 - (\bar{v}_1(0)t + \bar{x}_1(0)), \tilde{v}_1 = \bar{v}_1 - \bar{v}_1(0)$. Then the solution can be given by the following integral equation

$$\begin{pmatrix} \tilde{x}_1 \\ \tilde{v}_1 \end{pmatrix} = \int_0^t \begin{pmatrix} I_m & (t-s)I_m \\ 0_m & I_m \end{pmatrix} \times \begin{pmatrix} 0_m \\ -k^3(\alpha^T \otimes I_m)\bar{v}_2(s-r) - k^2(\alpha^T \otimes I_m)\bar{x}_2(s-r) \end{pmatrix} ds \tag{5.42}$$

As discussed before, the system (5.34) is uniformly asymptotically stable, which implies its exponential stability according to Lemma 5.3 of [319]. Thus, solution (5.42) has an exponential decay term with respect to time t and so it is convergent to zero as $t \to \infty$. Consequently, we have

$$\begin{pmatrix} \bar{x}_1 \\ \bar{x}_2 \end{pmatrix} - \begin{pmatrix} \bar{v}_1(0)t + \bar{x}_1(0) \\ 0_{m(n-1)} \end{pmatrix} \to 0_{mn}, \begin{pmatrix} \bar{v}_1 \\ \bar{v}_2 \end{pmatrix} \to \begin{pmatrix} \bar{v}_1(0) \\ 0_{m(n-1)} \end{pmatrix}, \text{ as } t \to \infty$$

From transformation (5.32), we have

$$x - (U \otimes I_m) \begin{pmatrix} \bar{v}_1(0)t + \bar{x}_1(0) \\ 0_{m(n-1)} \end{pmatrix} = x - 1 \otimes (\bar{v}_1(0)t + \bar{x}_1(0)) \to 0_{mn}$$

$$v = (U \otimes I_m)\bar{v} \to (U \otimes I_m) \begin{pmatrix} \bar{v}_1(0) \\ 0_{m(n-1)} \end{pmatrix} = 1 \otimes \bar{v}_1(0)$$

Therefore, $x_i - x_j \to 0_{mn}, v_i - v_j \to 0_{mn}$ for all $i, j \in I$ as $t \to \infty$. The conclusion follows.

$\Rightarrow$ It will be proved by contradiction. If G has no globally reachable node, then, from Lemma 5.5, there is at least two strong components of G having no neighbor sets. Thus, we renumber the nodes of G and L can be transformed to the following form

$$L = \begin{pmatrix} L_{11} & 0 & 0 \\ 0 & L_{22} & 0 \\ L_{31} & L_{32} & L_{33} \end{pmatrix} \tag{5.43}$$

where 0 denotes some zero matrices with appropriate dimensions and $L_{11} \in \Re^{k_1 \times k_1}$, $L_{22} \in R^{k_2 \times k_2}$ $(k_1 + k_2 < n)$ are the Laplacians associated with two strong components, respectively.

Then the system (5.29) can be expressed as

$$\begin{cases} \dot{x}_1 = v_1 \\ \dot{v}_1 = -k^3(L_{11} \otimes I_m)v_1(t-r) - k^2(L_{11} \otimes I_m)x_1(t-r) \end{cases} \tag{5.44}$$

$$\begin{cases} \dot{x}_2 = v_2 \\ \dot{v}_2 = -k^3(L_{22} \otimes I_m)v_2(t-r) - k^2(L_{22} \otimes I_m)x_2(t-r) \end{cases} \tag{5.45}$$

$$\begin{cases} \dot{x}_3 = v_3 \\ \dot{v}_3 = -k^3([L_{31} \quad L_{32} \quad L_{33}] \otimes I_m)v(t-r) \\ \qquad -k^2([L_{31} \quad L_{32} \quad L_{33}] \otimes I_m)x(t-r) \end{cases} \tag{5.46}$$

where

$$x = \begin{pmatrix} x_1 \\ x_2 \\ x_3 \end{pmatrix}, \quad v = \begin{pmatrix} v_1 \\ v_2 \\ v_3 \end{pmatrix}, \quad x_1, v_1 \in \Re^{mk_1}, \quad x_2, v_2 \in \Re^{mk_2}, \quad x_3, v_3 \in \Re^{m(n-k_1-K_1)}$$

Similar to the proof of sufficient condition, it is not hard to obtain that $x_1 - 1_{k_1} \otimes (\nu_1 t + \delta_1) \to 0_{mk_1}, v_1 \to 1_{k_1} \otimes \nu_1$ and $x_2 - 1_{k_2} \otimes (\nu_2 t + \delta_2) \to 0_{mk_2}, v_2 \to 1_{k_2} \otimes \nu_2$ for some constants $\nu_1, \delta_1, \nu_2, \delta_2$ as $t \to \infty$. Since ν_1, δ_1 and ν_2, δ_2 can be set arbitrarily, the system (5.29) cannot reach a consensus, which leads to a contradiction. □

Remark 5.10: *During the derivations of the bounds on k_1^* and τ_1^*, high-order inequalities and the bounds on matrix function need to be solved, many zoom techniques have to be applied and hence the results may be very conservative.*

5.2.4 Switched interconnection topology

We now consider the convergence of time-delay system (5.28) for the switched interconnection topology. Since it is hard to do this for switched interconnection topologies described by general digraphs, a special class of digraphs, that is, balanced digraphs, are considered in the subsequent stability analysis, see the appendix.

With a coordinate transformation

$$x = (V^* \otimes I_m)x, \quad \bar{v} = (V^* \otimes I_m)v \tag{5.47}$$

the system (5.28) becomes

$$\begin{cases} \dot{\bar{x}}_1 = \bar{v}_1 \\ \dot{\bar{v}}_1 = 0_m \end{cases} \tag{5.48}$$

$$\begin{cases} \dot{\bar{x}}_2 = \bar{v}_2 \\ \dot{\bar{v}}_2 = -k^3(H_\sigma \otimes I_m)\bar{v}_2(t-r) - k^2(H_\sigma \otimes I_m)\bar{x}_2(t-r) \end{cases} \tag{5.49}$$

where $\bar{x}_1, \bar{v}_1 \in C^m, \bar{x}_2, \bar{v}_2 \in C^{m(n-1)}$.

Consider (5.49), or equivalently

$$\dot{\varepsilon} = B\varepsilon(t) + E_\sigma \varepsilon(t-r) \tag{5.50}$$

where

$$B = \begin{pmatrix} 0_{(n-1)\times(n-1)} & I_{n-1} \\ 0_{(n-1)\times(n-1)} & 0_{(n-1)\times(n-1)} \end{pmatrix} \otimes I_m$$

$$E_\sigma = \begin{pmatrix} 0_{(n-1)\times(n-1)} & 0_{(n-1)\times(n-1)} \\ -k^2 H_\sigma & -k^3 H_\sigma \end{pmatrix} \otimes I_m$$

Based on Lemma 9.5 and the fact that the set I_Γ is finite, if the balanced digraph G_σ has a globally reachable node

$$\tilde{\lambda} = \min\{\text{eigenvalues of } H_\sigma + H_\sigma^*\} > 0$$

$$\tilde{\mu} = \max\{\text{eigenvalues of } H_\sigma H_\sigma^*\} > 0$$

can be well defined. Let $F_\sigma = B + E_\sigma$ and $\tilde{\lambda}_{\min}$ denotes the minimum eigenvalue of all possible $-(\tilde{P}F_\sigma + F_\sigma\tilde{P})$.

A result of the switched system (5.28) with time-varying delay is given as follows:

Theorem 5.6: *For system (5.28) with balanced interconnection topology G_σ, take*

$$k > k_2^* = \max\left\{\sqrt{\frac{1}{2\tilde{\lambda}} + 1}, \tilde{\mu} + 1\right\} \tag{5.51}$$

and assume that

$$\tau < \tau_2^* = \frac{\tilde{\lambda}_{\min}}{k^4 + k^6 + (k^{13} + k^{11})\tilde{\mu}^2 + 2q(k+1)} \tag{5.52}$$

with $q > 1$. Then the free consensus problem of the system (5.28) is solved if G_σ has a globally reachable node.

Proof. To obtain the result, we first consider (5.49). Take a Lyapunov–Razumikhin function

$$V(\varepsilon) = \varepsilon^* \tilde{P} \varepsilon \tag{5.53}$$

where

$$\tilde{P} = \begin{pmatrix} kI_{n-1} & I_{n-1} \\ I_{n-1} & kI_{n-1} \end{pmatrix} \otimes I_m$$

is positive definite for $k > 1$.

Similar to the proof of Theorem 5.5, we can obtain

$$\dot{V}|_{(33)} = \varepsilon^T (F_\sigma^T \tilde{P} + \tilde{P} F_\sigma)\varepsilon - 2\varepsilon^T \tilde{P} E_\sigma B \int_{-r}^{0} \varepsilon(t+s)ds$$

$$- 2\varepsilon^T \tilde{P} E_\sigma^2 \int_{-r}^{0} \varepsilon(t - r + s)ds$$

By a standard bounding inequality [30], it follows that

$$\begin{aligned}\dot{V}|_{(33)} \leq\ & \varepsilon^T(F_\sigma^T\tilde{P} + \tilde{P}F_\sigma)\varepsilon + r(k^4 + k^6)\varepsilon^*\varepsilon \\ & + \int_{-r}^{0} \varepsilon^*(t+s)\tilde{P}\varepsilon(t+s)ds + r\varepsilon^T\tilde{P}E_\sigma^2\tilde{P}^{-1}E_\sigma^{2^T}\tilde{P}\varepsilon \\ & + \int_{-r}^{0} \varepsilon^*(t-r+s)\tilde{P}\varepsilon(t-r+s)ds\end{aligned}$$

Set $\phi(s) = qs$ for some constant $q > 1$. In the case of

$$V(\varepsilon(t+\theta)) < qV(\varepsilon(t)), \quad -2\tau \leq \theta \leq 0 \tag{5.54}$$

we have

$$\dot{V} \leq -\varepsilon^T Q_\sigma \varepsilon + r\varepsilon^T((k^4 + k^6)I_{2m(n-1)} + \tilde{P}E_\sigma^2\tilde{p}^{-1}E_\sigma^2\tilde{p} + 2q\tilde{p})\varepsilon$$

where

$$\begin{aligned}Q_\sigma &= -(F_\sigma^T\Phi + \Phi F_\sigma) \\ &= \begin{pmatrix} k^2(H_\sigma^* + H_\sigma) & k^3(H_\sigma^* + H_\sigma) - kI_{n-1} \\ k^3(H_\sigma^* + H_\sigma) - kI_{n-1} & k^4(H_\sigma^* + H_\sigma) - 2I_n \end{pmatrix} \otimes I_m\end{aligned}$$

Clearly, Q_σ is positive definite and then $\dot{V}(\varepsilon)$ is negative definite if k is taken as (5.51) and

$$r < \tau \leq \tau_2^*$$

where τ_2^* is defined in (5.52). Thus, the subsystem (5.49) converges to $0_{2m(n-1)}$ according to the Lyapunov–Razumikhin theorem, see the appendix. On the other hand, for the system (5.48)

$$\bar{x}_1 = \bar{v}_1(0)t + \bar{x}_1(0), \quad \bar{v}_1 = \bar{v}_1(0)$$

with

$$\bar{x}_1(0) = (1/\sqrt{n})\sum_{i=1}^{n} x_i(0), \quad \bar{v}_1(0) = (1/\sqrt{n})\sum_{i=1}^{n} v_i(0)$$

Then with the transformation (5.47), we have

$$x - (V \otimes I_m)\begin{pmatrix}\bar{v}_1(0)t + \bar{x}_1(0) \\ 0_{m(n-1)}\end{pmatrix} = x - 1 \otimes \frac{1}{n}\sum_{i=1}^{n} x_i(0) \to 0_{mn} \tag{5.55}$$

$$v \to (V \otimes I_m)\begin{pmatrix}\bar{v}_1(0) \\ 0_{m(n-1)}\end{pmatrix} = 1 \otimes \frac{1}{n}\sum_{i=1}^{n} v_i(0) \tag{5.56}$$

which concludes the proof. □

Remark 5.11: *In Theorem 5.6, when we consider the stability of second-order time-delay system, by saying 'G_σ has a globally reachable node', we mean that every possible graph in the interconnection evolution has a globally reachable node. This is stronger than the usual requirement in [314]: there is $T \geq 0$, such that for all $t_0 > 0$ there is a node connected to all other nodes across $[t_0, t_0 + T]$, that is, the opposite graph, which is formed by changing the orientation of each arc of $g([t_0, t_0 + T]) = (V, \cup_{t\in[t_0,t_0+T]}E(t))$, has a globally reachable node.*

Remark 5.12: *In fact, with balanced graphs, the position x_i and the velocity v_i of agent $i(i = 1, \ldots, n)$ in the considered multiagent system converge to the average values of initial positions (i.e., $(1/n)\sum_{i=1}^{n} x_i(0)$) and initial velocities (i.e., $(1/n)\sum_{i=1}^{n} v_i(0)$), respectively, because of (5.55) and (5.56).*

In section 5.2.5, some numerical examples are provided to illustrate the theoretical results.

5.2.5 *Illustrative example 5.3*

Consider different number of agents are moving in a plane (i.e., $m = 2$) with ring-shaped interconnection G_1 and G_2, respectively (see Figure 5.4(a and b)).

It is not difficult to obtain two Laplacians associated with G_1 and G_2 as follows:

$$L_1 = \frac{1}{4}\begin{pmatrix} 1 & -1 & 0 & 0 \\ 0 & 1 & -1 & 0 \\ 0 & 0 & 1 & -1 \\ -1 & 0 & 0 & 1 \end{pmatrix}$$

$$L_2 = \frac{1}{4}\begin{pmatrix} 1 & -1 & 0 & 0 & 0 \\ 0 & 1 & -1 & 0 & 0 \\ 0 & 0 & 1 & -1 & 0 \\ 0 & 0 & 0 & 1 & -1 \\ -1 & 0 & 0 & 0 & 1 \end{pmatrix}$$

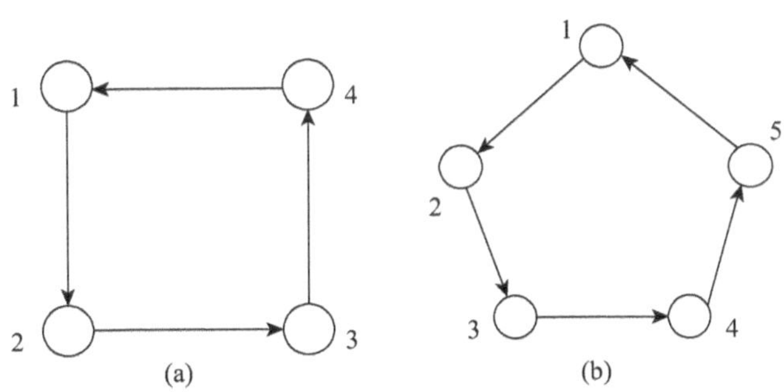

Figure 5.4 Interconnection topologies

Then, for the two interconnection topologies G_1, G_2, we can obtain $k_1^* = 1.5000$, $k_2^* = 2.0170$ and the upper bounds $\tau_1^* = 0.0104$, $\tau_2^* = 0.0011$, which shows that k^* will increase and τ^* will decrease as the number of agents become large.

5.2.6 Illustrative example 5.4

Consider the interconnection topology is switching between G_1 and G_3 (see Figures 5.4 and 5.5) while four agents are moving in a plane. Suppose that the time-varying interconnection delay is given by $r(t) = 0.01|\cos(t)|$ and the switching signal $\sigma(t) = \{G_1, G_3, G_1, G_3, \ldots\}$. The initial positions and velocities are given as follows:

$$x(0) = [0; 1; 0; 0; 1; 0; 1; 1]^T$$
$$v(0) = [-1; 1; 2:5; -1.5; -2; -1; 1; 2]^T$$

By using the proposed consensus algorithm (5.27) for four agents with time delay $r(t)$, the simulation results are shown in Figures 5.6 and 5.7. It can be seen that all the agents will reach consensus while the interconnection is dynamically changing and there exist time-varying delays in the local interactions between agents from

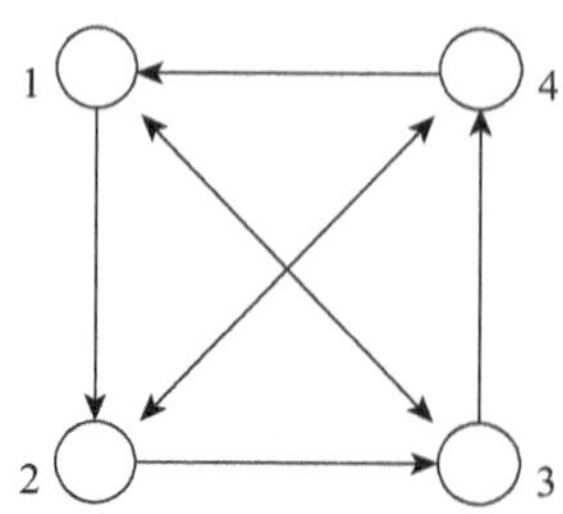

Figure 5.5 Interconnection topology G_3 for four agents: (a) G_1 for four agents; (b) G_2 for five agents

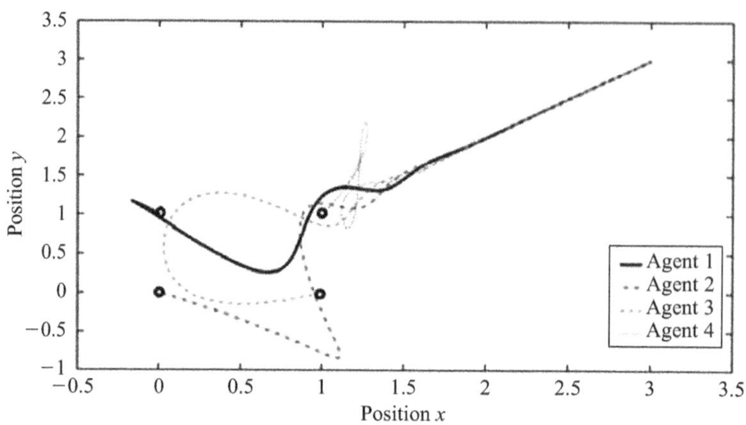

Figure 5.6 Position evolutions

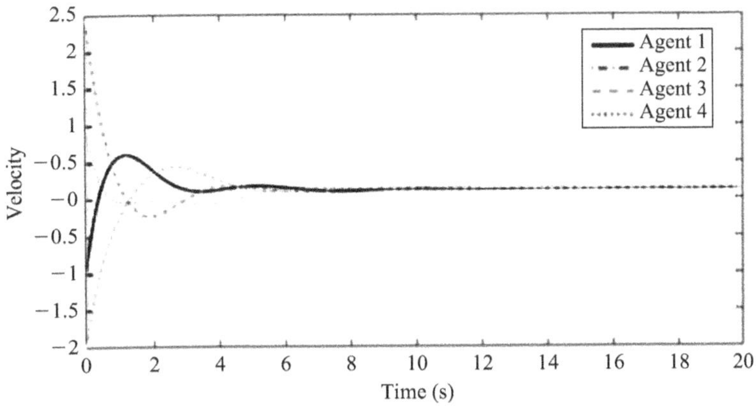

Figure 5.7 Velocity evolutions

Figure 5.6. From Figure 5.7, we also find that the velocity of each agent will approach the average of the initial values, that is, $v_i(t) \rightarrow 1/4\sum_{j=1}^{4} v_j(0) = v^* = [0.125, 0.125]^T$ as $t \rightarrow \infty$.

5.3 Robust consensus of multiagent systems

We are concerned with consensus problems for a class of multiagent systems with second-order dynamics. Some dynamic neighbor-based rules are adopted for the agents with the consideration of parameter uncertainties and external disturbances. Sufficient conditions are derived to make all agents asymptotically reach consensus while satisfying desired H_∞ performance. Finally, numerical simulations are provided to show the effectiveness of our theoretical results.

5.3.1 Introduction

As a new field of research, consensus problems in multiagent systems have attracted a great number of researchers with rather diverse background such as physics, biology, control theory, robotics, and computer science. Roughly speaking, consensus means that multiple agents reach an agreement on a common value that might be altitude in multispacecraft alignment, heading direction in flocking behavior, or average in distributed computation.

Numerous results have been obtained for consensus problems of multiagent systems [238,248,275,277–294,295–303,317,319–327]. For example, consensus problems for first-order multiagent systems were investigated under a mild connectivity assumption in the absence of centralized control in [276] and [323]. In [324] and [325], the results in [296–303,322] and [323] were extended to the nonuniform time-delay case where the time delays affect only the information that is being transmitted. Also, average consensus problems were studied with fixed/switching topology and time delays in [247,308–316,318]. Moreover, in [237] and [238], consensus-related problems for second-order multiagent systems such as flocking [238] and formation control of vehicles [302] were extensively studied. In practical applications, multiagent systems are often subjected to various disturbances such as actuator bias, measurement/calculation errors, and the variation of the communication topology. The existence of disturbances might destroy the convergence properties of multiagent systems. It is important to investigate their effects on the behavior of multiagent systems. Many studies have been conducted on this problem [328,329]. Ren [328] took into account the actuator saturation and the unavailability of the relative velocity information and proposed several consensus algorithms for second-order multiagent systems in the absence and presence of a group reference velocity. Lin *et al.* [330] studied robust H_∞ consensus problems for first-order multiagent systems with external disturbances and parameter uncertainties on fixed and switching topologies. Hong *et al.* [329] studied the tracking control for multiagent consensus with an active leader and gave a local controller together with a neighbor-based state-estimation rule.

5.3.2 Problem description

Suppose that the multiagent system under consideration consists of n agents that may be birds, airplanes, robots, and so on. Each agent is regarded as a node in a directed graph G. Each edge $(s_j, s_i) \in E$ corresponds to an available information channel from the agent s_i to the agent s_j. Moreover, each agent updates its current state based upon the information received from its neighbors.

Suppose the dynamics of the ith agent ($i \in I$) is

$$\begin{aligned} \dot{x}_i(t) &= v_i(t) \\ \dot{v}_i(t) &= u_i(t) \end{aligned} \tag{5.57}$$

where $x_i(t) \in R$ is the position state, $v_i(t) \in R$ is the velocity state, and $u_i(t) \in R$ is the control input (or protocol).

Given the dynamical system in (5.57), we say the protocol u_i asymptotically solves the consensus problem, if the states of agents satisfy

$$\lim_{t\to+\infty} [x_i(t) - x_j(t)] = 0, \quad \lim_{t\to+\infty} [v_i(t) - v_j(t)] = 0 \tag{5.58}$$

for all $i,j \in I$.

In practical applications, the multiagent systems often suffer from various disturbances, such as actuator bias, measurement/calculation errors, and the variation of the communication topology. The existence of disturbances might lead to oscillation or divergence of the multiagent systems. It is of significance to investigate their effects on the behavior of the multiagent systems and find appropriate protocols to make the multiagent systems robust to disturbances. In this section, we are interested in investigating the robust H_∞ performance of system (5.57) using the following consensus protocol:

$$\begin{aligned} u_i(t) &= k_0 v_i + k_1 \sum_{s_j \in N_i} (a_{ij} + \Delta a_{ij}(t))(x_j(t) - x_i(t)) \\ &\quad + k_2 \sum_{s_j \in N_i} (a_{ij} + \Delta a_{ij}(t))(v_j(t) - v_i(t)) + w_i(t) \end{aligned} \tag{5.59}$$

where $k_1 > 0, k_2 > 0$, and k_0 are protocol parameters

$$|\Delta a_{ij}(t)| = \begin{cases} \leq \psi_{ij} & i \neq j \text{ and } a_{ij} \neq 0 \\ 0 & \text{otherwise} \end{cases}$$

denotes the uncertainty of a_{ij}, which might be caused by actuator bias, for specified positive constants ψ_{ij}, and $w_i(t) \in L_2[0, \infty)$ is the external disturbance that might be caused by measurement/calculation errors. Here, it should be noted that there is no restricted condition imposed on the parameter k_0 and it might be positive, zero, or negative.

A natural way to combine the relative information is to define output functions $z_i(t) = [z_{i1}(t), z_{i2}(t)]^T \in R^2$ for $i \in I$ computed from an average of the relative displacements and velocities of all agents as follows:

$$\begin{aligned} z_{i1}(t) &= \frac{1}{n} \sum_{j=1}^{n} [x_i(t) - x_j(t)] = x_i(t) - \frac{1}{n} \sum_{j=1}^{n} x_j(t) \\ z_{i2}(t) &= \frac{1}{n} \sum_{j=1}^{n} [v_i(t) - v_j(t)] = v_i(t) - \frac{1}{n} \sum_{j=1}^{n} v_j(t) \end{aligned}$$

It is clear that consensus can be achieved if and only if

$$\lim_{t \to +\infty} z_i(t) = 0, \quad i = 1, \ldots, n$$

Denote

$$\begin{aligned} \xi &= [x_1, v_1, x_2, v_2, \ldots, x_n, v_n]^T \in \Re^{2n} \\ A &= \begin{bmatrix} 0 & 1 \\ 0 & k_0 \end{bmatrix} \in \Re^{2\times 2}, \quad B_1 = \begin{bmatrix} 0 & 1 \\ k_1 & k_2 \end{bmatrix} \in \Re^{2\times 2} \\ B_2 &= \begin{bmatrix} 0 \\ 1 \end{bmatrix} \in \Re^2 \\ C &= \begin{bmatrix} \frac{n-1}{n} & -\frac{1}{n} & \cdots & -\frac{1}{n} \\ -\frac{1}{n} & \frac{n-1}{n} & \ddots & -\frac{1}{n} \\ \vdots & \ddots & \ddots & \vdots \\ -\frac{1}{n} & -\frac{1}{n} & \cdots & \frac{n-1}{n} \end{bmatrix} \in \Re^{n\times n} \end{aligned}$$

With protocol (5.59), the network dynamics is

$$\begin{aligned}\dot{\xi}(t) &= [I_n \otimes A - (L + \Delta L) \otimes B_1]\xi(t) + (I_n \otimes B_2)w(t) \\ z(t) &= (C \otimes I_2)\xi(t)\end{aligned} \tag{5.60}$$

where $w(t) = [w_1(t), \ldots, w_n(t)]^T \in \Re^n, z(t) = [z_1(t), \ldots, z_n(t)]^T \in \Re^{2n}$, L is the Laplacian of the graph G and $\Delta L \in \Re^{n\times n}$ is the uncertainty Laplacian associated with the elements $\Delta a_{ij}(t)$.

By Lemma 9.4, ΔL can be transformed into $\Delta L = E_1\Sigma(t)E_2$, where $E_1 \in \Re^{n\times|E|}, E_2 \in \Re^{|E|\times n}$ are specified constant matrices and $\Sigma(t) \in \Re^{|E|\times|E|}$ is a diagonal matrix whose diagonal elements are the uncertainties of the edges, which is the non-zero $\Delta a_{ij}(t)$. It is easy to see that ΔL can be rewritten as $E_1\Sigma(t)E_2 = (E_1\Sigma_1)\tilde{\Sigma}(t)(\Sigma_2E_2)$, where Σ_1 and Σ_2 are adopted to validate $\tilde{\Sigma}(t)^T\tilde{\Sigma}(t) \le I$. Therefore, without loss of generality, we assume that $\Sigma^T(t)\Sigma(t) \le I$, that is $\psi_{ij} = 1$ for all $i = j$.

Define the following H_∞ performance index:

$$J = \int_0^\infty [z^T(t)z(t) - \gamma^2 w^T(t)w(t)]dt \tag{5.61}$$

where γ is a given positive constant. In [330], robust H_∞ consensus problems were studied for first-order multiagent systems. In this paper, our work focuses on extending the results of Lin *et al.* [330] to second-order multiagent systems and find proper values for the parameters k_0, k_1, and k_2 to make all agents reach consensus while satisfying the H_∞ performance index $J < 0$. However, due to the coupling of the two states of each agent, it is much harder to analyze the dynamics of the second-order multiagent systems especially when the communication graph is directed and the uncertainties are included. The approach adopted is to reduce the dimension of the whole system by separating out the agreement dynamics. A detailed discussion will be made in section 5.3.3.

5.3.3 Analytic results

Before presenting the main results, we need to first introduce some lemmas [330].

Lemma 5.7: *Consider the matrix C. The following statements hold:*

- *The eigenvalues of C are 1 with multiplicity $n-1$ and 0 with multiplicity 1. The vectors 1_n^T and 1_n are the left and the right eigenvectors of C associated with the zero eigenvalue, respectively.*
- *There exists an orthogonal matrix $U \in \Re^{n\times n}$ such that*

$$U^TCU = \begin{bmatrix} I_{n-1} & 0 \\ 0 & 0 \end{bmatrix}$$

and the last column is $1_n/\sqrt{n}$. Let $\Xi_1 \in \Re^{n\times n}$ be the Laplacian of any directed graph, then $U^T\Xi_1U = [\vartheta_1 \quad 0], \vartheta_1 \in \Re^{n\times(n-1)}$.

For convenience, denote

$$U = [U_1 \quad \bar{U}_1]$$

where $\bar{U}_1 = 1_n/\sqrt{n}$ *is the last column of* U *and* $U_1 \in \Re^{n\times(n-1)}$ *is the rest part.*

Theorem 5.7: *Consider a directed network of agents with fixed topology. The multiagent system (5.60) reach consensus while satisfying the* H_∞ *performance index* $J < 0$ *for a given constant* $\gamma > 0$, *if there exists a symmetric positive-definite matrix* $P \in \Re^{(2n-2)\times(2n-2)}$ *and a scalar* $\mu > 0$ *satisfying 5.62 where* $\bar{L} = U_1^T L U_1$ *and* $\bar{E} = (U_1^T E_2^T E_2 U_1) \otimes I_2$.

$$\begin{bmatrix} \begin{array}{l} P(I_{n-1}\otimes A - \bar{L}\otimes B_1) \\ \quad + (I_{n-1}\otimes A - \bar{L}\otimes B_1)^T P \\ \quad + \mu\bar{E} + I_{2n-2} \end{array} & P[(U_1^T E_1)\otimes B_1] & P(U_1^T \otimes B_2) \\ [(E_1^T U_1)\otimes B_1^T]P & -\mu I & 0 \\ (U_1 \otimes B_2^T)P & 0 & -\gamma^2 I \end{bmatrix} < 0 \tag{5.62}$$

Proof. Let

$$\begin{aligned} W(t) &= \frac{1}{n}\sum_{i=1}^{n}\int_0^t e^{A(t-s)}B_2 w_i(s)ds \\ \hat{\xi}(t) &= \xi(t) - 1_n \otimes W(t) \\ \delta(t) &= (U_1 \otimes I_2)^T \hat{\xi}(t) \\ \bar{\delta}(t) &= (\bar{U}_1 \otimes I_2)^T \hat{\xi}(t) \end{aligned} \tag{5.63}$$

where $W(t)$ describes the average effect of external disturbances on each agent; $\hat{\xi}(t)$ describes the states of all agents, which takes out the average of external disturbances; $\bar{\delta}(t)$ describes the average states of all agents; and $\delta(t)$ depicts the disagreement states of all agents.

Premultiplying the left-hand side of system (5.60) with the matrix $(U\otimes I_2)^T$ yields

$$\begin{aligned} (U\otimes I_2)^T \dot{\hat{\xi}}(t) &= \begin{bmatrix} \dot{\delta}(t) \\ \dot{\bar{\delta}}(t) \end{bmatrix} + (U\otimes I_2)^T (I_n \otimes \dot{W}(t)) \\ &= \begin{bmatrix} \dot{\delta}(t) \\ \dot{\bar{\delta}}(t) \end{bmatrix} + \begin{bmatrix} \begin{array}{l} \frac{1}{\sqrt{n}}\sum_{i=1}^{n}\int_0^t e^{A(t-s)}B_2 w_i(s)ds \\ \quad + \frac{1}{\sqrt{n}}\sum_{i=1}^{n} B_2 w_i(t) \end{array} \end{bmatrix} \\ &= \begin{bmatrix} \dot{\delta}(t) \\ \dot{\bar{\delta}}(t) \end{bmatrix} + \begin{bmatrix} 0 \\ \sqrt{n}AW(t) + \frac{1}{\sqrt{n}}\sum_{i=1}^{n} B_2 w_i(t) \end{bmatrix} \end{aligned} \tag{5.64}$$

Note that $L1_n = 0$ and $\Delta L1_n = 0$. Then premultiplying the right-hand side of system (5.60) with the matrix $(U \otimes I_2)^T$ yields

$$\begin{aligned}
(U \otimes I_2)^T \dot{\xi}(t) &= (U \otimes I_2)^T (I_n \otimes A)(\hat{\xi}(t) + 1_n \otimes W(t)) \\
&\quad - (U \otimes I_2)^T [(L + \Delta L) \otimes B_1](\hat{\xi}(t) \\
&\quad + 1_n \otimes W(t)) + (U^T \otimes B_2) w(t) \\
&= (U \otimes I_2)^T (I_n \otimes A)(U \otimes I_2) \begin{bmatrix} \delta(t) \\ \bar{\delta}(t) \end{bmatrix} \\
&\quad - (U \otimes I_2)^T [(L + \Delta L) \otimes B_1](U \otimes I_2) \begin{bmatrix} \delta(t) \\ \bar{\delta}(t) \end{bmatrix} \\
&\quad + (U \otimes I_2)^T (I_n \otimes A)(1_n \otimes W(t)) \\
&\quad + (U^T \otimes B_2) w(t) \\
&= \begin{bmatrix} I_{n-1} \otimes A & 0 \\ 0 & A \end{bmatrix} \begin{bmatrix} \delta(t) \\ \bar{\delta}(t) \end{bmatrix} \\
&\quad - \begin{bmatrix} (U_1^T L U_1 + U_1^T \Delta L U_1) \otimes B_1 & 0 \\ (\bar{U}_1^T L U_1 + \bar{U}_1^T \Delta L U_1) \otimes B_1 & 0 \end{bmatrix} \begin{bmatrix} \delta(t) \\ \bar{\delta}(t) \end{bmatrix} \\
&\quad + \begin{bmatrix} (U^T \otimes B_2) w(t) \\ \sqrt{n} A W(t) + B_2 \frac{1}{\sqrt{n}} \sum_{i=1}^{n} w_i(t) \end{bmatrix}
\end{aligned} \tag{5.65}$$

Denote $\overline{\Delta L} = U_1^T \Delta L U_1$. Then by (5.64) and (5.65), it is easy to see that

$$\begin{aligned}
\dot{\delta}(t) &= [I_{n-1} \otimes A - (\bar{L} + \bar{\Delta} L) \otimes B_1] \delta(t) + (U_1^T \otimes B_2) w(t) \\
\dot{\bar{\delta}}(t) &= [(\bar{U}_1^T L U_1 + \bar{U}_1^T \Delta L U_1) \otimes B_1] \delta(t) + A \bar{\delta}(t)
\end{aligned} \tag{5.66}$$

From (5.66), it is clear that $\delta(t)$ is independent of $\bar{\delta}(t)$ and $\bar{\delta}(t)$ is dependent on $\delta(t)$. Also, by Lemma 5.7

$$\begin{aligned}
z(t) &= (C \otimes I_2) \xi(t) = (C \otimes I_2)(U \otimes I_2)(U \otimes I_2)^T \xi(t) \\
&= diag\{U_1 \otimes I_2, 0\} \begin{bmatrix} \delta(t) \\ \bar{\delta}(t) \end{bmatrix}
\end{aligned} \tag{5.67}$$

It follows that $\lim_{t \to \infty} z(t) = 0$ when $\lim_{t \to \infty} \delta(t) = 0$. So, whether the multiagent system (5.60) can reach consensus is only related to the component $\delta(t)$. To investigate the H_∞ performance of the multiagent system (5.60), we can study the following system:

$$\begin{aligned}
\dot{\delta}(t) &= [I_{n-1} \otimes A - (\bar{L} + \bar{\Delta} L) \otimes B_1] \delta(t) + (U_1^T \otimes B_2) w(t) \\
z(t) &= (U_1 \otimes I_2) \delta(t)
\end{aligned} \tag{5.68}$$

Define a Lyapunov function for system (5.68) as follows:

$$V(t) = \delta^T(t)P\delta(t)$$

where $P \in \Re^{2(n-1)\times 2(n-1)}$ is a symmetric positive-definite matrix.

Calculating $\dot{V}(t)$ along the solution of system (5.68), it follows that

$$\begin{aligned}\dot{V}(t) &= 2\delta^T(t)P\dot{\delta}(t)\\ &= 2\delta^T(t)P[I_{n-1}\otimes A - (\bar{L} + \bar{\Delta}L)\otimes B_1]\delta(t)\\ &\quad + 2\delta^T(t)P(U_1^T \otimes B_2)w(t)\end{aligned}$$

For any $x, y \in \Re^n$ and any symmetric positive-definite matrix $R \in \Re^{n\times n}$

$$2x^Ty \le x^TR^{-1}x + y^TRy$$

And since $\Sigma^T(t)\Sigma(t) \le I_n$, it follows that

$$\begin{aligned}&-2\delta^T(t)P(\bar{\Delta}L\otimes B_1)\delta(t)\\ &\le \delta^T(t)P(U_1^TE_1\otimes B_1)\left(\frac{1}{\mu}I_{2n}\right)(U_1^TE_1\otimes B_1)^TP\delta(t)\\ &\quad + \delta^T(t)[(U_1^TE_2^T\Sigma^T)\otimes I_2](\mu I_{2n})[(\Sigma E_2U_1)\otimes I_2]\delta(t)\\ &\le \frac{1}{\mu}\delta^T(t)P(U_1^TE_1\otimes B_1)(U_1^TE_1\otimes B_1)^TP\delta(t) + \mu\delta^T(t)\bar{E}\delta(t)\end{aligned}$$

Consequently

$$\begin{aligned}\dot{V}(t) &\le 2\delta^T(t)P(I_{n-1}\otimes A - \bar{L}\otimes B_1)\delta(t) + 2\delta^T(t)P(U_1^T\otimes B_2)w(t)\\ &\quad + \frac{1}{\mu}\delta^T(t)P(U_1^TE_1\otimes B_1)(U_1^TE_1\otimes B_1)^TP\delta(t) + \mu\delta^T(t)\bar{E}\delta(t)\end{aligned}$$

Define the following cost performance index:

$$J_T = \int_0^T [z^T(t)z(t) - \gamma^2w^T(t)w(t)]dt$$

where $T > 0$ and $w(t) \in L_2[0,\infty)$. It is clear that $J = \lim_{T\to+\infty} J_T$. According to the linear superposition theorem, the response of a linear system can be decomposed into zero input response and zero state response. The former is caused by the non-zero initial condition while the latter is caused by the external input. So, to analyze the effects of the external disturbance $w(t)$ on system (5.68), suppose that all agents start from the consensus state, that is $\delta(0) = 0$. Clearly, $V(0) = 0$. Then it follows that

$$M = \begin{bmatrix} \begin{array}{c} P(I_{n-1} \otimes A - \bar{L} \otimes B_1) + (I_{n-1} \otimes A - \bar{L} \otimes B_1)^T P \\ + \dfrac{1}{\mu} P[(U_1^T E_1 E_1^T U_1) \otimes (B_1 B_1^T)]P + \mu \bar{E} + I_{2n-2} \end{array} & P(U_1^T \otimes B_2) \\ (U_1 \otimes B_2^T)P & -\gamma^2 I \end{bmatrix} < 0$$

$$\begin{aligned} J_T &= \int_0^T [z^T(t)z(t) - \gamma^2 w^T(t)w(t)]dt \\ &= \int_0^T [z^T(t)z(t) - \gamma^2 w^T(t)w(t) + \dot{V}(t)]dt - (V(T) - V(0)) \\ &\leq \int_0^T [\eta^T(t) M \eta(t)]dt - V(T) \end{aligned}$$

where $\eta(t) = [\delta^T(t) \quad w^T(t)]^T$.

By the Schur complements, $M < 0$ holds if and only if (5.62) holds. Note that $V(T) \geq 0$. Then, $J_T < 0$ when (5.62) holds. Let $T \to +\infty$. It follows that $J < 0$, that is $\int_0^\infty z^T(t)z(t)dt < \int_0^\infty \gamma^2 w^T(t)w(t)dt$. This completes the proof. □

Remark 5.13: *Transformation (5.63) plays a key role in our analysis. Applying (5.63), the agreement component is separated out from the dynamics of the multiagent system (5.60) and an equivalent system is obtained that describes the disagreement dynamics of (5.60). Moreover, it is shown that the agreement dynamics $\bar{\delta}(t)$ is independent of the external disturbance $w(t)$ but completely dependent on the disagreement dynamics $\delta(t)$.*

Remark 5.14: *It should be noted that a necessary condition for condition (5.62) is that the graph G has spanning trees. In fact, condition (5.62) implies that $P(I_{n-1} \otimes A - \bar{L} \otimes B_1) + (I_{n-1} \otimes A - \bar{L} \otimes B_1)^T P < 0$. Hence, $I_{n-1} \otimes A - \bar{L} \otimes B_1$ is Hurwitz. If the graph has no spanning trees, $\bar{L}$ must have zero eigenvalues, which implies $I_{n-1} \otimes A - \bar{L} \otimes B_1$ has zero eigenvalues and yields a contradiction.*

Condition (5.62) is a bilinear matrix inequality that can easily be solved by many approaches [331–333], when the number of the agents in the network, n, is not too large. But when n is very large, it becomes almost impossible to solve. In this case, we can consider assigning a specified form to the matrix P. In the following, discussion will be made about this on undirected graphs.

By the Schur complements, condition (5.62) is equivalent to

$$\begin{aligned} &P(I_{n-1} \otimes A - \bar{L} \otimes B_1) + (I_{n-1} \otimes A - \bar{L} \otimes B_1)^T P \\ &\quad + \frac{1}{\mu} P[(U_1^T E_1 E_1^T U_1) \otimes (B_1 B_1^T)]P + \mu \bar{E} + I_{2n-2} \\ &\quad + \gamma^{-2} P(I_{n-1} \otimes (B_2 B_2^T))P < 0 \end{aligned} \tag{5.69}$$

Note that E_1 and E_2 are constant matrices. Denote λ_{E_1} and λ_{E_2} as the largest eigenvalues of the matrices $U_1^T E_1 E_1^T U_1$ and $U_1^T E_2^T E_2 U_1$, respectively. Then (5.69) holds if

$$\begin{aligned} &P(I_{n-1} \otimes A - \bar{L} \otimes B_1) + (I_{n-1} \otimes A - \bar{L} \otimes B_1)^T P \\ &+ \frac{1}{\mu} P[I_{n-1} \otimes (\lambda_{E_1} B_1 B_1^T)]P + (\mu\lambda_{E_2} + 1)I_{2n-2} \\ &+ \gamma^{-2} P(I_{n-1} \otimes (B_2 B_2^T))P < 0 \end{aligned} \tag{5.70}$$

From Remark 5.14, the graph has spanning trees under condition (5.62). By Lemma 9.3, the eigenvalues of the Laplacian L can be denoted as $0 = \lambda_1 < \lambda_2 \leq \cdots \leq \lambda_n$. Moreover, from matrix theory, the matrices L and $U^T LU$ have the same eigenvalues. Then the eigenvalues of $\bar{L}$ are $\lambda_2, \lambda_3, \ldots, \lambda_n$ by Lemma 5.7. There must be an orthogonal matrix $H \in \Re^{(n-1)\times(n-1)}$ such that $H^T \bar{L} H = diag\{\lambda_2, \lambda_3, \ldots, \lambda_n\}$. Take $P = I_{n-1} \otimes \bar{P}$, where $\bar{P} \in \Re^{2\times 2}$ is an undermined positive-definite matrix. Pre- and post-multiplying the left-hand side of (5.70) with $H^T \otimes I_2$ and $H \otimes I_2$ yields

$$\begin{aligned} &I_{n-1} \otimes (\bar{P}A) - diag\{\lambda_2, \lambda_3, \ldots, \lambda_n\} \otimes (\bar{P}B_1) \\ &\quad + [I_{n-1} \otimes (\bar{P}A) \otimes diag\{\lambda_2, \lambda_3, \ldots, \lambda_n\} \otimes (\bar{P}B_1)]^T \\ &\quad + \frac{1}{\mu} I_{n-1} \otimes (\lambda_{E_1} \bar{P} B_1 B_1^T \bar{P}) + (\mu\lambda_{E_2} + 1)I_{2n-2} \\ &\quad + \gamma^{-2} I_{n-1} \otimes (\bar{P} B_2 B_2^T \bar{P}) < 0 \end{aligned} \tag{5.71}$$

It is easy to see that (5.71) holds if and only if

$$\begin{aligned} &\bar{P}A - \lambda_i, [\bar{P}A - \bar{P}B_1]^T + \frac{1}{\mu}\lambda_{E_1} \bar{P} B_1 B_1^T \bar{P} \\ &\quad + (\mu\lambda_{E_2} + 1)I_2 + \gamma^{-2} \bar{P} B_2 B_2^T \bar{P} < 0 \end{aligned} \tag{5.72}$$

for $2 \leq i \leq n$. This means that if the non-zero $n - 1$ eigenvalues of the Laplacian L all satisfy condition (5.72), the multiagent system (5.60) can reach consensus while satisfying the H_∞ performance index (5.61).

Remark 5.15: *As is well known, for a linear system, the asymptotic stability property is equivalent to the exponential stability property. Thus, under condition (5.62), system (5.68) is exponentially stable. That is, consensus can be achieved exponentially fast.*

In Theorem 5.7, the case with external disturbances is discussed. For better understanding the behavior of the multiagent system (5.60), let us consider the case without external disturbances. Set $w(t) \equiv 0$. System (5.66) becomes

$$\begin{bmatrix} \dot{\delta}(t) \\ \dot{\bar{\delta}}(t) \end{bmatrix} = \begin{bmatrix} I_{n-1} \otimes A - (\bar{L} + \bar{\Delta} L) \otimes B_1 & 0 \\ (\bar{U}_1^T L U_1 + \bar{U}_1^T \Delta L U_1) \otimes B_1 & A \end{bmatrix} \begin{bmatrix} \delta(t) \\ \bar{\delta}(t) \end{bmatrix} \tag{5.73}$$

It is clear that the system matrix of (5.73) has one eigenvalue at zero and one at k_0 and its other eigenvalues all have negative real parts under condition (5.62). And when

$k_0 \geq 0$, the first component of $\bar{\delta}(t)$ tends to infinity as $t \to +\infty$. From (5.63), it is easy to see that the average position state of the multiagent system tends to infinity as $t \to +\infty$. In what follows, it will be discussed whether the states of all agents stay bounded for $k_0 < 0$. Since the component $\delta(t)$ is independent of $\bar{\delta}(t)$, then $\delta(t)$ can be denoted as

$$\delta(t) = T(t, 0)\delta(0)$$

where $T(t, \cdot)$ is a continuous linear operator. Since $\delta(t)$ vanishes exponentially fast as $t \to +\infty$, there are two positive scalars r and α, for any $\Sigma(t)^T\Sigma(t) \leq I$, such that

$$\|T(t,s)\| \leq re^{-\alpha(t-s)}, \quad t \geq 0$$

Denote $\bar{\delta}(t) = [\bar{\delta}_1(t), \bar{\delta}_2(t)]^T$. From (5.73), it is easily obtained that

$$\begin{aligned}
\|\bar{\delta}_2(t)\| \leq & \int_0^t e^{k_0(t-s)}\|(\bar{U}_1^T LU_1 + \bar{U}_1^T \Delta LU_1)\| \\
& \otimes [k_1 \quad k_2]re^{-\alpha s}\|\delta(0)\|ds + \|e^{k_0 t}\bar{\delta}_2(0)\| \\
& \leq r\pi_1 \int_0^t e^{k_0 t-(k_0+\alpha)}\|\delta(0)\|ds + e^{k_0 t}\|\bar{\delta}_2(0)\| \\
& = \frac{r\pi_1}{k_0+\alpha}(-e^{-\alpha t} + e^{k_0 t})\|\delta(0)\| + e^{k_0 t}\|\bar{\delta}_2(0)\| \\
\|\bar{\delta}_1(t)\| = & \left\|\bar{\delta}_1(0) + \int_0^t \bar{\delta}_2(s)ds\right\| \\
& \leq \|\bar{\delta}_1(0)\| + \int_0^t \left\{\frac{r\pi_1}{k_0+\alpha}(-\tilde{e}^{-\alpha S} + e^{k_0 s})\|\delta(0)\| + e^{k_0 S}\|\bar{\delta}_2(0)\|\right\} ds \\
& = \left\|\bar{\delta}_1(0) + \frac{r\pi_1}{k_0+\alpha}\left[\frac{1}{\alpha}(e^{-\alpha t} - 1) + \frac{1}{k_0}(e^{k_0 t} - 1)\right] \|\delta(0)\| \right. \\
& \left. + \frac{1}{k_0}(e^{k_0 t} - 1)\|\bar{\delta}_2(0)\|\right.
\end{aligned}$$

where $\pi_1 = \max_t \|(\bar{U}_1^T LU_1 + \bar{U}_1^T \Delta LU_1) \otimes [k_1 \quad k_2]\|$. It is obvious that as $t \to +\infty$, each component of $\bar{\delta}(t)$ stays bounded.

By repeating the same argument of the proof of Theorem 5.7, the following proposition can be obtained.

Proposition 5.2: *Consider a directed network of agents with fixed topology in the absence of external disturbances. The multiagent system (5.60) can reach consensus if there exist a symmetric positive-definite matrix $P \in \Re^{(2n-2)\otimes(2n-2)}$ and a scalar $\mu > 0$ satisfying (5.74).*

In Theorem 5.7 and Proposition 5.2, we only discuss the case of fixed topology. In Theorem 5.8 and Proposition 5.3, we will discuss the case of switching topology.

To this end, we need to define a switching signal $\sigma = s(t) : [0,\infty) \rightarrow P = \{1,2,\ldots,N\}$ (N denotes the total number of the graphs of all possible topologies) that determines the topology. Such a function σ has a finite number of switching times and each time interval between every two consecutive switching times is assumed to be larger than a constant.

$$\begin{bmatrix} P(I_{n-1}\otimes A-\bar{L}\otimes B_1) \\ \quad +(I_{n-1}\otimes A-\bar{L}\otimes B_1)^T P & P[(U_1^T E_1)\otimes B_1] \\ \quad +\mu\bar{E} & \\ \quad [(E_1^T U_1)\otimes B_1^T]P & -\mu I \end{bmatrix} < 0 \tag{5.74}$$

$$\begin{bmatrix} P(I_{n-1}\otimes A-\bar{L}_\sigma\otimes B_1) & & \\ \quad +(I_{n-1}\otimes A-\bar{L}_\sigma\otimes B_1)^T P & P[(U_1^T E_{1\sigma})\otimes B_1] & P(U_1^T\otimes B_2) \\ \quad +\mu_\sigma\bar{E}_\sigma + I_{2n-2} & & \\ \quad [(E_{1\sigma}^T U_1)\otimes B_1^T]P & -\mu_\sigma I & 0 \\ \quad (U_1\otimes B_2^T)P & 0 & -\gamma^2 I \end{bmatrix} < 0 \tag{5.75}$$

$$\begin{bmatrix} P(I_{n-1}\otimes A-\bar{L}_\sigma\otimes B_1) \\ \quad +(I_{n-1}\otimes A-\bar{L}_\sigma\otimes B_1)^T P & P[(U_1^T E_{1\sigma})\otimes B_1] \\ \quad +\mu_\sigma\bar{E}_\sigma & \\ \quad [(E_{1\sigma}^T U_1)\otimes B_1^T]P & -\mu_\sigma I \end{bmatrix} < 0 \tag{5.76}$$

Theorem 5.8: *Consider a directed network of agents with switching topologies. The multiagent system (5.60) can reach consensus while satisfying the H_∞ performance index $J < 0$ for a given constant $\gamma > 0$, if there exist a common symmetric positive-definite matrix $P \in \Re^{(2n-2)\otimes(2n-2)}$ and a scalar $\mu_\sigma > 0$ for each possible communication graph G_σ satisfying (5.75), where $\bar{L}_\sigma = U_1^T L_\sigma U_1$ and $\bar{E}_\sigma = (U_1^T E_{2\sigma}^T E_{2\sigma} U_1)\otimes I_2$.*

Proof. This result can be proved following the lines of the proof of Theorem 5.7. However, it should be emphasized that all possible $I_{n-1}\otimes A - \bar{L}_\sigma \otimes B$ should share a common Lyapunov function $V(t) = \delta^T(t)P\delta(t)$ according to Theorem 2.1 in [334]. □

Remark 5.16: *As discussed previously, a necessary condition for the condition (5.64) is that each possible communication graph G_σ has spanning trees. Moreover, for undirected graphs, under condition (5.75), if each non-zero eigenvalue of all possible L_σ satisfies condition (5.72), then consensus can be achieved with desired H_∞ performance.*

Similar to Proposition 5.2, Proposition 5.3 can be easily obtained for the networks with switching topologies in the absence of external disturbances.

Proposition 5.3: *Consider a directed network of agents with switching topologies in the absence of external disturbances. The multiagent system (5.60) can* *reach consensus if there exists a common symmetric positive-definite matrix* $P \in R^{(2n-2)\otimes(2n-2)}$ *and a scalar* $\mu_\sigma > 0$ *for each possible communication graph* G_σ *satisfying, see (5.76).*

Remark 5.17: *In [302], decentralized coordination of second-order multiagent systems was studied with fixed topology. By the diagonalization of the Laplacian L, the multiagent system was decomposed into n subsystems and sufficient and necessary conditions were obtained for state consensus of the system. Also, in [327], second-order multiagent systems were studied with switching topology and time delay on undirected graphs. The approach used in [327] is to work out the average states of all agents and remove it from the dynamics of each agent. However, both approaches proposed in [302] and [327] cannot be applied to the models in this paper because there is no common transformation matrix that can diagonalize the matrix* $L + \Delta L(t)$ *and the average states of system (5.60) cannot be worked out due to the existence of uncertainties.*

Remark 5.18: *The agent models considered are in the form of second order. The proposed approach is promising and all the results might be extended to high-order multiagent systems with parameter uncertainties and external disturbances in the absence and presence of time delay.*

5.3.4 *Illustrative example 5.5*

Numerical simulations will be given to illustrate the theoretical results obtained in the previous sections. Figure 5.8 shows three different networks each with $n = 4$ agents. All directed graphs in this figure have spanning trees.

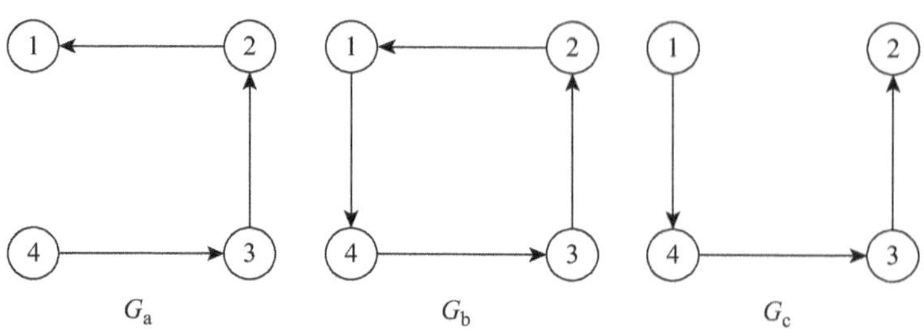

Figure 5.8 Three directed graphs

Suppose that the weight of each edge is 1, the uncertainty of each edge is

$$\Delta a_{ij} = \begin{cases} 0.01\sin[(i+j)t], & i \neq j \text{ and } a_{ij} \neq 0 \\ 0 & \text{otherwise} \end{cases}$$

and the initial condition of the multiagent system is zero. Moreover, the topology of the multiagent system switches every 0.01 s in the sequence of $\langle G_a, G_b, G_c, G_a \rangle$.

Choosing the performance index $\gamma = 1$, it is solved that (5.19) is feasible when $(k_0, k_1, k_2) = (-1, 5, 5)$, $(k_0, k_1, k_2) = (0, 5, 5)$ or $(k_0, k_1, k_2) = (0.3, 5, 5)$. According to Theorem 5.8, consensus can be achieved with desired H_∞ performance

$$\int_0^\infty z^T(t)z(t)dt < \int_0^\infty w^T(t)w(t)dt$$

when

$$(k_0, k_1, k_2) = (-1, 5, 5), \quad (k_0, k_1, k_2) = (0, 5, 5), \quad (k_0, k_1, k_2) = (0.3, 5, 5)$$

In practical situation, external disturbances usually occur in the form of pulse. So, take the external disturbance $w(t)$ as

$$w(t) = [1 \quad -2 \quad 0.5 \quad 3]^T \bar{\omega}(t)$$

where

$$\bar{\omega}(t) = \begin{cases} 1 & 0 \neq t \leq 1 \\ 0 & \text{otherwise} \end{cases}$$

is a pulse signal. Now, simulation results are presented for the consensus problems in directed networks of second-order multiagent systems with switching topology and external disturbance $w(t)$ for three cases:

(1) $(k_0, k_1, k_2) = (-1, 5, 5)$
(2) $(k_0, k_1, k_2) = (0, 5, 5)$
(3) $(k_0, k_1, k_2) = (0.3, 5, 5)$

Figures 5.9–5.14 show the position and velocity trajectories of all agents. It is observed that the pulse disturbance $w(t)$ makes the multiagent system diverge from consensus

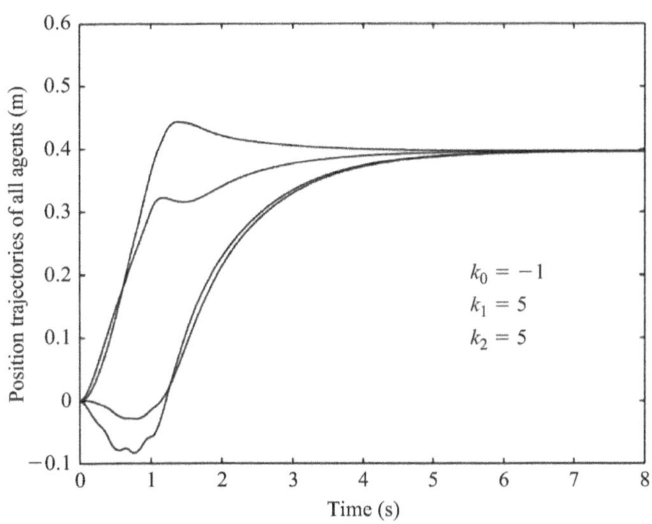

Figure 5.9 Position trajectories of all agents – case 1

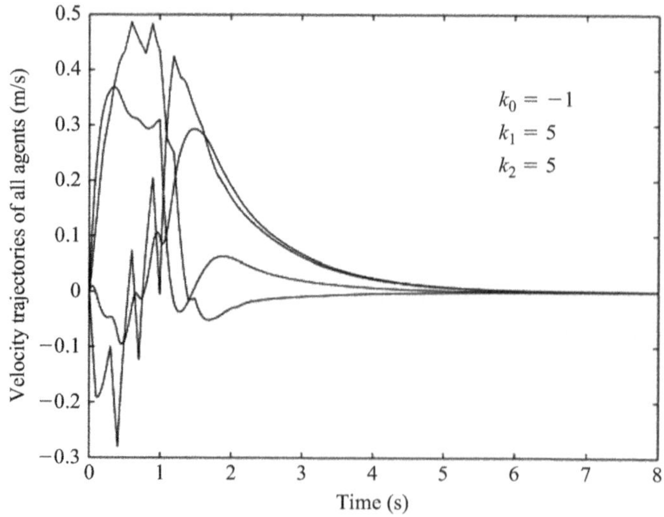

Figure 5.10 Velocity trajectories of all agents – case 1

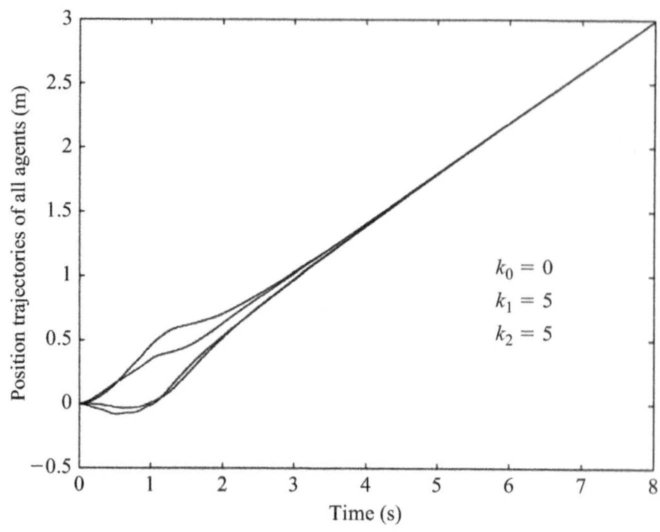

Figure 5.11 Position trajectories of all agents – case 2

in the first few seconds, but after a period of interaction between agents the multiagent system eventually return to consensus. Specifically, for $k_0 = -1$, all agents move to a common value; for $k_0 = 0$, all agents reach consensus and move with a common constant velocity; and for $k_0 = 0.3$, all agents reach consensus and move with a common constant acceleration. This suggests that the ultimate movement of

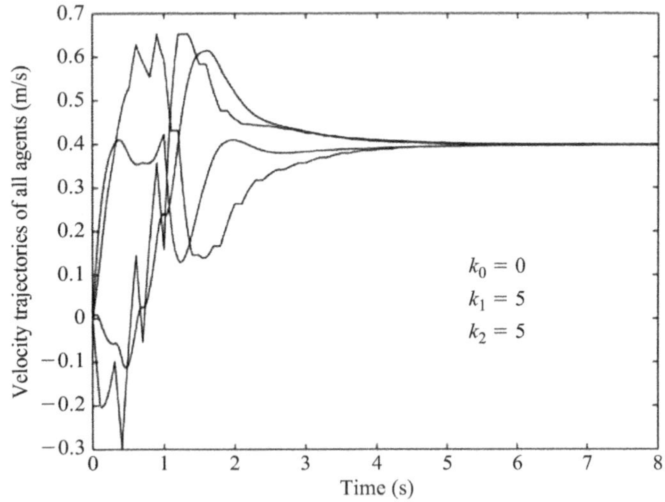

Figure 5.12 Velocity trajectories of all agents – case 2

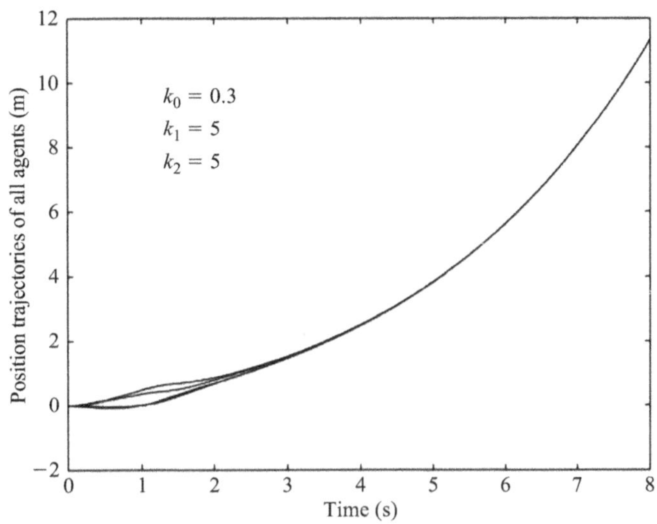

Figure 5.13 Position trajectories of all agents – case 1

the multiagent system is heavily dependent on the parameter k_0, which is consistent with [302].

Figures 5.15–5.20 describe the position error and the velocity error trajectories of all agents, that is $z_{i1}(t)$ and $z_{i2}(t)(i = 1, 2, \ldots, n)$, whereas Figures 5.14–5.16 show the energy trajectories of the output function $z(t)$ and the disturbance $w(t)$.

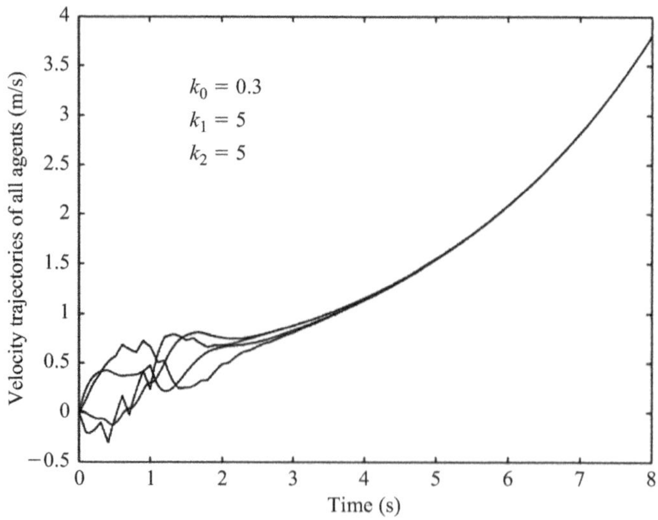

Figure 5.14 Velocity trajectories of all agents – case 1

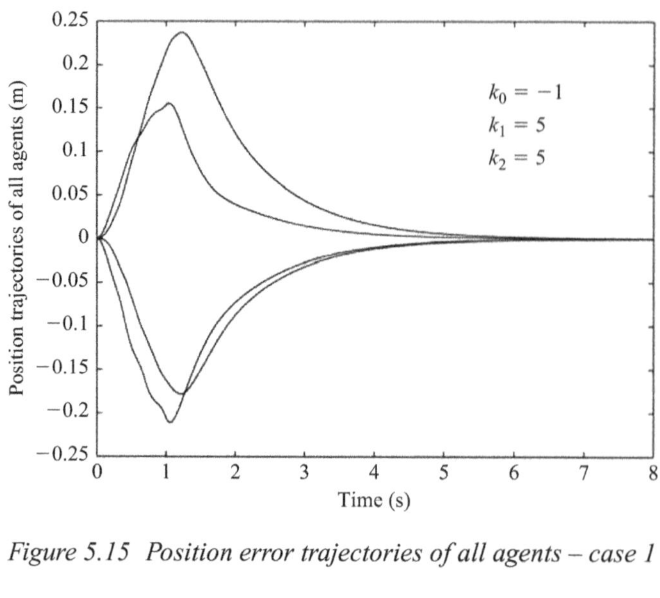

Figure 5.15 Position error trajectories of all agents – case 1

It is clear that the output function $z(t)$ vanishes $t \to +\infty$ and the desired H_∞ performance, $\int_0^\infty z^T(t)z(t)dt < \int_0^\infty w^T(t)w(t)dt$, is satisfied. Taken together, all simulations demonstrate that consensus can be achieved with desired H_∞ performance under the condition given by Theorem 5.8 (Figures 5.21–5.23).

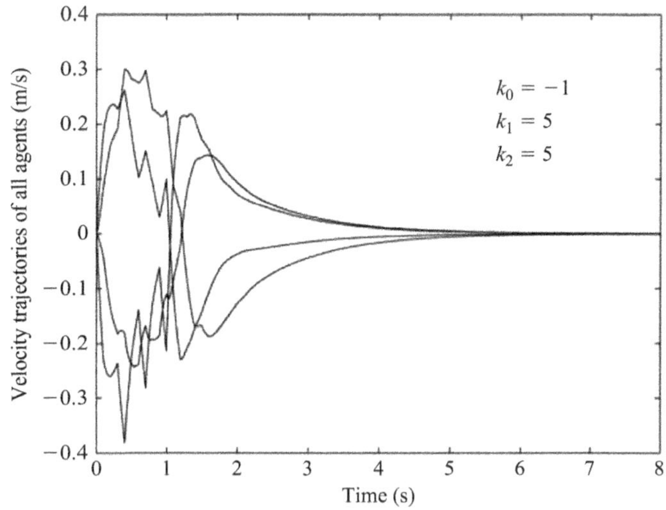

Figure 5.16 Velocity error trajectories of all agents – case 1

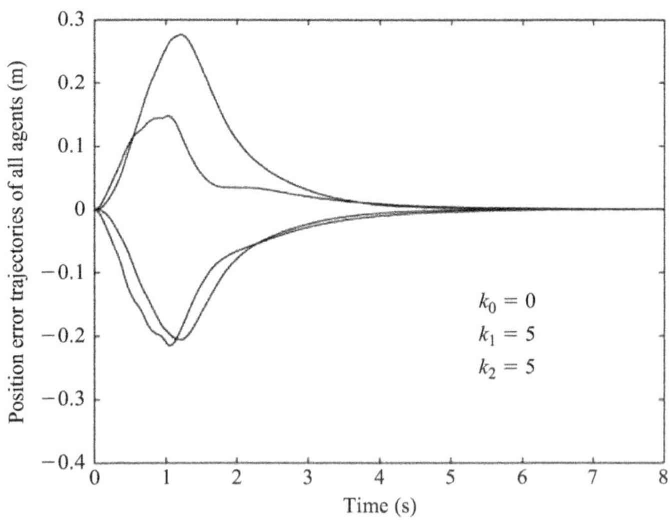

Figure 5.17 Position error trajectories of all agents – case 2

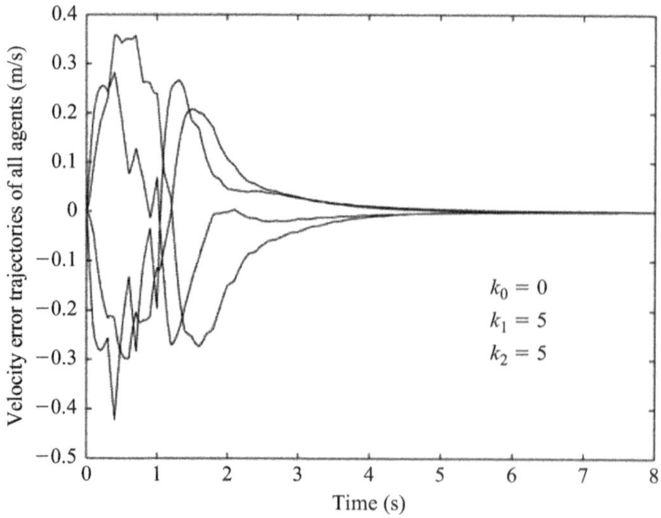

Figure 5.18 Velocity error trajectories of all agents – case 2

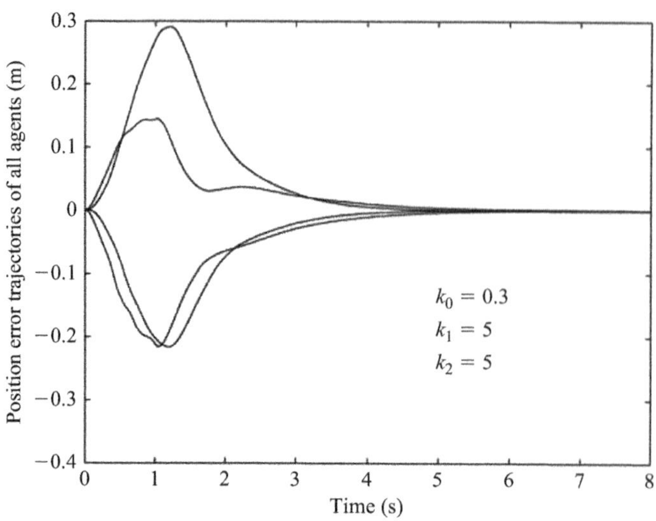

Figure 5.19 Position error trajectories of all agents – case 3

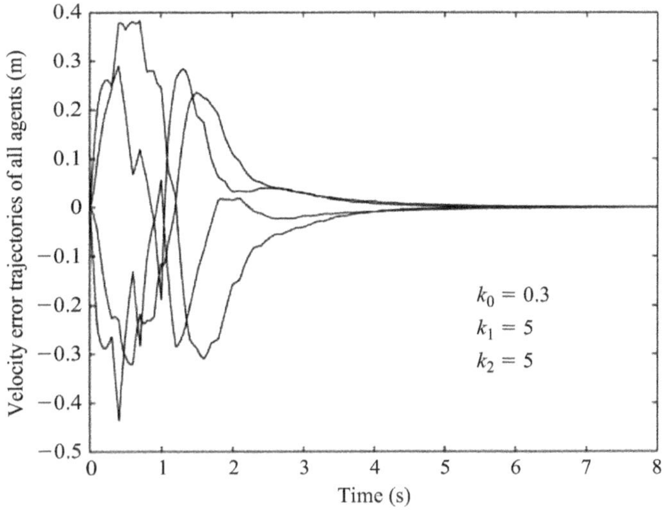

Figure 5.20 Velocity error trajectories of all agents – case 3

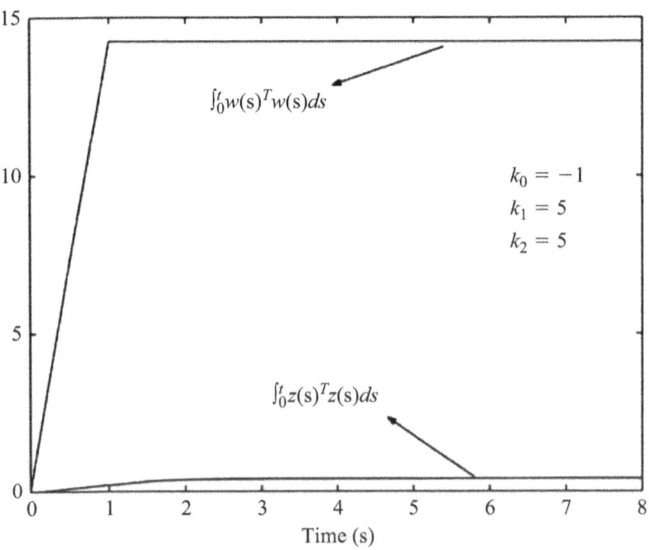

Figure 5.21 Energy trajectories of $z(t)$ and $w(t)$ – case 1

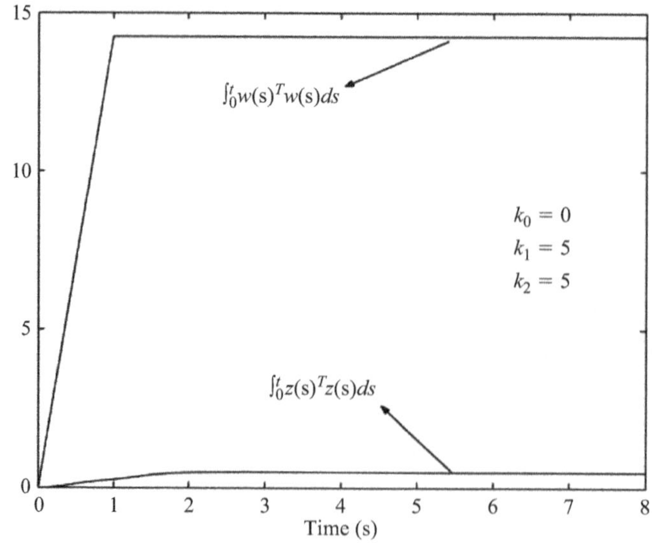

Figure 5.22 Energy trajectories of $z(t)$ and $w(t)$ – case 2

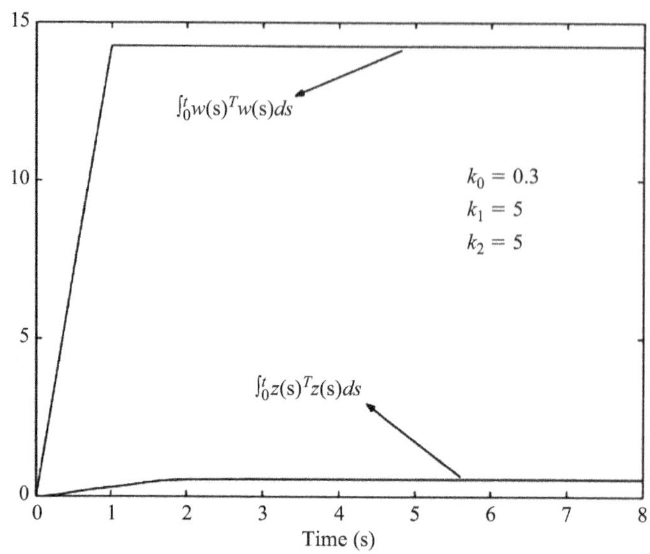

Figure 5.23 Energy trajectories of $z(t)$ and $w(t)$ – case 3

5.4 Notes

This chapter initially examined the dynamic consensus of a linear or linearized multi-agent system whose communication topology has a directed spanning tree. Based on the relative outputs of the neighboring agents, an observer-type protocol was adopted. It was shown, for neutrally stable agents, that there exists a protocol achieving consensus over a consensus region. This region is the entire open right-half plane if and only if each agent is stabilizable and detectable. Algorithms were developed to derive protocols to achieve consensus with or without convergence speed specification. It was established that the design procedures possess a computationally desirable decoupling property.

Next, delayed consensus problem of a group of leaderless multiple mobile agents with neighbor-based rule was addressed. The consensus stability was guaranteed with both fixed and switched interconnection topologies of the considered multiagent system. The dynamics of each agent was second-order with time-varying delays. A Lyapunov–Razumikhin function was employed in the stability analysis. For neutrally stable agents, it was shown that there exists a protocol achieving consensus and having a consensus region that is the entire open right-half plane if and only if each agent is stabilizable and detectable.

The consensus problem with a prescribed convergence speed was later investigated. A necessary and sufficient condition was then derived for the existence of a protocol that reaches consensus with a convergence speed larger than a given positive value and meanwhile yields an unbounded consensus region that means good robustness to the communication topology.

Later on, consensus problems were investigated in directed networks of second-order agents with uncertainties. Neighbor-based rules were adopted for each agent with the consideration of parameter uncertainties and external disturbances. Some conditions were derived under which all agents asymptotically reach consensus while satisfying desired H_∞ performance.

Chapter 6
Distributed estimation

In this chapter, an improved distributed estimation approach for sensor networks is initially developed. A distinct feature of this approach is toward characterizing the filter's performance limit and weights such that the variance of the estimation errors is minimized. Moreover, certain parameter optimization is alleviated with the application of a particular finite impulse response (FIR) filter. Next, the important topic of asynchronous multisensor systems is considered where data is missing and delays are likely to occur when performing state estimation at different rates. An algorithm to process both the delayed and missing measurements is developed, under the assumption that the data arrives between two consecutive sampling times. Local estimation is then implemented using a fusion algorithm.

6.1 Introduction

A sensor network (SN) is a network of autonomous devices that can sense their environment, make computations, and communicate with neighboring devices. SNs, and in particular wireless sensor networks (WSNs), have a growing domain of application in areas such as environmental monitoring, industrial automation, intelligent buildings, search and surveillance, and automotive applications [52,117,145, 146,159,162–173,175–181,335–341]. The characteristics of SNs motivate the development of new classes of distributed estimation and control algorithms, which explore these systems' limited power, computing, and communication capabilities. It is important that the algorithms have tuning parameters that can be adjusted according to the demands set by the applications. Therefore, we investigate a distributed estimation algorithm for tracking an unknown time-varying physical variable.

Distributed signal processing is a very active research area due to the recent developments in networking, computer, and sensor technologies. In-network computing thus differs from the traditional architecture where sensors simply provide raw data to a fusion center. By letting the network do the computations, it is possible to reach a scalable, fault tolerant, and flexible design. The drawback is that such a system is more difficult to analyze, as it is an asynchronous distributed computing system [342] with inputs and dynamics coupled to a physical environment. Despite current research activity and major progress, the theoretical understanding is far from satisfactory of these systems, exposed to link and node failures, packet drops, restricted power consumption, to name a few [511,512].

The new estimator for SNs presented in the sequel belongs to a class of recently developed filtering algorithms that exploit in-network computing [343,344,513]. The scalability of these algorithms is based on that node operates using only local information. Suitable cooperation between neighboring nodes improves the estimation quality considerably. Using sensor readings from more than one sensor, for example, can overcome intrinsic performance limitations due to uncertainty and noise present in individual devices.

In this chapter, we adopt the concept of minimum variance to develop an improved distributed estimation algorithm by considering the propagation delay of the physical quantity to be measured. This estimation algorithm has the following features:

- It incorporates an FIR filter aimed at simplifying the signal estimation and alleviating the parameter optimization.
- It tracks a time-varying signal, while the estimators of [247,345] are limited to averaging initial samples.
- Our approach does not require a model of the system that generates the signal to track, in contrast to model-based approaches [346,347].
- There is no pre-assigned coupling law imposed among the nodes as in [348].
- Compared to the methods of [247,346,349], our approach does not rely on the Laplacian matrix associated with the communication graph. Rather, it considers a more general model of the filter structure.
- Our filter parameters are computed through distributed algorithms, whereas, for example, the estimators of [350,351] rely on centralized algorithms for designing the filters.
- By extending the algorithms in [247,346,349], we design the filter weights such that the variance of the estimation errors is minimized, thereby improving the filter design considerably and we characterize the performance limit of the filter.

Today, an increasing number of applications demand remote control of plants over unreliable networks. The recent development of sensor web technology [335] enables the development of WSNs that can be immediately used for estimation and control. In these systems issues of communication delay, data loss and time-synchronization play critical roles. It is noted in [342] that several approaches have focused on diffusion mechanisms to have each node of the network obtain the average of the initial samples of the network nodes. Major progress has been made in understanding the convergence behavior of these consensus or state-agreement approaches. In [247], a scheme for sensor fusion based on a consensus filter is proposed. Here, each node computes a local weighted least-squares estimate that is shown to converge to the maximum likelihood solution for the overall network. An extension of this approach is presented in [350], where the authors study a distributed average computation of a time-varying signal, when the signal is affected by a zero-mean noise. A convex optimization problem is posed to compute the edge weights, which each node uses to minimize the least mean square deviation of the estimates. The same linear filter is also considered in [351], where the weights are computed off-line to speed up the computation of the averages. Further characterization of consensus filters for distributed sensor fusion is given in [349].

Another approach to distributed estimation is based on non-linear filters using self-synchronization and coupling functions, see [348,352–354]. In this case, the estimate of each node is provided by the state of a non-linear dynamical system. This system is coupled to some of the other nodes by a static coupling function. Some conditions on the coupling function that lead to asymptotic state synchronization are investigated in [348].

Distributed filtering using model-based approaches is studied in various wireless network contexts, see for example, [347,355–357]. Distributed Kalman filters and more recently a combination of the diffusion mechanism, discussed previously, with distributed Kalman filtering (DKF), see [337,358], have been proposed. A plausible approach is to communicate the estimates of the local Kalman filters, and then average these values using a diffusion strategy.

On another research front and regardless of the type of the signal (sound waves, infrared heat waves, or electromagnetic waves (EMW)), a time delay is always associated with wave propagation. This delay depends upon the

- propagation speed of the specific wave/signal,
- location of the source of signal, and
- geometry of the sensor nodes.

Ignoring this time delay is an obvious violation of the physics behind the propagation of signal, being envisioned to be best estimated from measurements of distantly placed sensor nodes.

6.2 Problem formulation

Consider $N > 1$ sensor nodes with known static positions in space. It is assumed the source of the signal is located at a known position and thus the radial r_i distance between an ith node and the signal source is also known. The time τ_i taken by a particular wavefront of the signal to travel from the source to an ith node can be obtained from the following relationship:

$$\tau_i = \frac{c}{r_i} \tag{6.1}$$

where c is the propagation speed of the signal, assumed to be known. An ith node measures a common scalar signal $d(t)$ affected by additive noise and delayed by τ_i:

$$u_{(\tau_i, v'_i)}(t) = d(t - \tau_i) + v'_i(t), \quad i = 1, \ldots, N$$

with $t \in N_0$ and where $v'_i(t)$ is zero-mean white noise.

With the knowledge of τ_i, data from an ith node and its neighbors can be appropriately *time adjusted* (either delayed or advanced, depending upon whether a neighbor node is closer to or farther from the source with respect to the ith node). We assume that this operation on the data is performed at the very beginning of the distributed estimation scheme under consideration and the resulting delay-adjusted noisy signal is denoted as $u_{v_i'}(t)$.

Since the signal spectrum is assumed to be known, a significant amount of noise can be filtered out with the application of a linear-phase FIR filter. This processing of the signal can be carried out at each individual node. As a result the signal will be left contaminated only with that part of the noise that has a spectrum overlapping with that of the signal of interest. Let the filtered noisy signal and vestige of the noise at node i are denoted as $u_i(t)$ and $v_i(t)$, respectively.

Let us collect measurements and noise variables in vectors

$$u(t) = (u_1(t), \ldots, u_N(t))^T, \quad v(t) = (v_1(t), \ldots, v_N(t))^T$$

so that we can rewrite the previous equation as

$$u(t) = d(t)1 + v(t), \quad t \in N_0$$

The covariance matrix of $v(t)$ is assumed to be diagonal $\Sigma = \sigma^2 I$, so $v_i(t)$ and $v_j(t)$, for $i \neq j$, are uncorrelated. The additive noise, in each node, can be averaged out only if nodes communicate measurements or estimates. The communication rate of the measurements and estimates should be just fast enough to track the variations of $d(t)$. Indeed, increasing the sampling rate, in general, is not beneficial because measurements might then be affected by auto-correlated noise.

It is convenient to model the communication network as an undirected graph $G = (V, E)$, where $V = \{1, \ldots, N\}$ is the vertex set and $E \subseteq V \times V$ the edge set. We assume that if $(i, j) \in E$ then $(j, i) \in E$, so that the direction of the edges can be dropped when representing the network. The graph G is said to be connected if there is a sequence of edges in E that can be traversed to go from any vertex to any other vertex.

In the sequel, we denote the set of neighbors of node $i \in V$ plus the node itself as

$$N_i = \{j \in V : (j, i) \in E\} \cup \{(i, i)\}$$

The estimation algorithm we propose is such that a node i computes an estimate $x_i(t)$ of $d(t)$ by taking a linear combination of neighboring estimates and measurements

$$x_i(t) = \sum_{j \in N_i} k_{ij}(t) x_j(t-1) + \sum_{j \in N_i} h_{ij}(t) u_j(t) \tag{6.2}$$

We assume that neighboring estimates and measurements are always successfully received, that is, there are no packet losses. This assumption is obviously plausible for wired connections, but it is still valid in wireless networks if certain assumptions hold true. More specifically, the designed estimator is suitable in wireless networks where the sampling time between measurements is long compared to the coherence time of the wireless channel (which is around some hundreds of milliseconds) and an Automatic Repeat Request (ARQ) protocol is used. Under such assumptions, if the wireless channel does not allow a packet to be successfully received at a given time instance, there is enough time to detect and retransmit erroneous packets until they are successfully received. These assumptions are representative of the

IEEE 802.11b and IEEE 802.11g [359], which have been actually used for distributed estimation and control algorithms of unmanned aerial vehicles [360].

We assume that for each node i, the algorithm is initialized with $x_j(0) = u_i(0)$, $j \in N_i$. In vector notation, we have

$$x(t) = K(t)x(t-1) + H(t)u(t) \tag{6.3}$$

where $K(t)$ and $H(t)$ can be interpreted as the adjacency matrices of two graphs with time-varying weights. These graphs are compatible with the underlying communication network represented by G.

Given an SN modeled as a connected graph G, we have the following design problem: find time-varying matrices $K(t)$ and $H(t)$, compatible with G, such that the signal $d(t)$ is consistently estimated and the variance of the estimate is minimized. Moreover, the solution should be distributed in the sense that the computation of $k_{ij}(t)$ and $h_{ij}(t)$ should be performed locally by node i.

6.3 Convergence of the centralized estimation error

In this section we derive conditions on $K(t)$ and $H(t)$ that guarantee the estimation error to converge. Define the estimation error $e(t) = x(t) - d(t)1$. Introduce $\delta(t) = d(t) - d(t-1)$, so that the error dynamics can be described as

$$\begin{aligned} e(t) &= K(t)e(t-1) + d(t)(K(t) + H(t) - I)1 \\ &\quad -\delta(t)K(t)1 + H(t)v(t) \end{aligned} \tag{6.4}$$

Taking the expected value with respect to the stochastic variable $v(t)$, we obtain

$$\begin{aligned} Ee(t) &= K(t)Ee(t-1) + d(t)(K(t) + H(t) - I)1 \\ &\quad -\delta(t)K(t)1 \end{aligned} \tag{6.5}$$

Proposition 6.1: *Consider system (6.4) and assume that*

$$(K(t) + H(t))1 = 1 \tag{6.6}$$

and that there exists $0 \leq \gamma_0 < 1$ *such that*

$$\gamma_{\max}(K(t)) \leq \gamma_0, \quad \forall\, t \in N_0 \tag{6.7}$$

(1) If $H(t)1 = 1$, *for all* $t \in N_0$, *then*

$$\lim_{t\to+\infty} [e(t)] = 0$$

(2) If $|\delta(t)| < \Delta$, *for all* $t \in N_0$, *then*

$$\lim_{t\to+\infty} \|\mathbb{E}[e(t)]\| \leq \frac{\sqrt{N}\Delta\gamma_0}{1-\gamma_0} \tag{6.8}$$

Proof. If $(K(t) + H(t))1 = 1$ then the system equation reduces to

$$\mathbb{E}[e(t)] = K(t)\mathbb{E}[e(t-1)] + \delta(t)(H(t) - I)1 \tag{6.9}$$

(1) If $H(t)1 = 1$, then (6.9) becomes $\mathbb{E}[e(t)] = K(t)\mathbb{E}[e(t-1)]$. Let us consider the function $V(t) = \|Ee(t)\|$. It follows that

$$V(t) \leq \|K(t)\|V(t-1) \leq \gamma_0 V(t-1) \leq \gamma_0^t V(0)$$

which implies that $\lim_{t\to+\infty} \|\mathbb{E}[e(t)]\| = 0$.

(2) In this case, we have $H(t)1 - 1 = -K(t)1$ and thus the system (6.4) becomes

$$\mathbb{E}[e(t)] = K(t)\mathbb{E}[e(t-1)] - \delta(t)K(t)1$$

With $V(t) = \|\mathbb{E}[e(t)]\|$, we have

$$\begin{aligned} V(t) &\leq \|K(t)\|V(t-1) + \|K(t)\|\sqrt{N}\Delta \\ &\leq \gamma_0 V(t-1) + \gamma_0\sqrt{N}\Delta \\ &\leq \gamma_0^t V(0) + \gamma_0 \frac{1-\gamma_0^t}{1-\gamma_0}\sqrt{N}\Delta \end{aligned}$$

Taking the limit for $t \to +\infty$ we obtained the result. □

Proposition 6.1 (i): *provides the condition $H(t)1 = 1$ under which the estimate is unbiased. It is possible to show that in this case the variance is minimized if $K(t) = 0$ and*

$$h_{ij}(t) = h_{ji}(t) = \begin{cases} \dfrac{1}{|N_i|} & \textit{if } j \in N_i \\ 0 & \textit{otherwise} \end{cases}$$

Note that nodes do not use any memory and that the error variance at each node is proportional to its neighborhood size. However, if $d(t)$ is slowly varying, then, under the assumptions of Proposition 6.1(ii), it is possible to guarantee that $\|\mathbb{E}[e(t)]\|$ tends to a neighborhood of the origin, but the estimate might be biased. Note also that $\|\mathbb{E}[e(t)]\|$ is a cumulative bias, that is, it is a function of the sum of the N biases of individual nodes.

The size of the cumulative bias can be kept small with respect to the signal to track by defining a proper value of γ_0. In particular, (6.8) can be related to the signal-to-noise ratio (SNR) of the average of the estimate as follows. Let P_d denote the average power of d and let P_b denote the desired power of the biases of the average of the estimates. Then, we define the desired SNR as $\text{SNR} = P_d/P_b$. Since there are N nodes, we consider the average SNR of each node as $\Upsilon = \text{SNR}/N$. Let us assume that we want the estimator to guarantee that the right-hand side of (6.8) be equal to this desired $\sqrt{SNR}$, i.e., that

$$\gamma_0 = \frac{\sqrt{\Upsilon}}{\sqrt{\Upsilon} + \Delta}$$

The right-hand side is useful in the tuning of the estimator. Hence, we denote it as $f(\Delta, \Upsilon)$. By choosing an appropriate Υ, we have a guaranteed convergence property of the estimator given by the corresponding $f(\Delta, \Upsilon)$. This function will allow us to relate the size of the bias of estimates with the variations of the signal to track, and the stability of the estimates, as we see in the next sections.

6.4 Minimum variance estimation design

In this section we describe how each node computes adaptive weights to minimize its estimation error variance. Note that, in order to guarantee that the estimation error of the overall sensor network, in average, converges to a neighbor of the origin, each node needs to locally compute the row-elements of $K(t)$ and $H(t)$ so that conditions in Proposition 6.1 are fulfilled. The condition $(K(t) + H(t))1 = 1$ is easily handled in a distributed way, as it states that the sum of the row-elements of $K(t)$ and $H(t)$ need to sum up to one. The bound on the maximum singular value of $K(t)$, however, requires some transformations so that new conditions on the row-elements of $K(t)$ fulfill $\gamma_{\max}(K(t)) \leq f(\Delta, \Upsilon)$. It turns out that it is possible to determine a local condition, $\sum_{j\in N_i} k_{ij}^2 \leq \psi_i$, where ψ_i is a constant that can be computed locally by the nodes. We then pose an optimization problem for finding optimal weights that minimize the error variance in each node, where the previous conditions are considered as constraints. An important aspect of the distributed optimal solution is that the weights depend on the error covariance matrix, which is not available at each node. We end this section by discussing a way of estimating it.

6.4.1 Distributed variance minimization

Let $M_i = |N_i|$, which denotes the number of neighbors of node i, including the node itself. Collect the estimation errors available at node i in the vector $\varepsilon_i \in \Re^{M_i}$. The elements of ε_i are ordered according to node indices:

$$\varepsilon_i = (e_{i1}, \ldots, e_{iM_i})^T, \quad i_1 < \cdots < i_{M_i}$$

Similarly, we introduce vectors $\kappa_i^T(t), \eta_i^T(t) \in \Re^{M_i}$ corresponding to the non-zero elements of row i of the matrices $K(t)$ and $H(t)$, respectively, and ordered according to node indices.

The expected value of the estimation error at node i can be written as

$$\mathbb{E}e_i(t) = \kappa_i^T(t)\mathbb{E}\varepsilon_i(t-1) - \kappa_i^T(t)\delta(t)1 \tag{6.10}$$

where we used the fact that $d(t) - d(t-1) = \delta(t)$ and that $(K(t) + H(t))1 = 1$. Note that the latter inequality is equivalent to $(\kappa_i(t) + \eta_i(t))^T 1 = 1$. We assume that $e_i(0) = u_i(0)$. Hence

$$\mathbb{E}[e_i(t) - \mathbb{E}[e_i(t)]]^2 = \kappa_i^T(t)\Gamma_i(t-1)\kappa_i(t) + \sigma^2\eta_i^T(t)\eta_i(t)$$

where $\Gamma_i(t) = \mathbb{E}[\varepsilon_i(t) - \mathbb{E}[\varepsilon_i(t)]][\varepsilon_i(t) - \mathbb{E}[\varepsilon_i(t)]]^T$. To minimize the variance of the estimation error in each node, we need to determine $\kappa_i(t)$ and $\eta_i(t)$ so that the previous expression is minimized at each time instant. We have the following optimization problem:

$$\mathrm{P}_1 : \min_{\kappa_i(t),\eta_i(t)} \kappa_i^T(t)\Gamma_i(t-1)\kappa_i(t) + \sigma^2\eta_i^T(t)\eta_i(t) \tag{6.11}$$

$$\text{such that} \quad (\kappa_i(t) + \eta_i(t))^T 1 = 1 \tag{6.12}$$

$$\gamma_{\max}(K(t)) \leq f(\Delta, \Upsilon) \tag{6.13}$$

The inequality constraint (6.13) is still global, since $\gamma_{\max}(K(t))$ depends on all $\kappa_i(t)$, $i = 1, \ldots, N$. We show next that it can be replaced by the local constraint

$$\|\kappa_i(t)\| \leq \psi_i, \quad t \in N_0 \tag{6.14}$$

where $\psi_i > 0$ is a constant that can be computed locally. The new constraint, however, even though ensures the stability of the estimation error, leads to a distributed solution that is in general different from the centralized one.

For $i = 1, \ldots, N$, define the set $\Theta_i = \{j \neq i : N_j \cap N_i \neq 0\}$, which is the collection of nodes located at two hops distance from node i plus neighbor nodes of i. We have the following result.

Proposition 6.2: *Suppose there exist $\psi_i > 0, i = 1, \ldots, N$, such that*

$$\psi_i + \sqrt{\psi_i}\sum_{j\in\Theta_i}\sqrt{\alpha_{i,j}^{(i)}\alpha_{i,j}^{(j)}\psi_j} \leq f^2(\Delta, \Upsilon) \tag{6.15}$$

where $\alpha_{i,j}^{(i)}\alpha_{i,j}^{(j)} \in (0, 1)$ are such that

$$\sum_{c\in N_j\cap N_i} k_{ic}^2 \leq \alpha_{i,j}^{(i)}\sum_{r=1}^{M_i}\kappa_{ii_r}^2, \qquad \sum_{c\in N_j\cap N_i} k_{jc}^2 \leq \alpha_{i,j}^{(j)}\sum_{r=1}^{M_j}\kappa_{ji_r}^2$$

If $\|\kappa_i(t)\|^2 \leq \psi_i$, $i = 1, \ldots, N$, then $\gamma_{\max}(K(t)) - f(\Delta, \Upsilon)$.

Proof. We use Gersgorin theorem to bound the eigenvalues of the matrix KK^T, that is, the singular values of K. The following relations hold: $[KK^T]_{ii} = \sum_{c=1}^{M_i} k_{ic}^2$ and $[KK^T]_{ij} = \sum_{c=1}^{N} k_{ic}k_{jc}$. By the Gersgorin theorem we know that for $r = 1, \ldots, N$

$$\lambda_r(KK^T) \in \cup_{i=1}^{N}\{z \in R : |z - [KK^T]_{ii}| \leq R_i(KK^T)\}$$

with

$$R_i(KK^T) = \sum_{j=1,j\neq i}^{N} |[KK^T]_{ij}| = \sum_{j=1,j\neq i}^{N}\left|\sum_{c=1}^{N} k_{ic}k_{jc}\right|$$

Now the inner sum in $R_i(KK^T)$ is non-zero only for $c \in N_j \cap N_i$. Thus,

$$R_i(KK^T) = \sum_{j\in\Theta_i}\left|\sum_{c\in N_j\cap N_i} k_{ic}k_{jc}\right|$$

Using the Cauchy–Schwartz inequality

$$\left| \sum_{c \in N_j \cap N_i} k_{ic} k_{jc} \right|^2 \leq \sum_{c \in N_j \cap N_i} k_{ic}^2 \sum_{c \in N_j \cap N_i} k_{jc}^2$$

Then,

$$\begin{aligned} \sum_{j=1, j \neq i}^{N} \left| \sum_{c=1}^{N} k_{ic} k_{jc} \right| &\leq \sum_{j \in \Theta_i} \sqrt{\sum_{c \in N_j \cap N_i} k_{ic}^2 \sum_{c \in N_j \cap N_i} k_{jc}^2} \\ &\leq \sum_{j \in \Theta_i} \sqrt{\alpha_{i,j}^{(i)} \sum_{r=1}^{M_i} \kappa_{ii_r}^2 \alpha_{i,j}^{(j)} \sum_{r=1}^{M_j} \kappa_{ji_r}^2} \\ &\times \sqrt{\sum_{r=1}^{M_i} \kappa_{ii_r}^2} \sum_{j \in \Theta_i} \sqrt{\alpha_{i,j}^{(i)} \alpha_{i,j}^{(j)} \sum_{r=1}^{M_j} \kappa_{jj_r}^2} \end{aligned}$$

Hence,

$$\lambda_r(KK^T) \in \bigcup_{i=1}^{N} \left\{ z \in C : \left| z - \sum_{r=1}^{M_i} \kappa_{jj_r}^2 \right| \leq \sqrt{\sum_{r=1}^{M_i} \kappa_{ii_r}^2} \sum_{j \in \Theta_i} \sqrt{\alpha_{i,j}^{(i)} \alpha_{i,j}^{(j)} \sum_{r=1}^{M_j} \kappa_{jj_r}^2} \right\}$$

From the hypothesis that $\|\kappa_i\| = \sum_{r=1}^{M_i} \kappa_{ii_r}^2 \leq \psi_i$ and (6.15), then

$$\sum_{r=1}^{M_i} \kappa_{ic}^2 + \sqrt{\sum_{r=1}^{M_i} \kappa_{ii_r}^2} \cdot \sum_{j \in \Theta_i} \sqrt{\sum_{r=1}^{M_j} \alpha_{i,j}^{(i)} \alpha_{i,j}^{(j)} \kappa_{jj_r}^2} \leq f^2(\Delta, \Upsilon)$$

Hence $\gamma_{\max}(K) \leq f(\Delta, \Upsilon)$. □

Proposition 6.2 provides a simple local condition on the filter coefficients such that $\gamma_{\max}(K) - f(\Delta, \Upsilon)$. We can expect that Proposition 6.2 is in general conservative, because no a priori knowledge of the network topology is used, the proof relies on the Gersgorin theorem and the Cauchy–Schwartz inequality. There are many other ways to bound the eigenvalues of a matrix by its elements than the one used in the proof above, for example, [361]. However, we do not know of any other bounds requiring only local information. Further, the Peron–Frobenius theory cannot be directly applied to bound the eigenvalues, because we make no assumption on the sign of the elements of $K(t)$.

The parameters $\alpha_{i,j}^{(i)}$ and $\alpha_{i,j}^{(j)}$ in Proposition 6.2 can all be set to one. However, this yields very conservative bounds on the maximum eigenvalue of KK^T. In section 6.4.7, we show how to choose these parameters to avoid bounds that are too conservative.

6.4.2 Optimal weights for variance minimization

Using previous results, we consider the following local optimization problem:

$$\mathrm{P}_2: \min_{\kappa_i(t),\eta_i(t)} \kappa_i(t)^T\Gamma_i(t-1)\kappa_i(t) + \sigma^2\eta_i(t)^T\eta_i(t) \tag{6.16}$$

$$\begin{aligned} s.t.\ & (\kappa_i(t)+\eta_i(t))^T 1 = 1 \\ & \|\kappa_i\|^2 \leq \psi_i \end{aligned} \tag{6.17}$$

We observe here that Problem P_2 has a different solution with respect to Problem P_1, because the constraint (6.13) has been replaced with (6.17). Problem P_2 is convex. In fact, the cost function is convex, as $\Gamma(t-1)$ is positive definite, since it represents the covariance matrix of Gaussian random variable, and the two constraints are also convex. The problem admits a strict interior point solution, corresponding to $\kappa_i(t)=0$ and $\eta_i(t)1=1$. Thus, Slater's condition is satisfied and strong duality holds [362]. The problem, however, does not have a closed-form solution: we need to rely on numerical algorithms to derive the optimal $\kappa_i(t)$ and $\eta_i(t)$. The following proposition provides a specific characterization of the solution.

Proposition 6.3: *For a given positive definite matrix $\Gamma_i(t-1)$, the solution to Problem P_2 is given by*

$$\kappa_i(t) = \frac{\sigma^2(\Gamma_i(t-1)+\xi_i I)^{-1}1}{\sigma^2 1^T(\Gamma_i(t-1)+\xi_i I)^{-1}1 + M_i} \tag{6.18}$$

$$\eta_i(t) = \frac{1}{\sigma^2 1^T(\Gamma_i(t-1)+\xi_i I)^{-1}1 + M_i} \tag{6.19}$$

with $\xi_i \in [0, \max(0, \sigma^2/\sqrt{M_i\psi_i} - \lambda_{\min}(\Gamma_i(t-1)))]$.

Proof. Since the problem is convex and Slater's condition holds, the KKT conditions are both necessary and sufficient for optimality. The primal and dual optimal points, (κ_i^*, η_i^*) and (λ_i^*, ν_i^*), respectively, need to satisfy

$$\begin{aligned} &(\kappa_i^*)^T\kappa_i^* - \psi_i \leq 0, && (\kappa_i^* + \eta_i)^T 1 - 1 = 0, \\ &\xi_i^* \geq 0, && \xi_i^*((\kappa_i^*)^T\kappa_i^* - \psi_i) = 0, \\ &2(\Gamma_i + \lambda_i^* I)\kappa_i^* + \nu_i^* 1 = 0, && 2\sigma^2\eta_i^* + \nu_i^* 1 = 0, \end{aligned}$$

where the last two KKT conditions follow from

$$\nabla_{\kappa_i} L(\kappa_i, \eta_i, \xi, \nu), \quad \nabla_{\eta_i} L(\kappa_i, \eta_i, \xi, \nu)$$

with the Lagrangian

$$L(\kappa_i, \eta_i, \xi, \nu) = \kappa_i^T\Gamma_i\kappa_i + \sigma^2\eta_i^T\eta_i + \xi_i(\kappa_i^T\kappa_i - \psi_i) + \nu_i((\kappa_i+\eta_i)^T 1 - 1)$$

Combining these two KKT conditions with the second KKT condition we obtain the optimal values. From the fourth KKT condition we have that either $\xi^* = 0$ or $(\kappa_i^*)^T\kappa_i^* = \psi_i$, where the second equality gives

$$\frac{\sigma^4 1^T(\Gamma_i + \xi_i^* I)^{-2}1}{(\sigma^1 1^T(\Gamma_i + \xi_i^* I)^{-1}1 + M_i)^2} = \psi_i$$

We are not able to provide a solution ξ^* in closed form. Instead we give a bound for the variable. From the previous equation, we can enforce a $\xi \geq 0$ such that

$$(\kappa_i^*)^T\kappa_i^* \leq \frac{\sigma^4\|(\Gamma_i + \xi I)^{-1}\|^2}{M_i} \leq \frac{\sigma^4}{M_i\lambda_{\min}^2(\Gamma_i + \xi I)} \leq \psi_i$$

from where we obtain

$$\xi \geq \frac{\sigma^2}{\sqrt{M_i\psi_i}} - \lambda_{\min}(\Gamma_i),$$

and for all these values of ξ the first KKT condition is always satisfied. This implies that the optimal value of ξ must be in the interval

$$[0, \max(0, \sigma^2/\sqrt{M_i\psi_i}) - \lambda_{\min}(\Gamma_i(t-1))]$$

and the theorem is proved. □

Proposition 6.3 gives an interval within which the optimal ξ_i can be found.

The first constraint in Problem P_2 is similar to the water-filling problem for power allocation in wireless networks [362]. Analogous to that problem, simple search algorithms such as a bisection algorithm, can be considered to solve for ξ_i numerically. Note that each node i needs to know the covariance matrix $\Gamma_i(t-1)$ to compute the weights. It is important to note that the Problem P_2 does not have the same solution as the Problem P_1, as the constraints (6.13) and (6.17) are not equivalent, although if (6.17) holds then (6.13) holds as well.

6.4.3 Bounds on the error variance

The optimal weights from Proposition 10 gives the following estimation error variance.

Proposition 6.4: *Let $\kappa_i(t)$ and $\eta_i(t)$ be an optimal solution given by (6.18) and (6.19). Then*

$$\mathbb{E}[e_i(0) - \mathbb{E}[e_i(0)]]^2 = \sigma^2$$

$$\mathbb{E}[e_i(t) - \mathbb{E}[e_i(t)]]^2 \leq \frac{\sigma^2}{M_i}, \quad t \in N_0\ \{0\}.$$

Proof. For $t=0$, $e_i(0)=u_i(0)=d(0)+v_i(0)$, so $\mathbb{E}[e_i(0)-\mathbb{E}[e_i(0)]]^2=\eta^2$. For $t>0$, the error variance of the ith node with the optimal values of $\kappa_i(t)$ and $\eta_i(t)$ is

$$\begin{aligned}\mathbb{E}[e_i(t)-\mathbb{E}[e_i(t)]]^2 &= \frac{\sigma^2}{M_i+\sigma^2 1^T(\Gamma_i(t-1)+\xi_i I)^{-1}1} \\ &\quad - \frac{\sigma^4\xi_i 1^T(\Gamma_i(t-1)+\xi_i I)^{-2}1}{(M_i+\sigma^2 1^T(\Gamma_i(t-1)+\xi_i I)^{-1}1)^2} \\ &= \frac{\sigma^2}{M_i+\sigma^2 1^T(\Gamma_i(t-1)+\xi_i I)^{-1}1}\end{aligned}$$

Since $\Gamma_i(t-1)$ is positive definite and $\xi_i\geq 0$, it holds that $1^T(\Gamma_i(t-1)+\xi_i I)^{-1}1>0$. Hence $\mathbb{E}[e_i(t)-\mathbb{E}[e_i(t)]]^2\leq\frac{\sigma^2}{M_i}$, $t\in N_0\{0\}$. This concludes the proof. □

A consequence of Proposition 6.4 is that the estimation error in each node is always upper-bounded by the variance of the estimator that computes the averages of the M_i measurements $u_i(t)$. The bound is obviously rather conservative, since we do not use any knowledge about the covariance matrix $\Gamma_i(t)$. Proposition 6.9 helps us to improve the bound in Proposition 6.4 as follows:

Corollary 6.1: *The optimal value of $\kappa_i(t)$ and $\eta_i(t)$ are such that the error variance at node i satisfies*

$$\mathbb{E}[e_i(t)-\mathbb{E}[e_i(t)]]^2 \leq \frac{\sigma^2}{M_i+\left(\sum_{j\in N_i}M_j^{-1}+(M_i\psi_i)^{-1/2}\right)^{-1}}$$

$$t\in N_0\ \{0\}.$$

Proof. Using the result in Proposition 6.4 we have that

$$Tr\Gamma_i(t-1)=\sum_{j\in N_i}\mathbb{E}[e_i(t-1)-\mathbb{E}[e_i(t-1)]]^2\leq\sum_{j\in N_i}\frac{\sigma^2}{M_j}$$

Thus

$$\begin{aligned}\lambda_{\max}(\Gamma_i(t-1)+\xi_i I) &\leq \sum_{j\in N_i}\frac{\sigma^2}{M_j}+\max\left(0,\frac{\sigma^2}{\sqrt{M_i\psi_i}}-\lambda_{\min}(\Gamma_i(t-1))\right) \\ &\leq \sum_{j\in N_i}\frac{\sigma^2}{M_j}+\frac{\sigma^2}{\sqrt{M_i\psi_i}}\end{aligned}$$

where we used the bound on ξ determined in Proposition 6.3. Since

$$1^T(\Gamma_i(t-1)+\xi_i I)^{-1}\geq\frac{1}{\lambda_{\max}(\Gamma_i(t-1)+\xi_i I)M_i}$$

we have that

$$\mathbb{E}[e_i(t) - \mathbb{E}[e_i(t)]]^2 \leq \frac{\sigma^2}{M_i + \sigma^2 1^T(\Gamma_i(t-1) + \xi_i I)^{-1} 1}$$
$$\leq \frac{\sigma^2}{M_i + \left(\sum_{j \in N_i} M_j^{-1} + (M_i \psi_i)^{-1/2}\right)^{-1}}$$

which completes the proof. □

6.4.4 Distributed computation of constraints

The choice of the constants $\psi_i, i = 1, \ldots, N$, in the local constraint of Problem P_2 is critical for the performance of the distributed estimator. Next we discuss how to compute good values of ψ_i. The intuition is that ψ_i has to be upper bounded to guarantee the estimation error to converge, but ψ_i should not be too small in order to put large enough weights on the estimates. Indeed, from the proof of Proposition 6.3 we see that if ψ_i is large then the Lagrangian multiplier ξ_i is small, since it must lie in the in-interval max $[0, \sigma^2/\sqrt{M_i \psi_i} - \lambda_{\min}(\Gamma_i(t-1))]$. From Proposition 6.4 (and Corollary 6.1) it is clear that the estimation error variance at the node i decreases as ξ_i decreases. Thus, larger the value of ψ_i lower the error variance.

The set of non-linear equations in Proposition 6.2 provides a tool to determine suitable values of ψ_i that guarantee stability. Since we are interested in determining the largest solution of the non-linear equations, we consider the following optimization problem:

$$\max_{\psi_1, \ldots, \psi_N} \sum_{i=1}^{N} \psi_i \tag{6.20}$$

$$\text{such that} \quad \psi_i + \sqrt{\psi} \sum_{j \in \Theta_i} \sqrt{\alpha_{i,j}^{(i)} \alpha_{i,j}^{(j)} \psi_j} \leq f^2(\Delta, \Upsilon)$$
$$\psi_i \geq 0 \tag{6.21}$$

with $i = 1, \ldots, N$. It is possible to show that previous problem has a unique solution, which is the solution to the following equations:

$$\psi_i + \sqrt{\psi} \sum_{j \in \Theta_i} \sqrt{\alpha_{i,j}^{(i)} \alpha_{i,j}^{(j)} \psi_j} = f^2(\Delta, \Upsilon), \quad i = 1, \ldots, N \tag{6.22}$$

Clearly, the solution of such system of non-linear equations is interesting in our setup if it can be solved in a decentralized fashion. The fact that in (6.22) only information from neighboring nodes is required, and not of the entire network, allows us to develop a decentralized algorithm to compute the solution. Following [342], we consider the iterative algorithm:

$$\psi(t+1) = T(\psi(t)) = (T_1(\psi(t)), \ldots, T_N(\psi(t))) \tag{6.23}$$

with initial condition $\psi(0) > 0$ and with $T : \Re_+^N \to \Re_+^N$ such that

$$T_i(\psi(t)) = \frac{1}{4}\left[\sqrt{\left(\sum_{j\in\Theta_i}\sqrt{\alpha_{i,j}^{(i)}\alpha_{i,j}^{(j)}\psi_j(t)}\right)^2 + 4f^2(\Delta,\Upsilon)} - \sum_{j\in\Theta_i}\sqrt{\alpha_{i,j}^{(i)}\alpha_{i,j}^{(j)}\psi_j(t)}\right]^2 \tag{6.24}$$

It is not difficult to show that $T_i(\psi)$ is a contractive function. The component solution method in [342] ensures that the fixed point solution at which the iteration converges is the solution of the non-linear equations (6.22). The computation of the iteration (6.23) can be done distributively. Note that node i does not need to know the thresholds $\psi_j, j \neq i$, of all the other nodes in the network, but those that concur in the definition of $T_i(\psi)$, that is, ψ_j that are associated to the nodes of the set Θ_i. Thresholds corresponding to nodes at two hops from node i can communicate to such node through its neighbors, with little communication overhead. Note that, the computation of the thresholds and the associated communication takes place before the nodes start to track the signal $d(t)$. Notice also that the convergence rate of the component solution method for block contraction converges geometrically to the fixed point.

6.4.5 Estimation of error covariance

Estimating the error covariance matrix is generally hard for the problem considered in this paper, because the estimator is a time-varying system and the stochastic process x, and thus e, is not stationary. However, if we consider the signals in the quasi-stationary sense, estimation based on samples guarantees to give good results. We have the following definition [56, p. 34].

Definition 6.1: A signal $s(t) : R \to R$ is said to be quasi-stationary if there exists a positive constant C and a function $R_s : R \to R$, such that s fulfills the following two conditions:

(1) $Es(t) = m_s(t), |m_s(t)| \leq C$ for all t
(2) $Es(t)s(r) = R_s(t,r), |R_s(t,r)| \leq C$ for all t and

$$\lim_{N\to+\infty}\frac{1}{N}\sum_{t=1}^{N} R_s(t,t-\tau) = R_s(\tau)$$

for all τ

It is easy to see that the time-varying linear system (6.3) is uniformly bounded-input bounded-output stable [363]. If a quasi-stationary signal is the input of such system, then its output is also quasi-stationary [364]. In our case, the measurement signal

$u(t)$ is (component-wise) stationary and ergodic and thus also quasi-stationary. This implies that also $x(t)$ is quasi-stationary, since it is the output of a uniformly exponentially stable time-varying linear system. Thus, we estimate the error covariance using the sample covariance. Specifically, we have that the mean $\mathbb{E}\varepsilon_i = m_{\varepsilon i}(t)$ and covariance $\Gamma_i(t)$ can be estimated from samples as

$$\hat{m}_{\varepsilon i}(t) = \frac{1}{t}\sum_{\tau=0}^{t}\hat{\varepsilon}_i(\tau) \tag{6.25}$$

$$\hat{\Gamma}_i(\tau) = \frac{1}{\tau}\sum_{\tau=0}^{t}(\hat{\varepsilon}_i(\tau) - \hat{m}_{\varepsilon_i}(\tau))(\hat{\varepsilon}_i(\tau) - \hat{m}_{\varepsilon_i}(\tau))^T \tag{6.26}$$

where $\hat{\varepsilon}_i(t)$ is the an estimate of the error. Thus, the problem reduces to design an estimator of $\varepsilon_i(t)$. Node i has estimates $x_{ij}(t)$ and measurements $u_{ij}(t), i_j \in N_i$, available. Let $x^{(i)}(t)$ and $u^{(i)}(t)$ denote the collection of all these variables. We can model this data set as

$$x^{(i)}(t) = d(t)1 + \beta(t) + w(t) \tag{6.27}$$

$$u^{(i)}(t) = d(t)1 + v(t) \tag{6.28}$$

where $\beta(t) \in \Re^{M_i}$ models the bias of the estimates and $w(t)$ is zero-mean Gaussian noise modeling the variance of the estimator. To *summarize*, node i has available $2M_i$ data values in which half of the data are corrupted by a small biased term $\beta(t)$ and a low variance noise $w(t)$ and the other half is corrupted by zero-mean Gaussian noise $v(t)$ with high variance. It is clear that using only $u^{(i)}(t)$ to generate an estimate $\hat{d}(t)$ of $d(t)$, which could then be used to estimate $\hat{\varepsilon}_i(t) = x^{(i)}(t) - \hat{d}(t)1$, would have the advantage of being unbiased. However, its covariance is rather large since M_i is typically small. Thus, using only measurements to estimate $d(t)$ would yield an overestimate of the error and this in turn would result in poor performance. On the other hand, using only $x^{(i)}(t)$ we obtain an underestimate of the error. This makes the weights $\eta_i(t)$ rapidly vanish and the signal measurements are discarded, thus tracking becomes impossible. From these arguments, in order to use both $x^i(t)$ and $u^i(t)$ we pose a linear least square problem as follows:

$$\min_{\hat{d},\hat{\beta}} \left\| \begin{bmatrix} x^i \\ u^i \end{bmatrix} - A\begin{bmatrix} \hat{d} \\ \hat{\beta} \end{bmatrix} \right\|^2$$

$$s.t.\quad \|B[\hat{d} \quad \hat{\beta}]\|^2 \le \rho$$

with $A \in \Re^{2M_i \times M_i+1}$ and $B \in \Re^{M_i \times M_i+1}$

$$A = \begin{bmatrix} 1 & I \\ 1 & 0 \end{bmatrix}, \quad B = [0 \quad I]$$

and ρ being the maximum value of the squared norm of the bias. However, the previous problem is very difficult to solve in a closed form, as it is a Quadratically Constrained Quadratic Problem and it typically requires heavy numerical algorithms to find the solution, such as the transformation into a **SDP** problem [362]. Note also that, in general, the value of ρ is not known in advance, being it a maximum value of the cumulative bias of M_i nodes. We thus consider the following regularized problem:

$$\min_{\hat{d},\hat{\beta}} \left\| \begin{bmatrix} x^i \\ u^i \end{bmatrix} - A \begin{bmatrix} \hat{d} \\ \hat{\beta} \end{bmatrix} \right\|^2 + \nu \left\| B \begin{bmatrix} \hat{d} \\ \hat{\beta} \end{bmatrix} \right\|^2 \tag{6.29}$$

where $\nu > 0$ is a parameter whose choice is typically rather difficult.

The solution of (6.29) is

$$(\hat{d}, \hat{\beta}) = [x^i, u^i]^T A[A^T A + \nu B^T B]^{-1}$$

The inverse of the matrix in the previous equation can be computed in closed form using the following result:

Proposition 6.5: *If* $\nu > 0$ *then*

$$[A^T A + \nu B^T B]^{-1} = \frac{1}{M_i(1+2\nu)} \begin{bmatrix} 1+\nu & -1^T \\ \bullet & \dfrac{M_i(1+2\nu)I + 11^T}{1+\nu} \end{bmatrix} \tag{6.30}$$

Proof. By Schur's complement we obtain

$$[A^T A + \nu B^T B]^{-1} = \begin{bmatrix} \left(2M_i - \dfrac{M_i}{1+\nu}\right)^{-1} & 1^T(11^T - 2M_i(1+\nu)I)^{-1} \\ \bullet & \left((1+\nu)I - \dfrac{11^T}{2M_i}\right)^{-1} \end{bmatrix}$$

From [365], it follows that

$$\left[(1+\nu)I - \frac{11^T}{2M_i}\right]^{-1} = \frac{I}{1+\nu} + \frac{11^T}{M_i(1+2\nu)(1+\nu)}$$

It is easy from here to show that the resulting matrix is (6.30).

Since we are interested in estimating $\varepsilon_i(t) = x(t) - d(t)1$ we observe that such an estimate is given by $\hat{\beta}$. From the solution of (6.29), we have

$$\hat{\beta} = \frac{x^i}{1+\nu} - \frac{\nu 1^T x^i (1+\nu) 1^T u^i}{M_i(1+2\nu)(1+\nu)} 1 \tag{6.31}$$

For the choice of the parameter ν we propose to use the Generalized Cross-Validation (GCV) method that consists of choosing

$$\nu = \arg\min \frac{\|(A^T A + \nu B^T B)^{-1} A^T (x^i, u^i)^T\|}{Tr(A^T A + \nu B^T B)^{-1}}$$

Typically, the GCV approach is computationally expensive since the trace of the matrix $(A^TA + \nu B^TB)^{-1}$ is difficult to compute, but in our case we have a closed-form representation of the matrix, and thus the computation is not difficult. However, it might be computationally difficult to carry out the minimization. Observing that

$$\begin{aligned}\nu &= \arg\min \frac{\|(A^TA + \nu B^TB)^{-1}A^T(x^i, u^i)^T\|}{tr(A^TA + \nu B^TB)^{-1}} \\ &\leq \arg\min \frac{\|(A^TA + \nu B^TB)^{-1}A^T\|}{tr(A^TA + \nu B^TB)^{-1}} \|(x^i, u^i)^T\| \end{aligned} \tag{6.32}$$

a sub-optimal value of ν can be computed solving the right-hand side of (6.32). Note that the first term on the right-hand side of (6.32) is a function of ν that can be computed off-line and stored in a look-up table at the node. Then, for different data, the problem becomes that of searching in the table.

Using (6.31) with the parameter ν computed from (6.32) we can then estimate the error mean and covariance matrix applying (6.25) and (6.26), respectively. □

6.4.6 Sub-optimal approximation of estimation bias

The previous section explains the estimation of error covariance matrix, which in turn requires computation of two parameters naming, $\beta(t)$ and $\nu(t)$. The computation of these parameters causes an increase in the overall computational complexity of the whole distributed estimation scheme. We propose an alleviation of computing $\beta(t)$ by utilizing the fact that in most of the real world data acquisition systems, the contaminating noise lies in frequency bands beyond the signal's spectrum. Since the data is low-pass filtered prior to estimation, we can assume that most of the noise power has been removed with the application of this filter, and thus (6.27) can be re-written as

$$x^{(i)}(t) = d(t)1 + \tilde{\beta}(t) \tag{6.33}$$

where

$$\tilde{\beta}(t) = \tilde{\beta}_i(t)1$$

and

$$\tilde{\beta}_i(t) = u_{\nu_i'}(t) - u_i(t)$$

Though the accuracy of $\tilde{\beta}_i$ is less than $\hat{\beta}_i$, the reduced computational efforts justify this approximation in the practical estimation systems.

6.4.7 Numerical results

This section presents the estimator structure and the algorithmic implementation followed by some numerical results.

6.4.8 Estimator structure and implementation

Figure 6.1 summarizes the structure of the estimator implemented in each node. The estimator has a cascade structure with two subsystems: the one to the left is an adaptive filter that produces the estimate of d; the one to the right computes an estimate of the error covariance matrix Γ_i. In the following, we discuss in some detail a pseudo-code implementation of the blocks in the figure.

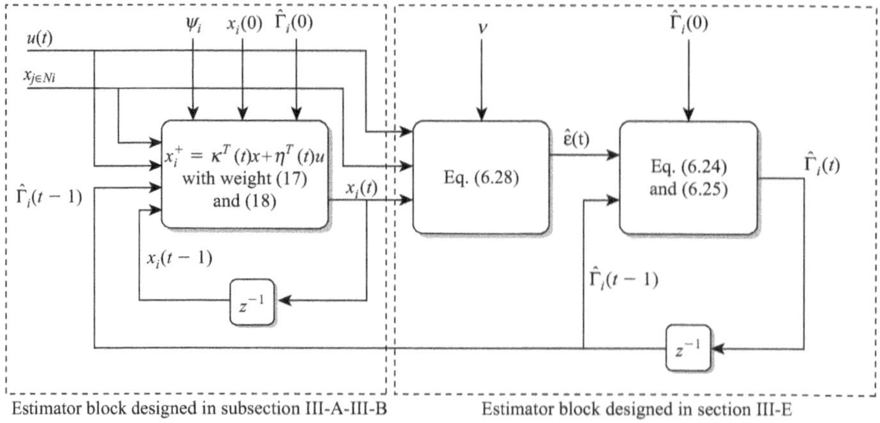

Figure 6.1 Block diagram of the proposed estimator

The estimator is presented by the following algorithm:

Algorithm 6.1 (Estimation algorithm for node i):

STATE $t := 0$
STATE $\psi_i(t-1) = 0$
STATE $\psi_i(t) = 1/M_i$
WHILE $|\psi_i(t) - \psi_i(t-1)| \geq \bar{\omega} = 10^{-10}$
STATE $\psi_i(t+1) = T_i(\psi(t))$
STATE *Collect thresholds from nodes in* Θ_i
STATE $t := t+1$
ENDWHILE
STATE $\tau_i = \dfrac{c}{r_i}$
STATE $\tau_j = \dfrac{c}{r_j}, \quad j \in M_i$
STATE $t := \max(\tau_i), \quad i = 1, 2, 3, \ldots, N$
STATE *If* $r_j < r_i$
STATE $u_{(v_j')}(t) = u_{(\tau_j, v_j')}(t - \tau_j)$
STATE *Else if* $r_i < r_j$
STATE $u_{(v_j')}(t) = u_{(\tau_j, v_j')}(t + \tau_j)$

16. *STATE* *End*
17. *STATE* $u_i(t) = Filter(u_{(v_i')}(t))$
18. *STATE* $u_j(t) = Filter(u_{(v_j')}(t)), \quad j \in M_i$
19. *STATE* $t := \max(\tau_i), \quad i = 1, 2, 3, \ldots, N$
20. *STATE* $\hat{m}_{e_i}(t) := 0$
21. *STATE* $\hat{\Gamma}_i(t) := \sigma^2 I$
22. *STATE* $x_i(t) := u_i(t)$
23. *WHILE* *forever*
24. *STATE* $M_i := |N_i|$
25. *STATE* $t := t + 1$
26. *STATE* $\xi_i = \text{bisection}(\max[0, \sigma^2/\sqrt{M_i \psi_i} - \lambda_{\min}(\Gamma_i(t-1))])$
27. *STATE* $\kappa_i(t) := \dfrac{\sigma^2(\hat{\Gamma}_i(t-1) + \xi_i I)^{-1} 1}{M_i + \sigma^2 1^T (\hat{\Gamma}_i(t-1) + \xi_i I)^{-1} 1}$
28. *STATE* $\eta_i(t) := \dfrac{1}{M_i + \sigma^2 1^T (\hat{\Gamma}_i(t-1) + \xi_i I)^{-1} 1}$
29. *STATE* $x_i(t) := \sum_{j \in N_i} \kappa_{i_j} x_j(t-1) + \sum_{j \in N_i} \eta_{i_j}(t) u_j(t)$
30. *STATE* $\hat{\beta} := \dfrac{x^i}{1+\nu} - \dfrac{\nu 1^T x^i + (1+\nu) 1^T u^i}{M_i(1+2\nu)(1+\nu)} 1$
31. *STATE* *Or*
32. *STATE* $\tilde{\beta}_i(t) := u_{v_i'}(t) - u_i(t)$
33. *STATE* $\hat{\varepsilon}_i := \hat{\beta}$
34. *STATE* *Or*
35. *STATE* $\hat{\varepsilon}_i := \tilde{\beta}$
36. *STATE* $\hat{m}_{e_i}(t) := \dfrac{t-1}{t} \hat{m}_{e_i}(t-1) + \dfrac{1}{t} \hat{\varepsilon}_i(t)$
37. *STATE* $\hat{\Gamma}_i(t) := \dfrac{t-1}{t} \hat{\Gamma}_i(t-1) + \dfrac{1}{t} (\hat{\varepsilon}_i(t) - \hat{m}_{e_i}(t))(\hat{\varepsilon}_i(t) - \hat{m}_{e_i}(t))^T$
38. *ENDWHILE*

Initially, the distributed computation of the threshold is performed (lines 1–8): node i updates its threshold ψ_i until a given precision $\bar{\omega}$ is reached. In the computations of ψ_i, we chose $\alpha_{i,j}^{(i)} = |N_j \cap N_i|/(M_i - 1)$ and $\alpha_{i,j}^{(j)} = |N_j \cap N_i|/(M_j - 1)$. This works well in practice because $k_{ii_r}, i_r = 1, \ldots, M_i$, are of similar magnitude. Indeed, the stability of the average of the estimation error established in section 6.4.6, and the bounds on the error variance in section 6.4.3, ensure that estimates among nodes have similar performance.

Numerical results show that the while-loop (lines 4–8) converges after about 10–20 iterations. Lines 9 and 10 calculate the time delay of node i and all of its neighbor nodes based on the knowledge of propagation speed c. Since there is delaying and advancing of the signal involved, the farthest node is taken as the reference point and the corresponding time delay τ_i is taken as the starting time of the algorithm (line 11). The input signal at each node is then delayed or advanced, depending upon its geometrical location relative to the node i (lines 12–16). These delay-adjusted signals are then filtered to remove the higher frequency noise (lines 17 and 18). The estimators for the local mean estimation error and the local covariance matrix are then initialized (lines 19 and 20). The main loop of the estimator is lines 23–38. Lines 24–29 are related to the left subsystem of Figure 6.1.

The optimal weights are computed using (6.18) and (6.19) (lines 27 and 28). Note that the optimal Lagrangian multiplier ξ_i is computed using the function bisection that takes as argument the interval $\max(0, \sigma^2\sqrt{M_i\psi_i} - \lambda_{\min}(\Gamma_i(t-1)))$ where the optimal value lays. Note that, if the nodes have limited computational power, so that the minimum eigenvalue of the matrix $\Gamma_i(t-1)$ cannot be exactly computed, an upper-bound based on Gersgorin Theorem can be used instead. The estimate of $d(t)$ is computed in line 29. Lines 30–38 are related to the right subsystem of Figure 6.1. These lines implement the error covariance estimation by solving the constrained least-squares minimization problem described in section 6.4.5 and 6.4.6. Sample mean and covariance of the estimation error are updated in lines 36 and 37. These formulas correspond to recursive implementation of (6.25) and (6.26). As an option $\tilde{\beta}_i(t)$ can be computed using (6.34). With regards to the inversions of the estimated error covariance matrix $\hat{\Gamma}_i$ in lines 27 and 28. In general, the dimension of $\hat{\Gamma}_i$ is not a problem because we consider cases when the number of neighbors is small. Precautions have still to be taken, because even though the error covariance matrix Γ_i is always positive definite, its estimate $\hat{\Gamma}_i$ may not be positive definite before sufficient statistics are collected. In our implementation, we use heuristics to ensure that $\hat{\Gamma}_i$ is positive definite.

6.4.9 *Simulation results*

Numerical simulations have been carried out in order to validate performance of the proposed distributed estimator. We have simulated a scenario of $N = 10$, with the structure as shown in Figure 6.2, where each vertex is the location of a node. Figure 6.3 show the original data $d(t)$ and input data $u_{(\tau_i, v_i')}(t)$ at $u_{\tau_1, v_1'}(t)$. Figures 6.4 and 6.5 show the estimations x_1 to x_{10}.

6.5 Asynchronous multirate multismart sensors

In asynchronous multisensor systems, data missing and delays are likely to occur when performing state estimation at different rates. In this paper, an algorithm to process both the delayed and missing measurements is developed, under the assumption that the data arrives between two consecutive sampling times. Local estimation is then implemented using a fusion algorithm. This study extends the currently available state estimation for asynchronous multirate multisensor that only considers random measurements missing with the ability to remove delay effect. Simulation experiments are performed to illustrate the effectiveness of the developed theory.

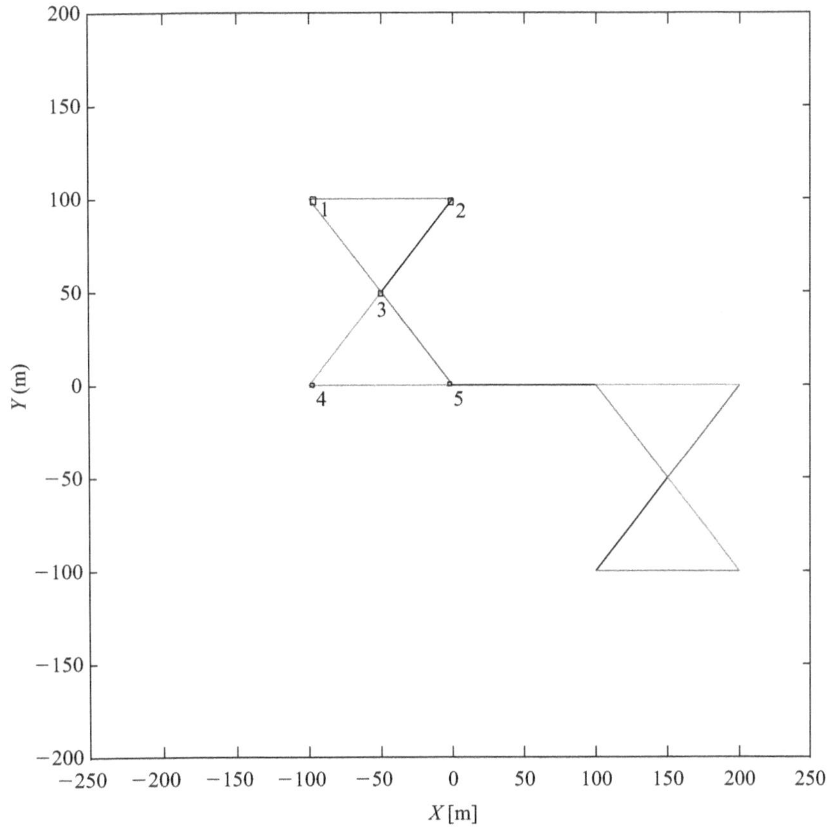

Figure 6.2 Structure of sensor network

6.5.1 Introduction

It becomes increasingly apparent that in multisensor systems there is a possibility to incorporate all of the measurements coming from these sensors to get better estimation of dynamical system. In this regard, state fusion is the process to make combination of information coming from different sources of state estimates. Formerly, the sensor only sent the raw data, such as the value of its measurement, but with the advancement of electronics, the new generation of sensor, so called *smart sensor*, can do more complex tasks, such as signal conditioning, calibration, diagnostic tests, and communication. The smart sensor transforms the raw sensor signal to a standardized digital representation, checks and calibrates the signal, and finally transmits this digital signal to the outside world via a standardized interface using a standardized communication protocol.

Applications for multisensor data fusion are widespread. Military applications include automated target recognition (e.g., for smart weapons), guidance for autonomous vehicles, remote sensing, battlefield surveillance, and automated threat

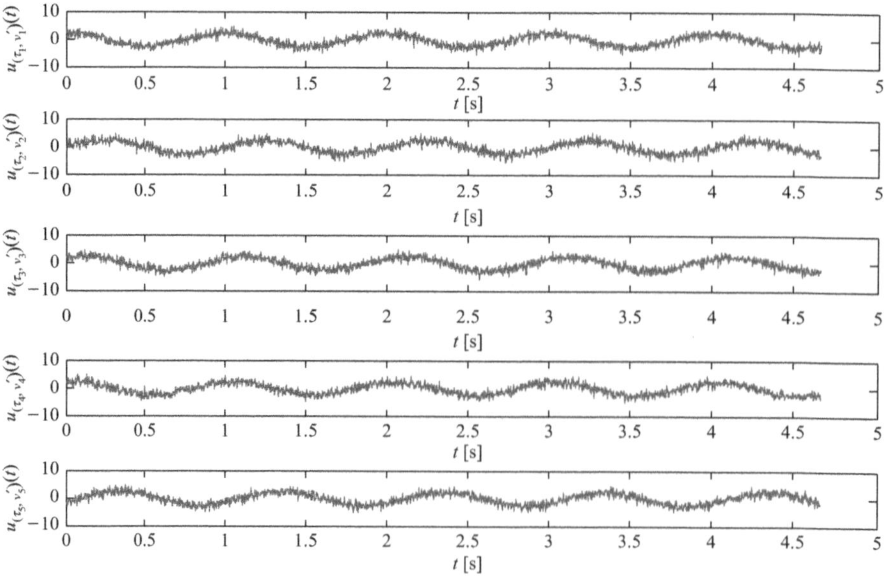

Figure 6.3 The original signal $d_{(t)}$

recognition systems, such as identification-friend-foe-neutral (IFFN) systems. Non-military applications include monitoring of manufacturing processes, condition-based maintenance of complex machinery, robotics, and medical applications.

Although there is a vast research related to multisensor fusion, only less work has been done in asynchronous multirate multisensor. General asynchronous sensor-to-sensor linear distributed estimation problem is solved in [366–368]. On the theoretical level, the results of [366] solved three problems in one step: fusing asynchronous tracks without synchronization, dealing with out-of-sequence track, and accounting for latent tracks. In [369], the authors presented exact state update equation for such out-of-sequence problem using retrodicted state estimation. Recently, in [370], the authors considered the problem of removal of out-of-sequence (actually, 'old') measurements (OOSM) from tracks. The optimal solution and a one-step solution have been presented using the Information Filter (IF-)-based equivalent measurement.

Development of multisensor optimal information fusion criterion weighted by matrices in the linear minimum variance sense is presented in [371]. Based on this optimal fusion criterion, a general multisensor optimal information fusion decentralized Kalman filter with a two-layer fusion structure is given for discrete-time linear stochastic control systems with multiple sensors and correlated noises. Some related works are reported in [372,373]. Recent results on multisensor data fusion can be found in [374,375].

The problem of state estimation using multiple sensor with incomplete measurement is found in several works [376–381]. In [376], Kalman filtering was studied

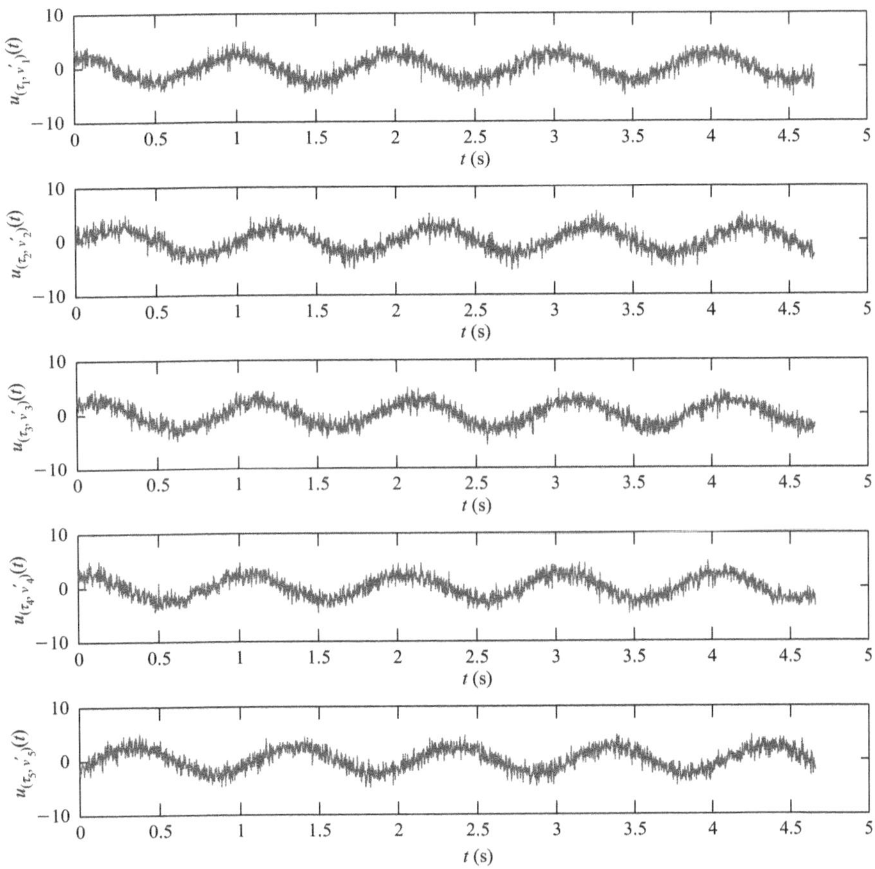

Figure 6.4 The input $u_{\tau_i,v_i'}(t)$ at node i, for i = 1, 2, . . . 5

with intermittent observations using modeling of the arrival of the observation as a random process. Later, state estimations for asynchronous multisensor system were developed in [374]. In their approach, measurement losses are detected based on statistical hypothesis.

It appears from the published results that data missing and delays are likely to occur in asynchronous multisensor systems when performing state estimation at different rates. In this paper, an algorithm to process both the delayed and missing measurements is developed, under the assumption that the data arrives between two consecutive sampling times. Local estimation is then implemented using a fusion algorithm. Looked at in this light, our study extends the currently available state estimation for asynchronous multirate multisensor that only considers random measurements missing with the ability to remove delay effect.

On related research fronts, a pair-wise decomposition and coordinated control for complex interconnected systems with any information structure constraints is

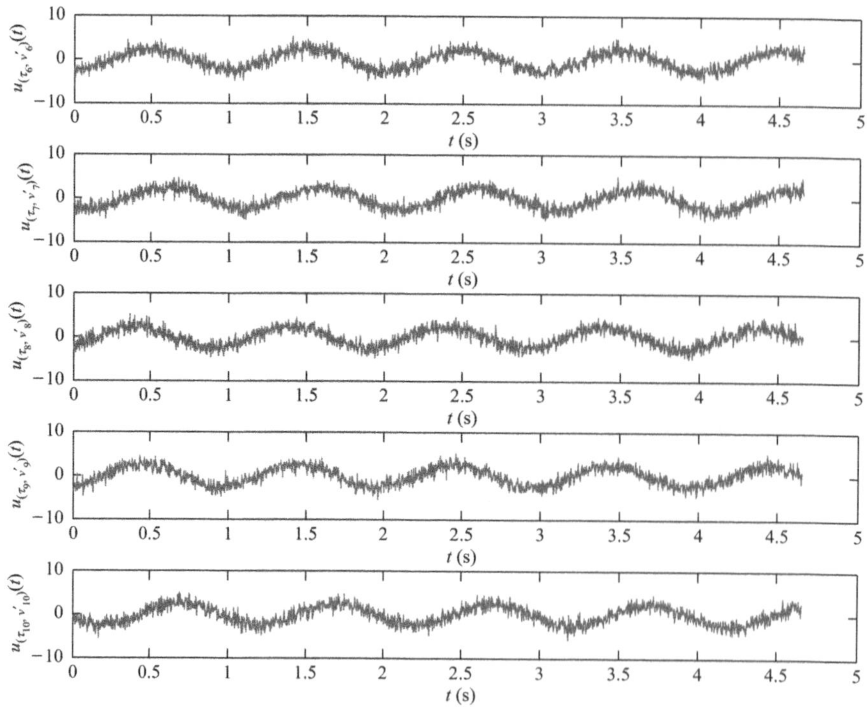

Figure 6.5 The input $u_{\tau_i,v_i'}(t)$ at node i, for $i = 6, 7, \ldots, 10$

proposed in [382] based on a permuted inclusion principle. The authors in [383] introduce swarm intelligence as a relatively novel field in order to address the study of the collective behaviors of systems made by many components that coordinate using decentralized controls and self-organization. A robust anti-windup fuzzy control approach for uncertain non-linear time-delay systems with actuator saturations is developed in [384] using the Takagi–Sugeno (T–S) fuzzy model where the nonlinearity of input saturation was characterized by a specific sector condition to facilitate the design process. The problem of $\ell_2 - \ell_\infty$ filter design for discrete-time singular systems with Markovian jump and time-varying delays is investigated in [385] using the delay partitioning technique where a delay-dependent condition is established to guarantee the filtering error systems to be stochastically admissible and achieve a prescribed $\ell_2 - \ell_\infty$ performance index. In a parallel development, a robust $\mathcal{H}_\infty$ control problem for a class of T–S fuzzy systems with time delays using delta operator approach is presented in [385].

In comparison with the available results, the present work uses a priori estimation to handle the case of measurement losses. The optimality of fusion algorithm is also presented, neglecting the effect of cross-covariance among the local state estimation errors. In this regard, our approach extends the work of [374] by allowing delays to occur between two consecutive sampling time and by assuming that the

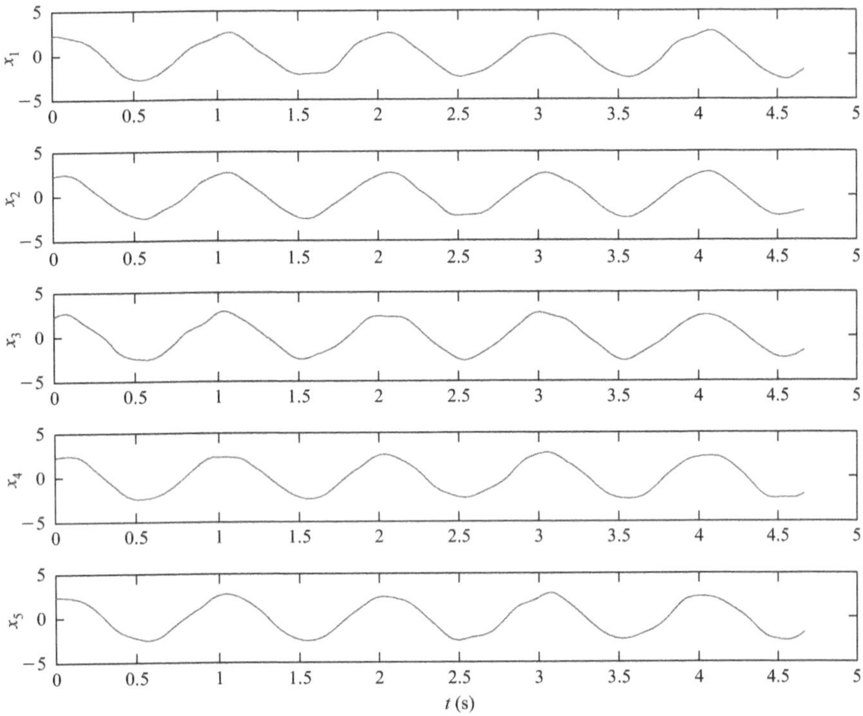

Figure 6.6 Estimates of $d(t)$ at node 1 to node 5

measurement loss detection is done as basic communication protocols feature using data validity checking mechanism (Figures 6.6 and 6.7).

6.5.2 Problem formulation

In the sequel, the following assumptions are recalled:

- Fusion center are assumed to have capability to determine measurement-loss resulting communication mechanism.
- Sensors measurements are time stamped.
- Sensors sampling rates are assumed to be integer number multiplication from it fastest rate.
- Delay can happen on measurement, but it is assumed here that the sensor delay is distributed uniformly and is less than corresponding VS sampling time.
- It is assumed that the sensor does not compute the estimation locally, the estimation process is done at fusion center.

Suppose that there is N sensor at different sampling rate. Because we allow sensor to measure asynchronously, the measurement is assumed to come only at the same

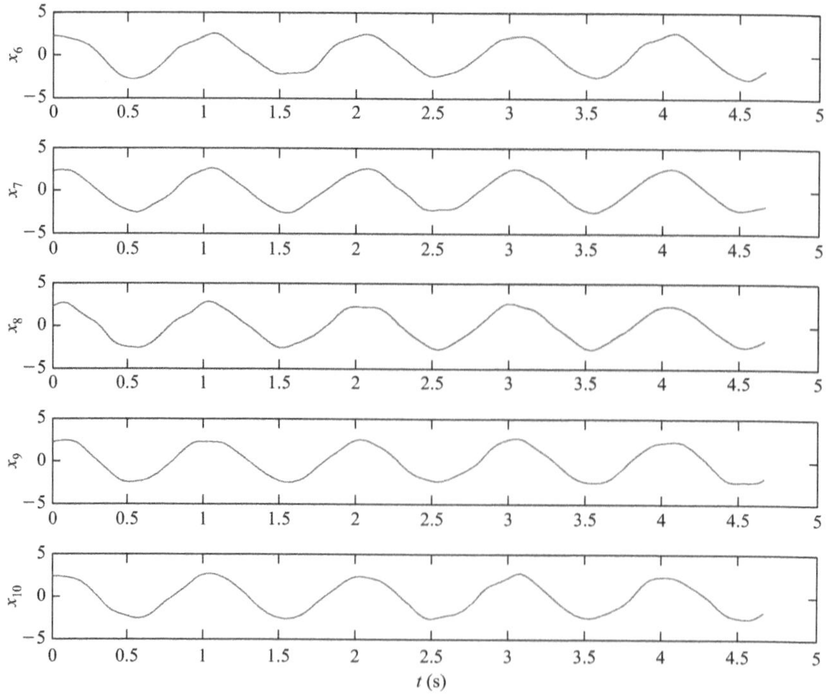

Figure 6.7 Estimates of $d(t)$ at node 6 to node 10

number during a period, where the time distance between two measurements is not uniform. For every measurement coming from these sensors, which is received in fusion center, there is a corresponding estimation, based solely on one sensor that is so called *virtual sensor* (VS). Suppose at VS, last estimate update is at t_{k_i}, and the next estimate time is $t_{k_i+1} = t_{k_i} + T_{s_i}$, where T_{s_i} is a constant sampling time at VS chosen based on sampling rate of corresponding sensor. Suppose that during the time interval $[t_{k_i}, t_{k_i+1}]$, a number of asynchronous measurement arrive at the fusion center. Because of communication delays, some of the measurements are generated during the time interval $[t_{k_i}, t_{k_i+1}]$ and some may have been generated during the interval $[t_{k_i-1}, t_{k_i}]$ and arrive during the interval $[t_{k_i}, t_{k_i+1}]$. Before the VS estimation process takes place, some of the measurements may be removed. If more than one measurement generated by the same sensor is received, only the most recent measurement is used for fusion. If only one measurement is received from a given sensor platform and was taken during the time interval $[t_{k_i-1}, t_{k_i}]$, a decision has to be made whether to keep such track or not. This mainly depends on the data rate of the sensor, the sensor accuracy, as well as how old that measurement is. The measurement that is kept for track fusion is called valid measurement. Because of the difference in communication delays, some of the validated tracks may arrive out-of-sequence, which is considered as measurement loss.

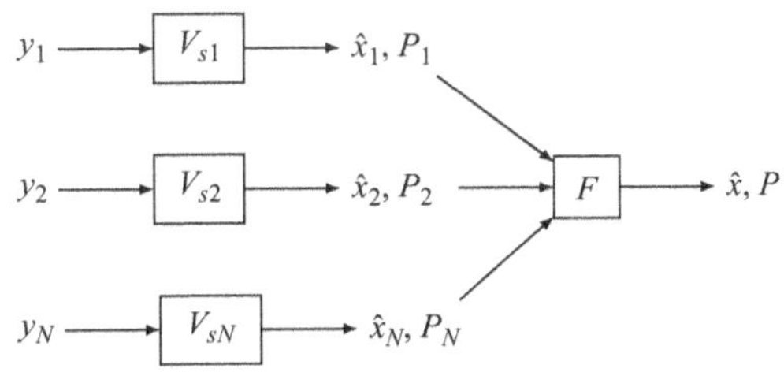

Figure 6.8 Overall proposed fusion design

Every estimation from single VS is then processed through the fusion algorithm to get optimal estimation of the state. The overall diagram of fusion process using multiple sensor can be seen in Figure 6.8.

6.5.3 Lossy channel

A lossy channel modeled by a stochastic process $\gamma_k \in \{0, 1\}, \forall k \in \mathbb{N}$ with the understanding that $\gamma_k = 1$ signifies that the measurement y_k sent at time k reaches its destination and that $\gamma_k = 0$ when it does not. Whether the measurement has reached the sensor or failed is assumed to be known to the sensor from the network data validity checking mechanism. It is generally assumed that the dropout process $\gamma(k)$ is statistically independent of x_0, w_k, and v_k. This estimate can be computed recursively using the following time-varying Kalman filter (TVKF):

$$\begin{aligned}\hat{x}(0|-1) &= 0\\ \hat{x}(k|k-1) &= A_i\hat{x}(k|k)\\ \hat{x}(k|k) &= \hat{x}(k|k-1) + \gamma(k)K(k)(y(k) - C\hat{x}(k|k-1))\end{aligned} \tag{6.34}$$

6.5.4 Dealing with delay

As VS in the central data fusion read the measurement from each sensor in a uniform time step t_{k_i}, each measurement received could be the measurement from t_ς, where $t_{k_i} < t_\varsigma < t_{k_i+1}$. Since state estimation from each sensor will be counted at t_{k_i+1}, by incorporating the measurement that arrive before that, it should be an intermediate propagation from the former time sampling instance t_{k_i}. Consider that the measurement time stamp is at t_ς, then from former explanation

$$\varepsilon = \frac{t_\varsigma - t_{k_i}}{T_{s_i}}, \qquad \tau = 1 - \varepsilon \tag{6.35}$$

where it is assumed, for simplicity in exposition, that the dynamics of system is linear time-invariant and the system modeling is in the continuous-time domain. Then it is discretized in view of the fact that the measurements are obtained in discrete time.

$$\begin{aligned}\dot{x}(t) &= A_c x(t) + w_c(t)\\ x_i(k+1) &= A_i x_i(k) + w_i(k), \qquad \text{where } A_i = e^{A_c T_{s_i}}\end{aligned} \tag{6.36}$$

The propagation is taken in two steps: initially from t_{k_i} to t_ζ then from t_ζ to t_{k_i+1}.

step 1 at time $k_i + \varepsilon$

$$\begin{aligned}
\hat{x}_i(k_i + \varepsilon|k_i) &= A_i^\varepsilon \hat{x}_i(k_i|k_i) \\
P_i(k_i + \varepsilon|k_i) &= A_i^\varepsilon P_i(k_i|k_i) A_i^{\varepsilon,T} + Q_{i,\varepsilon} \\
K_i(k_i + \varepsilon) &= P_i(k_i + \varepsilon|k_i) C_i^T \left[C_i P_i(k_i + \varepsilon|k_i) C_i^T + R_{i,\varepsilon} \right] \\
\hat{x}_i(k_i + \varepsilon|k_i + \varepsilon) &= \hat{x}_i(k_i + \varepsilon|k_i) + \gamma_i(k_i) K_i(k_i + \varepsilon) \tilde{y}(k_i + \varepsilon) \\
\tilde{y}(k_i + \varepsilon) &= (y(k_i + \varepsilon) - C_i \hat{x}_i(k_i + \varepsilon|k))
\end{aligned} \tag{6.37}$$

step 2 at time $k_i + 1$

$$\hat{x}_i(k_i + 1|k_i + \varepsilon) = A_i^\tau \hat{x}_i(k_i + \varepsilon|k_i + \varepsilon)$$

Here both $Q_{i,\varepsilon}$ and $R_{i,\varepsilon}$ are the process and measurement noises of the corresponding sensors related to ε step. While $R_{i,\varepsilon}$ will be readily set to R_i since there is no lumped effect on the measurement, $Q_{i,\varepsilon}$ on the other hand, will be different from Q_i.

Easy inspection of (6.37) shows that it essentially reduces to the ordinary Kalman filter, when $\varepsilon \to 0$. When measurement losses occur, it follows that (6.37) reduces to one step a priori state estimation leading to $\hat{x}_i(k_i + 1) = A_i \hat{x}_i(k_i)$. At this stage, one might have doubt of whether $A^\varepsilon \in \mathbb{R}$. This can be easily answered by carefully looking into (6.36) and $A^\varepsilon = e^{A_c \varepsilon T_{s_i}}$.

Let us assume that there is an integer $p > 0$ such that one time sampling of corresponding sensor can be divided into p parts, and there is also an integer $m > 0$ such that $\varepsilon = m/p$. Based on this fact, one can write an expression for state propagation in ε step as below:

$$\begin{aligned}
x_i(k_i) &= \mathbf{x}_i(pk_i) \\
x_i(k_i + \varepsilon) &= \mathbf{x}_i(pk_i + m) \\
&= \mathbf{A}_i \mathbf{x}_i(pk_i + m - 1) + \mathbf{w}_i(pk_i + m - 1) \\
&= \mathbf{A}_i^m \mathbf{x}_i(pk_i) + \sum_{j=0}^{m-1} \mathbf{A}_i^j \mathbf{w}_i(pk_i + m - 1 - j) \\
&= \mathbf{A}_i^m \mathbf{x}_i(pk_i) + \mathbf{w}_{i,m}(pk_i) \\
&= A^\varepsilon x_i(k_i) + w_{i,\varepsilon}(k_i)
\end{aligned} \tag{6.38}$$

Compared with the original Kalman filter, only one additional computation has to be computed here, no matter how big the value p is. Higher p does not affect the computational effort at every step. Rather, it only affects the way of computing the A^ε and process noise variance $Q_{i,\varepsilon}$ at step ε. One can then express $Q_{i,\varepsilon}$ as below:

$$Q_{i,\varepsilon} = \sum_{j=0}^{m-1} \mathbf{A}_i^j \mathbf{Q} \mathbf{A}_i^{j,T} = Q_i - \sum_{j=m}^{p} \mathbf{A}_i^j \mathbf{Q} \mathbf{A}_i^{j,T} \tag{6.39}$$

Interestingly enough, we solve (6.39) by approximation as follows:

$$
\begin{aligned}
Q_i &= \sum_{j=0}^{p} \mathbf{A}_i^j \mathbf{Q} \mathbf{A}_i^{j,T} \\
\lambda_M(Q_i) &= \lambda_M \left(\sum_{j=0}^{p-1} \mathbf{A}_i^j \mathbf{Q} \mathbf{A}_i^{j,T} \right) \\
&\simeq \lambda_M(\mathbf{Q}) \lambda_M \left(\sum_{j=0}^{p-1} \mathbf{A}_i^j \mathbf{A}_i^{j,T} \right) \\
&\simeq \lambda_M(\mathbf{Q}) \sum_{j=0}^{p-1} \lambda_M(\mathbf{A}_i^j \mathbf{A}_i^{j,T})
\end{aligned}
\tag{6.40}
$$

This leads to

$$
\lambda_M(\mathbf{Q}) \simeq \frac{\lambda_M(Q_i)}{\sum_{j=0}^{p-1} \lambda_M(\mathbf{A}_i^j \mathbf{A}_i^{j,T})}
$$

which in turn yields

$$
\lambda_M(Q_{i,\varepsilon}) \simeq \frac{\lambda_M(Q_i) \sum_{j=0}^{m-1} \lambda_M(\mathbf{A}_i^j \mathbf{A}_i^{j,T})}{\sum_{j=0}^{p-1} \lambda_M(\mathbf{A}_i^j \mathbf{A}_i^{j,T})}
$$

Thus, we have

$$
\frac{\lambda_M(Q_{i,\varepsilon})}{\lambda_M(Q_i)} \simeq \left(\frac{\sum_{j=0}^{m-1} \lambda_M(\mathbf{A}_i^j \mathbf{A}_i^{j,T})}{\sum_{j=0}^{p-1} \lambda_M(\mathbf{A}_i^j \mathbf{A}_i^{j,T})} \right)
$$

Letting

$$
\delta(m,p) \equiv \frac{\lambda_M(Q_{i,\varepsilon})}{\lambda_M(Q_i)}
$$

we finally arrive at

$$
Q_{i,\varepsilon} \simeq \delta(m,p) U_{m,p} Q_i U_{m,p}^T, \quad U_{m,p} = \frac{\mathbf{A}_i^m}{|\mathbf{A}_i^m|} \tag{6.41}
$$

The problem therefore turns into how to select appropriate p for an acceptable delay tolerance. As soon as p is determined, $\delta(m,p)$ can be counted. For example, if p is 64 then one has a collection of 64 values of δ. These values can be counted once at initialization, then it can be used at every time step. One can even have more precise value of $Q_{i,\varepsilon}$ using a simple well-known Newton–Raphson, using $Q_{i,\varepsilon}$ obtained in (6.41) as initial guess. All procedures for state estimation using delayed measurements described hereafter are named as *Intermediate Step Kalman Filter* (ISKF). Figure 6.9 shows a comparison between approximated result of $\delta_{m,p}$ processed with (6.41) along with the associated true value. Rough estimates can be taken $Q_{i,\varepsilon} \simeq \varepsilon Q_i$ as given in [369].

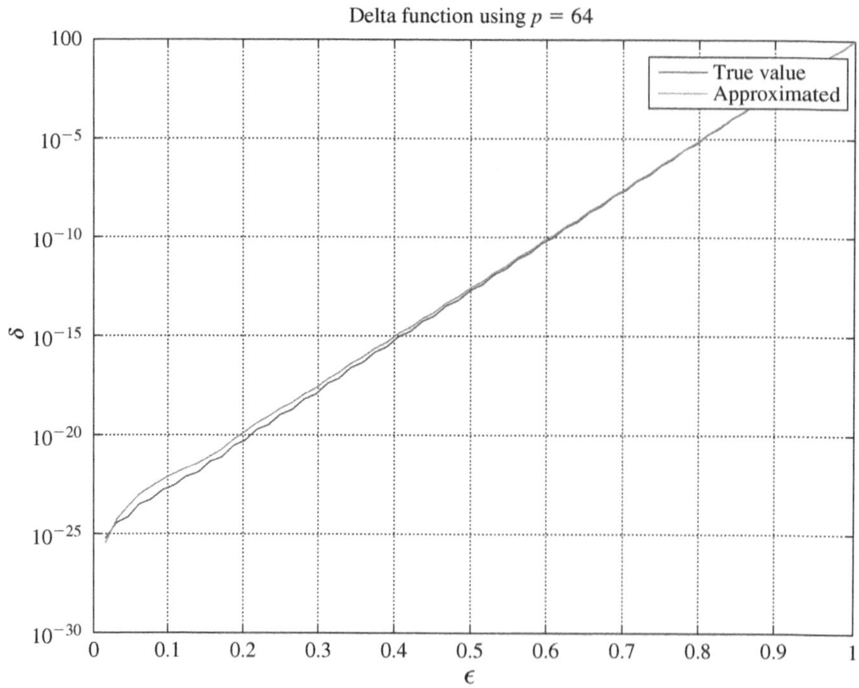

Figure 6.9 Comparison of approximated and true δ

6.5.5 Fusion algorithm

When an estimate of the state is available, the problem now turns to how to combine these different estimations to get the optimal result. In the sequel, an approach purposed in [374] is incorporated effectively. The dynamics of the system is first defined on the lowest sampling-time (VS) by

$$\begin{aligned} x(k+1) &= Ax(k) + w(k) \\ y_i(k_i) &= \gamma_i(k_i)C_i x_i(k_i) + v_i(k_i), \quad i = 1, 2, \cdots N \end{aligned} \tag{6.42}$$

where nonsubindexed variable belongs to the lowest sampling-time sensor that the system dynamic is defined with, A is the system matrix, and w is supposed to be zero-mean white Gaussian noise with variance Q. Also y_i is the measurement observed by sensor i, x_i is the state whose dynamics is described by A_i, which will be defined subsequently, C_i is the measurement matrix, and v_i is supposed to be zero-mean white Gaussian noise with variance R_i, $i = 1, 2, \ldots, N$.

In the sequel, it is assumed that each sensor follows

$$n_i = \frac{S_N}{S_i}, \quad n_i \in \mathbb{N} \tag{6.43}$$

where S_i is sampling rate on sensor i, and S_N is sampling rate at the highest sensor sampling rate. Equivalently at VS

$$n_i = \frac{T_{si}}{T_s}, \quad n_i \in \mathbb{N} \tag{6.44}$$

where T_s is sampling time of the lowest sampling time VS, and T_{s_i} is VS i sampling time.

Fused estimation based on this series of sensors can then be computed every T_s. Obviously, the fused estimation $\hat{x}(k|k)$ is no more than a weighted estimation coming from each sensor $\hat{x}_i(k|k)$. One can intuitively think that, the estimation from sensor i, that have less error variance should be weighted more, compared to those that have higher error variance. It can be easily shown that the foregoing weighting selection of state estimations coming from corresponding VS gives the optimal result in sense of error variance under the assumption that there is no cross-covariance among VS's estimations [374].

$$\begin{aligned}
\hat{x}(k|k) &= \sum_{i=1}^{N} \alpha_i(k)\hat{x}_{N|i}(k|k) \\
P(k|k) &= \left[\sum_{i=1}^{N} P_{N|i}^{-1}(k|k)\right]^{-1} \\
\alpha_i(k) &= P(k|k)P_{N|i}^{-1}(k|k)
\end{aligned} \tag{6.45}$$

where $\hat{x}_{N|i}(k|k)$ is state estimation at the highest sample rate based on estimation from VS i and $P_{N|i}(k|k)$ is its error covariance with both as given below:

$$\hat{x}_{N|i}(k|k) = \begin{cases} \left[\sum_{m=0}^{n_i-1} A^- m\right]^{-1} \hat{x}_i(l|l), & k = n_i l \\ n_i A^p \left[\sum_{m=0}^{n_i-1} A^- m\right]^{-1} \hat{x}_i(l|l), & k = n_i l + p \end{cases} \tag{6.46}$$

$$P_{N|i}(k|k) = \begin{cases} n_i^2 \left[\sum_{m=0}^{n_i-1} A^- m\right]^{-1} P_i(l|l) \left[\sum_{m=0}^{n_i-1} A^- m\right]^{-1,T}, & k = n_i l \\ n_i^2 A^p \left[\sum_{m=0}^{n_i-1} A^- m\right]^{-1} P_i(l|l) \left[\sum_{m=0}^{n_i-1} A^- m\right]^{-1,T} A^{p,T} \\ \quad + \sum_{m=0}^{p-1} A^- m^{-1} A^l Q A^{l,T}(l|l), & k = n_i l + p \end{cases} \tag{6.47}$$

The terms $P_i(l|l)$ and $\hat{x}_i(l|l)$ in (6.47) and (6.46) are computed using ISKF as described in previous section, with system matrix A_i and process noise covariance Q_i in corresponding sensor i defined below:

$$A_i = A^{n_i}$$
$$Q_i = \frac{1}{n_i^2}\left[\sum_{m=0}^{n_i-1} A^{-m}\right]\left[\sum_{m=0}^{n_i-1} A^m Q A^{m,T}\right]\left[\sum_{m=0}^{n_i-1} A^{-m}\right]^t \tag{6.48}$$

From (6.45) it can be verified that

$$P(k|k) \leq P_{N|i}(k|k) \tag{6.49}$$

which means that the fused estimation error from estimation of different sensors is always less or equal to the estimation error of each sensor.

In what follows, the purposed algorithm will be tested with a system using Kinetic Model using discrete Wiener acceleration process (DWAP) as in [367]. Given the true system as

$$x(k+1) = \begin{bmatrix} 1 & T_s & T_s^2/2 \\ 0 & 1 & T_s \\ 0 & 0 & 1 \end{bmatrix} x(k) + \Gamma w(k)$$
$$Q = \sigma_w^2 \Gamma \Gamma^t \tag{6.50}$$
$$\Gamma = \begin{bmatrix} T_s^2/2 \\ T_s \\ 1 \end{bmatrix}$$

Here $T_s = 0.01$ and $\sigma_w = 0.8$. The simulations experiments are divided into two parts: in the first part, ISKF will be compared to ordinary Kalman filter. In second part, the fusion algorithm based on ISKF is implemented.

6.5.6 Single-sensor simulations

Given position sensor, with its VS $T_s = 5.12$ s and with measurement noise of 0.9. At the beginning, measurement loss probability is assumed zero here to allow comparison between ISKF and Kalman filter. Next, measurement loss probability is set as 0.3. Several selections of delay partition are presented here to demonstrate the effect of ISKF using different delay partitions and on comparison with Kalman filter.

In the simulation results, we the following notations are applied:

- Black dotted is the measurement, and blue dotted line is the measurement that arrive at Fusion center.
- Kalman estimation tend to follow the delayed measurement.
- ISKF, despite it was using small partition number, like in Figure 6.2, is able to move state estimation closer to the true value.

Simulation results of single-sensor estimations without measurement loss and different partitions are displayed in Figures 6.10–6.13.

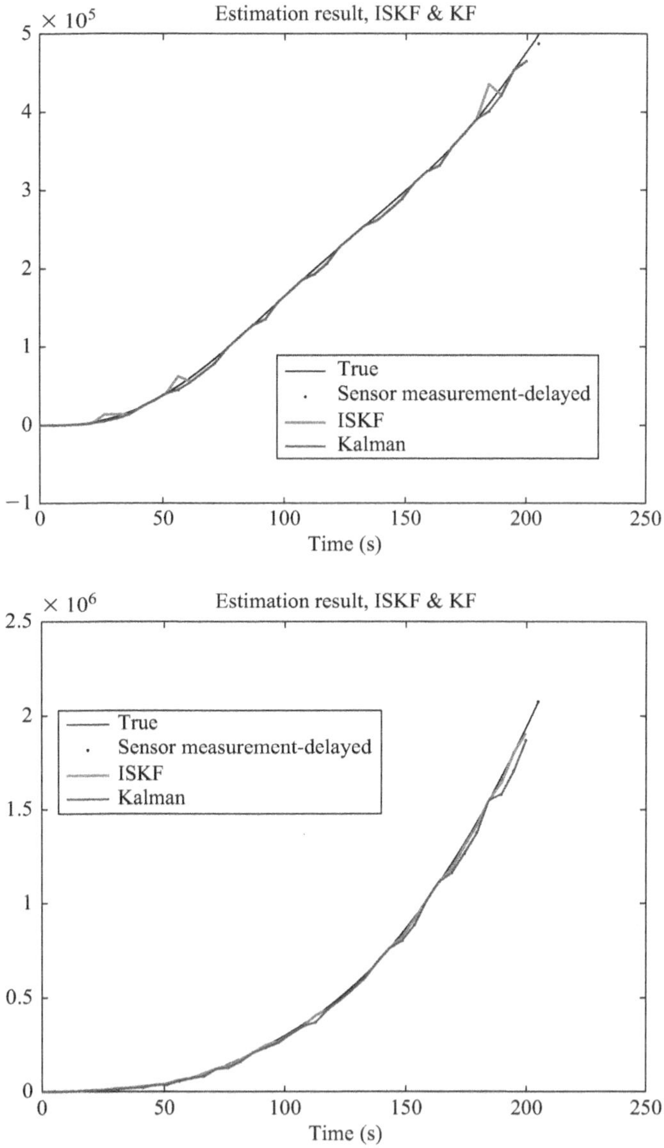

Figure 6.10 Single-sensor estimations without measurement loss: Part 1

Remark 6.1: *The following points pertaining to the simulation results are in order:*

- *When there is no loss in measurements, ISKF is shown to be able to give nearly the 'non-delayed' estimate from delayed measurement. We observe that the more the partition delay used in ISKF, the lower estimation error is obtained. It can be verified that, if delay partition number is equal to one, ISKF reduces to Kalman*

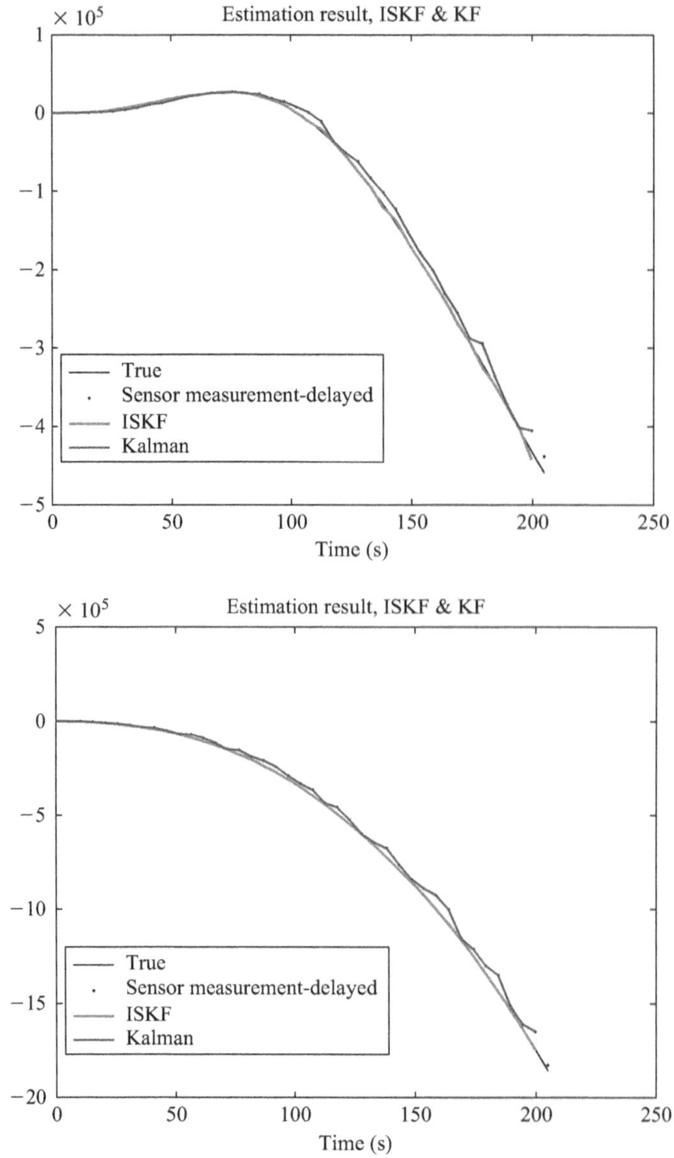

Figure 6.11 Single-sensor estimations without measurement loss: Part 2

filter, however with delays in its estimations. It can be seen clearly that Kalman filter estimation follows delayed estimations, where ISKF nearly follows true measurement, especially if partition number goes higher, see Figures 6.10–6.13.

- *When measurement loss occurs, ISKF tends to give a better a priori estimation compared to Kalman filter. This can be readily seen on several sharp peaks on*

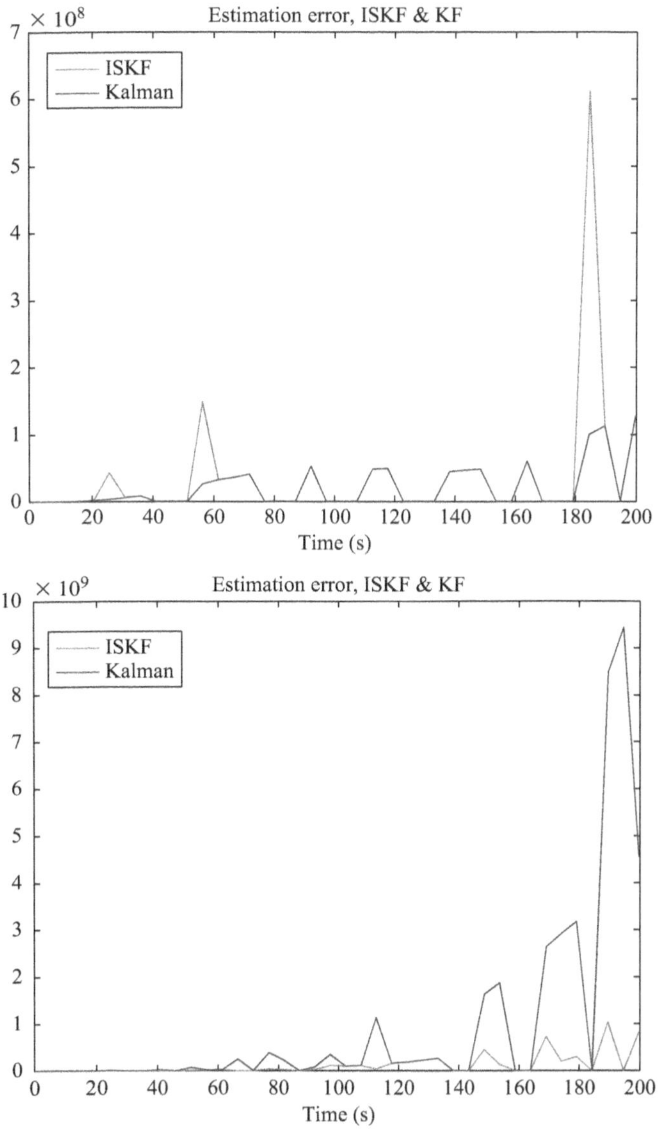

Figure 6.12 Single-sensor estimations without measurement loss: Part 1

Kalman filter estimation results. Measurement losses are the main source for the occurrence of such sharp peaks on Kalman estimation. When measurements are lossy, Kalman filter propagates its state estimation using its last estimate that already had an error due to delayed measurement. Upon propagation, the state estimate tends to go to false directions, till the arrival of the next measurement. The peak becomes higher if measurement loss happened in consecutive measurements.

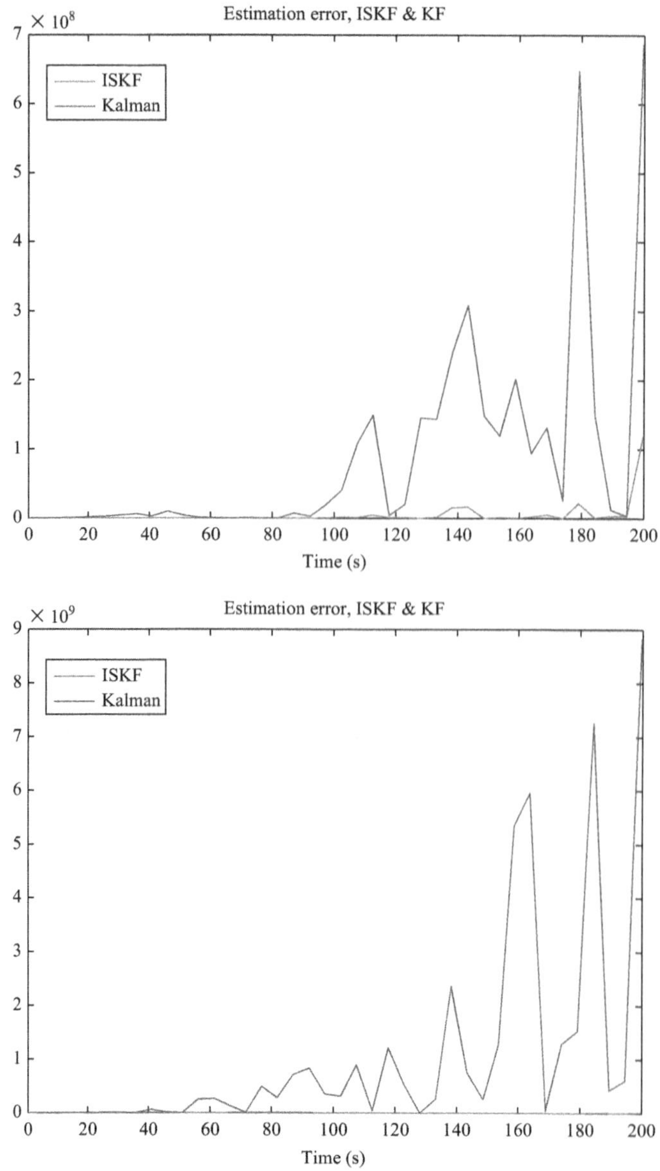

Figure 6.13 Single-sensor estimations error without measurement loss: Part 2

Similar things also happen in ISKF, except that due to the closer estimate before the loss occurs. In such case, the state estimation, although it also propagated into the wrong direction, the peak is not as high as in Kalman filter estimations. The peak even reduced significantly if the number of partition goes higher, see Figures 6.14 and 6.17.

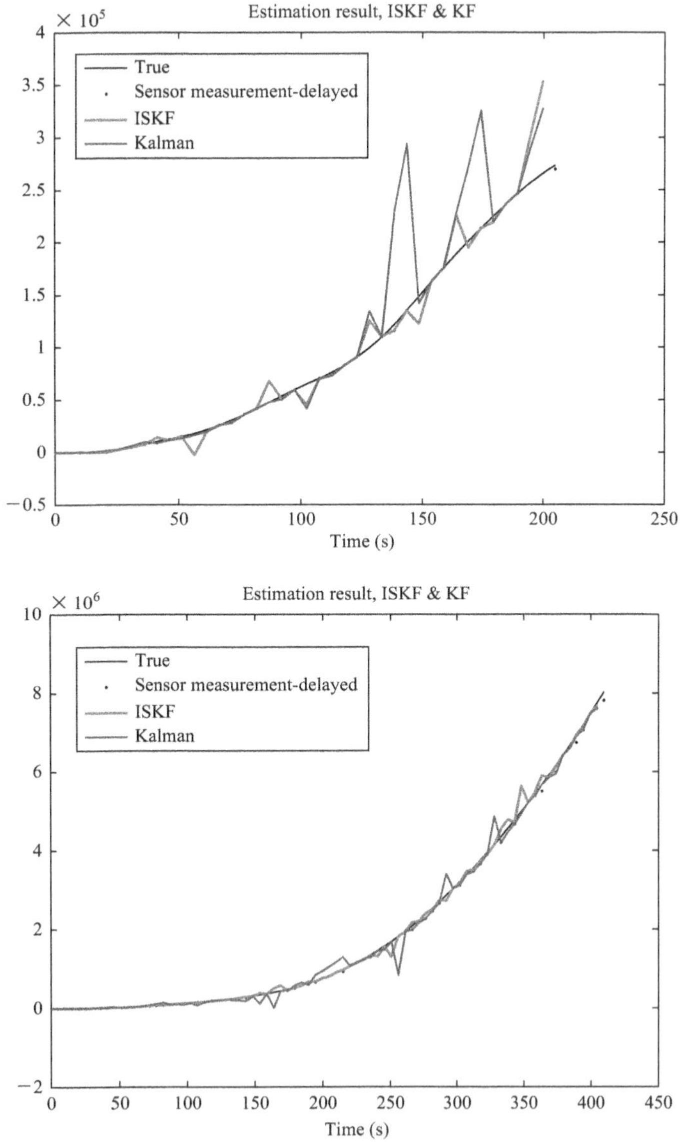

Figure 6.14 Single-sensor estimations with measurement loss: Part 1

- *As have been noted earlier, since higher p does not influence the computational effort, one can easily select the highest number that is reasonable. The difference will be only on how many matrices will be stored at the initialization phase of the algorithm in addition to the convergence of Newton Raphson algorithm used to compute $Q_{i,\varepsilon}$, and the precision of $Q_{i,\varepsilon}$ matrices that can be reached.*

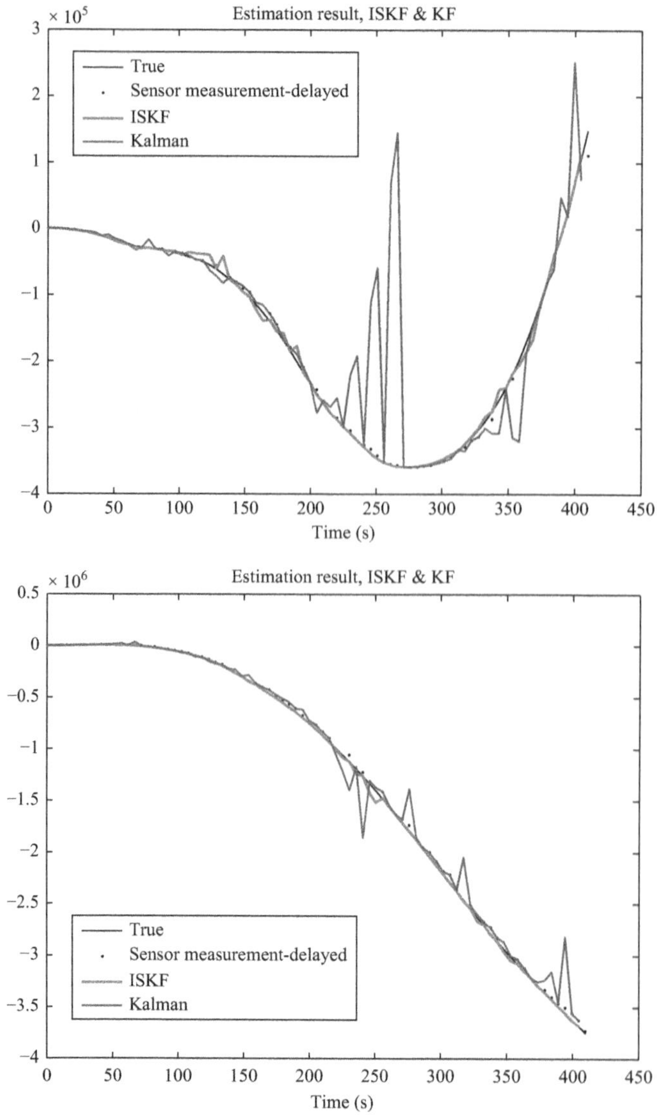

Figure 6.15 Single-sensor estimations with measurement loss: Part 2

The corresponding simulation results of single-sensor estimations with measurement loss and different partitions are displayed in Figures 6.14–6.17.

6.5.7 Multiple sensor simulations

Using the same model with previous subsection, fusion algorithm is simulated using position, velocity, and acceleration sensors. The highest sampling rate is acceleration

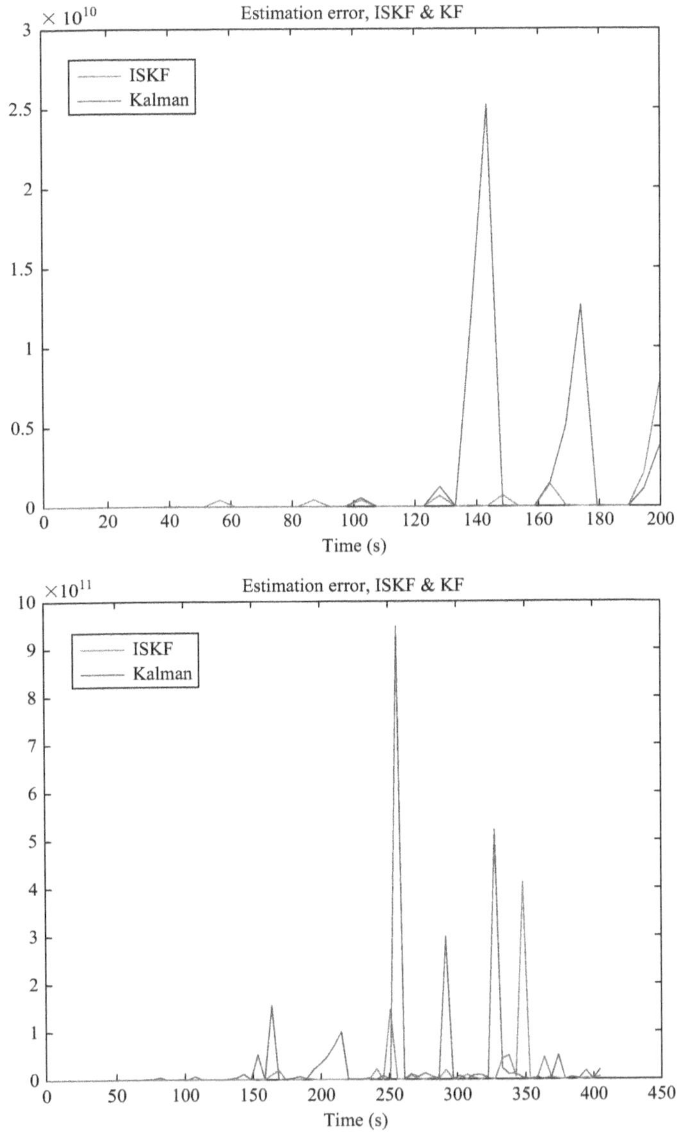

Figure 6.16 Single-sensor estimations with measurement loss: Part 1

sensor, with $T_s = 5.12$ at its VS, drop probability = 0.6, and noise variance = 0.1. Velocity sensors have half the sensor sampling rate, and 0.3 drop probability, with 1.9 noise variance. Position sensor has the slowest sensor, at one-third of sampling rate of acceleration sensor, and 0.1 drop probability, with 2.8 noise variance.

It is expected that the fusion algorithm can overtake the measurement loss problem, by considering that during the event of measurement loss in one sensor, other

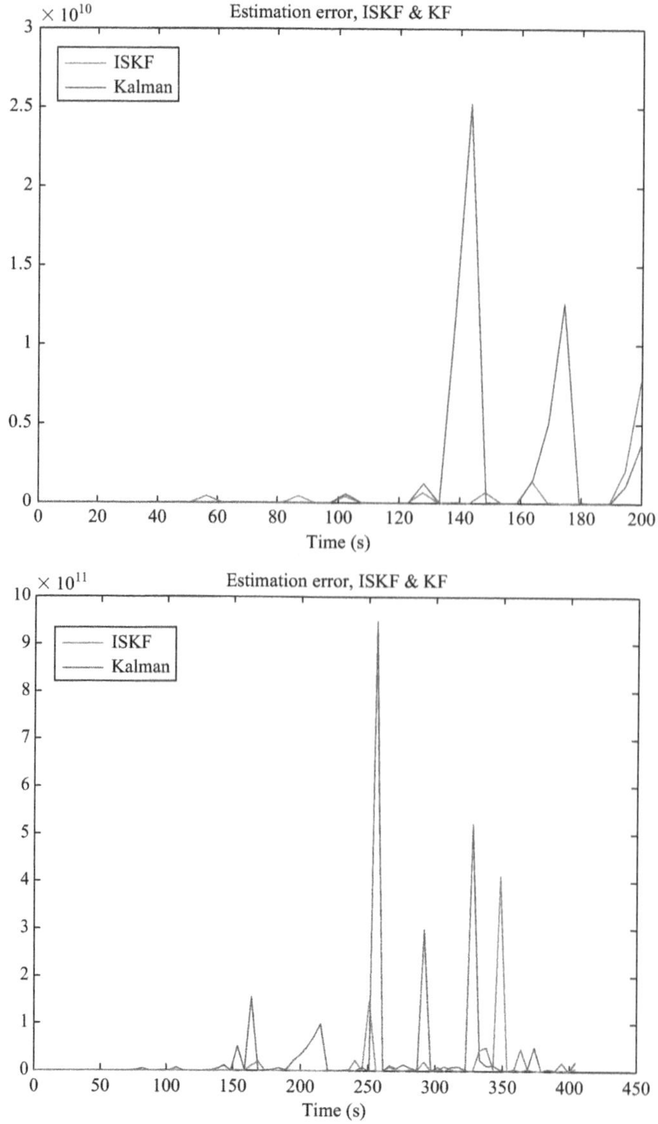

Figure 6.17 Single-sensor estimations with measurement loss: Part 2

sensors still get the measurement, so the peak arising in the estimation pattern due to the measurement loss can be essentially removed. It is also expected that variance of state estimation error would be less than state estimation error arising from single-sensor measurement.

Simulation results of multisensor estimations with measurement loss and different partitions are displayed in Figures 6.18–6.20.

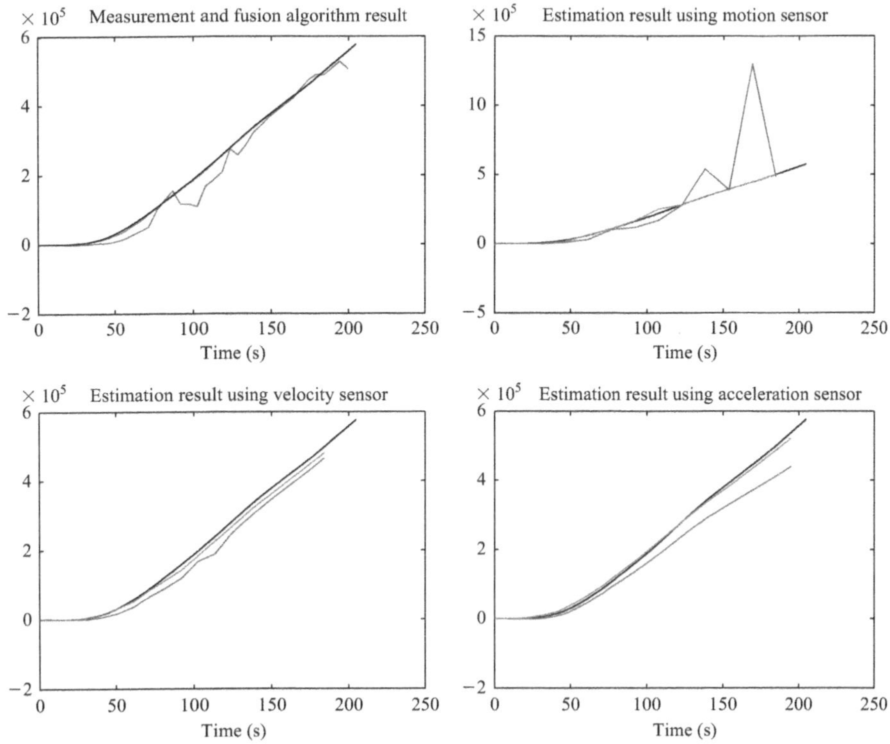

Figure 6.18 Multiple sensor estimation: position state

Remark 6.2: *Estimations arising from single measurement (VS's estimation) using ISKF – light grey line – generally perform better than Kalman filter – dark grey line. This can be seen in Figures 6.18 and 6.19. Estimations of position and velocity based on position and velocity sensor based on ISKF are closer to true value compared to Kalman Filter estimation. ISKF estimation also have less peak on each VS estimation compared to KF.*

6.6 Distributed non-linear estimation

This section is concerned with the distributed $\mathcal{H}_\infty$ filtering problem for sensor networks with repeated scalar non-linearities and multiple probabilistic packet losses. The class of non-linear systems is represented by a discrete time, state-space model involving repeated scalar non-linearities that cover several types of frequently investigated non-linearities as special cases. A number of stochastic variables, all of which are mutually independent but satisfy a certain probabilistic distribution in the interval [0, 1], are introduced to account for the packet dropout phenomena occurring in the channels from the original system to the networked sensors. The concept of

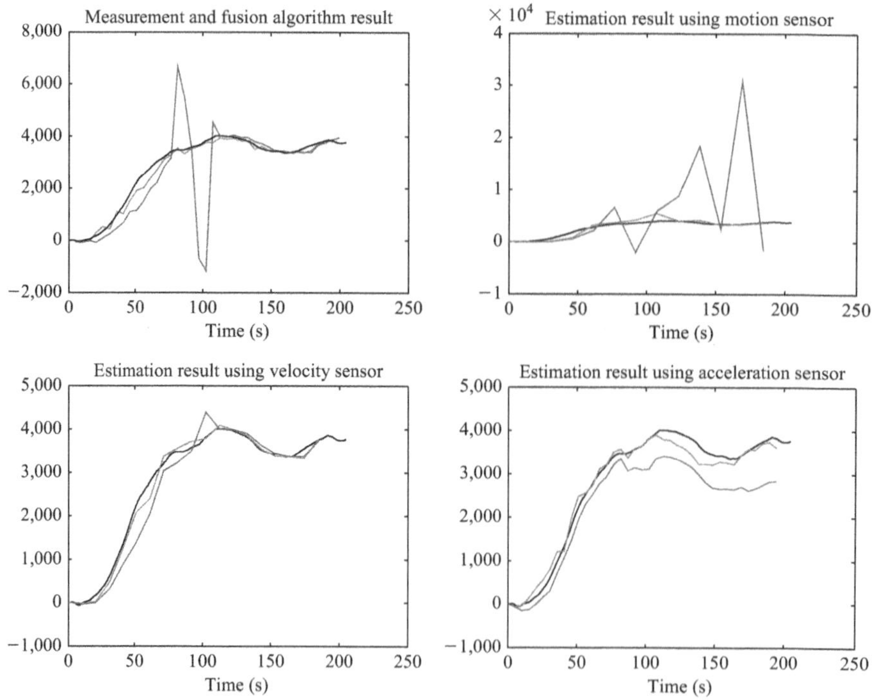

Figure 6.19 Multiple sensor estimation: velocity state

average $\mathcal{H}_\infty$ index is first introduced to measure the overall performance of the sensor networks. Then, by utilizing available measurement information from not only each individual sensor but also its neighboring sensors according to a given topology, stability analysis is carried out to obtain sufficient conditions for ensuring stochastic stability as well as the prescribed average $\mathcal{H}_\infty$ performance constraint. The determination of the parameters of the distributed filters is attained through the feasibility of a convex optimization problem.

6.6.1 Introduction

In the past decade, sensor networks have been attracting increasing attention from many researchers in different disciplines owing to the extensive applications of sensor networks in many areas including surveillance, environment monitoring, information collection, industrial automation, and wireless networks [238,360,386–391]. A sensor network typically consists of a large number of sensor nodes and also a few control nodes, all of which are distributed over a spatial region. The distributed filtering or estimation, as an important issue for sensor networks, has been an area of active research for many years. Different from the traditional filtering for a single sensor [392–394], the information available for the filter algorithm on an individual node of the sensor network is not only from its own measurement but also from its

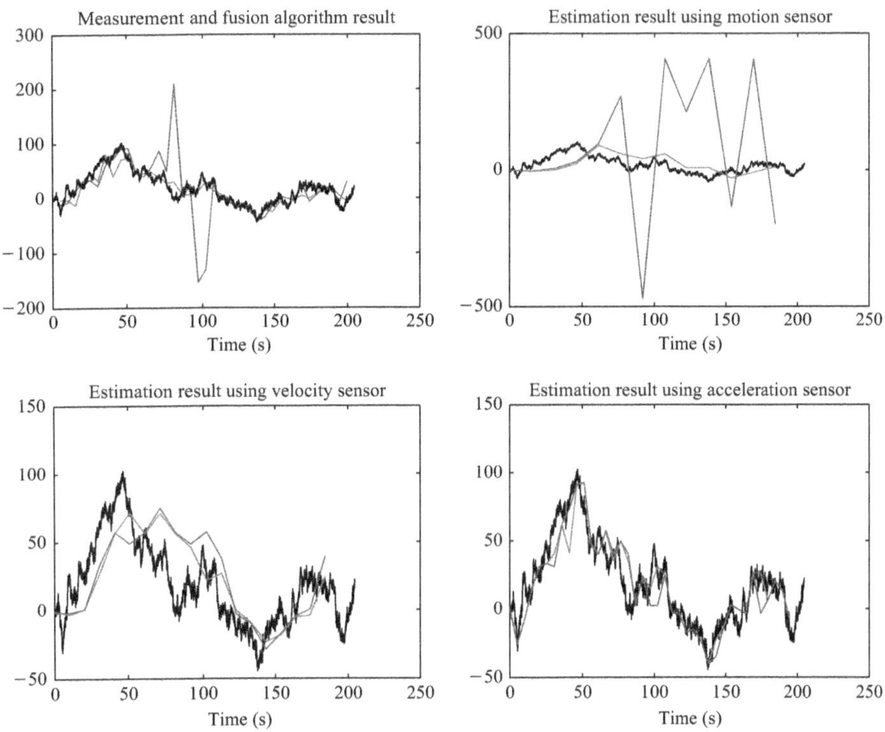

Figure 6.20 Multiple sensor estimation: acceleration state

neighboring sensors' measurements according to the given topology. As such, the objective of filtering based on a sensor network can be achieved in a distributed yet collaborative way. It is noticed that one of the main challenges for distributed filtering lies in how to handle the complicated coupling issues between one sensor and its neighboring sensors.

In recent years, the distributed filtering problem for sensor networks has received considerable research interest and a lot of research results have been available in the literature, see [337,360,390,395–397]. The distributed diffusion filtering strategy has been established in [386,395] for the design of distributed Kalman filters and smoothers, where the information is diffused across the network through a sequence of Kalman iterations and data aggregation. A DKF algorithm has been introduced in [397] through which a crucial part of the solution is used to estimate the average of n signals in a distributed way. Furthermore, three novel DKF algorithms have been introduced in [396] with the first one being a modification of the previous DKF algorithm [397]. Also, a continuous-time DKF algorithm has been rigorously derived and analyzed in [396], and the corresponding extension to the discrete time setting has been conducted in [390] that includes the optimality and stability analysis.

Most of the reported distributed filter algorithms for sensor networks have thus far been mainly based on the traditional Kalman filtering theory that requires exact information about the plant model. In the presence of unavoidable parameter drifts and external disturbances, a desired distributed filtering scheme should be made as robust as possible. However, the robust performance of the available distributed filters has not yet been thoroughly studied, and this would inevitably restrict the application potential in practical engineering. Therefore, it is of great significance to introduce the $\mathcal{H}_\infty$ performance requirement with the hope to enhance the disturbance rejection attenuation level of designed distributed filters. Note that some initial efforts have been made to address the robustness issue. Very recently, a new distributed $\mathcal{H}_\infty$-consensus performance has been proposed in [398] to quantify the consensus degree over a finite horizon and the distributed filtering problem has been addressed for a class of linear time-varying systems in the sensor network, and the filter parameters have been designed recursively by resorting to the difference linear matrix inequalities. In [399], an $\mathcal{H}_\infty$-type performance measure of disagreement between adjacent nodes of the network has been included and a robust filtering approach has been proposed to design the distributed filters for uncertain plants. On the other hand, since non-linearities are ubiquitous in practice, it is necessary to consider the distributed filtering problem for target plants described by non-linear systems.

A sensor network possesses spatially distributed autonomous sensors to monitor physical conditions in cooperative environment where wireless communication is typically adopted among the sensors. As the network scale increases, distributed filtering in a sensor network unavoidably suffers from the limited communication induced by the channel bandwidth constraints. Among others, the network-induced data-packet-loss phenomenon takes place quite often in wireless communication channels between nodes because of fading, interference [400–405]. In most relevant literature, the binary random distribution model has been widely employed due to its simplicity in describing the packet-loss phenomenon. In such a model, probability 0 is usually used to represent the missing of an entire signal and probability 1 denotes the intactness, that is, there is no signal missing at all. Recently, due to various reasons such as sensor aging and sensor temporal failure, the traditional model of binary random distribution has been generalized in [406] to account for the partial data missing problem where the missing probability is defined over the interval of [0,1].

6.6.2 Problem formulation

The filtering problem in sensor networks is shown in Figure 6.21, where each sensor can derive the information from both the plant and its neighboring sensors according to the sensor-networked topology. The information received by sensor i from the plant is transmitted via communication cables that are of limited capacity, and therefore may encounter the phenomena of data-packet losses.

In the sequel, we assume that the n sensor nodes are distributed in space according to a fixed network topology represented by a directed graph $G = (V, \xi, A)$ of

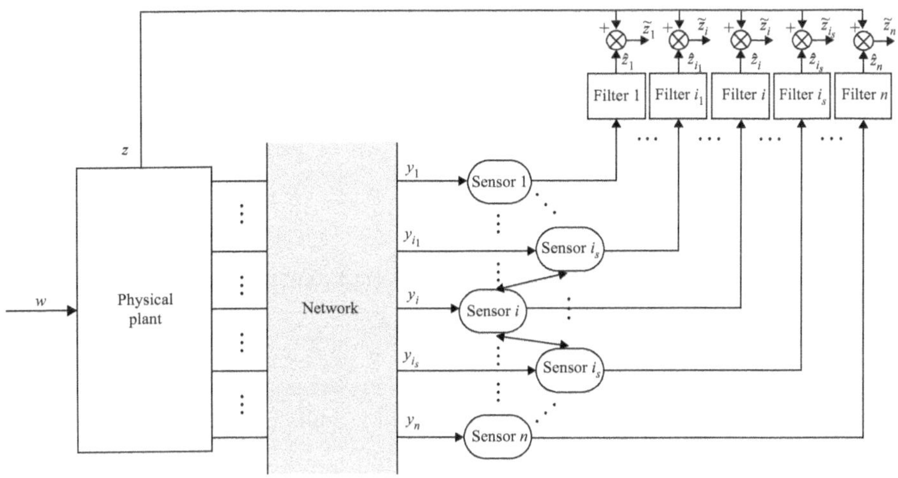

Figure 6.21 A sensor networks filtering problem

the order n with the set of nodes $V = 1, 2, \ldots, n$, the set of edges $\xi \in V \times V$ and the weighted adjacency matrix $A = [a_{ij}]$ with non-negative adjacency element a_{ij}. An edge of G is denoted by ordered pair (i, j). The adjacency elements associated with the edges of the graph are positive, that is, $a_{ij} > 0 \Leftrightarrow (i, j) \in E$, which means that sensor i can obtain information from sensor j. Also, we assume that $a_{ii} = 1$ for all $i \in V$, and therefore (i, i) can be regarded as an additional edge. The set of neighbors of node $i \in V$ plus the node itself are denoted by $N_i = \{j \in V : (i, j) \in E\}$. With reference to Figure 6.21, the plant is described by the following discrete-time non-linear system:

$$\begin{aligned} x(k+1) &= Af(x(k)) + Bw(k), \\ z(k) &= Mf(x(k)) \end{aligned} \tag{6.51}$$

where $x(k) \in \Re^{n_x}$ represents the state vector that cannot be observed directly; $z(k) \in \Re^{n_z}$ is the output to be estimated; $w(k) \in \Re^{n_w}$ is the disturbance input belonging to $l_2[0, \infty)$; and f is the non-linear function. In the sequel, for the vector $x = [x_1\, x_2 \ldots x_{n_x}]^T$, we denote

$$f(x) := [g(x_1)g(x_2)\ldots g(x_{n_x})]^T \tag{6.52}$$

and $g : R \rightarrow R$ is a piecewise continuous identical non-linearity acting on every component of the state x as specified in the following assumption [407]:

Assumption 6.1: *The non-linear function $g : R \rightarrow R$ in system (6.51) satisfies*

$$\forall a, b \in R \quad |g(a) + g(b)| \leq |a + b| \tag{6.53}$$

The model (6.51) is called a system with repeated scalar non-linearity since it is analogous to an upper normal fractional transformation (NFT) with respect to a repeated scalar block, see [407–409]. It is implied from Assumption 6.1 that f encapsulates some typical classes of non-linearities, such as

- the standard saturation
$$\begin{cases} s, & |s| \leq 1 \\ sgn(s), & |s| > 1 \end{cases}$$
- the hyperbolic tangent function that has been extensively used for activation function in neural networks;
- the sine function, etc.

The target plant (6.51) can be applied in many areas including marketing and production control systems and n-stand cold-rolling mills [410] and manufacturing systems for decision making [411].

The n sensors are modeled in what follows by

$$y_i(k) = \beta_i(k)C_i f(x(k)) + D_i w(k), \quad i = 1, 2, \ldots, n, \tag{6.54}$$

where $y_i(k) \in \Re^{n_y}$ is the measurement output measured by sensor i from the plant and $\beta_i(k)$ $(i = 1, 2, \ldots, n)$ are n unrelated and independent stochastic variables. It is assumed that $\beta_i(k)$ has the probabilistic density function $q_i(s)$ $(i = 1, 2, \ldots, n)$ on the interval [0, 1] with mathematical expectation μ_i and variance σ_i^2. Moreover, the matrices mentioned above, that is, A, B, C_i, D_i, and M are all known matrices with appropriate dimensions.

Remark 6.3: *In the sensor model (6.54), the stochastic variable $\beta_i(k)$ corresponding to the ith sensor is introduced to cater for the probabilistic packet-loss problem. It is worth noting that, in most literature concerning the packet-loss phenomena, the Bernoulli distributed model is arguably the most popular one in which 0 is used to represent the missing of signals and 1 denotes the intactness. However, there is indeed a case that the data losses at one moment might be partialy owing to sensor aging and sensor temporal failure, and therefore the missing probability may not be appropriately described by 0 or 1. In (6.54), $\beta_i(k)$ can take value on the interval* [0, 1] *and is therefore more general than the traditional Bernoulli variables.*

The key point in the distributed filter design for sensor networks is how to fuse the information available for the filter on the sensor node i both from the sensor i itself and from its neighbors. Keeping such a fact in mind, the following filter structure is adopted in this article on sensor node i:

$$\begin{aligned} \hat{x}_i(k+1) &= \sum_{j \in N_i} a_{ij} K_{ij} f(\hat{x}_j(k)) + \sum_{j \in N_i} a_{ij} H_{ij} y_j(k), \\ \hat{z}_i(k) &= M f(\hat{x}_i(k)) \end{aligned} \tag{6.55}$$

where $\hat{x}_i(k) \in \Re^{n_x}$ is the state estimate on sensor node i and $\hat{z}_i(k) \in \Re^{n_z}$ the estimate of $z(k)$ from the filter on sensor node i. Here, matrices K_{ij} and H_{ij} are the parameters of filter on node i to be determined.

For convenience of later analysis, we denote

$$\begin{aligned}
\hat{x}(k) &= [\hat{x}_1^T(k)\hat{x}_2^T(k)\ldots\hat{x}_n^T(k)]^T \\
\bar{x}(k) &= 1_N \otimes x(k), \quad \bar{C}_i = (e_i e_i^T) \otimes C_i \\
\tilde{z}(k) &= [(z(k)-\hat{z}_1(k))^T(z(k)-\hat{z}_2(k))^T \ldots (z(k)-\hat{z}_n(k))^T]^T \\
C_\mu &= diag\{\mu_1 C_1, \mu_2 C_2, \ldots, \mu_n C_n\} \\
\bar{A} &= I_n \otimes A, \quad \bar{M} = I_n \otimes M \\
\bar{B} &= I_N \otimes B, \quad \bar{D} = [D_1^T D_2^T \ldots D_n^T]^T \\
\tilde{M} &= [\bar{M} \quad -\bar{M}], \quad \tilde{A} = \begin{bmatrix} \bar{A} & 0 \\ \bar{H}C_\mu & \bar{K} \end{bmatrix} \\
\tilde{C}_i &= \begin{bmatrix} 0 & 0 \\ \bar{H}\bar{C}_i & 0 \end{bmatrix}, \quad \tilde{D} = [\bar{B}^T \quad (\bar{H}\bar{D})^T]^T
\end{aligned} \tag{6.56}$$

where

$$\begin{aligned}
\bar{K} &= [\bar{K}_{ij}]_{n\times n} \quad \text{with} \\
\bar{K}_{ij} &= \begin{cases} a_{ij}K_{ij}, & i = 1,2,\ldots,n; j \in N_i \\ 0, & i = 1,2,\ldots,n; j \neq N_i \end{cases} \\
\bar{H} &= [\bar{H}_{ij}]_{n\times n} \\
\bar{H}_{ij} &= \begin{cases} a_{ij}H_{ij}, & i = 1,2,\ldots,n; j \in N_i \\ 0, & i = 1,2,\ldots,n; j \neq N_i \end{cases}
\end{aligned} \tag{6.57}$$

Obviously, since $a_{ij} = 0$ when $j \neq N_i$, $\bar{K}$ and $\bar{H}$ are two sparse matrices that can be expressed as

$$\bar{K} \in T_{n_x \times n_x}, \quad \bar{H} \in T_{n_x \times n_y} \tag{6.58}$$

where $T_{p\times q} = \{\bar{U} = [U_{ij}] \in \Re^{np\times nq} | U_{ij} \in \Re^{p\times q}; U_{ij} = 0 \text{ if } j \neq N_i\}$.

Introducing $\tilde{x}(k) = [\bar{x}^T(k) \quad \hat{x}^T(k)]^T$, the following augmented system that governs the filtering dynamics for the sensor network is obtained:

$$\begin{aligned}
\tilde{x}(k+1) &= [\tilde{A} + sum_{i=1}^n(\beta_i(k) - \mu_i)\tilde{C}_i]f(\tilde{x}(k)) + \tilde{D}w(k) \\
\tilde{z}(k) &= \tilde{M}f(\tilde{x}(k))
\end{aligned} \tag{6.59}$$

Before proceeding further, we introduce the following definitions.

Definition 6.2: The solution $\tilde{x}(k) = 0$ of the augmented system in (6.59) with $w(k) \equiv 0$ is said to be stochastically stable if, for any $\varepsilon > 0$, there exists a $\delta > 0$ such that

$$\mathbb{E}\{\|\tilde{x}(k)\|\} < \varepsilon \tag{6.60}$$

whenever $k \in I^+$ and $\|\tilde{x}(0)\| < \delta$.

Definition 6.3: A square matrix $P := [p_{ij}] \in \Re^{n\times n}$ is called diagonally dominant if for all $i = 1, \ldots, n$

$$\forall i, |p_{ii}| \geq \sum_{j\neq i} |p_{ij}|. \tag{6.61}$$

The objective now is to design a filter of the form in (6.55) on each node i of the sensor network for system (6.51). In other words, we are going to find the filter parameters K_{ij} and H_{ij} such that the following two requirements are satisfied simultaneously:

1. (Stochastic stability) The zero-solution of the augmented system (6.59) with $w(k)0$ is stochastically stable.
2. (Average $\mathcal{H}_\infty$ performance) Under the zero-initial condition, for a given disturbance attenuation level $\gamma > 0$ and all non-zero $w(k)$, the filtering error $\tilde{z}(k)$ from (6.59) satisfies the following condition, which will be called in the sequel *average $\mathcal{H}_\infty$ performance constraint*:

$$J = \frac{1}{n}\|\tilde{z}\|_{E_2}^2 - \gamma^2\|w\|_2^2 < 0, \tag{6.62}$$

where

$$\|\tilde{z}\|_{E_2} = \mathbb{E}\left\{\sqrt{\sum_{k=0}^{\infty}\tilde{z}^T(k)\tilde{z}(k)}\right\}, \quad \|w\|_2 = \sqrt{\sum_{k=0}^{\infty}w^T(k)w(k)} \tag{6.63}$$

Remark 6.4: *The average $\mathcal{H}_\infty$ performance presented in (6.62) is a generalization of the classical $\mathcal{H}_\infty$ performance constraint. It means that the average energy gains (from the average energy of all disturbance on the system and sensor networks to the average energy of all estimation errors) for filters in the sensor network should be less than a given disturbance attenuation level γ.*

6.6.3 Analytic results

In this section, let us investigate both the distributed $\mathcal{H}_\infty$ filtering performance analysis and filter design problems for system (6.51) with n sensors whose topology is determined by the given graph $G = (V, E, A)$. The following lemmas will be needed in establishing our main results.

Lemma 6.1 [409]: *Let f be defined as in (6.52). If $P > 0$ is diagonally dominant, then the following inequality holds for any vector ζ :*

$$f^T(\zeta)Pf(\zeta) \le \zeta^T P\zeta \tag{6.64}$$

Lemma 6.2: *If there exist a Lyapunov function $V(\tilde{x}(k))$ and a function $\phi(r) \in \Xi$ satisfying the following conditions:*

$$V(0) = 0 \tag{6.65}$$

$$\phi(\|\tilde{x}(k)\|) \le V(\tilde{x}(k)) \tag{6.66}$$

$$\mathbb{E}\{V(\tilde{x}(k+1))\} - \mathbb{E}\{V(\tilde{x}(k))\} < 0, \quad k \in I^+ \tag{6.67}$$

then the solution $\tilde{x}(k) = 0$ of the filtering augmented system in (6.59) with $w(k) \equiv 0$ is stochastically stable.

The following theorem gives a sufficient condition under which the stochastic repeated scalar non-linear system governed by (6.59) is stochastically stable in the sense of Definition 6.2 with $\mathcal{H}_\infty$ performance constraint given in (6.62).

Theorem 6.1: *Consider the repeated scalar non-linear system (6.51) with given filter parameters K_{ij}, H_{ij}, and a prescribed $\mathcal{H}_\infty$ index $\gamma > 0$. If there exists a positive diagonally dominant matrix $P > 0$ satisfying*

$$\phi = \begin{bmatrix} \left\{ \begin{array}{c} \tilde{A}^T P \tilde{A} + \sum_{i=1}^{n} \sigma_i^2 \tilde{C}_i^T P \tilde{C}_i \\ -P + \frac{1}{n} \tilde{M}^T \tilde{M} \end{array} \right\} & * \\ \tilde{D}^T P \tilde{A} & \tilde{D}^T P \tilde{D} - \gamma^2 I \end{bmatrix} < 0 \tag{6.68}$$

then the filtering dynamics in (6.59) is stochastically stable and also satisfies the $\mathcal{H}_\infty$ performance constraint (6.62).

Proof. Choose the following Lyapunov function for system (6.59):

$$V(\tilde{x}(k)) = \tilde{x}^T(k) P \tilde{x}(k) \tag{6.69}$$

with $P > 0$ being a matrix to be determined. Then, along the trajectory of system (6.59) with $w(k) = 0$, we have

$$\begin{aligned} &\mathbb{E}\{\tilde{x}^T(k+1) P \tilde{x}(k+1) - \tilde{x}^T(k) P \tilde{x}(k)\} \\ &= \mathbb{E}\left\{ \left(\left(\tilde{A} + \sum_{i=1}^{n} (\beta_i(k) - \mu_i) \tilde{C}_i \right) f(\tilde{x}(k)) \right)^T \right. \\ &\quad \left. \times P \left(\tilde{A} + \sum_{i=1}^{n} (\beta_i(k) - \mu_i) \tilde{C}_i \right) f(\tilde{x}(k)) - \tilde{x}^T(k) P \tilde{x}(k) \right\} \\ &= \mathbb{E}\left\{ f^T(\tilde{x}(k)) \left(\tilde{A}^T P \tilde{A} + \sum_{i=1}^{n} \sigma_i^2 \tilde{C}_i^T P \tilde{C}_i \right) f(\tilde{x}(k)) - \tilde{x}^T(k) P \tilde{x}(k) \right\} \end{aligned} \tag{6.70}$$

According to Lemma 6.1, we have

$$\begin{aligned} &\mathbb{E}\{\Delta V(\tilde{x}(k))\} \\ &\leq \mathbb{E}\left\{ f^T(\tilde{x}(k)) \left(\tilde{A}^T P \tilde{A} + \sum_{i=1}^{n} \sigma_i^2 \tilde{C}_i^T P \tilde{C}_i - P \right) f(\tilde{x}(k)) \right\} \end{aligned} \tag{6.71}$$

Note that (6.68) in Theorem 6.1 implies that

$$\tilde{A}^T P \tilde{A} + \sum_{i=1}^{n} \sigma_i^2 \tilde{C}_i^T P \tilde{C}_i - P < 0$$

which results in

$$\mathbb{E}\{V(\tilde{x}(k+1))\} - E\{V(\tilde{x}(k))\} < 0$$

Taking $\phi(r) = \lambda_{\min}(P)\Re^2$ such that $\phi(r) \in \Xi$, we obtain

$$\begin{aligned}\phi(\|\tilde{x}(k)\|) &= \lambda_{\min}(P)\|\tilde{x}(k)\|^2 \\ &= \lambda_{\min}(P)\tilde{x}^T(k)\tilde{x}(k) \leq \tilde{x}^T(k)P\tilde{x}(k) = V(\tilde{x}(k))\end{aligned}$$

which satisfies (6.66). Considering $V(0) = 0$, it follows readily from Lemma 6.2 that the filtering dynamics in (6.59) with $w(k) = 0$ is stochastically stable.

To establish the $\mathcal{H}_\infty$ performance, we introduce the following index:

$$\begin{aligned}\bar{J} &= \mathbb{E}\{V(\tilde{x}(k+1))\} + E\left\{\frac{1}{n}\tilde{z}^T(k)\tilde{z}(k)\right\} - \gamma^2 w^T(k)w(k) \\ &\quad -\mathbb{E}\left\{f^T(\tilde{x}(k))Pf(\tilde{x}(k))\right\} \qquad (6.72)\end{aligned}$$

From (6.59), one has

$$\begin{aligned}\bar{J} &= \mathbb{E}\left\{\left[\left(\tilde{A} + \sum_{i=1}^{n}(\beta_i(k) - \mu_i)\tilde{C}_i\right)f(\tilde{x}(k)) + \tilde{D}w(k)\right)^T\right. \\ &\quad \times P\left(\left(\tilde{A} + \sum_{i=1}^{n}(\beta_i(k) - \mu_i)\tilde{C}_i\right)f(\tilde{x}(k)) + \tilde{D}w(k)\right] \\ &\quad \left. +f^T(\tilde{x}(k))\left(\frac{1}{n}\tilde{M}^T(k)\tilde{M} - P\right)f(\tilde{x}(k)) - \gamma^2 w^T(k)w(k)\right\} \\ &= \mathbb{E}\{\eta^T(k)\Phi\eta(k)\} \qquad (6.73)\end{aligned}$$

where $\eta(k) = [f^T(\tilde{x}(k)) \quad w^T(k)w(k)]^T$. It follows from (6.68) in Theorem 6.1 that $\bar{J} < 0$.

By summing up (6.72) on both sides from 0 to ∞ with respect to k and taking mathematical expectation, we arrive at

$$\begin{aligned}\sum_{k=0}^{\infty}\bar{J} &= \sum_{k=0}^{\infty} E\{V(\tilde{x}(k+1))\} - \sum_{k=0}^{\infty} E\{f^T(\tilde{x}(k))Pf(\tilde{x}(k))\} \\ &\quad +\frac{1}{n}\|\tilde{z}\|_{E_2}^2 - \gamma^2\|w\|_2^2 < 0 \qquad (6.74)\end{aligned}$$

Hence, the $\mathcal{H}_\infty$ performance index defined in (6.62) is given by

$$J = \sum_{k=0}^{\infty}\bar{J} - \left(\sum_{k=0}^{\infty} E\{V(\tilde{x}(k+1))\} - \sum_{k=0}^{\infty}\{f^T(\tilde{x}(k))Pf(\tilde{x}(k))\}\right) \qquad (6.75)$$

By Lemma 6.1 and noting that $\bar{J} < 0$ together with zero-initial condition, we have

$$J \leq \sum_{k=0}^{\infty} \bar{J} - \left(\sum_{k=0}^{\infty} E\{V(\tilde{x}(k+1))\} - \sum_{k=0}^{\infty} E\{\tilde{x}^T(k)P\tilde{x}(k)\} \right)$$
$$\leq \sum_{k=0}^{\infty} \bar{J} - E\{VV(\tilde{x}(\infty))\} < 0$$

which is equivalent to (6.62), and the proof is complete. □

We now deal with the problem of designing distributed $\mathcal{H}_\infty$ filters for the repeated scalar non-linear systems. To tackle the difficulty of dilating a diagonally dominant positive definite matrix P, that is, letting P remain in its original form (instead of being partitioned) in solving the filter design problems, we rewrite the parameters $\tilde{A}, \tilde{D}$, and $\tilde{C}_i$ in Theorem 6.1 in the following form:

$$\tilde{A} = \tilde{A}_0 + EK_hR_1, \quad \tilde{D} = \tilde{D}_0 + EK_hR_2, \quad \tilde{C}_i = EK_hR_{3i} \tag{6.76}$$

where

$$\tilde{A}_0 = \begin{bmatrix} \bar{A} & 0 \\ 0 & 0 \end{bmatrix}, \quad E = \begin{bmatrix} 0 \\ I \end{bmatrix}, \quad R_1 = \begin{bmatrix} 0 & I \\ C_\mu & 0 \end{bmatrix}$$
$$R_{3i} = \begin{bmatrix} 0 & 0 \\ \bar{C}_i & 0 \end{bmatrix}, \quad \tilde{D}_0 = [\bar{B}^T \quad 0]^T \tag{6.77}$$
$$R = [0 \quad \bar{D}^T]^T, \quad K_h = [\bar{K} \quad \bar{H}]$$

Note that the matrix K_h consists of all desired filter parameters. By the Schur complement and noting (6.76), it follows that (6.68) is equivalent to

$$\begin{bmatrix} \frac{1}{n}\tilde{M}^T(k)\tilde{M} - P & \bullet & \bullet & \bullet \\ 0 & -\gamma^2 I & \bullet & \bullet \\ \tilde{A}_0 + EK_hR_1 & \tilde{D}_0 + EK_hR_2 & -P^{-1} & \bullet \\ \bar{X} & 0 & 0 & -\bar{P}^{-1} \end{bmatrix} < 0 \tag{6.78}$$

where

$$\bar{X} = [\sigma_1 \Re_{31}^T (EK_h)^T \ldots \sigma_n \Re_{3n}^T (EK_h)^T]^T, \quad \bar{P}^{-1} = I_n \otimes P^{-1} \tag{6.79}$$

Theorem 6.2 provides the existence condition of the desired distributed H_∞ filters.

Theorem 6.2: *Given a positive scalar $\gamma > 0$. The filtering dynamics in (6.59) is stochastically stable and satisfies the distributed H_∞ performance constraint (6.62) if there exist matrices $0 < P := [p_{ij}] \in \Re^{2nn_x \times 2nn_x}, 0 < Q := [q_{ij}] \in \Re^{2nn_x \times 2nn_x}$, $L > 0, \bar{K} \in T_{n_x \times n_x}$ and $\bar{H} \in T_{n_x \times n_y}$ satisfying*

$$\begin{bmatrix} \frac{1}{n}\tilde{M}^T(k)\tilde{M} - P & \bullet & \bullet & \bullet \\ 0 & -\gamma^2 I & \bullet & \bullet \\ \tilde{A}_0 + EK_hR_1 & \tilde{D}_0 + EK_hR_2 & -L & \bullet \\ \bar{X} & 0 & 0 & -\bar{L} \end{bmatrix} < 0 \tag{6.80}$$

$$p_{ii} - \sum_{j \neq i} (p_{ij} + 2q_{ij}) \geq 0 \tag{6.81}$$

$$q_{ij} \geq 0 \ \forall i \neq j \tag{6.82}$$

$$p_{ij} + q_{ij} \geq 0 \ \forall i \neq j \tag{6.83}$$

$$PL = I \tag{6.84}$$

where K_h is defined in (6.77) and

$$\bar{L} = I_n \otimes L \tag{6.85}$$

Moreover, if the above inequalities (with one equality constraint) are feasible, the desired filter parameters K_{ij} and H_{ij} ($i = 1, 2, \ldots, n, j \in N_i$) can be derived from (6.57).

Proof. By setting $P^{-1} = L$, inequality (6.78) follows from (6.80) immediately. In addition, from (6.81) to (6.83), we have

$$p_{ii} \geq \sum_{j \neq i} (p_{ij} + 2q_{ij}) = \sum_{j \neq i} (|p_{ij} + q_{ij}| + |-q_{ij}|) \geq \sum_{j \neq i} |p_{ij}|$$

which guarantees the positive definite matrix P to be diagonally dominant. The proof is complete. □

Due to the presence of the matrix equality in Theorem 6.2, condition (6.80) will give rise to a nonconvex feasible set. In the following, a cone complementarity linearization (CCL) method is proposed to solve the problem.

In [409], it is reported that the basic idea in the CCL algorithm is that if $\Omega(P, L) = \left[\begin{smallmatrix} P & I \\ I & L \end{smallmatrix}\right] \geq 0$ is feasible in the $2nn_x \times 2nn_x$ matrix variables $L > 0$ and $P > 0$, then $tr(PL) \geq 2nn_x$, and $tr(PL) = 2nn_x$ if and only if $PL = I$.

Based on such a statement, we put forward the following CCL problem instead of the original nonconvex feasibility problem formulated in Theorem 6.2.

CCL Problem:

$$\min\ tr(PL) \text{subject to (6.80) (6.83)}, \quad \Omega(P, L) \geq 0$$

If a solution exists and min $tr(PL) = 2nn_x$, then the conditions in Theorem 6.2 are solvable.

The following design procedure is proposed to implement the distributed $\mathcal{H}_\infty$ filters for the target plant with repeated scalar non-linearity and random packet losses.

Distributed $\mathcal{H}_\infty$ filter design algorithm

Step 1: Given the $\mathcal{H}_\infty$ performance index γ and the positive weighting matrix R. Find a feasible set $(P_{(0)}, L_{(0)}, Q_{(0)}, \bar{K}_{(0)}, \bar{H}_{(0)})$ satisfying (6.80)–(6.83) and $\Omega(P, L) \geq 0$. Set $q = 0$. If there are none, exit.

Step 2: According to (6.80)–(6.83) and $\Omega(P, L) \geq 0$, solve the following problem: min $tr(PL_{(q)} + P_{(q)}L)$.

Step 3: Substitute the obtained matrix variables $(P, L, Q, \bar{K}, \bar{H})$ into (6.68). If condition (6.68) is satisfied with $|tr(PL) - 2nn_x| < \delta$ for some sufficiently small scalar $\delta > 0$, then output the feasible solutions $(P, L, Q, \bar{K}, \bar{H})$. Exit.

Step 4: If $q > N$, where N is the maximum number of iterations allowed, exit.

Step 5: Set $q = q + 1, (P_{(q)},$

$$L_{(q)}, Q_{(q)}, \bar{K}_{(q)}, \bar{H}_{(q)}) = (P, L, Q, \bar{K}, \bar{H})$$

and go to Step 2.

6.6.4 *Illustrative example 5.1*

To demonstrate the applicability of the proposed filtering techniques, following [410], we consider a factory that produces two kinds of products ($j = 1, 2$) sharing common resources and raw materials, like color TV and black/white TV, PC and laptop computer, for instance. During the kth period (quarter or season), we define

- $s_j(k)$ amount of sales of product j;
- $a_j(k)$ advertisement cost spent for product j;
- $i_j(k)$ amount of inventory of product j;
- $m_j(k)$ production of product j,

where $j = 1, 2$. Let

$$x(k) = \begin{bmatrix} s_1(k) \\ s_2(k) \\ i_1(k) \\ i_2(k) \end{bmatrix}, \quad u_k = \begin{bmatrix} m_1(k+1) \\ m_2(k+1) \\ a_1(k) \\ a_2(k) \end{bmatrix}$$

Then, the effect of advertisements on sales in the marketing process and the interlink between inventory and production in the production process can be expressed dynamically by the following form:

$$x(k+1) = Af(x(k)) + Bw(k) + Eu(k)$$

where $f(x(k))$, which accounts for the phenomenon of state-dependent perturbations, is a saturation non-linearity function with the following form:

$$f(x(k)) := [g(x_1(k)) \quad g(x_2(k)) \quad g(x_3(k)) \quad g(x_4(k))]^T \tag{6.86}$$

$$g(x_i(k)) = \begin{cases} x_i(k), & -1 \leq x_i(k) \leq 1 \\ 1, & x_i(k) > 1 \\ -1, & x_i(k) < -1 \end{cases} \tag{6.87}$$

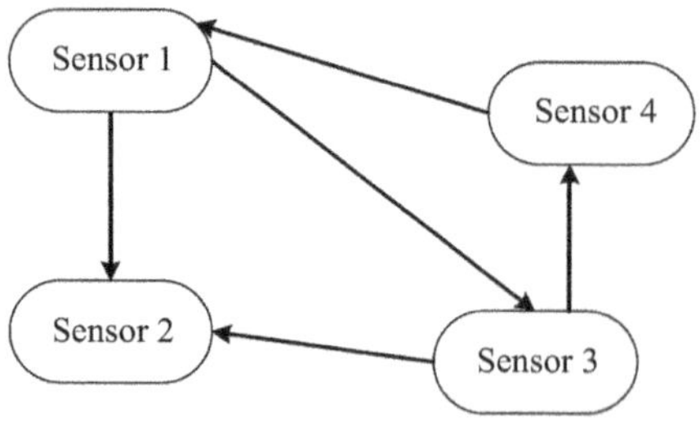

Figure 6.22 Topological structure of the sensor network

Let us assume that no production and advertisement are available in this example, hence $E = 0$. Also, we assume that the information transmission for the measured amounts of sales and inventory of the products is conducted through a sensor network that are subject to possible packet losses. The sensor network shown in Figure 6.22 is represented by a directed graph $G = (V, E, A)$ with the set of nodes $V = \{1, 2, 3, 4\}$, set of edges $E = \{(1,1), (1,4), (2,1), (2,2), (2,3), (3,1), (3,3), (4,3), (4,4)\}$ and the following adjacency matrix:

$$A = \begin{bmatrix} 1 & 0 & 0 & 1 \\ 1 & 1 & 1 & 0 \\ 1 & 0 & 1 & 0 \\ 0 & 0 & 1 & 1 \end{bmatrix}$$

To employ the proposed distributed filtering scheme, we now consider a specific example of the problem discussed above. The corresponding system parameters are listed as follows:

$$A = \begin{bmatrix} 0.4 & 0 & 0 & 0 \\ 0 & 0.5 & 0 & 0 \\ -0.4 & 0 & 0.6 & 0 \\ 0 & -0.5 & 0 & 0.6 \end{bmatrix}$$

$$B = \begin{bmatrix} 0 & 0.3 & 0 & 0 \\ 0.2 & 0 & 0 & 0 \\ 0 & -0.4 & 0 & 0 \\ -0.2 & 0 & 0 & 0 \end{bmatrix}$$

$$C_1 = [1 \quad 0 \quad 0 \quad 0], \quad C_2 = [0 \quad 1 \quad 0 \quad 0],$$

$$C_3 = [1 \quad 1 \quad 0 \quad 0], \quad C_4 = [1 \quad 0 \quad 1 \quad 0],$$

$$D_1 = [0.5 \quad 0 \quad 0 \quad 0], \quad D_2 = [0.2 \quad 0.1 \quad 0 \quad 0]$$

$$D_3 = [0 \quad 0 \quad 0.2 \quad 0], \quad D_4 = [0.1 \quad 0 \quad 0 \quad 0.1]$$

$$M = [0 \quad 1 \quad 0 \quad 0], \quad E = 0$$

Let the probabilistic density functions of $\beta_1(k)$, $\beta_2(k)$, $\beta_3(k)$, and $\beta_4(k)$ in $[0, 1]$ be described by

$$
q_1(s_1) = \begin{cases} 0, & s_1 = 0 \\ 0.1, & s_1 = 0.5 \\ 0.9, & s_1 = 1 \end{cases}, \quad q_2(s_2) = \begin{cases} 0.1, & s_2 = 0 \\ 0.1, & s_2 = 0.5 \\ 0.8, & s_2 = 1 \end{cases}
$$
$$
q_3(s_3) = \begin{cases} 0, & s_3 = 0 \\ 0.2, & s_3 = 0.5 \\ 0.8, & s_3 = 1 \end{cases}, \quad q_4(s_4) = \begin{cases} 0, & s_4 = 0 \\ 0.4, & s_4 = 0.5 \\ 0.6, & s_4 = 1. \end{cases} \tag{6.88}
$$

It follows that the expectations and variances are calculated as $\mu_1 = 0.95, \mu_2 = 0.85, \mu_3 = 0.9, \mu_4 = 0.8, \sigma_1^2 = 0.15, \sigma_2^2 = 0.32, \sigma_3^2 = 0.2$, and $\sigma_4^2 = 0.245$. The disturbance attenuation level and the positive definite matrices are given as $\gamma = 1.2$ and $R_1 = R_2 = R_3 = R_4 = 2I_8$, respectively. The purpose is to design a distributed $\mathcal{H}_\infty$ filter in order to estimate the amount of sales of product. By applying Theorem 6.2 with help from the design procedure, we can obtain the following parameters of the desired distributed filters:

$$
K_{11} = \begin{bmatrix} 0.1137 & -0.1625 & 0.0158 & -0.0531 \\ -0.1915 & 0.4749 & -0.0877 & -0.1100 \\ 0.2716 & -0.1200 & 0.3754 & -0.2617 \\ -0.0188 & 0.0202 & 0.1044 & 0.6081 \end{bmatrix}
$$

$$
K_{14} = \begin{bmatrix} 0.2439 & 0.0750 & -0.0961 & -0.0938 \\ 0.0458 & 0.1037 & 0.0736 & 0.0247 \\ -0.2167 & 0.3131 & -0.1808 & -0.0713 \\ 0.0102 & -0.1909 & 0.0602 & 0.0352 \end{bmatrix}
$$

$$
K_{21} = \begin{bmatrix} -0.0852 & 0.0193 & 0.0280 & -0.0121 \\ -0.1592 & 0.0814 & 0.1600 & 0.0569 \\ 0.0110 & 0.0287 & 0.2255 & 0.0179 \\ -0.0409 & -0.0854 & -0.0536 & 0.0567 \end{bmatrix}
$$

$$
K_{22} = \begin{bmatrix} 0.5380 & -0.0176 & -0.0566 & 0.1805 \\ -0.2530 & 0.4950 & -0.0272 & -0.0026 \\ -0.3970 & 0.0877 & 0.4407 & -0.2361 \\ 0.3020 & 0.0255 & -0.0732 & 0.4410 \end{bmatrix}
$$

$$
K_{23} = \begin{bmatrix} 0.0175 & 0.0216 & -0.0046 & 0.0136 \\ 0.1703 & -0.0971 & -0.1353 & 0.1793 \\ 0.2161 & -0.0226 & -0.1292 & 0.1740 \\ -0.0519 & 0.0367 & 0.0420 & -0.1516 \end{bmatrix}
$$

$$
K_{31} = \begin{bmatrix} 0.0690 & -0.0771 & -0.1142 & 0.0979 \\ 0.2841 & 0.0122 & 0.0060 & -0.0909 \\ 0.2946 & 0.0726 & 0.0485 & -0.0467 \\ -0.1500 & -0.0464 & 0.1253 & 0.1185 \end{bmatrix}
$$

$$
K_{33} = \begin{bmatrix} 0.3091 & 0.0055 & 0.1363 & 0.0433 \\ -0.0245 & 0.6476 & 0.0959 & 0.0404 \\ 0.3798 & 0.3682 & 0.4620 & 0.2560 \\ 0.2610 & -0.0723 & 0.0302 & 0.6325 \end{bmatrix}
$$

$$K_{43} = \begin{bmatrix} -0.1472 & 0.0150 & 0.0801 & 0.0331 \\ -0.1411 & -0.0687 & 0.1451 & -0.2717 \\ -0.2472 & -0.2496 & 0.3624 & -0.1996 \\ -0.0143 & -0.0310 & -0.0192 & 0.0628 \end{bmatrix}$$

$$K_{44} = \begin{bmatrix} 0.5910 & 0.0532 & -0.1676 & -0.1281 \\ -0.0105 & 0.5823 & -0.0889 & -0.0401 \\ -0.3628 & 0.1010 & 0.4647 & 0.0711 \\ -0.1753 & 0.0505 & 0.0249 & 0.4736 \end{bmatrix}$$

$$H_{11} = \begin{bmatrix} 0.0373 \\ -0.1116 \\ -0.8248 \\ 0.1619 \end{bmatrix}, \quad H_{14} = \begin{bmatrix} 0.1336 \\ -0.0340 \\ 0.5162 \\ 0.0074 \end{bmatrix}$$

$$H_{21} = [0.0278 \quad -0.0884 \quad -0.0630 \quad 0.0845]^T$$

$$H_{22} = [0.3096 \quad -0.3339 \quad -0.6263 \quad 0.4021]^T$$

$$H_{23} = [0.0180 \quad 0.0047 \quad 0.0066 \quad -0.0897]^T$$

$$H_{31} = [0.6074 \quad 0.0560 \quad -0.8735 \quad 0.0408]^T$$

$$H_{23} = [0.0180 \quad 0.0047 \quad 0.0066 \quad -0.0897]^T$$

$$H_{33} = [-0.2468 \quad -0.0439 \quad 0.2344 \quad 0.1118]^T$$

$$H_{43} = [-0.0235 \quad -0.0300 \quad -0.1328 \quad 0.0966]^T$$

$$H_{44} = [0.1702 \quad 0.0700 \quad 0.4929 \quad -0.0385]^T$$

For simulation purpose, we first assume that $w(k) = 0$ and the initial conditions $x(0) = [-0.4\ 0.5\ 00.2]^T$ and $\hat{x}_i(0) = [0\ 0.1\ 0\ 0]^T (i = 1, 2, 3, 4)$. Figure 6.23

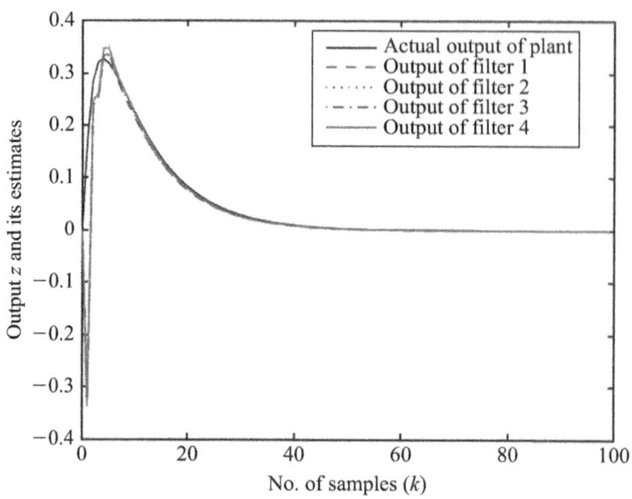

Figure 6.23 Output $z(k)$ and its estimates when $w(k) = 0$

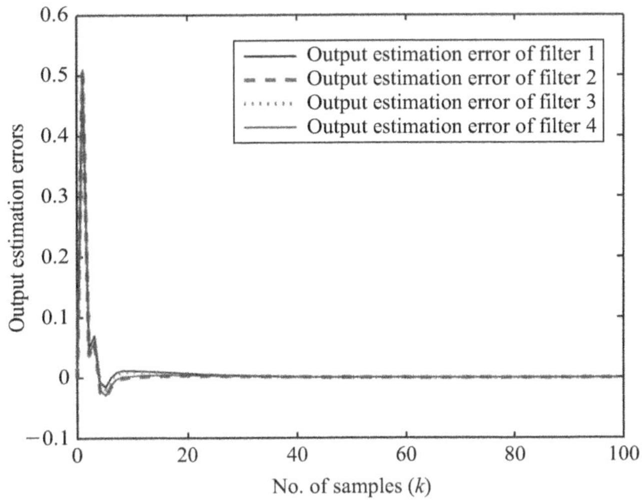

Figure 6.24 Filtering errors $z(k) - \hat{z}_i(k)$ $(i = 1, 2, 3, 4)$

shows the output $z(k)$ and its estimates from the filters 1, 2, 3, and 4, and Figure 6.24 gives the simulation results of estimation error $\tilde{z}(k)$, which confirms that the augmented systems in (6.59) is stochastically stable. To further illustrate the performance of the designed distributed filters, we now assume the zero-initial conditions and set the exogenous disturbance input to be $w(k) = \exp(-k/40)$. The simulation results are shown in Figures 6.25 and 6.26. Figure 6.25 plots the output $z(k)$ and its estimates from the filters 1, 2, 3, and 4. Figure 6.26 shows the estimation errors

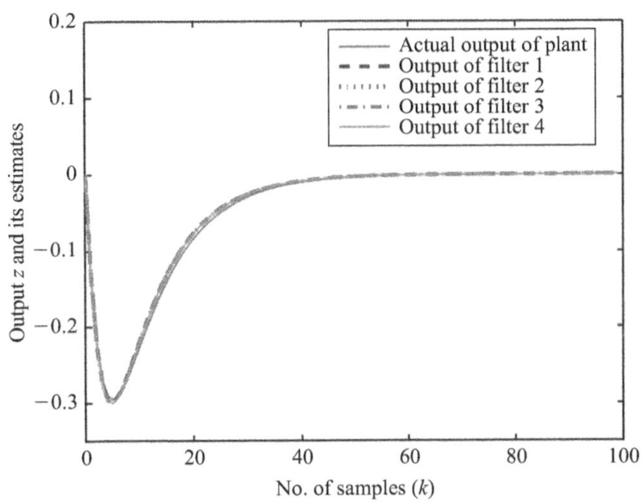

Figure 6.25 Output $z(k)$ and its estimates

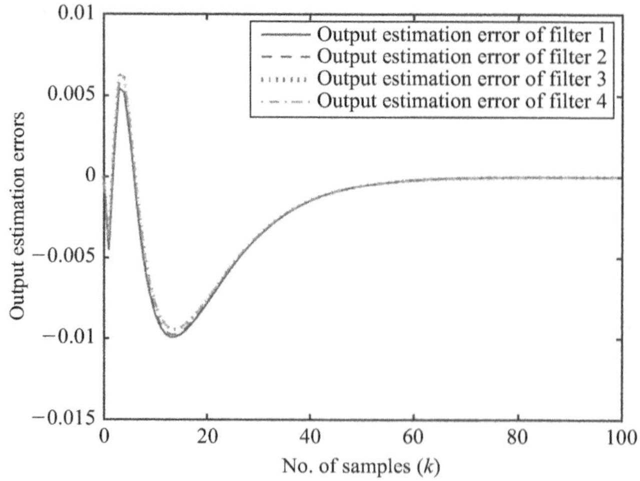

Figure 6.26 Filtering errors $z(k) - \hat{z}_i(k)(i = 1, 2, 3, 4)$

$z(k) - \hat{z}_i(k)$ $(i = 1, 2, 3, 4)$. It can be calculated that $\frac{\frac{1}{n}\|\tilde{z}\|_{E_2}}{\|w\|_2} = 0.8933$ that stays below the prescribed upper bound $\gamma = 1.2$.

6.7 Notes

In this chapter, an improved distributed minimum variance estimation algorithm was first developed by considering the propagation delay of the physical quantity to be measured. It was established that this estimation algorithm provide the following features:

- It incorporated an FIR filter that aims at simplifying the signal estimation and alleviating the parameter optimization.
- It had a time-varying signal tracking capability without a pre-assigned coupling law imposed among the nodes.
- It considered a more general model of the filter structure.
- The parameters were computed through distributed algorithms.
- The filter weights were designed such that the variance of the estimation errors is minimized, thereby improving the filter design considerably and we characterize the performance limit of the filter.

Next, we developed new algorithm to deal with asynchronous multisensor measurements. The developed algorithm, ISKF, have been shown to be able to recover undelayed estimations from delayed measurement. ISKF has also been demonstrated to perform better compared with the regular Kalman filter if the measurement loss happens. Incorporating its estimation in fusion algorithm, the fused estimations tend to have less estimation error compared to single-sensor estimations.

Then, an improved distributed estimation method was developed based on the concept of minimum variance with a consideration of time delay appearing in the measurements of different sensor nodes. This was attributed to the time taken by physical quantity to be measured (e.g., temperature, acoustic waves, to name few) to propagate from one point in the space to another one. In this regard, it is assumed that source of the data is located at a known position and the spatial ordinates of the sensor nodes are known. This knowledge is applied to calculate the time delay between the source of the data and any node in the network. The spectrum of the signal of interest is assumed to be known a priori. Based on this knowledge the data is first passed through a low-pass FIR filter. The signal is then jointly tracked by a SN, in which each node computes an estimate as a delay-adjusted weighted sum of its own and its neighbors measurements and estimates. The filter weights are time varying and updated locally and it has a cascade structure comprised of two loops: an inner loop producing the state estimate and an outer loop producing an estimate of the error covariance. The state estimate is thus obtained as the solution of an optimization problem with quadratic cost function and quadratic constraints.

Finally, we dealt with the distributed $\mathcal{H}_\infty$ filtering problem for systems with repeated scalar non-linearities in the sensor network. The phenomena of packet losses occurring between the system and each sensor have been taken into account. The distributed filters were designed such that the filtering dynamics system to be stochastically stable and the estimation errors to satisfy the average $\mathcal{K}_\infty$ performance constraint. Furthermore, the explicit expression of the desired filter gains have been derived. Finally, an illustrative example highlighting the usefulness of the presented results have been discussed. It is also noted that the proposed design method for sensor networks can be easily extended to other networks, like NCS [412], gene regulatory networks, complex networks, and neural networks, to name a few.

Chapter 7
Distributed Kalman filtering

In this chapter, we provide the theory and analytical development of distributed Kalman filtering based on various methods. In one method, the weighted averages, based on the Bayesian point of view, yield the best possible estimate of the state at a particular time instant that can be computed from the conditional probability distribution of state x given all past measurements up to the particular time.

7.1 Introduction

As electrical and processing power of nodes in sensor networks increase the possibility of more intelligent estimation schemes become more and more important. The use of sensor networks was first driven by military applications, but with cheaper technology many other areas could make use of sensor networks, see for example, [413,414]. Advantages with wireless sensor networks typically include more robustness, because more than one unit is performing the same task with increased flexibility. However, decentralized estimation is a far more complex task than traditional centralized estimation. At least two big questions arise, what to send and how to make use of the received information. The first question has two obvious candidates, measurements or estimates.

In the case when measurements are transmitted the main problem is that all nodes usually do not have the possibility to communicate with all other nodes, thus the measurements have to be routed through other nodes. These nodes might experience a very high communication load, thus limiting the size of the network. This is especially evident when the number of nodes exceed the size of the state vector to be estimated. Once the routing problem is solved, generating an estimate based on measurements is straight forward, at least if the system is linear and subject to Gaussian disturbances.

The case where estimates are communicated has been given great attention in the literature. In [415] a decentralized Kalman filter was proposed. However, this algorithm requires every node to be able to communicate with every other node, which might not be possible. An alternative approach is to only allow nodes to communicate with their neighbors. As opposed to the case where measurements are communicated no routing is required when estimates are used as information carriers.

Without direct communication between all nodes a new problem is introduced, namely how to combine estimates from just neighboring nodes. To optimally combine two estimates one has to know the mutual information between the estimates. Computing this quantity for a general communication graph is a difficult task that requires global knowledge of the topology. In the case of a loop-free graph the problem was solved in [416] by introducing a channel filter. This approach was used in a coordinated search strategy application, see [417]. If the state is assumed constant, the problem can be viewed as a distributed average problem that has been studied in [245]. The problem was generalized to time-varying states in [418,419] using consensus filters. A closely related area is how to combine estimates from a number of noncommunicating estimators, so called track-to-track fusion algorithms. This problem differs in that the combined estimate is not used to compute a new estimate in the individual nodes. The fused estimate is only communicated to a central node to be used for some task. This problem was studied in [420, 421,518].

7.2 Self-tuning Kalman filtering

Multisensor data fusion is a synergistic approach used in industry, which involved the mechanism of fusing data from various local resources, for example, local sensors and individual sub-units. The data thus collected is being centralized and is being used for monitoring the health of the plant, controlling various scenarios, deciding on instant decisions based on local conditions. One of the key issues in developing a multisensor data fusion is to combine and collect the information from various local resources and then obtain an accurate joint estimate of the system state vector. The two basic fusion architectures are centralized and decentralized (or distributed) [422]. Both architectures can be used in specific scenarios. When the processing center can receive all measurements from the local sensors in time, the centralized Kalman filtering (CKF) can be used there. This can be used in mission critical scenarios, where every local sensor is important with its local information, this local information then transmits the processed data – local state estimate – to a fusion center. The main advantage of the centralized fusion method is that it involves minimum information loss. Under some regularity conditions, in particular, the assumption of cross-independent sensor noises, an optimal Kalman filtering fusion was proposed in [423–426], which was proved to be equivalent to the CKF using all sensor measurements; therefore, such fusion is optimal. However, it may result in high computational load due to overloading of the filter with more than it can handle. On the other hand, the distributed-fusion method can give the globally optimal or suboptimal state estimation by combining or weighting the local state estimators placed at ith sensors [427]. Also, then there is no feedback from the fusion center to local sensors, we propose a distributed Kalman filtering fusion formula under a mild condition, and prove that the fused state estimate is equivalent to the CKF using all sensor measurements.

7.2.1 Introduction

In [489], preliminary analysis of information fusion estimation of noise statistics for multisensor systems was presented. Parameter estimation on linear time-varying systems was the subject of [429,519]. Some related results on fault detection and diagnosis are reported in [430–432]. Optimal filtering over linear observations with unknown parameters was treated in [433] and recent results on H_∞ filtering for systems with non-linear perturbations are found in [434,435].

In this section, by the information matrix approach, we build on the results of [436,437] and examine carefully an approach of self-tuning Kalman filter fusion (STKFF) with sensor noises cross-correlated for distributed recursive state estimators of dynamic systems. In [437], the idea was implemented using an ordinary Kalman filter with no conditions of feedback from the fusion center to local sensors and the corresponding condition of fusion formula with feedback. It has been shown that the corresponding self-tuning Kalman fusion without feedback is, as the fusion with feedback, exactly equivalent to the corresponding CKF fusion formula using all sensor measurements. The results have been evaluated by testing on a quadruple-tank system.

It has been shown that an efficient modified self-tuning Kalman filter approach has no requirement of providing a feedback loop in view of its embedded information matrix. The condition of feedback from fusion center to the local sensors has helped in the implementation of self-tuning Kalman filter. Through lab simulation on a quadruple-tank system, the modified filter has performed without feedback in a much better way as compared to the regular Kalman with feedback.

7.2.2 Problem formulation

Consider the l-sensor distributed dynamic system

$$\begin{aligned} x_{k+1} &= \phi_k x_k + \upsilon_k, \quad k = 0, 1, \ldots && (7.1)\\ y_k^i &= H_k^i x_k + w_k^i, \quad i = 1, \ldots, \ell && (7.2) \end{aligned}$$

where ϕ_k is a matrix of order $(r \times r)$, $\hat{x}_k$, $v_k \in \mathcal{R}^r$, $H_i^k \in \mathcal{R}^{N_i \times r}$, y_k^i, $w_i^k \in \mathcal{R}^{N_i}$. The process noise υ_k and measurement noise w_i^k are both zero-mean random variables independent of each other temporally but w_i^k and w_j^k may be cross-correlated for $i \neq j$ at the same time instant k. To compare performances between the centralized and distributed filtering fusion, the stacked measurement equation is written as follows:

$$y_k = H_k x_k + w_k \tag{7.3}$$

where

$$\begin{aligned} y_k &= (y_k^{1^t}, \ldots\ldots, y_k^{l^t})^t, \quad H_k = (H_k^{1^t}, \ldots\ldots, H_k^{l^t})^t \\ w_k &= (w_k^{1^t}, \ldots\ldots, w_k^{l^t})^t, \quad R = diag[R_1 \ldots\ldots R_L] \end{aligned} \tag{7.4}$$

and the covariance of the noise w_k is given by

$$Cov(w_k) = R_k, \quad Cov(w_k^i) = R_k^i, \quad i = 1, \ldots, \ell \tag{7.5}$$

and R_k and R_k^i are both invertible for all i. According to the optimal distributed fusion Kalman filter [436, 437], the local steady-state optimal Kalman filter is described by

$$\hat{x}_{k/k}^i = \hat{\Psi}^i \hat{x}_{k-1/k-1}^i + \hat{K}_k^i y_k^i \tag{7.6}$$

$$\hat{\Psi}^i = \hat{P}_{k/k}^i \hat{P}_{k/k-1}^{-1} \hat{\Phi}_k \tag{7.7}$$

$$\hat{K}_k^i = \hat{P}_{k/k}^i H_k^{iT} \hat{R}_k^{i^{-1}} \tag{7.8}$$

with covariance of filtering error given by

$$\hat{P}_{k/k}^{i^{-1}} = \hat{P}_{k/k-1}^{i^{-1}} + H_k^{it} \hat{R}_k^{i^{-1}} H_k^i \tag{7.9}$$

where

$$\begin{aligned} \hat{x}_{k/k-1}^i &= \hat{\phi}_k \hat{x}_{k-1/k-1}^i \\ \hat{P}_{k/k}^i &= E[(\hat{x}_{k/k}^i - \hat{x}_k)(\hat{x}_{k/k-1}^i - \hat{x}_k)^t] \\ \hat{P}_{k/k-1}^i &= E[(\hat{x}_{k/k-1}^i - \hat{x}_k)(\hat{x}_{k/k-1}^i - \hat{x}_k)^t] \end{aligned}$$

From (7.1) and (7.3), we have the steady-state optimal centralized fusion Kalman filter

$$\hat{x}_{k/k} = \hat{\Psi} \hat{x}_{k-1/k-1} + P \sum_{i=1}^{L} H_k^T \hat{R}_k^{-1} y_k \tag{7.10}$$

Putting values, we get

$$x\hat{x}_{k/k} = \Psi \hat{x}_{k-1/k-1} + P \sum_{i=1}^{L} [\hat{P}_{k/k}^{-1} \hat{x}_{k/k} - \hat{P}_{k/k-1}^{-1} \hat{\Phi} \hat{x}_{k-1/k-1}] \tag{7.11}$$

$$\hat{K}_k = \hat{P}_{k/k} H_k^{i^t} \hat{R}_k^{-1} \tag{7.12}$$

$$\hat{P}_{k/k}^{-1} = \hat{P}_{k/k-1}^{-1} + H_k^t \hat{R}_k^{-1} H_k \tag{7.13}$$

$$\hat{\Psi} = \hat{P}_{k/k} \hat{P}_{k/k-1}^{-1} \Phi \tag{7.14}$$

where

$$\begin{aligned} \hat{x}_{k/k-1} &= \hat{\phi}_k \hat{x}_{k-1/k-1}, \\ \hat{P}_{k/k} &= E[(\hat{x}_{k/k} - \hat{x}_k)(\hat{x}_{k/k-1} - \hat{x}_k)^T] \\ \hat{P}_{k/k-1} &= E[(\hat{x}_{k/k-1} - \hat{x}_k)(\hat{x}_{k/k-1} - \hat{x}_k)^t] \end{aligned}$$

7.2.3 Self-tuning distributed Kalman fusion filter

Using the system identification method [438], we can obtain the consistent estimators.

Assumption 7.1: The model parameter and noise variance estimators are consistent, that bis, we assume that

$$\hat{\theta} \to \theta, \hat{R}_i \to R_i, \text{ as } k \to \infty, i.a.r \tag{7.15}$$

where symbol 'i.a.r' denotes the convergence 'in a realization'.

The basic principle of the self-tuning fusion filter (STFF) is that substituting the online estimates of the model parameters and noise variances into the optimal fusion filter will yield a STFF. It can be obtained by the following three steps [428, 139]:

Step 1. Using the system identification method and the correlation function method, the estimators of model parameters and noise variances are obtained.

Step 2. Substituting the estimates $\hat{\Phi}_k$ and $\hat{R}_i$ into (7.6)–(7.9) yields the self-tuning local Kalman filters.

$$\hat{x}^{S_i}_{k/k} = \hat{P}^{S_i}_{k/k} \hat{P}^{S_i^{-1}}_{k/k-1} \hat{\Phi}_k \hat{x}^{S_i}_{k-1/k-1} + \hat{K}^{S_i}_k y^{S_i}_k \tag{7.16}$$

$$\hat{K}^{S_i}_k = \hat{P}^{S_i}_{k/k} H^{i^T}_k \hat{R}^{S_i^{-1}}_k \tag{7.17}$$

$$\hat{P}^{S^{-1}}_{k/k} = \hat{P}^{S^{-1}}_{k/k-1} + H^{i^T}_k \hat{R}^{S_i^{-1}}_k H^i_k \tag{7.18}$$

Step 3. Substituting $\hat{P}^{S_i}_{k/k}$, $\hat{P}^{S^{-1}}_{k/k-1}$, and $\hat{x}^S_{k/k}$ into the steady-state optimal distributed fusion Kalman filter (7.11)–(7.14) yields the self-tuning distributed fusion Kalman filter

$$\hat{x}^S_{k/k} = \hat{P}^S_{k/k} \hat{P}^{S^{-1}}_{k/k-1} \hat{\Phi}_k \hat{x}^S_{k-1/k-1} + \hat{P}^S_{k/k} \sum_{i=1}^{L} [\hat{P}^{S^{-1}}_{k/k} \hat{x}^S_{k/k} - \hat{P}^{S^{-1}}_{k/k-1} \hat{\Phi}_k \hat{x}^S_{k-1/k-1}] \tag{7.19}$$

$$\hat{P}^{S^{-1}}_{k/k} = \hat{P}^{S^{-1}}_{k/k-1} + \sum_{i=1}^{L} [\hat{P}^{S_i^{-1}}_{k/k} - \hat{P}^{S_i^{-1}}_{k/k-1}] \tag{7.20}$$

with the initial value $\hat{x}^S(0 \mid 0) = \hat{x}^S_0$, $\hat{P}(0 \mid 0) = \hat{P}_0$. When sensor noises are cross-dependent, clearly

$$H^T_k \hat{R}^{S^{-1}}_k H_k = \sum_{i=1}^{l} H^{i^T}_k \hat{R}^{S_i^{-1}}_k H^i_k \tag{7.21}$$

To express the centralized self-tuning filter $\hat{x}^S_{k/k}$ in terms of the local filtering, we use (7.6)–(7.8), then we have

$$H^{i^T}_k \hat{R}^{S_i^{-1}}_k y^{S_i}_k = \hat{P}^{S_i^{-1}}_{k/k} \hat{x}^{S_i}_{k/k} - \hat{P}^{S_i^{-1}}_{k/k} \hat{\Phi}_k \hat{x}^{S_i}_{k-1/k-1} \tag{7.22}$$

Thus, substituting (7.22) into (7.19) yields

$$\hat{P}^{S^{-1}}_{k/k}\hat{x}^{S}_{k/k} = \hat{P}^{S^{-1}}_{k/k}\hat{\Phi}_k\hat{x}^{S}_{k-1/k-1} + \sum_{i=1}^{l}(\hat{P}^{S_i^{-1}}_{k/k}\hat{x}^{S_i}_{k/k} - \hat{P}^{S_i^{-1}}_{k/k-1}\hat{\Phi}_k\hat{x}^{S_i}_{k/k-1}) \tag{7.23}$$

That is to say that the centralized self-tuned filtering (7.23) and error matrix (7.20) are explicitly expressed in terms of the local filtering. It was claimed in [436] that although the feedback cannot improve the performance at the fusion center, the feedback does reduce the covariance of each local tracking error, and they proceeded further from here on in proving that. Also it was claimed that when sensor noises were cross-correlated, it was still possible to express equivalently the CKF in terms of the local filtering, and derive a similar performance analysis for the distributed Kalman filtering with feedback. In the following sections, it is shown that the self-tuning Kalman filter approach can out-perform even without the feedback, as it gives the same results without the feedback, which was given here with the feedback.

7.2.4 Distributed STKFF without feedback

To answer the question above, assume all sensor measurement matrices H_i^k are of full row rank. In many practical applications, for example, in the tracking problem *cf*, this assumption is fulfilled very often. Thus, we know $H_i^k(H_k^{i'})^+ = I$, where the superscript '+' stands for pseudo-inverse.

Similar to the derivation of (7.20) and (7.23), the CKF with cross-correlated sensor noises is given by

$$\begin{aligned}\hat{x}^{S}_{k/k} &= \Psi\hat{x}^{S}_{k-1/k-1} + \hat{P}_{k/k}H_k^t\sum_{i=1}^{l}\hat{R}_k^{-1}(*i)y_k^i \\ K_k &= \hat{P}_{k/k}H_k^t\hat{R}_k^{-1}\end{aligned} \tag{7.24}$$

Also, in (7.24)

$$\hat{K}_k^S y_k = \hat{P}_{k/k}H_k^t\sum_{i=1}^{l}\hat{R}_k^{-1}(*i)y_k^i \tag{7.25}$$

where $\hat{R}_k^{-1}(*i)$ is the ith submatrix column of $\hat{R}_k^{-1}$, with covariance of filtering error given by

$$\begin{aligned}\hat{P}^{-1}_{k/k} &= \hat{P}^{-1}_{k/k-1} + H_k^t\hat{R}_k^{-1}H_k \\ &= \hat{P}^{-1}_{k/k-1} + H_k^t\sum_{i=1}^{l}\hat{R}_k^{-1}(*i)\hat{R}_k^i(H_k^{i'}) + H_k^{i'}\hat{R}_k^{-i}H_k^i\end{aligned} \tag{7.26}$$

or

$$\hat{P}_{k/k} = (I - \hat{K}_kH_k)\hat{P}_{k/k-1} \tag{7.27}$$

Using (7.9) and (7.26), the estimation error covariance of the CKF is given by using the estimation error covariances of all local filters:

$$\hat{P}_{k/k}^{-1} = \hat{P}_{k/k-1}^{-1} + H_k^t \sum_{i=1}^{l} \hat{R}_k^{-1^i} \hat{R}_k^i (H_k^{i'}) + (\hat{P}_{k/k}^{i-1} - \hat{P}_{k/k-1}^{i-1}) \tag{7.28}$$

Using (7.24), (7.4), and (7.5), we have

$$\begin{aligned} K_k y_k &= \hat{P}_{k/k} H_k^t \sum_{i=1}^{l} \hat{R}_k^{-1}(^*i) y_k^i \\ &= \hat{P}_{k/k} H_k^t \sum_{i=1}^{l} \hat{R}_k^{-1}(^*i) \hat{R}_k^i (H_k^{i'}) + H_k^{i'} \hat{R}_k^{-i} H_k^i \end{aligned} \tag{7.29}$$

To express the centralized filtering $\hat{x}_{k/k}^S$ in terms of the local filtering, we use (7.6) and (7.29) to eliminate y_k from (7.24). From (7.6) we have

$$H_k^{i^T} \hat{R}_k^{i^{-1}} y_k^i = [\hat{P}_{k/k}^{i-1} \hat{x}_{k/k}^{S_i} - \hat{P}_{k/k-1}^{i-1} \Phi \hat{x}_{k/k-1}^{S_i}] \tag{7.30}$$

Thus, substituting (7.27), (7.29), and (7.30), into (7.24) yields

$$\begin{aligned} P_{k/k}^{-1} x_{k/k} &= P_{k/k-1}^{-1} \Phi x_{k/k-1} + H_k^t \sum_{i=1}^{l} R_k^{-1}(^*i) R_k^i (H_k^{i'})^+ \\ &\quad \cdot (P_{k/k}^{i-1} x_{k/k}^i - P_{k/k-1}^{i-1} \Phi x_{k/k-1}^i) \end{aligned} \tag{7.31}$$

That is to say that the centralized filtering (7.31) and error matrix (7.28) are explicitly expressed in terms of the local filtering. Hence, the performance of the distributed Kalman filtering fusion given in (7.25) and (7.31) is the same as that of the centralized fusion.

Remark 7.1: *It is not difficult to see from (7.24) that the centralized Kalman is different from the one in [315]. Moreover, in (7.31), the information vector is the main source of carrying the information.*

7.2.5 Optimality of STKFF with feedback

When there is feedback, the fusion center broadcasts its latest estimate to the local sensors. Thus, for all i, the following local and global one-stage predictions were modified naturally (146) as

$$x_{k/k-1}^i = \Phi_k \hat{x}_{k-1/k-1} = x_{k/k-1} \tag{7.32}$$

$$\hat{P}_{k/k-1}^i = \hat{P}_{k/k-1} \tag{7.33}$$

Naturally, the following filtering fusion with feedback was suggested:

$$\hat{P}_{k/k}^{-1} = \hat{P}_{k/k-1}^{-1} + H_k^t \sum_{i=1}^{l} R_k^{-1}(^*i)R_k^i(H_k^{i'}) + (\hat{P}_{k/k}^{i-1} - \hat{P}_{k/k-1}^{-1}) \tag{7.34}$$

$$\hat{P}_{k/k}^{-1}x_{k/k} = \hat{P}_{k/k-1}^{-1}\Phi x_{k/k-1} + H_k^t \sum_{i=1}^{l} R_k^{-1}(^*i)R_k^i(H_k^{i'})^+ \cdot (\hat{P}_{k/k}^{i-1}x_{k/k}^i - \hat{P}_{k/k-1}^{i-1}\Phi x_{k/k-1}^i) \tag{7.35}$$

where the superscript *hat* here is to show the corresponding estimates with feedback. Here it should be noted that (4.22) and (4.23) is equal to (7.28) and (7.31), respectively.

In what follows, a rigorous analysis for the feedback filtering fusion with cross-correlated sensor noises is presented:

$$\hat{x}_{k/k} = x_{k/k}, \quad \hat{P}_{k/k} = P_{k/k} \tag{7.36}$$

$$\hat{P}_{k/k}^i = E[(\hat{x}_{k/k}^i)(\hat{x}_{k/k}^i - x_k)^t] \tag{7.37}$$

The main advantage of the feedback filtering fusion is to reduce the covariance of every local estimate error, that is

$$\hat{P}_{k/k}^i \leq P_{k/k}^i, \quad i = 1, 2, \ldots, l \tag{7.38}$$

7.2.6 Global optimality of the feedback filtering fusion

We use mathematical induction to show no global performance difference between two types of filtering fusion: with feedback and without feedback. Clearly, it is reasonable to have the same initial values at both the local and the global filters no matter feedback exists or not.

$$\begin{aligned} \hat{x}_{0/0} &= x_{0/0} = x_{0/0}^i = \hat{x}_{0/0}^i \\ \hat{P}_{0/0} &= P_{0/0} = P_{0/0}^i = \hat{P}_{0/0}^i \end{aligned} \tag{7.39}$$

Suppose that

$$\hat{x}_{k-1/k-1} = x_{k-1/k-1}, \quad \hat{x}_{k/k-1} = x_{k/k-1} \tag{7.40}$$

$$\hat{P}_{k-1/k-1} = P_{k-1/k-1}, \quad \hat{P}_{k/k-1} = P_{k/k-1} \tag{7.41}$$

Therefore, at the kth stage, using the above equations and (7.34), we have

$$\hat{P}_{k/k}^{-1} = \hat{P}_{k/k-1}^{-1} + H_k^t \sum_{i=1}^{l} R_k^{-1}(^*i)R_k^i(H_k^{i'}) + (\hat{P}_{k/k}^{i-1} - \hat{P}_{k/k-1}^{-1}) \tag{7.42}$$

Note the definition of $\hat{P}_{k/k}^i$ similar to that in (7.9), we get

$$\hat{P}_{k/k}^{-1} = P_{k/k-1}^{-1} + H_k^{i'} R_k^{i-1} H_k^i \tag{7.43}$$

Substituting (7.43) into (7.42) and comparing it with (7.16)–(7.18) yields

$$\hat{P}^{-1}_{k/k} = P^{-1}_{k/kl} \tag{7.44}$$

Then using (7.15)–(7.18) in the feedback case, and (7.43), we have

$$\begin{aligned}
\hat{P}^{i^{-1}}_{k/k}\hat{x}^{i}_{k/k} &= \hat{P}^{i^{-1}}_{k/k}\Phi x_{k/k-1} + H^{i'}_{k}R^{i^{-1}}_{k}(y^{i}_{k} - H^{i}_{k}x_{k/k-1}) \\
&= \hat{P}^{i^{-1}}_{k/k}\Phi x_{k/k-1} + H^{i'}_{k}R^{i^{-1}}_{k}y^{i}_{k} - H^{i'}_{k}R^{i^{-1}}_{k}H^{i}_{k}x_{k/k-1} \\
&= \hat{P}^{i^{-1}}_{k/k}\Phi x_{k/k-1} + H^{i^{-1}}_{k}R^{i^{-1}}_{k}y^{i}_{k} - \hat{P}^{i^{-1}}_{k/k}x_{k/k-1} + P^{-1}_{k/k-1}x_{k/k-1} \\
&= H^{i'}_{k}R^{i^{-1}}_{k}y^{i}_{k} + P^{-1}_{k/k-1}x_{k/k-1} + \hat{P}^{-1}_{k/k-1}(\Phi - 1)x_{k-1/k-1}
\end{aligned} \tag{7.45}$$

Remark 7.2: *Note that (7.45) shows that the feedback loop is employed as followed by the steps of [437]. Obviously, it is different from the nonfeedback case. However, it will be shown below that when feedback is not applied here in (7.45), it gives the similar answer that was given in [437] with the application of feedback. Thus*

$$\hat{P}^{i^{-1}}_{k/k}\hat{x}^{i}_{k/k} = \hat{P}^{i^{-1}}_{k/k}\Phi x_{k/k-1} + H^{i'}_{k}R^{i^{-1}}_{k}(y^{i}_{k}) \tag{7.46}$$

Similarly, from (7.19) and (7.20), we have

$$\begin{aligned}
P^{-1}_{k/k}x_{k/k} &= H^{t}_{k}R^{-1}_{k}y_{k} + P^{-1}_{k/k-1}\Phi x_{k/k-1} \\
&= P^{-1}_{k/k-1}\Phi x_{k/k-1} + H^{t}_{k}\sum_{i=1}^{l}R^{-1}_{k}(^{*}i)R^{i}_{k}(H^{i'}_{k}) + H^{i'}_{k}R^{i^{-1}}_{k}y^{i}_{k}
\end{aligned} \tag{7.47}$$

Hence, using (7.35), (7.41), and (7.44)–(7.47), we have

$$\begin{aligned}
\hat{x}_{k/k} &= P_{k/k}P^{-1}_{k/k-1}\Phi x_{k/k-1} + P_{k/k}H^{t}_{k}\sum_{i=1}^{l}R^{-1}_{k}(^{*}i)R^{i}_{k}(H^{i'}_{k}) + H^{i'}_{k}R^{i^{-1}}_{k}y^{i}_{k} \\
&= x_{k/k}
\end{aligned} \tag{7.48}$$

Now, calculating the local estimates without feedback with self-tuning Kalman filter, $\hat{P}^{i^{-1}}_{k/k}$ has been defined in (7.43), we need to verify if it is still the covariance of local estimate error. From the last equality of (7.45) and (7.43), we know

$$\begin{aligned}
\hat{x}^{i}_{k/k} &= \hat{P}^{i}_{k/k}H^{i'}_{k}R^{i^{-1}}_{k}y^{i}_{k} + \hat{P}^{i}_{k/k}\hat{P}^{i^{-1}}_{k/k}x_{k/k-1} \\
&= \hat{P}^{i}_{k/k}H^{i'}_{k}R^{i^{-1}}_{k}y^{i}_{k} + x_{k/k-1} \\
&= \hat{P}^{i}_{k/k}H^{i'}_{k}R^{i^{-1}}_{k}y^{i}_{k} + x^{i}_{k/k} - \hat{P}^{i}_{k/k}H^{i'}_{k}R^{i^{-1}}_{k}y^{i}_{k} \\
&= x_{k/k}
\end{aligned} \tag{7.49}$$

It follows that

$$= E[(\hat{x}^{i}_{k/k})(\hat{x}^{i}_{k/k})'] = \hat{P}^{i}_{k/k} \tag{7.50}$$

The result shows that with the self-tuning Kalman filter, no feedback is required that was practiced in [437]. The self-tuning Kalman filter without feedback is good enough to run and prove compatible with the ordinary Kalman filter with feedback.

7.2.7 Evaluation and testing

In what follows, we demonstrate the effectiveness of the distributed STKFF algorithm as applied to a typical dynamic process. The process is a quadruple-tank system and it consists of four interconnected water tanks and two pumps. Its manipulated variables are voltages to the pumps and the controlled variables are the water levels in the two lower tanks. The quadruple-tank process is being built by considering the concept of two double-tank processes. The quadruple-tank system presents a multiinput–multioutput (MIMO) system. This system is a real-life control problem prototyped to experiment on, and try to solve in the most efficient way, since it deals with multiple variables, thus it gives a reflection for the large systems in industry. The schematic description of the four-tank system can be visualized as in Figure 7.1. The system has two control inputs (pump throughputs) that can be manipulated to control the water level in the tanks. The two pumps are used to transfer water from a sump into four overhead tanks. By adjusting the bypass valves of the system, the proportion of the water pumped into different tanks can be changed to adjust the degree of interaction between the pump throughputs and the water levels. Thus, each pump output goes to two tanks, one lower and another upper, diagonally opposite and the ratio of the split up is controlled by the position of the valve. Because of the large water distribution load, the pumps have been supplied 12 V each. The mathematical modeling of the quadruple-tank process can be obtained by using Bernoulli's law. Combining all the equations for the interconnected four-tank system we obtain the physical system. A fault model can then be constructed by

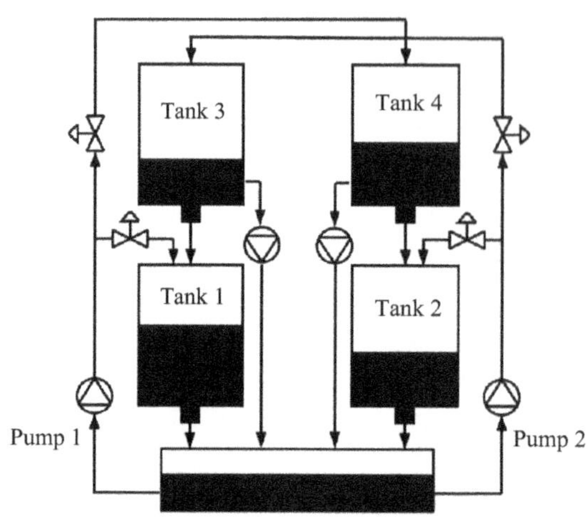

Figure 7.1 Schematic diagram of a quadruple-tank system

adding extra holes to each tank. The mathematical model of the faulty quadruple-tank system can be given as

$$\begin{aligned}
\frac{dh_1}{dt} &= -\frac{a_1}{A_1}\sqrt{2gh_1} + \frac{a_3}{A_1}\sqrt{2gh_3} + \frac{\gamma_1 k_1}{A_1}\nu_1 + \frac{d}{A_1} - \frac{a_{leak1}}{A_1}\sqrt{2gh_1} \\
\frac{dh_2}{dt} &= -\frac{a_2}{A_2}\sqrt{2gh_2} + \frac{a_4}{A_2}\sqrt{2gh_4} + \frac{\gamma_2 k_2}{A_2}\nu_2 - \frac{d}{A_2} - \frac{a_{leak2}}{A_2}\sqrt{2gh_2} \\
\frac{dh_3}{dt} &= -\frac{a_3}{A_3}\sqrt{2gh_3} + \frac{(1-\gamma_2)k_2}{A_3}\nu_2 - \frac{a_{leak3}}{A_3}\sqrt{2gh_3} \\
\frac{dh_4}{dt} &= -\frac{a_4}{A_4}\sqrt{2gh_4} + \frac{(1-\gamma_1)k_1}{A_4}\nu_1 - \frac{a_{leak4}}{A_4}\sqrt{2gh_4} \\
\frac{d\nu_1}{dt} &= -\frac{\nu_1}{\tau_1} + \frac{1}{\tau_1}u_1 \\
\frac{d\nu_2}{dt} &= -\frac{\nu_2}{\tau_2} + \frac{2}{\tau_2}u_2
\end{aligned} \tag{7.51}$$

By carefully evaluating the ensuing results of the quadruple-tank system for three scenarios:

1. the regular Kalman filter fusion without feedback,
2. the regular Kalman filter fusion with feedback,
3. the efficient distributed self-tuning Kalman filter implementation without feedback, see Figure 7.2,

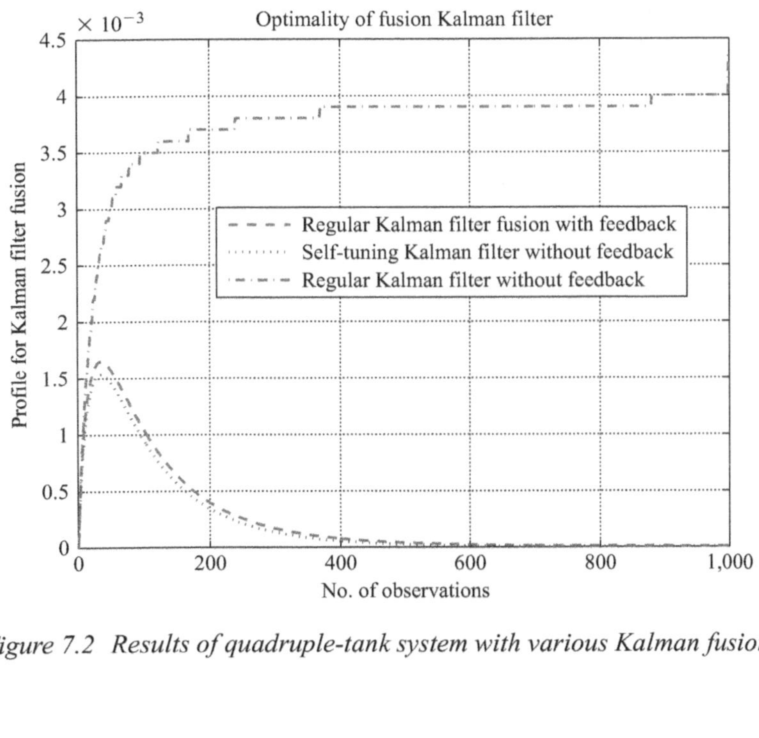

Figure 7.2 Results of quadruple-tank system with various Kalman fusions

it can be readily seen that what is achieved by [437] with feedback is achieved by self-tuning Kalman without feedback. This is due to the well performance of the structure of the self-tuning Kalman filter with information matrix.

7.3 Kalman filtering with intermittent communications

In this section, we employ the forward–backward (FB) Kalman filtering approach to tackle the distributed Kalman filtering fusion with passive packet loss or initiative intermittent communications from local estimators to a fusion center while the process noise does exist. We present a specific condition – properly short intermittent communication period/packet loss under which a proposed optimal distributed algorithm can be equivalent to the corresponding CKF fusion with complete communications even if the process noise does exist. When this condition is not satisfied, using some recent results, a suboptimal distributed-fusion algorithm is motivated. Experimental results on coupled tank system data indicate that the FB Kalman filter fusion is strong enough to provide better results than the regular Kalman. It works well and significantly better than similar existing schemes in view of the smoothing property of the FB Kalman filter.

7.3.1 Introduction

In hi-tech environment, a strict surveillance unit is required for an appropriate supervision. It often utilizes a group of distributed sensors that provide information of the local targets. Comparing with the CKF, which can be used in mission critical scenarios, where every local sensor is important with its local information, the distributed fusion architecture has many advantages. There is no second thought that in certain scenarios, centralized Kalman filter plays a major role, and it involves minimum information loss. Under some regularity conditions, in particular, the assumption of cross-independent sensor noises, an optimal Kalman filtering fusion was proposed in [324,423,425,426], which was proved to be equivalent to the CKF using all sensor measurements; therefore, such fusion is optimal. However, it may result in high computational load due to overloading of the filter with more than it can handle. For the distributed Kalman filter, lets' consider the case with packet loss or intermittent communications from local sensors/estimators to fusion center. In [439], it has been notified that given the target state the optimality of the fusion equations in reproducing the centralized estimates depends on the conditional independence of the measurements. Thus, the techniques implemented in [424] cannot be directly implemented here.

However, comparing with the original multisensor system, the centralized filtering fusion performance of the modified multisensor system may be reduced in some degree since sensor noises of the modified multisensor system are cross-correlated. Thus, as shown in [440,441], such fusion algorithm is suboptimal. Since the proposed suboptimal algorithm takes advantage of sensor estimates obviously more than the standard distributed Kalman filtering fusion with lost sensor estimates replaced by their predictions, it can be expected that performance of the former should be better than that of the latter. In the literature, there are several issues related to Kalman filtering and relevant to the present work [117,145,159,170–173,175–178,339], whereas filtering techniques within sensor networks are found in [442,443,520,521] and their references.

In this section, we build on [444] and extend it further by developing an improved approach of FB Kalman filtering to tackle the distributed Kalman filtering fusion with passive packet loss or initiative intermittent communications from local estimators to a fusion center and in the presence of process noise. It is shown that the FB Kalman filter fusion is strong enough to provide better results than the regular Kalman filter and it also performs better than the proposed scheme in [444]. This is due to the smoothing property of the FB Kalman filter. An evaluation of the modified filter on coupled tank system further ensures its effectiveness. The remaining part of the paper is organized as follows: problem formulation and the proposed solution formulation is presented in section 7.2, distributed FB Kalman filtering with correlation to process noise and sensor noise in section 7.3, followed by Kalman filtering fusion with packet loss/intermittent communications in section 7.4, followed by the evaluation and testing in section 7.5. Finally, some concluding remarks are given in section 7.6.

7.3.2 Problem formulation

Consider an ℓ-sensor distributed dynamic system described by

$$x_{k+1} = \phi_k x_k + \upsilon_k, \quad k = 0, 1, \ldots \tag{7.52}$$

$$y_k^i = H_k^i x_k + w_k^i, \quad i = 1, \ldots, \ell \tag{7.53}$$

where ϕ_k is a matrix of order $r \times r$, x_k, $v_k \in \Re^r$, $H_i^k \in \Re^{N_i \times r}$, y_k^i, and $w_i^k \in \Re^{N_i}$. The process noise υ_k and measurement noise w_i^k are both zero-mean random variables independent of each other temporally but w_i^k and w_j^k may be cross-correlated for $i \neq j$ at the same time instant k.

To compare performances between the centralized and distributed filtering fusion, the stacked measurement equation is written as

$$y_k = H_k x_k + w_k \tag{7.54}$$

where

$$\begin{aligned} y_k &= (y_k^{1'}, \ldots, y_k^{\ell'})' \\ H_k &= (H_k^{1'}, \ldots, H_k^{\ell'})', \\ w_k &= (w_k^{1'}, \ldots, w_k^{\ell'})' \end{aligned} \tag{7.55}$$

and the covariance of the noise w_k is given by

$$Cov(w_k) = R_k, \quad R_k^i = Cov(w_k^i), \qquad i = 1, \ldots, \ell \tag{7.56}$$

where R_k and R_k^i are both invertible for all i. According to the standard results in FB Kalman filtering, the local Kalman filtering at the ith sensor is expressed as

Forward run: for $(k = 0; k < T; +k)$

$$\hat{K}_k^i = \hat{P}_{k/k}^i H_k^{i^t} \hat{R}_k^{i^{-1}} \tag{7.57}$$

$$\hat{x}_{k/k}^i = \hat{x}_{k/k-1}^i + \hat{K}_k^i(y_k^i - H_k^i \hat{x}_{k/k-1}^i) \tag{7.58}$$

$$\hat{P}_{k/k}^i = \hat{P}_{k/k-1}^i - \hat{K}_k^i H_k \hat{P}_{k/k-1}^i \tag{7.59}$$

Backward run: for $(k = T-1; t \geq 0; -k)$

$$\hat{\lambda}_{k-1/T}^i = (I_{P+N} - H_k^{i^*} \hat{K}_{f,i}^*)\hat{\Phi}_k^i \hat{\lambda}_{k+1/T}^i + H_k^i \hat{R}_{e,i}^{-1}(y_k^i - H_k^i \hat{x}_{k/k-1}^i) \tag{7.60}$$

$$\hat{x}_{k-1/T}^i = \hat{x}_{k-1/k-1}^i + \hat{P}_{k/k-1}^i \hat{\lambda}_{k-1/T}^i \tag{7.61}$$

$$\hat{P}_{k-1/T}^i = \hat{P}_{k-1/k-1}^i \tag{7.62}$$

where, the covariance of filtering error can be stated as

$$\hat{P}_{k/k}^{i^{-1}} = \hat{P}_{k/k-1}^{i^{-1}} + H_k^{it} \hat{R}_k^{i^{-1}} H_k^i \tag{7.63}$$

where

$$\begin{aligned}
\hat{x}_{k/k-1}^i &= \hat{\Phi}_k \hat{x}_{k-1/k-1}^i, \\
\hat{P}_{k/k}^i &= E[(\hat{x}_{k/k}^i - \hat{x}_k)(\hat{x}_{k/k-1}^i - \hat{x}_k)^t] \\
\hat{P}_{k/k-1}^i &= E[(\hat{x}_{k/k-1}^i - \hat{x}_k)(\hat{x}_{k/k-1}^i - \hat{x}_k)^t]
\end{aligned}$$

Similarly, the centralized FB Kalman filtering with all sensor data is given by

Forward run: for $(k = 0; k < T; +k)$

$$\hat{K}_k = \hat{P}_{k/k} H_k^t \hat{R}_k^{-1} \tag{7.64}$$

$$\hat{x}_{k/k} = \hat{x}_{k/k-1} + \hat{K}_k(y_k - H_k \hat{x}_{k/k-1}) \tag{7.65}$$

$$\hat{P}_{k/k} = \hat{P}_{k/k-1} - \hat{K}_k H_k \hat{P}_{k/k-1} \tag{7.66}$$

Backward run: for $(k = T-1; t \geq 0; -k)$

$$\hat{\lambda}_{k-1/T} = (I_{P+N} - H_k^* \hat{K}_{f,i}^*)\hat{\Phi}_k \hat{\lambda}_{k+1/T} + H_k \hat{R}_{e,i}^{-1}(y_k^i - H_k \hat{x}_{k/k-1}) \tag{7.67}$$

$$\hat{x}_{k-1/T} = \hat{x}_{k-1/k-1} + \hat{P}_{k/k-1} \hat{\lambda}_{k-1/T} \tag{7.68}$$

$$\hat{P}_{k-1/T} = \hat{P}_{k-1/k-1} \tag{7.69}$$

where, the covariance of filtering error can be described as

$$\hat{P}_{k/k}^{-1} = \hat{P}_{k/k-1}^{-1} + H_k{}^t \hat{R}_k^{-1} H_k \tag{7.70}$$

where

$$\begin{aligned}
\hat{x}_{k/k-1} &= \hat{\Phi}_k \hat{x}_{k-1/k-1} \\
\hat{P}_{k/k} &= E[(\hat{x}_{k/k} - \hat{x}_k)(\hat{x}_{k/k-1} - \hat{x}_k)^t] \\
\hat{P}_{k/k-1} &= E[(\hat{x}_{k/k-1} - \hat{x}_k)(\hat{x}_{k/k-1} - \hat{x}_k)^t]
\end{aligned}$$

It is quite clear when the sensor noises are cross-dependent that

$$H_k^t \hat{R}_k^{-1} H_k = \sum_{i=1}^{l} H_k^{i^t} \hat{R}_k^{i^{-1}} H_k^i$$

Likewise, it has already been proved in [7,29,70] that the centralized filtering and error matrix could be explicitly expressed in terms of the local filtering and error matrices as follows:

$$\hat{P}_{k/k}^{-1} = \hat{P}_{k/k-1}^{-1} + \sum_{i=1}^{l} (\hat{P}_{k/k}^{i^{-1}} - \hat{P}_{k/k-1}^{i^{-1}}) \tag{7.71}$$

and

$$\hat{P}_{k/k}^{-1}\hat{x}_{k/k} = \hat{P}_{k/k-1}^{-1} + \sum_{i=1}^{l} (\hat{P}_{k/k}^{i^{-1}}\hat{x}_{k/k}^{i} - \hat{P}_{k/k-1}^{i^{-1}}\hat{x}_{k/k-1}^{i}) \tag{7.72}$$

Also,

$$H_k^{i'} \hat{R}_k^{i^{-1}} y_k^i = \hat{P}_{k/k}^{i^{-1}}\hat{x}_{k/k}^{i} - \hat{P}_{k/k-1}^{i^{-1}}\hat{x}_{k/k-1}^{i} \tag{7.73}$$

In what follows, we are going to deal with the practical situation in which the local sensors may fail to send their estimates to the fusion center. In this case, the measurement equation of the corresponding centralized multisensor system has to be modified, that is, the original multiple individual observations should be stacked as a modified single observation. This modification process is made by FB Kalman filter smoother. It will be shown that these modification equations will offer better results than [423] in cases when

- The process noise and sensor noise are one-step mutually correlated.
- The sensor noises are cross-correlated, which both obviously violate the condition for the global optimality of the distributed filtering fusion.

Before proceeding further, we present a preliminary result on globally optimal distributed Kalman filtering fusion for dynamic systems with process noise one-step correlated to measurement noise and sensor noises cross-correlated.

Consider a discrete dynamic process

$$\hat{x}_{k+1} = \hat{\Phi}_k \hat{x}_k + \hat{\upsilon}_k, \quad k = 0, 1, \ldots \tag{7.74}$$

$$\hat{y}_k^i = \hat{H}_k^i \hat{x}_k + \hat{w}_k^i, \quad i = 1, \ldots, \ell \tag{7.75}$$

where the process noises $\hat{\upsilon}_k$ and $\hat{w}_k$ are both zero-mean random variables independent of each other temporally. But $\hat{\upsilon}_k$ and $\hat{w}_k$ are one-step correlated here, that is

$$\mathbb{E}(\hat{\upsilon}_k) = 0, \quad Cov(\hat{\upsilon}_k) = \hat{R}\upsilon_k \tag{7.76}$$

$$\mathbb{E}(\hat{w}_k^i) = 0, \quad Cov(\hat{\upsilon}_k^i) = \hat{R}_k^i \tag{7.77}$$

$$\mathbb{E}(\hat{\upsilon}_k \hat{w}_j^{i'}) = \hat{M}_k^i \delta_{k-j+1} \tag{7.78}$$

where δ_{k-j+1} is the Kronecker delta function. Recall that the stacked measurement equation remains the same as (7.54) and (7.55).

Since, $\hat{\upsilon}_k$ and $\hat{w}_k$ are one-step correlated, we have the following, also with the covariance of $\hat{w}_k$

$$\mathbb{E}(\hat{\upsilon}_k \hat{w}_j^t) = \hat{M}_k = (\hat{M}_k^1, \ldots, \hat{M}_k^l)\delta_{k-j+1} \tag{7.79}$$

$$Cov\hat{w}_k = \hat{R}_k \tag{7.80}$$

However, the local measurement noises $\hat{\upsilon}_k$ and $\hat{w}_k$ are not necessarily uncorrelated here.

We are now in a position to prove the following results:

Theorem 7.1: *1. The FB Kalman filtering of the ith sensor given by*

Forward run: *for* $(k = 0; k < T; +k)$

$$\begin{aligned}
\hat{K}_k^i &= \hat{P}_{k/k}^i(\hat{H}_k^{i'} + \hat{P}_{k/k-1}^{i-1}\hat{M}_k^i)(\hat{R}_k^{i'} - \hat{M}_k^{i'}\hat{P}_{k/k-1}^{i-1}\hat{M}_k^i)^{-1} \\
\hat{x}_{k/k}^i &= \hat{x}_{k/k-1}^i + \hat{K}_k^i(y_k^i - H_k^i\hat{x}_{k/k-1}^i) \\
\hat{P}_{k/k}^i &= \hat{P}_{k/k-1}^i - \hat{K}_k^i(\hat{H}_k^{i'} + \hat{P}_{k/k-1}^{i-1}\hat{M}_k^i)\hat{P}_{k/k-1}^i
\end{aligned}$$

Backward run: *for* $(k = T-1; t \geq 0; -k)$

$$\begin{aligned}
\hat{\lambda}_{k-1/T}^i &= (I_{P+N} - (\hat{H}_k^{i*} + \hat{P}_{k/k-1}^{i-1}\hat{M}_k^i)\hat{K}_{f,i}^*)\hat{\Phi}_k^i\hat{\lambda}_{k+1/T}^i \\
&\quad + (\hat{H}_k^{i'} + \hat{P}_{k/k-1}^{i-1}\hat{M}_k^i)(\hat{R}_k^{i'} - \hat{M}_k^{i'}\hat{P}_{k/k-1}^{i-1}\hat{M}_k^i)^{-1}(y_k^i - H_k^i\hat{x}_{k/k-1}^i) \\
\hat{x}_{k-1/T}^i &= \hat{x}_{k-1/k-1}^i + \hat{P}_{k/k-1}^i\hat{\lambda}_{k-1/T}^i \\
\hat{P}_{k-1/T}^i &= \hat{P}_{k-1/k-1}^i
\end{aligned} \tag{7.81}$$

with covariance of filtering error for the forward run as

$$\begin{aligned}
\hat{P}_{k/k}^{i-1} &= \hat{P}_{k/k-1}^{i-1} + (\hat{H}_k^{i'} + \hat{P}_{k/k-1}^{i-1}\hat{M}_k^i) \\
&\quad \times (\hat{R}_k^{i'} - \hat{M}_k^{i'}\hat{P}_{k/k-1}^{i-1}\hat{M}_k^i)^{-1}(\hat{H}_k^{i'} + \hat{P}_{k/k-1}^{i-1}\hat{M}_k^i)^t
\end{aligned} \tag{7.82}$$

where

$$\begin{aligned}
\hat{x}_{k/k-1}^i &= \hat{\Phi}_{k-1}\hat{x}_{k-1/k-1}^i \\
\hat{P}_{k/k-1}^i &= E[(\hat{x}_{k/k-1}^i - \hat{x}_k)(\hat{x}_{k/k-1}^i - \hat{x}_k)^t \,|\, \hat{y}_0^i, \ldots, \hat{y}_{k-1}^t]
\end{aligned}$$

is globally optimal in the MSE sense.

2. When the row dimensions of all matrices $\hat{H}_k^i + (\hat{P}_{k/k-1}^{i-1}\hat{M}_k^i)^t$ *to be less than or equal to the dimension of the state and all of them to be of full row rank, the centralized filtering can be explicitly expressed in terms of local filtering.*

Proof 1. It suffices to show point 2, using point 1. The centralized FB Kalman filtering with all sensor data is given by:

Forward run: for $(k = 0; k < T; +k)$

$$\hat{K}_k = \hat{P}_{k/k}(\hat{H}_k^t + \hat{P}_{k/k-1}^{-1}\hat{M}_k)(\hat{R}_k^t - \hat{M}_k^t\hat{P}_{k/k-1}^{-1}\hat{M}_k)^{-1} \tag{7.83}$$

$$\begin{aligned}\hat{x}_{k/k} &= (I - \hat{K}_k\hat{H}_k)\hat{x}_{k/k-1} + \hat{K}_k\hat{y}_k \\ &= (I - \hat{K}_k\hat{H}_k)\hat{x}_{k/k-1} + \hat{P}_{k/k}(\hat{H}_k^t + \hat{P}_{k/k-1}^{-1}\hat{M}_k) \\ &\quad \times \sum_{i=1}^{l}(\hat{R}_k - \hat{M}_k^t)^{-1}(\times i)\hat{y}_k \end{aligned} \tag{7.84}$$

$$\hat{P}_{k/k} = \hat{P}_{k/k-1} - \hat{K}_k(\hat{H}_k^t + \hat{P}_{k/k-1}^{-1}\hat{M}_k^i)\hat{P}_{k/k-1} \tag{7.85}$$

Backward run: for $(k = T - 1; t \geq 0; -k)$

$$\begin{aligned}\hat{\lambda}_{k-1/T} &= (I_{P+N} - (\hat{H}_k^* + \hat{P}_{k/k-1}^{-1}\hat{M}_k)\hat{K}_{f,i}^*) \\ &\quad \times \hat{\Phi}_k\hat{\lambda}_{k+1/T} + (\hat{H}_k^t + \hat{P}_{k/k-1}^{-1}\hat{M}_k) \\ &\quad \times (\hat{R}_k^t - \hat{M}_k^t\hat{P}_{k/k-1}^{-1}\hat{M}_k)^{-1}(y_k - H_k\hat{x}_{k/k-1}) \end{aligned} \tag{7.86}$$

$$\hat{x}_{k-1/T} = \hat{x}_{k-1/k-1} + \hat{P}_{k/k-1}\hat{\lambda}_{k-1/T} \tag{7.87}$$

$$\hat{P}_{k-1/T} = \hat{P}_{k-1/k-1} \tag{7.88}$$

with covariance of filtering error for the forward run can also be given by

$$\begin{aligned}\hat{P}_{k/k}^{i^{-1}} &= \hat{P}_{k/k-1}^{i^{-1}} + (\hat{H}_k^{i'} + \hat{P}_{k/k-1}^{i^{-1}}\hat{M}_k^i) \\ &\quad \times (\hat{R}_k^{i'} - \hat{M}_k^{i'}\hat{P}_{k/k-1}^{i^{-1}}\hat{M}_k^i)^{-1}(\hat{H}_k^{i'} + \hat{P}_{k/k-1}^{i^{-1}}\hat{M}_k^i)^t \end{aligned} \tag{7.89}$$

In the sequel, we let $(\hat{R}_k - \hat{M}_k^t\hat{P}_{k/k-1}^{-1}\hat{M}_k)^{-1}$ $(\times i)$ to be the ith submatrix column of $(\hat{R}_k - \hat{M}_k^t\hat{P}_{k/k-1}^{-1}\hat{M}_k)^{-1}$.

Since we assume the row dimensions of all matrices $\hat{H}_k^i + (\hat{P}_{k/k-1}^{i^{-1}}\hat{M}_k^i)^t$ to be less than or equal to the dimension of the state, and all of them to be of full row rank. In addition, we know $(\hat{H}_k^i + (\hat{P}_{k/k-1}^{i^{-1}}\hat{M}_k^i)^t)(\hat{H}_k^i + (\hat{P}_{k/k-1}^{i^{-1}}\hat{M}_k^i)^t)^+ = I$, where the superscript '+' stands for pseudo-inverse. Using (7.83), we get

$$\begin{aligned}\hat{K}_k\hat{y}_k &= \hat{P}_{k/k}(\hat{H}_k^t + \hat{P}_{k/k-1}^{-1}\hat{M}_k)\sum_{i=1}^{l}(\hat{R}_k - \hat{M}_k^t\hat{P}_{k/k-1}^{-1}\hat{M}_k)^{-1}(\times i)\hat{y}_k^i \\ &= \hat{P}_{k/k}(\hat{H}_k^t + \hat{P}_{k/k-1}^{-1}\hat{M}_k)\sum_{i=1}^{l}(\hat{R}_k - \hat{M}_k^t\hat{P}_{k/k-1}^{-1}\hat{M}_k)^{-1}(\times i) \\ &\quad \times (\hat{R}_k^i - \hat{M}_k^{i^t}\hat{P}_{k/k-1}^{-1}\hat{M}_k)(\hat{H}_k^{i^t} + \hat{P}_{k/k-1}^{i^{-1}}\hat{M}_k^i)^+ \\ &\quad \times (\hat{H}_k^{i^t} + \hat{P}_{k/k-1}^{i^{-1}})(\hat{R}_k^i - \hat{M}_k^{i^t}\hat{P}_{k/k-1}^{i^{-1}}\hat{M}_k^i)^{-1}\hat{y}_k^i \end{aligned} \tag{7.90}$$

Proceeding to express the FB centralized filtering $\hat{x}_{k/k}$ in terms of the local filtering and using (7.81), we get

$$
\begin{aligned}
&(\hat{H}_k^{i'} + \hat{P}_{k/k-1}^{i-1})(\hat{R}_k^i - \hat{M}_k^{i'}\hat{P}_{k/k-1}^{i-1}\hat{M}_k^i)^{-1}\hat{y}_k^i \\
&\quad = \hat{P}_{k/k}^{i-1}\hat{K}_k^i\hat{y}_k^i \\
&\quad = \hat{P}_{k/k}^{i-1}[\hat{x}_{k/k}^i - (I - \hat{K}_k^i\hat{H}_k^i)\hat{x}_{k/k-1}^i]
\end{aligned} \tag{7.91}
$$

Using (7.90) and (7.91), it yields

$$
\begin{aligned}
\hat{x}_{k/k} &= (I - \hat{K}_k\hat{H}_k)\hat{x}_{k/k-1} + \hat{P}_{k/k}(\hat{H}_k^t + \hat{P}_{k/k-1}^{-1}\hat{M}_k) \\
&\quad \times \sum_{i=1}^{l}(\hat{R}_k - \hat{M}_k^t\hat{P}_{k/k-1}^{-1}\hat{M}_k)^{-1}(\times i) \\
&\quad \times (\hat{R}_k^i - \hat{M}_k^{i'}\hat{P}_{k/k-1}^{i-1}\hat{M}_k^i)(\hat{H}_k^{i'} + \hat{P}_{k/k-1}^{i-1}\hat{M}_k^i)^+ \\
&\quad \times \hat{P}_{k/k}^{i-1}[\hat{x}_{k/k}^i - (I - \hat{K}_k^i)\hat{x}_{k/k-1}^i].
\end{aligned} \tag{7.92}
$$

and

$$
\begin{aligned}
\hat{\lambda}_{k-1/T} &= I_{P+N} - \Bigg(\hat{H}_k^t\Bigg[\hat{P}_{k/k}(\hat{H}_k^t + \hat{P}_{k/k-1}^{-1}\hat{M}_k) \\
&\quad \times \sum_{i=1}^{l}(\hat{R}_k - \hat{M}_k^t\hat{P}_{k/k-1}^{-1}\hat{M}_k)^{-1}(\times i) \\
&\quad \times (\hat{R}_k^i - \hat{M}_k^{i'}\hat{P}_{k/k-1}^{-1}\hat{M}_k)(\hat{H}_k^{i'} + \hat{P}_{k/k-1}^{i-1}\hat{M}_k^i)^+ \\
&\quad \times (\hat{H}_k^{i'} + \hat{P}_{k/k-1}^{i-1})(\hat{R}_k^i - \hat{M}_k^{i'}\hat{P}_{k/k-1}^{i-1}\hat{M}_k^i)^{-1} \\
&\qquad -H_k^iR_{e_i}^{-1}H_k^i\hat{x}_{k/k-1}^i\Bigg]\Bigg)\hat{\Phi}_k\hat{\lambda}_{k+1/T} + \hat{P}_{k/k}(\hat{H}_k^t + \hat{P}_{k/k-1}^{-1}\hat{M}_k) \\
&\quad \times \sum_{i=1}^{l}(\hat{R}_k - \hat{M}_k^t\hat{P}_{k/k-1}^{-1}\hat{M}_k)^{-1}(\times i)(\hat{R}_k^i - \hat{M}_k^{i'}\hat{P}_{k/k-1}^{-1}\hat{M}_k) \\
&\quad \times (\hat{H}_k^{i'} + \hat{P}_{k/k-1}^{i-1}\hat{M}_k^i)^+(\hat{H}_k^{i'} + \hat{P}_{k/k-1}^{i-1}) \\
&\quad \times (\hat{R}_k^i - \hat{M}_k^{i'}\hat{P}_{k/k-1}^{i-1}\hat{M}_k^i)^{-1}\hat{y}_k^i - H_k^iR_{e_i}^{-1}H_k^i\hat{x}_{k/k-1}^i \\
\hat{x}_{k-1/T} &= \hat{x}_{k/k-1} + P_{k/k-1}\lambda_{k/T}
\end{aligned} \tag{7.93}
$$

Simplifying, it is easy to show that the FB centralized filtering can be explicitly expressed in terms of the local filtering in the following form:

Forward run:

$$\begin{aligned}\hat{x}_{k/k} &= (I - \hat{K}_k\hat{H}_k)\hat{x}_{k/k-1} + \hat{P}_{k/k}(\hat{H}_k^t + \hat{P}_{k/k-1}^{-1}\hat{M}_k)\\ &\times \sum_{i=1}^{\ell}(\hat{R}_k - \hat{M}_k^t\hat{P}_{k/k-1}^{-1}\hat{M}_k)^{-1}(\times i)(\hat{R}_k^i - \hat{M}_k^{i^t}\hat{P}_{k/k-1}^{i-1}\hat{M}_k^i)\\ &\times (\hat{H}_k^{i^t} + \hat{P}_{k/k-1}^{i-1}\hat{M}_k^i)^{+}\hat{P}_{k/k}^{i-1}[\hat{x}_{k/k}^i - (I - \hat{K}_k^i)\hat{x}_{k/k-1}^i]\end{aligned}$$

Backward run:

$$\begin{aligned}\hat{\lambda}_{k-1/T} &= \hat{\Phi}_k\hat{\lambda}_{k+1/T} - \hat{P}_{k/k}^{i-1}\hat{K}_k^i(\hat{H}_k^t\hat{P}_{k/k}\hat{\Phi}_k\hat{\lambda}_{k+1/T} - \hat{y}_k^i) - \hat{H}_k^t\hat{x}_{k/k-1}^i\\ \hat{x}_{k-1/T} &= \hat{x}_{k/k-1} + P_{k/k-1}\lambda_{k/T}\end{aligned}$$

Hence, the performance of the distributed Kalman filtering fusion with process noise one-step correlated to measurement noise and sensor noises cross-correlated is better than the centralized fusion because of the smoothing effect. □

7.3.3 *Impact of packet loss/intermittent communications*

In this section, and for notational and presentation simplicity, we consider the dynamic system (7.52) and (7.53) with synchronous intermittent communications from sensors to fusion center. The packet loss from sensors to fusion center is actually passive asynchronous intermittent communications. Then, we can easily extend the result in this section to asynchronous intermittent communications/packet loss case.

We now assume the intermittent communication is periodical and the periods of all the local sensors are the same n in this section, that is, the fusion center only receives $x_{N/N}^i$ $(N = kn, k = 1, 2, \ldots)$ from the local sensor.

7.3.4 *Modified centralized multisensor system*

Firstly, by stacking n-step sensor observations and some simple algebraic operations, we will convert the n-step dynamic system to the compact form with one-step correlated process and measurement noises in the following proposition.

Proposition 7.1: *Assume Φ_k to be invertible. The n steps of the original local dynamic system*

$$\begin{aligned}x_{kn+1} &= \Phi_{kn}x_{kn} + \upsilon_{kn}\\ y_{kn+1}^i &= H_{kn+1}^i x_{kn+1} + w_{kn+1}^i\\ &\cdots\\ x_{(k+1)n} &= \Phi_{(k+1)n} - 1x_{(k+1)n} - 1 + \upsilon_{(k+1)n} - 1\\ y_{(kn+1)n}^i &= H_{(k+1)n}^i x_{(k+1)n} + w_{(k+1)n}^i\end{aligned}$$

which can be rewritten in a compact form

$$\begin{aligned} \tilde{x}_{k+1} &= \tilde{\Phi}_k \tilde{x}_k + \tilde{\upsilon}_k, \quad k = 0, 1, \ldots \\ \tilde{y}^i_{(k+1)} &= \tilde{H}^i_{(k+1)} \tilde{x}_{(k+1)} + w^i_{(k+1)}, \quad i = 1, \ldots, l \end{aligned} \tag{7.94}$$

$$\tilde{w}^i_{k+1} = \begin{bmatrix} w^i_{kn+1} - H^i_{kn+1}(\Pi^{n-1}_{j=1}\Phi_{kn+j})^{-1}\sum_{l=2}^{n-1}(\Pi^{n-1}_{j=1}\Phi_{kn+j})\upsilon_{kn+l-1} \\ \quad -H^i_{kn+1}(\Pi^{n-1}_{j=1}\Phi_{kn+j})^{-1}\upsilon_{(k+l)n-1} \\ \vdots \\ w^i_{(k+1)n-1} - H_{(kn+1)n-1}\Phi^{-1}_{(k+1)n-1}\upsilon_{(k+1)n-1} \\ w^i_{(k+1)n} \end{bmatrix}$$

where

$$\begin{aligned} \tilde{x}_k &= \tilde{x}_{kn} \\ \tilde{y}^i_k &= (y_{(k-1)n+1}, y^t_{(k-1)n+2}, \ldots y^t_{kn})^t \\ \tilde{\Phi}_k &= \Pi^{n-1}_{j=0}\Phi_{kn+j} \end{aligned} \tag{7.95}$$

$$\tilde{\upsilon}_k = \sum_{l=1}^{n-1}(\Pi^{n-1}_{j=l})\upsilon_{kn+l-1} + \upsilon_{(k+l)n-1} \tag{7.96}$$

$$\tilde{H}^i_{k+1} = \begin{bmatrix} H^i_{kn+1}(\Pi^{n-1}_{j=1}\Phi_{kn+j})^{-1} \\ \vdots \\ H_{(kn+1)n-1}\Phi^{-1}_{(k+1)n-1} \\ H^i_{(k+1)n} \end{bmatrix} \tag{7.97}$$

Similar to (7.94) and (7.95), the n-step stacked measurement equations

$$\begin{aligned} y_{kn+1} &= H_{kn+1}x_{kn+!} + w_{(kn+1)} \\ \ldots, & \\ y_{(k+1)n} &= H_{(k+1)n}x_{(k+1)n} + w_{(k+1)n} \end{aligned} \tag{7.99}$$

where $(m = 1, \ldots, n)$

$$\begin{aligned} y_{kn+m} &= (y^{1^t}_{kn+m}, \ldots, y^{\ell^t}_{kn+m})^t \\ H_{kn+m} &= (H^{l^t}_{kn+m}, \ldots, H^{\ell^t}_{kn+m})^t \\ w_{kn+m} &= (w^{l^t}_{kn+m}, \ldots, w^{\ell^t}_{kn+m})^t \end{aligned}$$

can then be rewritten as

$$\tilde{y}_{k+1} = \tilde{H}_{k+1}\tilde{x}_{k+1} + \tilde{w}_{k+1} \tag{7.100}$$

where

$$\begin{aligned} y_{k+1} &= (y_{k+1}^{1^t}, \ldots, y_{k+1}^{\ell^t})^t \\ H_{k+1} &= (H_{k+1}^{1^t}, \ldots, H_{k+1}^{\ell^t})^t \\ w_{k+1} &= (w_{k+1}^{1^t}, \ldots, w_{k+1}^{\ell^t})^t \end{aligned} \tag{7.101}$$

In addition to the correlation of sensor noises as mentioned above, the process and measurement noises of the modified centralized system are also one-step correlated, that is

$$Cov(\tilde{w}_k) = \tilde{R}_k, \quad \mathbb{E}(\tilde{\upsilon}_k \tilde{w}_j^t) = \tilde{M}_k \delta_{k-j+1} \tag{7.102}$$

In Section 7.3, we assumed *the row dimensions of all matrices* $\hat{H}_k^i + (\hat{P}_{k/k-1}^{i^{-1}} \hat{M}_k^i)^t$ *to be less than or equal to the dimension of the state, and all of them to be of full row rank*. According to whether or not this assumption is satisfied, the Kalman filtering with packet loss/intermittent communications problem can be divided into two cases to discuss.

Remark 7.3: *In view of (7.96) and, (7.98), since the original process noise* υ_k *is now involved in the new process and measurement noises,* υ_k *and* w_k^i *are one-step correlated and sensor noises* w_k^i *and* w_k^j *are cross-correlated here, that is*

$$\begin{aligned} \mathbb{E}(\tilde{\upsilon}_k) &= 0, \quad Cov(\tilde{\upsilon}_k) = \tilde{R}\upsilon_k \\ \mathbb{E}(\tilde{w}_k^i) &= 0, \quad Cov(\tilde{w}_k^i) = \tilde{R}_k^i \\ \mathbb{E}(\tilde{w}_k^i \tilde{w}_k^{j'}) &\neq 0, \quad i \neq j; \qquad \mathbb{E}(\tilde{v}_k \tilde{w}_j^{i'}) = \tilde{M}_k^i \delta_{k-j+1}, \quad \forall k, i, j \end{aligned}$$

Obviously, the larger the variance of υ_k *is, the larger is the correlation of sensor noises* $\tilde{w}_k^i$ *and* $\tilde{w}_k^{j'}$.

In the sequel, two important cases are considered: case 1 implies that the row dimensions of all matrices $\hat{H}_k^i + (\hat{P}_{k/k-1}^{i^{-1}} \hat{M}_k^i)^t$ to be less than or equal to the dimension of the state, and all of them to be of full row rank. Case 2 treats the situation in which such an assumption does not hold.

Case 1: The assumption is satisfied. Since we have already converted the local sensor's n-step dynamic equations (7.94) to the compact form with one-step correlated process and measurement noises (7.94). According to **Theorem** 3.1(a), the local filtering of (7.94) at the ith sensor is

Forward run:

$$\begin{aligned} \tilde{x}_{k/k-1}^i &= \tilde{\Phi}_k \tilde{x}_{k-1/k-1}^i \\ \tilde{P}_{k/k-1}^i &= \tilde{\Phi}_k \tilde{P}_{k-1/k-1}^i \Phi_k^t + \tilde{R}\upsilon_k \\ \tilde{P}_{k/k}^{-1} &= \tilde{P}_{k/k-1}^{i^{-1}} + (\tilde{H}_k^{i^t} + \tilde{P}_{k/k-1}^{i^{-1}} \tilde{M}_k^i)(\tilde{R}_k^i - \tilde{M}_k^{i^t} \tilde{P}_{k/k-1}^{i^{-1}} \tilde{M}_k^i)^{-1} (\tilde{H}_k^{i'} + \tilde{P}_{k/k-1}^{i^{-1}} \tilde{M}_k^i)^t \\ \tilde{K}_k^i &= \tilde{P}_{k/k}^i (\tilde{H}_k^{i'} + \tilde{P}_{k/k-1}^{i^{-1}} \tilde{M}_k^i)(\tilde{R}_k^i - \tilde{M}_k^{i'} \tilde{P}_{k/k-1}^{i^{-1}} \tilde{M}_k^i)^{-1} \\ \tilde{x}_{k/k}^i &= \tilde{x}_{k/k-1}^i + \tilde{K}_k^i (y_k^i - \tilde{H}_k^i \tilde{x}_{k/k-1}^i) \end{aligned}$$

Backward run:

$$\begin{aligned}
\tilde{\lambda}^i_{k-1/T} &= (I_{P+N} - (\tilde{H}^{i'}_k + \tilde{P}^{i-1}_{k/k-1}\tilde{M}^i_k)K^*_{f,i})\Phi_k\lambda_{k+1/T} + (\tilde{H}^{i'}_k + \tilde{P}^{i-1}_{k/k-1}\tilde{M}^i_k) \\
&\quad \times (\tilde{R}^i_k - \tilde{M}^{i'}_k\tilde{P}^{i-1}_{k/k-1}\tilde{M}^i_k)^{-1}(y_k - (\tilde{H}^{i'}_k + \tilde{P}^{i-1}_{k/k-1}\tilde{M}^i_k)\tilde{x}_{k/k-1}) \\
\tilde{x}^i_{k-1/T} &= \tilde{x}^i_{k-1/k-1} + \tilde{P}^i_{k/k-1}\tilde{\lambda}^i_{k/T} \\
\tilde{P}^i_{k-1/T} &= \tilde{P}^i_{k-1/k-1}
\end{aligned}$$

By part 1, of Theorem 3.1, the optimality of (7.103) is the same as the standard Kalman filtering (i.e., globally optimal in the MSE sense). Combining this with the equivalence of dynamic systems (7.94) and (7.95), the state estimate $\tilde{x}^i_{k/k}$ of n-step observation Kalman filtering with one-step correlated process and measurement noises is equal to the standard Kalman filtering state estimate $x^i_{kn/kn}$ of (7.28), that, $\tilde{x}^i_{k/k} = x^i_{kn/kn}$.

Since the primitive local measurement matrix $H^i_k \in R^{N_i \times r}$, $x_k \in R^r$, and the communication period of all the local sensor is the same n. If $N_i n \leq r$, the row dimensions of all matrices $\tilde{H}^i_k + (\tilde{P}^{i-1}_{k/k-1}\tilde{M}^i_k)^t$ equals $N_i n$ and to be less than or equal to the dimension of the state. In many practical applications, $\tilde{H}^i_k + (\tilde{P}^{i-1}_{k/k-1}\tilde{M}^i_k)^t$ is of full row rank under the condition $N_i n \leq r$. Then by the assumption in Theorem 3.1, part 2 is satisfied. Thus, we give the filtering at the fusion center as follows:

Forward run:

$$\begin{aligned}
\tilde{x}_{k/k-1} &= \tilde{\Phi}_k\tilde{x}_{k-1/k-1} \\
\tilde{P}_{k/k-1} &= \tilde{\Phi}_k\tilde{P}_{k-1/k-1}\Phi^t_k + \tilde{R}\upsilon_k \\
\tilde{P}^{-1}_{k/k} &= \tilde{P}^{-1}_{k/k-1} + (\tilde{H}^t_k + \tilde{P}^{-1}_{k/k-1}\tilde{M}_k)(\tilde{R}_k - \tilde{M}^t_k\tilde{P}^{-1}_{k/k-1}\tilde{M}_k)^{-1}(\tilde{H}^t_k + \tilde{P}^{-1}_{k/k-1}\tilde{M}_k)^t \\
\tilde{K}_k &= \tilde{P}_{k/k}(\tilde{H}^t_k + \tilde{P}^{-1}_{k/k-1}\tilde{M}_k)(\tilde{R}_k - \tilde{M}^t_k\tilde{P}^{-1}_{k/k-1}\tilde{M}_k)^{-1} \\
\tilde{x}_{k/k} &= (I - \tilde{K}_k\tilde{H}_k)\tilde{x}_{k/k-1} + \tilde{P}_{k/k}(\tilde{H}^t_k + \tilde{P}^{-1}_{k/k-1}\tilde{M}_k) \\
&\quad \times \sum_{i=1}^{l}(\tilde{R}_k - \tilde{M}^t_k\tilde{P}^{-1}_{k/k-1}\tilde{M}_k)^{-1}(\times i) \\
&\quad \times (R^i_k - \tilde{M}^{i'}_k\tilde{P}^{i-1}_{k/k-1}\tilde{M}^i_k)(\tilde{H}^{i'}_k + \tilde{P}^{i-1}_{k/k-1}\tilde{M}^i_k)^+ \\
&\quad \times \tilde{P}^{i-1}_{k/k}[\tilde{x}^i_{k/k} - (I - \tilde{K}^i_k)\tilde{x}^i_{k/k-1}]
\end{aligned}$$

Backward run:

$$\begin{aligned}
\tilde{\lambda}_{k-1/T} &= \tilde{\Phi}_k\tilde{\lambda}_{k+1/T} - \tilde{P}^{i-1}_{k/k}\tilde{K}^i_k(\tilde{H}^t_k\tilde{P}_{k/k}\tilde{\Phi}_k\tilde{\lambda}_{k+1/T} - \tilde{y}^i_k) - \tilde{H}^t_k\tilde{x}^i_{k/k-1} \\
\tilde{x}_{k-1/T} &= \tilde{x}_{k/k-1} + P_{k/k-1}\tilde{\lambda}_{k/T} \\
\tilde{P}_{k-1/T} &= \tilde{P}_{k-1/k-1}
\end{aligned}$$

Remark 7.4: *It has already been proved in Theorem 3.1(b) that under certain assumptions (roughly speaking, flat measurement matrix in local sensors and short packet loss/intermittent communications period), the centralized filtering* $\tilde{x}_{k/k}$

in (7.38) can be explicitly expressed in terms of the local filtering $\tilde{x}^i_{k/k}$. *Since* $\tilde{x}^i_{k/k} = x^i_{kn/kn}$, *the centralized filtering* $\tilde{x}^i_{k/k}$ *in (7.38), can be explicitly expressed in terms of* $x^i_{kn/kn}$. *This is to say the performance of the estimation at fusion center does not change with packet loss/intermittent communications from local sensors to the fusion center even if the process noise does exist. Since there is no process noise, it is just a sufficient but not necessary condition for reconstructing global optimality of the algorithm with packet loss/intermittent communications.*

Case 2: The assumption is not satisfied. It occurs in practice, when the measurement matrix is not flat or the packet loss/intermittent communications period is long, that the row dimension of $\tilde{H}^i_k + (\tilde{P}^{i-1}_{k/k-1}\tilde{M}^i_k)'$ is greater than the dimension of the state. In this case, our assumption is not satisfied. To motivate a suboptimal distributed fusion algorithm, we compress the local sensor observation first so that the compressed measurement matrix is of row full rank, then the proposed optimal distributed fusion algorithm in section 3.3 can be used to this modified system. We will demonstrate that the compressed sensor observations make the local sensor estimate same as that of uncompressed sensor observations case, but comparing with the original whole multisensor system, the filtering performance of the compressed system may be reduced in some degree since sensor noises of the latter become cross-correlated therefore the block-diagonal compression matrix consisting of all sensor lossless compression matrices is not a lossless compression matrix for the original whole multi-sensor system, see for example [4,52]. Thus, the motivated algorithm is suboptimal.

The following lemma proposes a way to reduce the row dimension of $\tilde{H}^i_k + (\tilde{P}^{i-1}_{k/k-1}\tilde{M}^i_k)^t$ to the rank of $\tilde{H}^i_k + (\tilde{P}^{i-1}_{k/k-1}\tilde{M}^i_k)^t$ that is less than or equal to the dimension of the state. This dimensionality reduction does not change the estimation performance at each local sensor.

Lemma 7.1: *Suppose a dynamic system is given by*

$$x_{k+1} = \Phi_k x_k + \upsilon_k, \quad k = 0, 1, \ldots \tag{7.103}$$

$$y_k = H_k x_k + w_k \tag{7.104}$$

where $\Phi_k \in R^{r\times r}$, x_k, $\upsilon_k \in R^r$, $H_k \in R^{N\times r}$, y_k, $w_k \in R^N$, $Cov(w_k) = R_k$, $E(\upsilon_k w_j) = M_k\ \delta_{k-j+1}$, *and* $N > r$. *Then one finds a dimensionality reduction matrix* S_k, *for example, see (7.42), such that the dimension-reduced measurement equation*

$$z_k = S_k y_k = S_k H_k x_k + S_k w_k \tag{7.105}$$

replaces (7.104) but does not change the performance of Kalman filtering.

Proof 2. Let $H_k + (P^{-1}_{k/k-1}M_k)^t = T_k G_k$ be full-rank decompose of matrix $H_k + (P^{-1}_{k/k-1}M_k)^t$, where T_k and G_k are full column rank and full row rank matrices, respectively, and their ranks are equal to the rank of $H_k + (P^{-1}_{k/k-1}M_k)^t$. We will show that

$$S_k = T^t_k(R_k - M^t_k P^{-1}_{k/k-1} M_k)^{-1} \tag{7.106}$$

is a lossless dimensionality reduction matrix.

Denote the estimate error covariances using y_k and z_k by $P^{-1}_{k/k}(y_k)$ and $P^{-1}_{k/k}(z_k)$, respectively. Recalling (21), we have

$$\begin{aligned} P^{-1}_{k/k}(y_k) &= P^{-1}_{k/k-1} + (H^t_k + P^{-1}_{k/k-1}M_k) \\ &\quad \times (R_k - M^t_k P^{-1}_{k/k-1} M_k)^{-1} (H^t_k + P^{-1}_{k/k-1}M_k)^t \end{aligned} \tag{7.107}$$

Denoting $H_k = S_k H_k, \hat{w}_k = S_k w_k$. Then

$$\mathbb{E}(\upsilon_k \hat{w}^t_j) = \hat{M}_k \delta_{k-j+1}, \quad Cov(\hat{w}_k) = \hat{R}_k = S_k R_k S^t_k$$

By assumption

$$\begin{aligned} \hat{H}_k + (P^{-1}_{k/k-1}\hat{M}_k)^t &= S_k H_k + (P^{-1}_{k/k-1} M_k S^t_k)^t \\ &= S_k(H_k + (P^{-1}_{k/k-1}M_k)^t) \\ &= T^t_k (R_k - M^t_k P^{-1}_{k/k-1} M_k)^{-1} T_k G_k \end{aligned}$$

Obviously, $T^t_k(R_k - M^t_k P^{-1}_{k/k-1}M_k)^{-1}T_k$ is an invertible matrix and then $\hat{H}_k + (P^{-1}_{k/k-1}\hat{M}_k)^t = T^t_k(R_k - M^t_k P^{-1}_{k/k-1}M_k)^{-1}T_k G_k$ is a full row rank matrix.

This leads to

$$\begin{aligned} P^{-1}_{k/k}(z_k) &= P^{-1}_{k/k-1} + (\hat{H}^t_k + P^{-1}_{k/k-1}\hat{M}_k) \\ &\quad \times (\hat{R}_k - \hat{M}^t_k P^{-1}_{k/k-1}\hat{M}_k)^{-1}(\hat{H}^t_k + P^{-1}_{k/k-1}\hat{M}_k)^t \\ &= P^{-1}_{k/k-1} + (T^t_k(R_k - M^t_k P^{-1}_{k/k-1}M_k))^{-1} \\ &\quad \times (\hat{H}_k + (P^{-1}_{k/k-1}\hat{M}_k)^t)^t (T^t_k(R_k - M^t_k P^{-1}_{k/k-1}M_k))^{-1} \\ &\quad \times (R_k - M^t_k P^{-1}_{k/k-1}M_k)(R_k - M^t_k P^{-1}_{k/k-1}M_k)^{-1}T^{-1}_k \\ &\quad \times (T^t_k(R_k - M^t_k P^{-1}_{k/k-1}M_k))^{-1}(H_k + (P^{-1}_{k/k-1}M_k)^t) \\ &= P^{-1}_{k/k-1} + (T^t_k(R_k - M^t_k P^{-1}_{k/k-1}M_k))^{-1}(\hat{H}_k + (P^{-1}_{k/k-1}\hat{M}_k)^t)^t \\ &\quad \times (T^t_k((R_k - M^t_k P^{-1}_{k/k-1}M_k))^{-1}T_k)^{-1} \\ &\quad \times (T^t_k(R_k - M^t_k P^{-1}_{k/k-1}M_k))^{-1}(H_k + (P^{-1}_{k/k-1}M_k)^t) \\ &= P^{-1}_{k/k-1} + (H_k + (P^{-1}_{k/k-1})^t)^t (R_k - M^t_k P^{-1}_{k/k-1}M_k)^{-1} \\ &\quad \times T_k(T^t_k(R_k - M^t_k P^{-1}_{k/k-1}M_k)^{-1}T_k)^{-1}T^t_k(R_k - M^t_k P^{-1}_{k/k-1}M_k)^{-1}T_k G_k \\ &= P^{-1}_{k/k-1} + (H_k + (P^{-1}_{k/k-1}M_k)^t)^t (R_k - M^t_k P^{-1}_{k/k-1}M_k)^{-1}T_k G_k \\ &= P^{-1}_{k/k-1} + (H_k + (P^{-1}_{k/k-1}M_k)^t)^t \\ &\quad \times (R_k - M^t_k P^{-1}_{k/k-1}M_k)^{-1}(H_k + (P^{-1}_{k/k-1}M_k)^t) \\ &= P^{-1}_{k/k}(y_k) \end{aligned}$$

which completes the proof. □

By compressing the sensor observation, the local filtering at the ith sensor could be rewritten as

Forward run:

$$
\begin{aligned}
\bar{x}^i_{k/k-1} &= \tilde{\Phi}_k \bar{x}^i_{k-1/k-1} \\
\bar{P}^i_{k/k-1} &= \tilde{\Phi}_k \bar{P}^i_{k-1/k-1} \tilde{\Phi}^t_k + \tilde{R}\upsilon_k \\
\bar{H}^i_k &= S^i_k \tilde{H}^i_k, \bar{M}^{i^t}_k = \tilde{M}^{i^t}_k S^{i^t}_k, \quad \bar{R}^i_k = S^i_k \tilde{R}^i_k S^{i^t}_k \\
\bar{P}^{i-1}_{k/k} &= \bar{P}^{i-1}_{k/k-1} + (\bar{H}^{i^t}_k + \bar{P}^{i-1}_{k/k-1} \bar{M}^i_k)(\bar{R}^i_k - \bar{M}^{i^t}_k \bar{P}_{k/k-1})^{-1} (\bar{H}^{i^t}_k + \bar{P}^{i-1}_{k/k-1} \bar{M}^i_k)^t \\
\bar{K}^i_k &= \bar{P}^i_{k/k} (\bar{H}^{i^t}_k + \bar{P}^{i-1}_{k/k-1} \bar{M}^i_k)(\bar{R}^i_k - \bar{M}^{i^t}_k \bar{P}_{k/k-1})^{-1} (\bar{H}^{i^t}_k + \bar{P}^{i-1}_{k/k-1}) \\
\bar{x}^i_{k/k} &= \bar{x}^i_{k/k-1} + \bar{K}^i_k (S^i_k \bar{y}^i_k - \bar{H}^i_k \bar{x}^i_{k/k-1})
\end{aligned}
$$

Backward run:

$$
\begin{aligned}
\bar{\lambda}^i_{k-1/T} &= (I_{P+N} - (\bar{H}^{i^*}_k + \bar{P}^{i-1}_{k/k-1} \bar{M}^i_k) \bar{K}^*_{f,i}) \bar{\Phi}^i_k \bar{\lambda}^i_{k+1/T} + (\bar{H}^{i^t}_k + \bar{P}^{i-1}_{k/k-1} \bar{M}^i_k) \\
&\quad \times (\bar{R}^{i^t}_k - \bar{M}^{i^t}_k \bar{P}^{i-1}_{k/k-1} \bar{M}^i_k)^{-1} (S^i_k y^i_k - H^i_k \bar{x}^i_{k/k-1}) \\
\bar{x}^i_{k-1/T} &= \bar{x}^i_{k-1/k-1} + \bar{P}^i_{k/k-1} \bar{\lambda}^i_{k-1/T} \\
\bar{P}^i_{k-1/T} &= \bar{P}^i_{k-1/k-1}
\end{aligned}
$$

where S^i_k is the dimensionality reduction matrix given in Lemma 7.1, that is

$$\bar{x}^i_{k/k} = \tilde{x}^i_{k/k} = x^i_{kn/kn}$$

Now the row dimensions of all new matrices $\bar{H}^i_k + (\bar{P}^{i-1}_{k/k-1} \bar{M}^i_k)$ to be less than or equal to the dimension of the state. Our assumption in part 2 of Theorem 3.1 is thus satisfied. Then the proposed optimal distributed fusion algorithm above can be used to the new compressed system. Let $S = diag(S^1_k, \ldots, S^l_k)$, the filtering at the fusion center is

Forward run:

$$
\begin{aligned}
\bar{x}_{k/k-1} &= \tilde{\Phi}_k \bar{x}_{k-1/k-1} \\
\bar{P}_{k/k-1} &= \tilde{\Phi}_k \bar{P}_{k-1/k-1} \tilde{\Phi}^t_k + \tilde{R}\upsilon_k \\
\bar{H}^i_k &= S_k \tilde{H}_k, \bar{M}_k = \tilde{M}_k S^t_k, \bar{R}_k = S \tilde{R}_k S^t \\
\bar{P}^{-1}_{k/k} &= \bar{P}^{-1}_{k/k-1} + (\bar{H}^t_k + \bar{P}^{-1}_{k/k-1} \bar{M}_k)(\bar{R}_k - \bar{M}^t_k \bar{P}_{k/k-1})^{-1} (\bar{H}^t_k + \bar{P}^{-1}_{k/k-1} \bar{M}_k)^t \\
\bar{K}_k &= \bar{P}_{k/k} (\bar{H}^t_k + \bar{P}^{-1}_{k/k-1} \bar{M}_k)(\bar{R}_k - \bar{M}^t_k \bar{P}^{-1}_{k/k-1} \bar{M}_k)^{-1} \\
\bar{x}^i_{k/k} &= (I - \bar{K}_k \bar{H}_k) \bar{x}_{k/k-1} + \bar{P}_{k/k} (\bar{H}^t_k + \bar{P}^{-1}_{k/k-1} \bar{M}_k) \\
&\quad \times \sum_{i=1}^{l} (\bar{R}_k - \bar{M}^t_k \bar{P}^{-1}_{k/k-1} \bar{M}_k)^{-1} (\times i) \\
&\quad \times (\bar{R}^i_k - \bar{M}^{i^t}_k \bar{P}^{i-1}_{k/k-1} \bar{M}^i_k)(\bar{H}^{i^t}_k + \tilde{P}^{i-1}_{k/k-1} \bar{M}^i_k)^+ \\
&\quad \times \bar{P}^{i-1}_{k/k} [\bar{x}^i_{k/k} - (I - \bar{K}^i_k \bar{H}^i_k) \bar{x}^i_{k/k-1}]
\end{aligned}
$$

Backward run:

$$\begin{aligned}
\hat{\lambda}_{k-1/T} &= \hat{\Phi}_k\hat{\lambda}_{k+1/T} - \hat{P}^{i-1}_{k/k}\hat{K}^i_k(\hat{H}^t_k\hat{P}_{k/k}\hat{\Phi}_k\hat{\lambda}_{k+1/T} - \hat{y}^i_k) - \hat{H}^t_k\hat{x}^i_{k/k-1} \\
\hat{x}_{k-1/T} &= \hat{x}_{k/k-1} + P_{k/k-1}\lambda_{k/T} \\
&\quad \bar{P}^i_{k-1/T}\bar{P}^i_{k-1/k-1}
\end{aligned}$$

In general, however, the fusion estimate $\bar{x}_{k/k}$ in (7.108) is not globally optimal except some special case, for example, the process noise υ_k is equal to zero (therefore, the converted sensor noises are mutually uncorrelated). Thus, when the assumption for the matrices $H^i_k + (\tilde{P}^{i-1}_{k/k-1}\tilde{M}^i_k)'$ is not satisfied, (44) and (45) provide a suboptimal algorithm for the packet loss or intermittent communications problem. The numerical example in the next section shows that this suboptimal algorithm still works well.

Remark 7.5: *In this section, it is assumed the packet loss or intermittent communications from local sensors to fusion center are synchronous. In fact, it is easy to extend the results to the distributed Kalman filtering fusion with asynchronous intermittent communications or packet loss. In the asynchronous communications case, we first synchronize the distributed system by regarding every possible communication instant from sensors to the fusion center as a point of the time subscript set of the whole dynamic system. Then, at each time instant, the fusion center only fuses the sensor estimates that are received by the fusion center using the proposed distributed filtering fusion algorithms above. More precisely speaking, if, for example,* $\bar{x}^i_{k-1/k-1}$ *is not received by the fusion center, view* $\bar{x}^i_{k/k} = \bar{x}^i_{k/k-1}$ *and* $\bar{K}^i_k\bar{H}^i_k = 0$ *in (45). Since Kalman filtering is a recursive minimal MSE estimator using all received observations up to now, such extension does not change the performance analysis in section 7.4 at all.*

7.3.5 Process data collection

The benchmarked laboratory-scale two-tank process control system has been used to collect data at a sampling rate of 50 ms. Process data has been generated through an experimental setup as shown in Figure 7.3. The main objective of the benchmarked dual-tank system is to reach a reference height of 200 ml of the second tank. During this process, several faults have been introduced such as the leakage faults, sensor faults, and actuator faults. Leakage faults have been introduced through the pipe clogs of the system, knobs between the first and the second tank, etc. Sensor faults have been introduced by introducing a gain in the circuit as if there is a fault in the level sensor of the tank. Actuator faults have been introduced by introducing a gain in the setup for the actuator that comprises of the motor and pump. A proportional integral (PI) controller works in a closed loop configuration to reach the desired height of the second tank. Due to the inclusion of faults, the controller was finding it difficult to reach the desired level. For this reason, the power of the motor has been increased from a scale of 0–5 V to scale of 5–18 V in order to provide it the maximum throttle to reach the desired level. In doing so, the actuator performed well

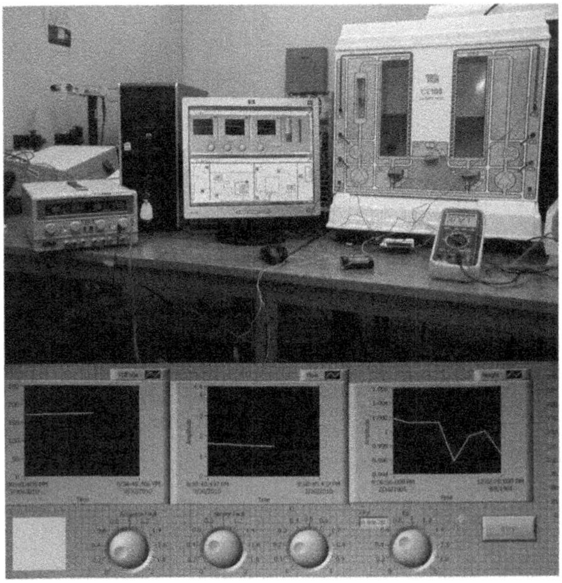

Figure 7.3 *(a) The two tank system interfaced with the LabView through a data acquisition (DAQ) and the amplifier for the magnified voltage; (b) the labview setup of the apparatus including the circuit window and the block diagram of the experiment*

in achieving its desired level but it also suppressed the faults of the system. So, it made the task of detecting the faults even more difficult. After the collection of data, techniques such as settling time, steady-state value, and coherence spectra can help us to give an insight of the fault.

In this section, in particular, leakage fault has been considered. Hydraulic height and liquid output flow rate of the second tank are the inputs while leakage fault level on a discrete scale of 1–4 is the considered output. Data is collected by introducing leakage fault in the closed loop system.

7.3.6 Model of the coupled tank system

The physical system under evaluation is formed of two tanks connected by a pipe. The leakage is simulated in the tank by opening the drain valve. A DC motor-driven pump supplies the fluid to the first tank and a PI controller is used to control the fluid level in the second tank by maintaining the level at a specified level, as shown in Figure 7.4. A step input is applied to the DC motor-pump system to fill the first tank. The opening of the drainage valve introduces a leakage in the tank. Various types of leakage faults are introduced and the liquid height in the second tank, H_2, and the inflow rate, Q_i, are both measured. The National Instruments LABVIEW package is employed to collect these data. A benchmark model of a cascade connection of a DC motor and a pump relating the input to the motor, u, and the flow, Q_i, is

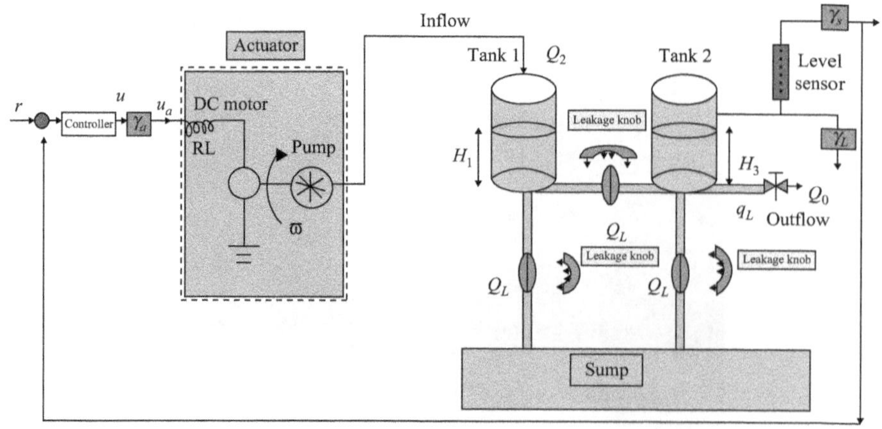

Figure 7.4 Process control system: a lab-scale two-tank system

a first-order system:

$$\dot{Q}_i = -a_m Q_i + b_m \phi(u) \tag{7.108}$$

where a_m and b_m are the parameters of the motor–pump system and $\phi(u)$ is a dead-band and saturation type of non-linearity. It is assumed that the leakage Q_ℓ occurs in tank 1 and is given by (see (7.109)):

$$Q_\ell = C_{d\ell}\sqrt{2gH_1} \tag{7.109}$$

With the inclusion of the leakage, the liquid level system is modeled by (see (7.110)):

$$A_1 \frac{dH_1}{dt} = Q_i - C_{12}\varphi(H_1 - H_2) - C_\ell \varphi(H_1) \tag{7.110}$$

$$A_2 \frac{dH_2}{dt} = C_{12}\varphi(H_1 - H_2) - C_0 \varphi(H_2) \tag{7.111}$$

where $\varphi(\cdot) = sign(\cdot)\sqrt{2g(\cdot)}$, $Q_\ell = C_\ell \varphi(H_1)$ is the leakage flow rate, $Q_0 = C_0\varphi(H_2)$ is the output flow rate, H_1 is the height of the liquid in tank 1, H_2 is the height of the liquid in tank 2, A_1 and A_2 are the cross-sectional areas of the 2 tanks, $g = 980$ cm/s^2 is the gravitational constant, C_{12} and C_0 are the discharge coefficient of the inter-tank and output valves, respectively.

The model of the two-tank fluid control system, shown above in Figure 7.4, is of a second order and is non-linear with a smooth square-root type of non-linearity. For design purposes, a linearized model of the fluid system is required and is given in (7.112) and (7.113):

$$\frac{dh_1}{dt} = b_1 q_i - (a_1 + \alpha)h_1 + a_1 h_2 \tag{7.112}$$

$$\frac{dh_2}{dt} = a_2 h_1 - (a_2 - \beta)h_2 \tag{7.113}$$

where h_1 and h_2 are the increments in the nominal (leakage-free) heights H_1^0 and H_2^0 (see (105)–(106)):

$$b_1 = \frac{1}{A_1}, \quad a_1 = \frac{C_{db}}{2\sqrt{2g(H_1^0 - H_2^0)}}, \quad \beta = \frac{C_0}{2\sqrt{2gH_2^0}},$$

$$a_2 = a_1 + \frac{C_{do}}{2\sqrt{2gH_2^0}}, \quad \alpha = \frac{C_{d\ell}}{2\sqrt{2gH_1^0}}$$

and the parameter α indicates the amount of leakage.

A PI controller, with gains k_p and k_I, is used to maintain the level of the Tank 2 at the desired reference input r as (see (7.114)):

$$\begin{aligned} \dot{x}_3 &= e = r - h_2 \\ u &= k_p e + k_I x_3 \end{aligned} \tag{7.114}$$

The linearized model of the entire system formed by the motor, pump, and the tanks is given by (see (7.115)):

$$\dot{x} = Ax + Br \quad y = Cx \tag{7.115}$$

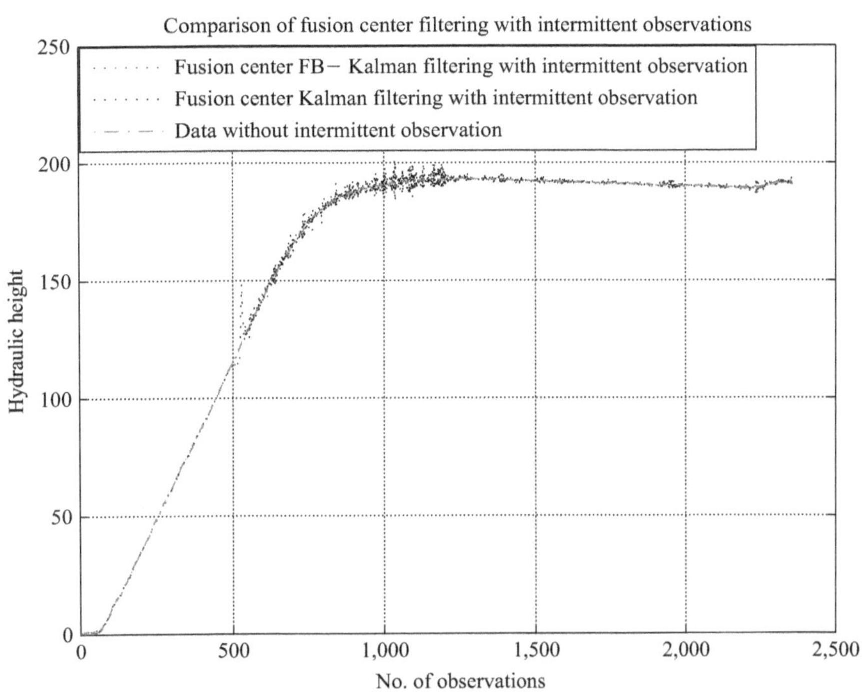

Figure 7.5 Implementation results of fusion center-based Kalman and FB Kalman on the intermittent observations

where

$$x = \begin{bmatrix} h_1 \\ h_2 \\ x_3 \\ q_i \end{bmatrix}, \quad A = \begin{bmatrix} -a_1 - \alpha & a_1 & 0 & b_1 \\ a_2 & -a_2 - \beta & 0 & 0 \\ -1 & 0 & 0 & 0 \\ -b_m k_p & 0 & b_m k_I & -a_m \end{bmatrix} \tag{7.116}$$

$$B = [0 \quad 0 \quad 1 \quad b_m k_p]^T, \quad C = [1 \quad 0 \quad 0 \quad 0]$$

where q_i, q_ℓ, q_0, h_1, and h_2 are the increments in Q_i, Q_ℓ, Q_0, H_1^0, and H_2^0, respectively, the parameters a_1 and a_2 are associated with linearization whereas the parameters α and β are, respectively associated with the leakage and the output flow rate, that is $q_\ell = \alpha h_1$, $q_o = \beta h_2$.

7.3.7 Evaluation of results

The distributed FB Kalman filtering fusion with intermittent observation has been employed here on a coupled tank system. The intermittent observations have been created in the height profile of the water in the coupled tank system. Three profiles have been shown in Figures 7.5 and 7.6 respectively, first, with the ordinary hydraulic height without the intermittent observations, second, with the fusion center Kalman

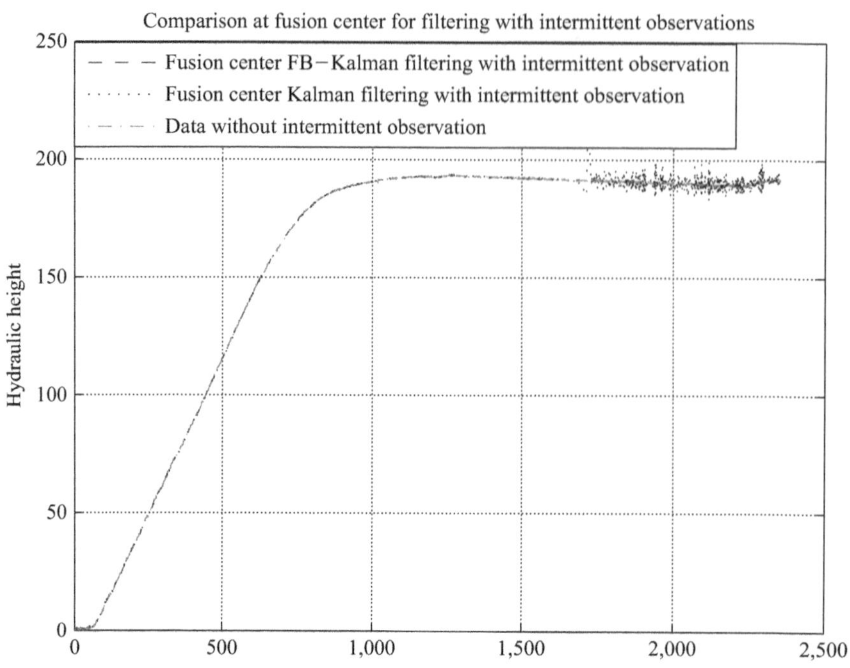

Figure 7.6 Implementation results of fusion center-based Kalman and FB Kalman on the intermittent observations

filter with intermittent observation, third, with the FB Kalman fusion center. It has been shown that the FB Kalman fusion performs well with its habitual instincts of smoothing and providing better results as compared to the ordinary Kalman filter fusion.

Also, the mean square error at the fusion center with ordinary Kalman filter for the first scenario at Figure 7.5 is 0.0017, whereas the mean square error at the fusion center with FB Kalman filter is 0.00019144. This shows that by calculating the values in the backward-loop, we are able to minimize the error. Likewise for Figure 7.6, it is 0.0027 for ordinary Kalman filter, whereas it is 0.00012147 for the FB Kalman filter.

7.4 Notes

It was shown that self-tuning Kalman filter approach has no requirement of providing a feedback loop in view of its embedded information matrix. The condition of feedback from fusion center to the local sensors has helped in the implementation of self-tuning Kalman filter. The developed scheme has been successfully evaluated on quadruple-tank system to demonstrate its effectiveness practical implementation.

Then, a FB Kalman filter fusion approach is implemented on the developed scheme on regular Kalman filter. It has been shown that FB Kalman filter has a smoothing property that provides better results than the regular Kalman filter. Moreover, the conditions with intermittent observations are handled in a satisfactory manner with equivalent corresponding to the loss communication. The proposed scheme has been successfully evaluated on coupled tank system, thus underpinning the proposed scheme with its practical implementation.

Chapter 8
Experimental setups

Distributed control systems (DCSs) based on communication network are becoming a common interest in the contemporary control research community. This chapter presents simulation and experimental investigations for such systems. We investigate the available simulation environments and introduce appropriate experimental setup to conduct control designs in distributed-networked control systems (DNCSs).

8.1 Introduction

Networked control system has been a hotspot in the research fields of control theory and control engineering applications at home and abroad. Due to major advancements in the area of networking over the past decade, a new paradigm of control systems has emerged, namely networked control systems. Such systems differ from classical control systems in that their control loops are closed around communication networks.

Simulation and experimental setups play vital role in the analysis and design of improved methods in control systems. DNCSs offer challenging problems due to the communication in feedback loop. These problems are related to packet rate, asynchronous sampling, random network delay, packet drop, bit rate, and quantization in addition to the structural complexity of decentralized system and its largeness. This requires a comprehensive need of simulation tools and experimental setups to conduct research in the area of DNCS. The aim of this section is to study and conduct simulations and experiments for advanced research in the field of DNCS. Highlights of the topics to be explored include DCSs with communication. Aim is to achieve best balance between the industrial and academic approach to the topic. Several elements from state of the art plant environment like PLC, DCS, and SCADA systems in view of topic will be explored. Several of the industrial communication networks like actuator–sensor interface, HART Protocol, Foundation FieldBus, Profibus, etc., will also be explored. Academic research with emphasis on experimental and simulation setups in the field of DNCS will be explored.

The *TrueTime* toolbox [504] in the MATLAB environment provided very good platform for the research of networked control system, and the dynamic process of distributed real-time control system, control task execution, and co-simulation environment of network interaction can be built by *TrueTime* toolbox (Figure 8.1).

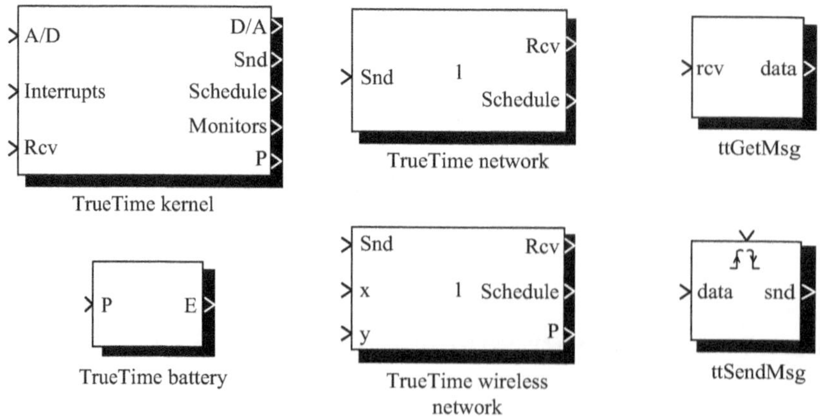

Figure 8.1 The TrueTime 1.5 library blocks

The advantages given by wireless technology are several. First of all, it permits to carry the capability of wired networks to areas that cables cannot reach. Considering industrial plants, wireless technologies can significantly facilitate deployment and reconfiguration by eliminating the need for installing and maintaining cabling, reducing both cost and time. However, the lack of maturity, the failure to provide real-time performance, and the lack of reliability metrics comparable to those of wired networks do not help the use of this technology in industrial environments.

The field is growing in setting up the appropriate simulation environment and experimental setup. The objectives of this work are:

- To review simulation and experimental setups for decentralized and networked control systems.
- To conduct simulations for networked control systems.
- To establish experimental setup for networked control systems.

Our ultimate aim is to provide a technical description on distributed-networked simulation and experimental setup to support the analysis and control research in DNCSs. In addition, we seek to cover the knowledge of simulation and experimental tools and methods of advanced research in DNCSs.

8.2 Overview of related work

A MATLAB/Simulink-based, event-based simulator for real-time control systems termed (*TrueTime*) was developed in [445]. In effect, *TrueTime*, as a co-design package, makes it possible to simulate the temporal (true) behavior of multitasking real-time kernels including controller tasks and to critically examine the effects of central processing unit (CPU) and network scheduling on the performance of dynamic control systems. Basically, the simulated real-time kernel can handle external interrupts as well as fine-grained details including context switches. One of the basic features of the simulator is that arbitrary scheduling policies may be defined, and the control tasks may be implemented using **C**-functions, **M**-functions, or Simulink block diagrams. The recent applications of distributed mobile agents and wireless sensor/actuator networks have called for co-simulation as a necessary tool during system development. In [446], a simulation environment for mobile wireless networked embedded systems that facilitates simulation of computer architecture (nodes) and communication networks interacting with the continuous-time dynamics of the real world was presented. Basic features of the simulator include interrupt handling, task scheduling, wired and wireless communication, local clocks, dynamic voltage scaling, and battery-driven operation. The modeling and implementation of the IEEE 802.11b standard for wireless communication within the *TrueTime* framework were provided. An effort describing simulation of wireless control using the WirelessHART standard was given in [447]. The impact of clock-drift on predictive controller performance where no synchronization exist between the controller and the wireless network was investigated. The simulations were done using an extension of the Simulink package *TrueTime*.

Two MATLAB-based tools, *Jitterbug* and *TrueTime*, for the analysis and simulation of how the process of timing affects control performance were described in [448]. Control system is described by a signal model and a timing model in Jitterbug simulation. On one hand, *Jitterbug* was used to determine the sensitivity of a control system against slow sampling, delay, and jitter in a quick way. On the other hand, *TrueTime* was preferred for more detailed analysis from overall system perspectives, where the modeling targeted at a combination of computational nodes and network blocks along with describing any dynamical system. In this way, a complete networked control system environment was prepared.

Connecting a spatially distributed system with sensors, actuators, and controllers as a networked control scheme (NCS) by a shared data network can reduce the wiring and cost remarkably. Excellent results incurred when NCS was utilized in remote operation of linear systems. However, the presence of non-linearities was considered the major barrier in implementing a NCS. In [449], with focus on the control of induction motors, a sliding mode flux observer was used to linearize the induction motor model, such that the application of NCS became feasible. A fuzzy logic speed controller with state predictor was developed to adapt various network conditions due to the variable quality of service (QoS). Simulations were conducted by employing *TrueTime* toolbox to demonstrate the effectiveness of the adaptive fuzzy logic PI speed controller.

In [450], a network-based cascade control system on a real laboratory pilot plant using two alternative FieldBus (FB) and NCS control approaches was designed and implemented. The proposed network platform was designed so that a Smar foundation FB controller (DFI-302) could be linked to a remote Siemens programmable logic controller (S7-315-2DP) through an industrial Ethernet. The networked-based configuration was maintained operational using the Smar OPC Tag browser and Siemens WinCC OPC C servers' facilities. Experimental observations indicated that the network data transmission delay has no serious effect on the resulting control performance in the FB implementation approach. However, the increase in the network transmission delay degrades the control performances in the NCS implementation approach. To improve the control performance due to the variable transmission delay degradation effect, a fuzzy PID control approach was proposed. In [451], experimental platform for NCS was designed and implemented. The system used control, video, and ftp/web data over Controller Area Network (CAN) and Switched Ethernet networks.

Recently, wireless networked control systems (WNCSs) became an interesting area in research community, and in particular, designing WNCS over mobile ad hoc network (MANET) brought new challenges to the researchers. In ad hoc systems, strict guarantee for topology, delay, etc. were not practical. A smart integration of communication network, computing, and control was therefore required for quality performance of WNCS. In [452], the network, control, and computing co-design issues were reviewed and guidelines for successful implementation were established. Some co-simulation tools for such systems were addressed. Important ideas for subsequent development were identified, including online co-simulation using MATLAB for control/computing, NS2/OPNET for communication networks and development of co-design theory and co-simulation tools for control, computing, and communication.

The sampling period scheduling of NCSs with multiple control loops was addressed in [453]. The generalized exponential function was employed to describe integral absolute error performance versus sampling period based on *TrueTime* toolbox where the sampling periods were scheduled to obtain the optimal integrated performance. Algorithms were developed to guarantee the stability of every control loop and the bandwidth on available network resource. In [454], the structure of NCS for immune PID controller was designed. The immune PID controller constituted by the combination of biological immune principles and traditional PID control had simple structure and algorithm, and with the features of strong robustness and adaptability, which can improve the control performance of the system further. Several results pertaining to the analysis and design of NCS were provided based on sampled data, model-based, and hybrid models. Also, some of the most critical problems in the design of NCS such as packet loss, network-induced time delays, and limited communication in the control loop were presented.

With the emerging IP-based technologies, the quality of network traffic of a control system is becoming much more challenging. Operators must constantly monitor the state of traffic flow in the control system and decide whether the system is functioning properly. Obviously, the characteristics of normal traffic and extract boundary conditions, for example, thresholds at which the QoSs begin to deteriorate to intolerable levels, must be known. This will help distinguish safe traffic from an undesirable one. In [455], a network simulation technique was used to analyze the traffic behavior of IP-based control systems and construct a flexible simulation architecture. With focus on the message jitter, as it is embedded within the control loop, the implementation of NCS was considered in [456] where a strong impact on the control execution delay was recognized. By comparing the timing properties of both CAN and PROFIBUS control networks, the capability of the CAN network was assessed to fulfill control application requirements.

A new wireless communication architecture is specified in [457] that provides the required reliability, safety, security, and real-time parameters for wireless sensor networks. The primary goals of the work are:

- The implementation of a WirelessHART network simulator. A new wireless protocol that aims to establish a new communication standard for process automation applications. Essentially, the simulator was written in C++ and interfaced with MATLAB through an **S**-function, to have a friendly use of it. The possibility to work with this simulator was added in the *TrueTime* library giving to the user a very powerful tool to simulate different wireless scenarios.
- The study of delay compensation in a WirelessHART network. This protocol provides for the synchronization between all the devices of the network. If the controller is not part of the wireless network, no synchronization with the other devices is guaranteed. In this condition, if the controller clock is affected by drift, a varying delay is introduced in the control loop. This effect degrades the performance desired and, in some critical cases, causes instability to the whole system.

A framework for NCS simulation to enable the analysis of the influence of network transmissions on the performance of control systems was presented in [458]. An agent-based design was introduced to simulate data packet transmissions over the network. The network simulator 2.26 (NS-2.26) release was extended by modifying the user data protocol (UDP) common header in order to support transmission of application data. Then, modifying the network topology parameters, networked control system simulations are analyzed for different parameter changes, such as the network bandwidth, the number of plant nodes, and the sampling period. The off-line identification problem for open-loop stable LTI processes working in the networked environment was extensively studied in [459] with emphasis on the effects of random network-induced delays and packet dropouts based on event-driven actuators (D/A conversion) and a general nonuniformly nonsynchronized sampled data. A modified version of the simplified refined instrumental variable method was developed to solve this problem, and was validated in a networked identification experiment using the simulator *TrueTime*. The TCP/IP communication issue of data stream in NCS was investigated in [460]. Analytic model of TCP/IP was presented according to slow start theory. The simulation model was established based on OPNET software. By applying TCP or UDP to different data streams, the simulation results showed that real time and reliability of most data streams can be guaranteed.

8.3 Simulation environments

It turns out from the foregoing discussions that designing a real-time control system is essentially a co-design problem since the choices made will affect the control

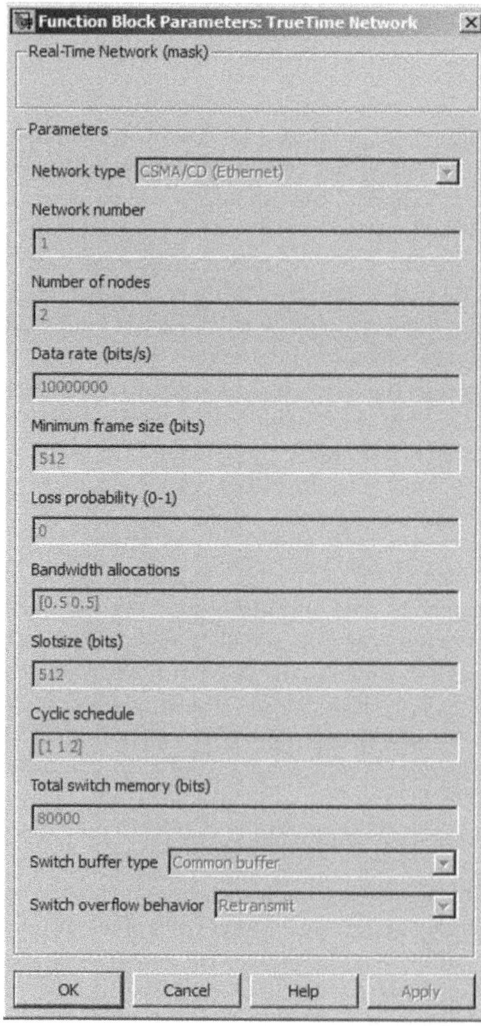

Figure 8.2 Parameters of network block in TrueTime

design and vice versa. Deciding on a particular network protocol will induce certain delay distributions that must be taken into account in the controller design. Alternatively, bandwidth requirements in the control loops will influence the choice of CPU and network speed. Therefore, using analysis tool like *Jitterbug*, one can compute a quadratic performance criterion for a linear control system under various timing conditions and model representations [461]. A stochastic timing model with random delays can be invoked to describe the execution of the system. Guided by this information, the user can proceed using a simulation tool such as *TrueTime* to perform event-based co-simulation of a multitasking real-time kernel containing controller tasks and the continuous dynamics of controlled plants. The simulations capture the true, timely behavior of real-time controller tasks and communication networks, and dynamic control and scheduling strategies can be evaluated from a control performance perspective. The controllers can be implemented as MATLAB **M**-functions, C++ functions, or ordinary discrete-time Simulink blocks. The following subsections shed light of the simulation tools.

8.3.1 TrueTime

TrueTime is a MATLAB/Simulink-based simulator that effectively offers co-simulation of controller, network transmission, and plant dynamics. Introduced in 2002, *TrueTime* is still evolving and currently a beta version 2 is available, although version 1.5 is considered as stable software. Main feature of *TrueTime* is the use of real-time kernels in which controller tasks are executed. This environment can be used to study standard and distributed control, scheduling, overrun handling, synchronization, control over wireless networks, and wireless ad hoc routing using AODV. The *TrueTime* blocks can be connected with other Simulink blocks to form a real-time distributed (networked) control system. The *TrueTime* blocks additionally require to initialize the kernel and network blocks. This also requires creating and interrupting handlers, timers, events, monitors, etc. for the simulation environment to work.

TrueTime network block simulates Medium Access and Packet Transmission in a local area network (LAN) environment. A node uses ttSendMsg primitive function to transmit a message over network. This causes a triggering signal to the network block on the corresponding channel of node on the network. Version 1.5 of *TrueTime* supports simple models of networks including: CSMA/CD (e.g., Ethernet), CSMA/AMP (e.g., CAN), Round Robin (e.g., Token Bus), FDMA, TDMA (e.g., TTP), and Switched Ethernet. The models ignore propagation delay since it is very small in LAN. Models support packet-level simulation. Figure 8.2 illustrates the network parameters including: network number, number of nodes connected to network, data rate in bits/second (i.e., speed of network), minimum frame size (bits), and loss probability from 0 to 1 (i.e., probability that message is lost during transmission), etc.

8.4 Simulation examples

In what follows, we provide two representative examples.

8.4.1 Illustrative example 8.1

Control performance versus loss probability is studied in this example. *Loss probability* is the probability that a network message is lost during transmission. Lost

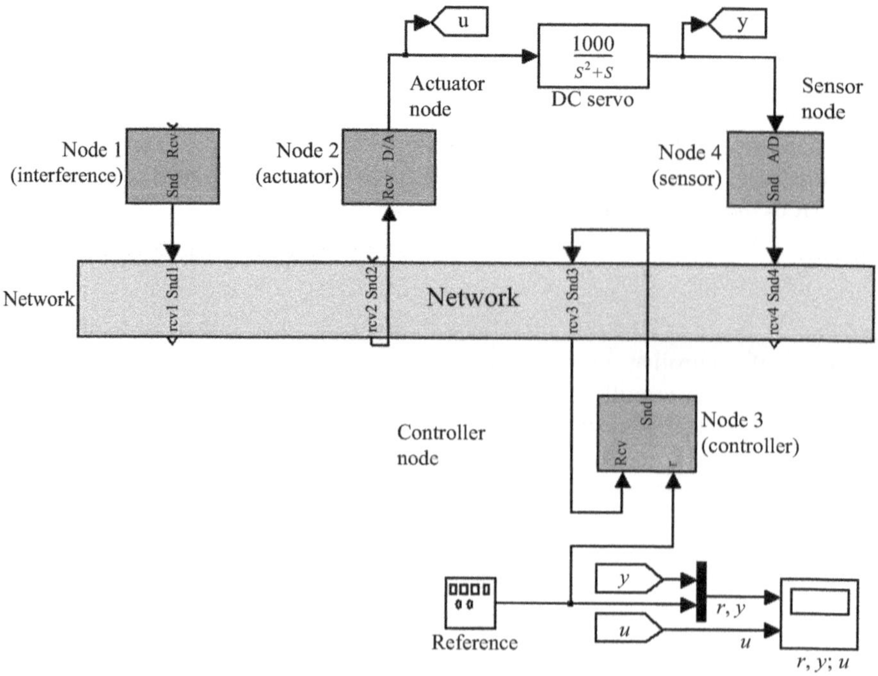

Figure 8.3 Simulation example 1

messages will consume network bandwidth, but will never arrive at the destination. Example 6.1 is shown in Figure 8.3.

Network settings are:

- Network Type: CSMA/CD (Ethernet)
- Data rate: 80,000 bits/s
- Minimum frame size: 40 bits
- Loss probability: (0–1)

Investigation of control is observed for different values of loss probability in Figures 8.4–8.7. It is found that the loss of control increased as the loss probability (packet loss) is increased beyond 0.2.

8.4.2 Illustrative example 8.2

Investigation of the system shown in Figure 8.3 is carried out for different networks providing same loss probability of 0.1. Figure 8.8 shows the tracking signal with respect to reference in case of different networks under similar conditions. The error $(y - r)$ is shown in Figure 8.9.

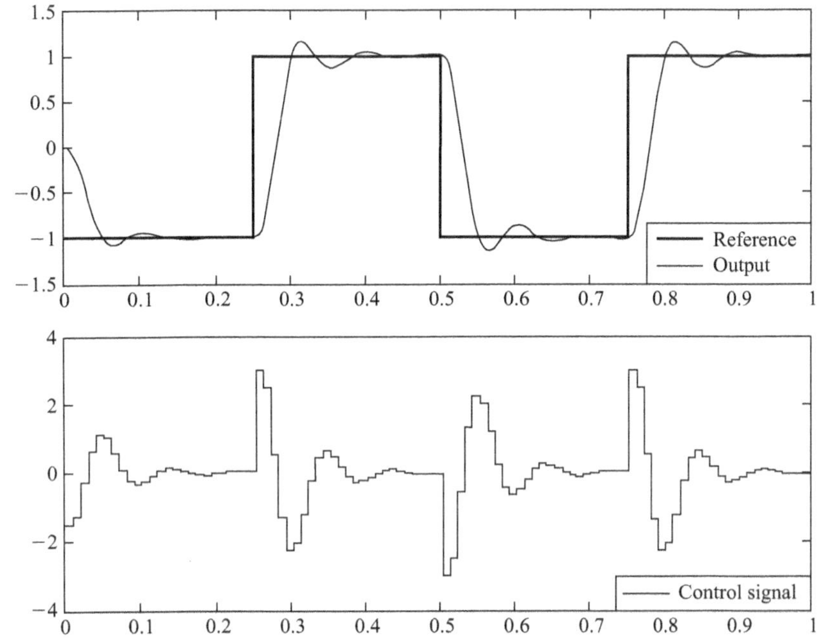

Figure 8.4 Loss probability = 0

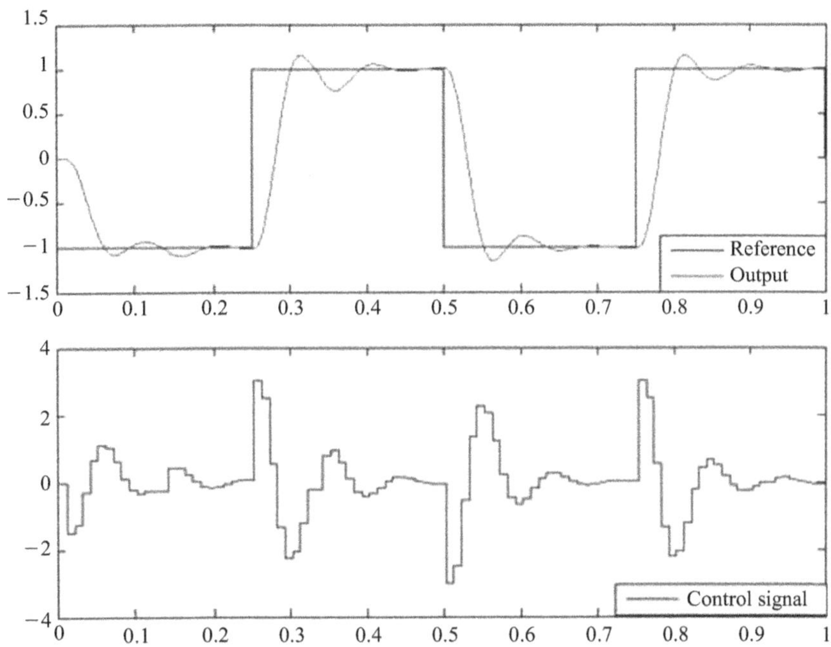

Figure 8.5 Loss probability = 0.01

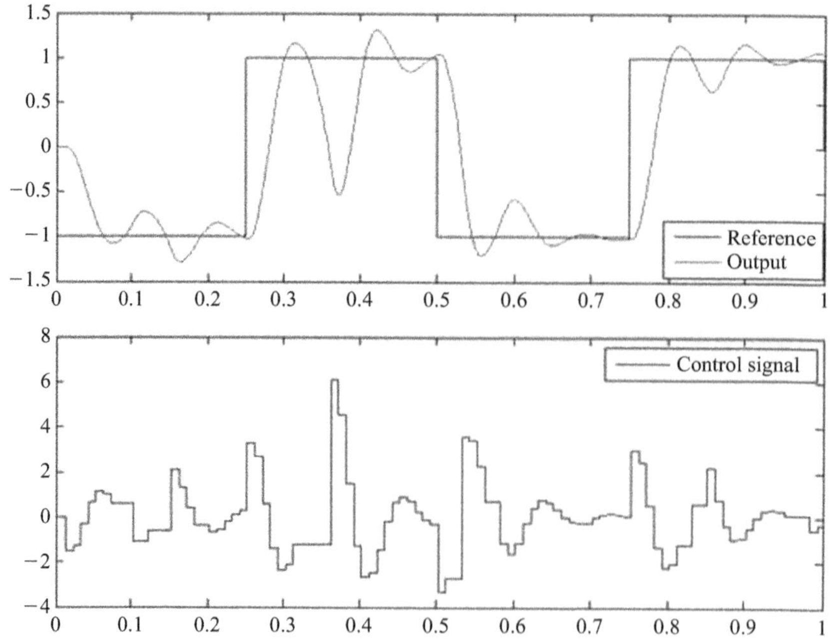

Figure 8.6 Loss probability = 0.1

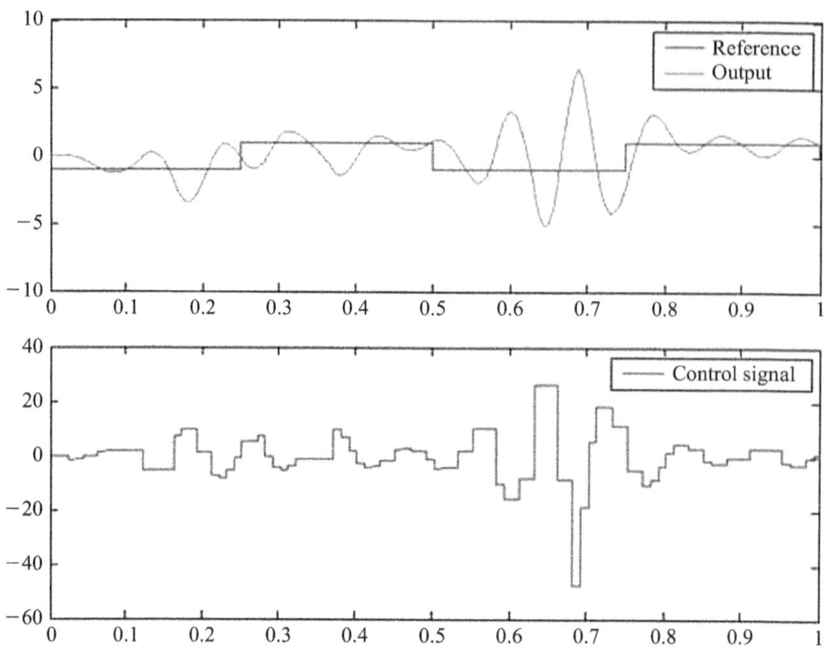

Figure 8.7 Loss probability = 0.2

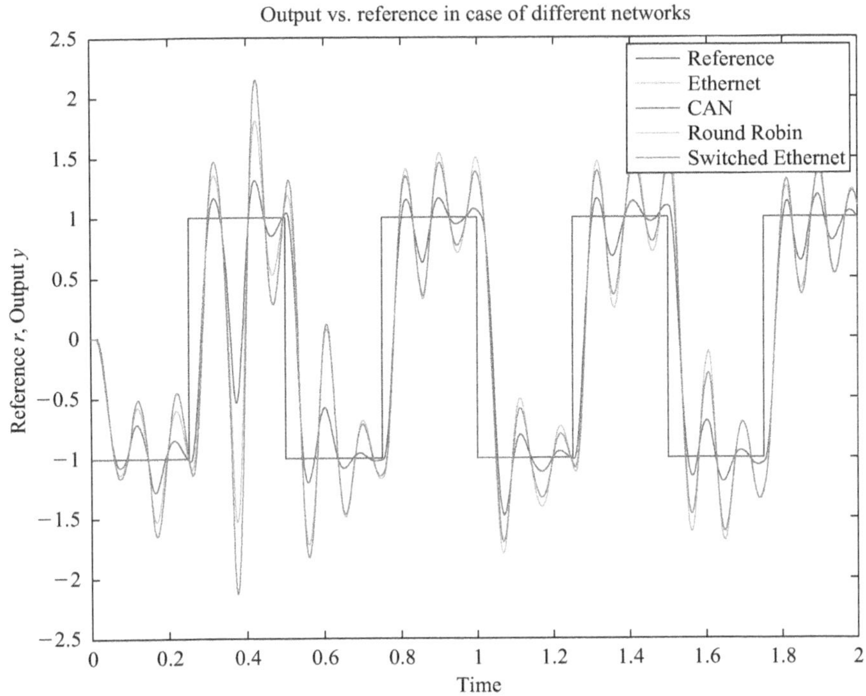

Figure 8.8 Output vs. reference for different networks under similar condition and Loss probability of 0.1

8.5 Experimental setup

A coupled two-tank physical system is used to establish a network control system. Sensors and actuator on the physical system are connected to a Plant-PC via USB-based Data Acquisition card. This Plant-PC is connected on LAN (known as CCSE-Network in KFUPM). The Sensor information is transmitted to remote PC (which is Monitoring and Controller PC) on the network via TCP/IP communication protocol. Data Acquisition, TCP/IP communication, and Graphical User Interface are built in LABVIEW software from National Instruments. Graphical block diagram of this experimental set-up is shown in Figure 8.10.

Plant PC is connected to the Two Tank. Analog Input Channel is connected to the Level Sensor (0–10 V), and Analog Output is connected to the Pump. National Instruments USB-DAQ is used for the Interfacing. TCP/IP connection is established between the Two PCs (i.e., Plant PC and Controller PC) over CCSE-Network, which is the LAN of the University Campus. A Manual knob through User Interface on Controller PC is used to send Control Signal to Plant PC and is applied to the Two-Tank system via USB-DAQ. Control Signal from remote Controller PC is also displayed on the Plant PC's Front Panel. Reference Signal is set from Front Panel of Plant PC and

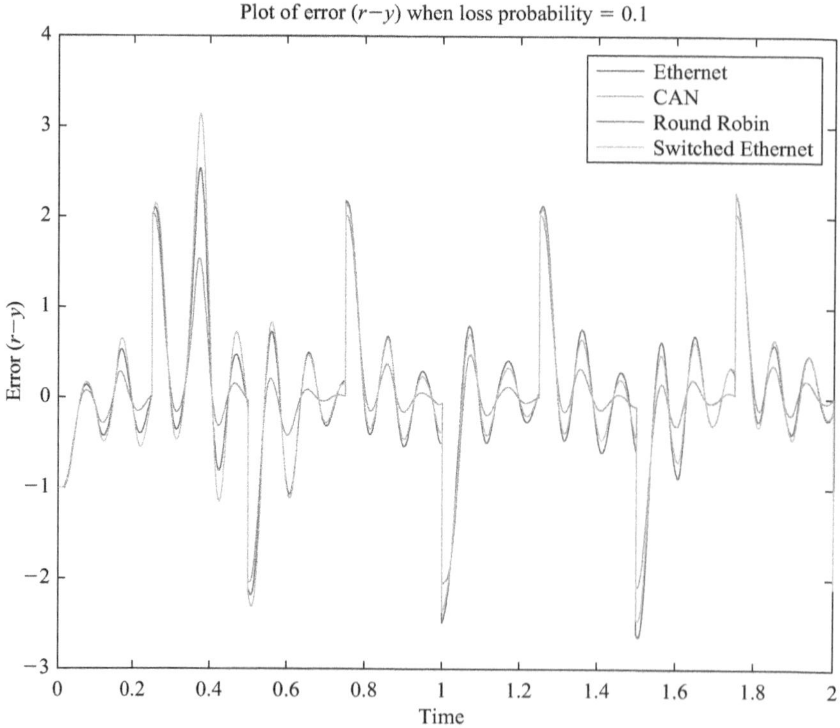

Figure 8.9 Error (r–y) in case of different networked under similar condition and loss probability of 0.1

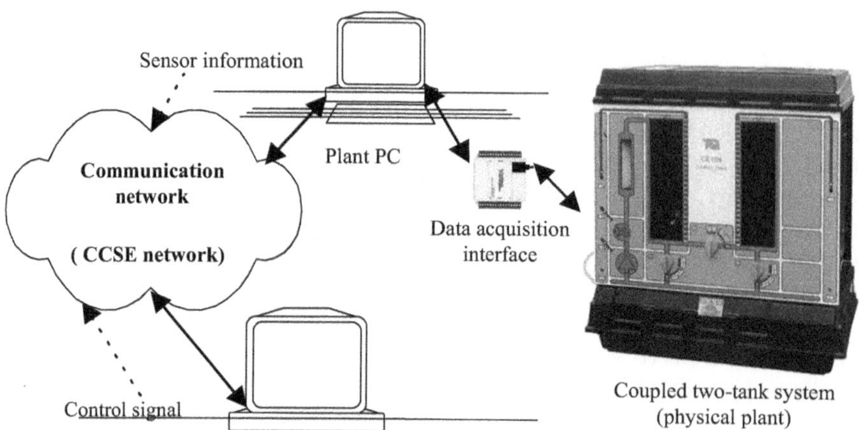

Figure 8.10 Networked two-tank system

can be sent from the Controller PC. TCP/IP connection is successfully established between Plant PC and Controller PC. Control Action received from Networked Controller PC is implemented to the Two-Tank System (Practical System) connected to Plant PC (Figures 8.11–8.14).

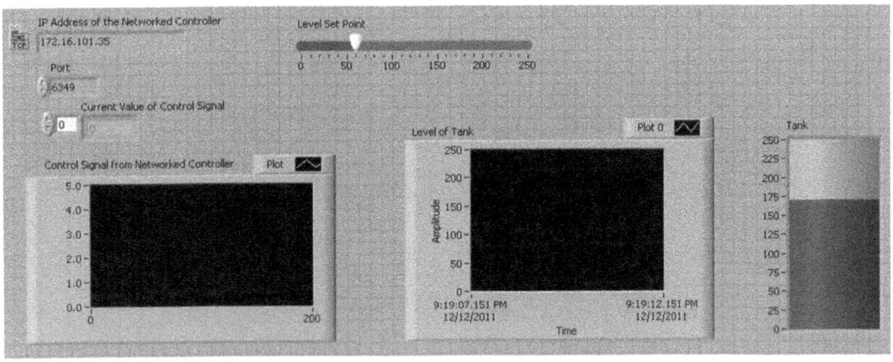

Figure 8.11 Front panel of plant PC

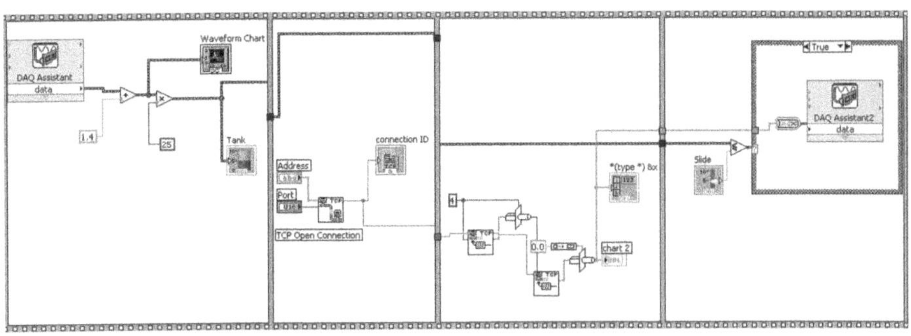

Figure 8.12 Block diagram of plant PC

Programming Language used in conducting this experimental set-up is LABVIEW, which is the Graphical Language from National Instruments. Any LABVIEW program consists of two parts: (1) Front Panel (that can be used for User Interaction), and (2) Block Diagram (where programming is build through function blocks and data flow). Front Panel and Block Diagram for Plant PC are shown in section 8.6.

8.6 Multiagent modeling approach

One approach for automating industrial processes is a DCS based on a hierarchical architecture. In this hierarchy, each level is characterized by a group of different tasks to be carried out within the control system and by using and generating different information. In this article, a reference model for the development of Intelligent

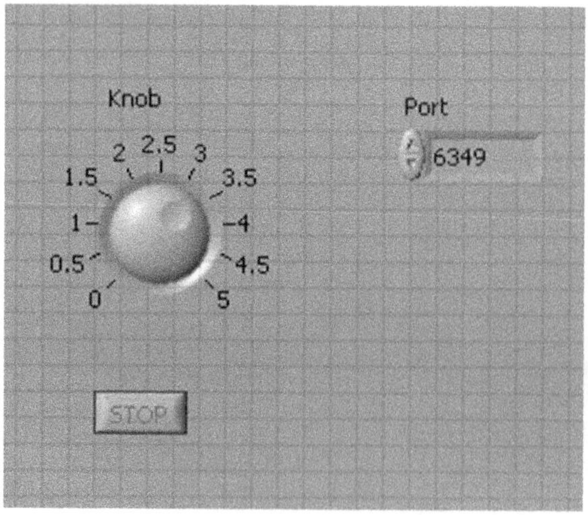

Figure 8.13 Front panel of networked controller PC

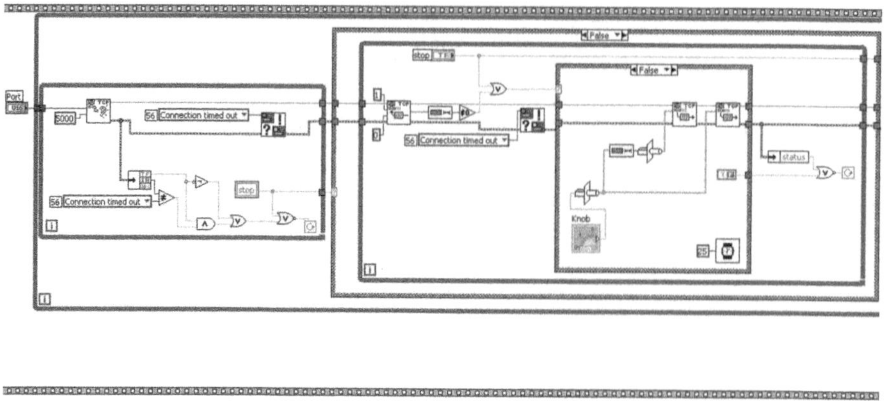

Figure 8.14 Block diagram of networked controller PC

Distributed Control Systems based on Agents (IDCSA) inspired by this hierarchical architecture is presented. The agents of this model are classified in two categories: control agents and service management environment agents. To define these agents we have used an extension of the MAS-Common KADS methodology for Multi-Agent Systems (MAS).

8.6.1 Introduction

Presently, there is great interest in the development of integrated automation systems that permit monitoring different plant operation variables in a broad and dynamic way and to transform such variables in control commands that are later integrated into the plant through actuators. Moreover, they should take into account production and economic criteria that can be applied as control commands or as part of a plant programming function. The integrated structure should permit the flow of information at all levels (management, operation, etc.) concerning the plant, the products obtained, and all the relevant information. On the other hand, a MAS can be defined as a network of problem solvers that work together to solve problems that could not be worked out individually [462]. The main preoccupation of the MAS is the coordination between the groups of autonomous agents, perhaps intelligent, to coordinate their objectives, skills, knowledge, tasks, etc.

In this section, a reference model for Intelligent Distributed Control System (IDCS) based on Agents is proposed. Our model will be made up of entities called agents that work together dynamically to satisfy the control system's local and global objectives and whose design could be developed and made completely independent of the system. The description of the agents in this reference model is based on the MASINA methodology [463] that has an extension of MAS-Common KAD methodology to incorporate other characteristics of agents such as emerging behavior, the reasoning, and the possibility of using intelligent techniques (expert systems, artificial neuronal networks, genetic algorithms, fuzzy logic, etc.) for carrying out their jobs.

8.6.2 *A reference model*

In the sequel, we describe a reference model for IDCSA. The control jobs and information management needed in automation processes can be distributed and expressed through a hierarchical logical structure. Figure 8.15 shows hierarchical

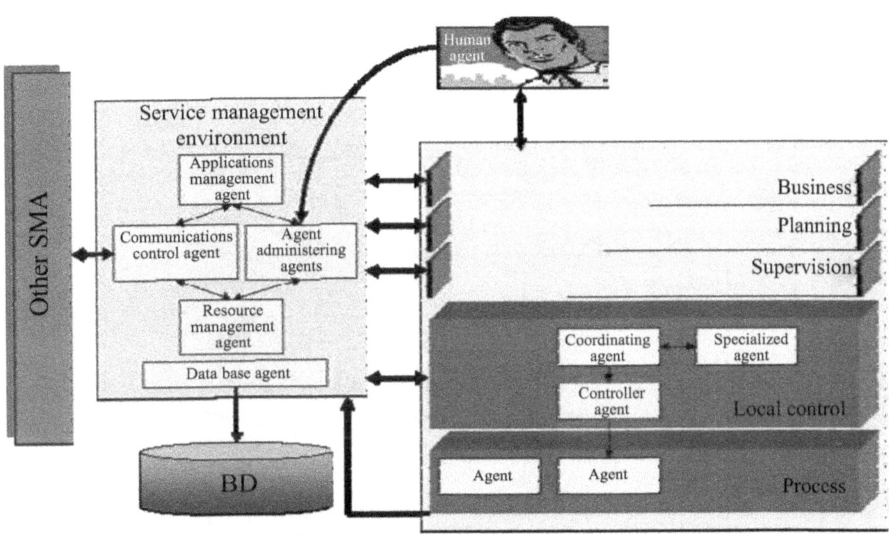

Figure 8.15 Reference model for distributed control systems

reference architecture to develop a distributed control system (DCS) that permits the automation of an industrial plant [464,522].

At the business and planning levels decisions are made at the managerial level, and the control process jobs are carried out at the lowest levels. The Supervisory level adjusts the parameters of the controllers, the control signal is obtained at the Local Control level, to be incorporated later at the plant at the Process level. This DCS model is complemented with a group of agents in each one of the levels of the hierarchy, this is the IDCSA model shown in Figure 8.16. These agents carry out diverse jobs looking toward reaching the specific control objective. The incorporation of the MAS to the reference model permits the control to emerge from the interactions of those entities (agents). The IDCSA is seen as a network of autonomous agents, with different responsibilities according to the level to which it belongs. As such, the details of each agent (objectives, communication, jobs, intelligence, etc.) come according to the levels to which the given agents belong in the IDCSA model. This way, the agents are distributed through the control hierarchy and can be geographically dispersed. The dynamic interaction of the multiple agents happens between levels and in the interior of the levels. In the IDCSA model two categories of agents exist:

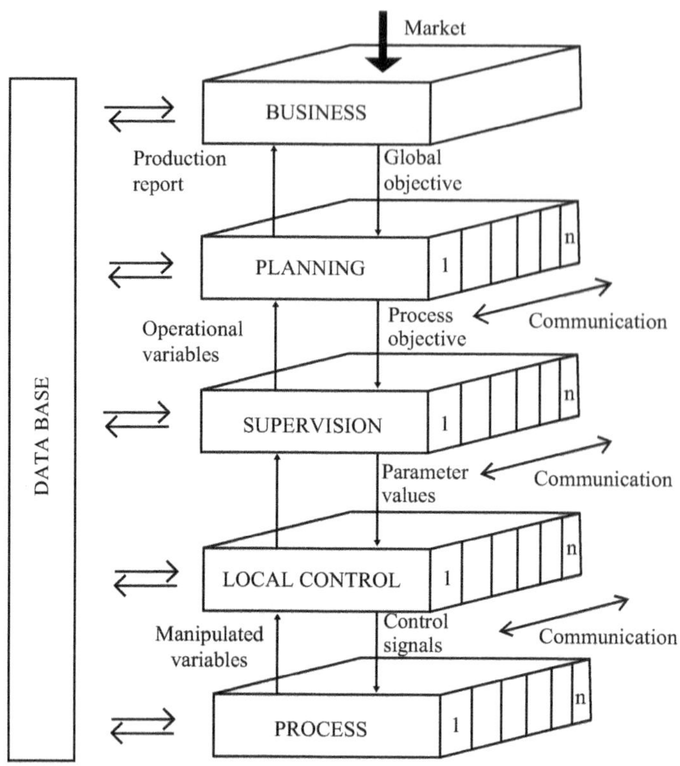

Figure 8.16 IDCSA reference model

Control Agents: Carry out the jobs of control, measurement, control decision making, and putting the decisions into practice, among others. The agents are:

- **Coordinating agent:** make decisions and plan control jobs.
- **Specialized agent:** carry out specific jobs that serve to support the coordinating agent.
- **Controller agent:** obtains control action.
- **Actuator agent:** executes the control action.
- **Observation agent:** measures and processes the variables of the plant.
- **Human agent:** supervises the SCIDA.

Agents of the service management environment (SME): It is the base of the distribution system given that it manages the communication of the IDCSA model and permits the distribution of the control system and heterogeneity among geographically dispersed agents. The agents of the SME are [465]:

- **Agent administering agents:** coordinates the multiagent society.
- **Resource management agent:** administers the use of the resources.
- **Applications management agent:** administers the use of applications (software) of the system.
- **Data base agent:** manages the information storage means.
- **Communication control agent:** establishes communication with other MAS.

A detailed specification of the IDCSA model is presented below. The submodels proposed in the MASINA methodology have been used according to the needs of IDCSA [463].

8.6.3 Organization model

The main objective of the organization model is to specify the environment in which IDCSA will be developed. In this stage of modeling, a human organization is analyzed to determine the areas susceptible to the introduction of the IDCSA model. The model is conceived in three stages:

1. **Modeling of the organization environment:** Describes the environment where we are going to introduce the IDCSA model.
2. **Evaluating the organization environment:** The viability of introducing the IDCSA model to the organization is evaluated.
3. **Model of the multiagent society:** In this stage, the following activities are developed:
    i. *Identify the IDCSA levels in the organization*: some applications can consider the levels proposed in the IDCSA; others only require a group of them [466].
    ii. *Agent's proposal at each IDCSA level*: after identifying the levels needed, the group of IDCSA agents adapted to the functions that are carried out at each level is defined.

iii. *Identify the environment objects*: in each one of the levels, the surrounding objects with which agents will interact to achieve their objective should be identified.

8.6.4 Agent model

The agent model serves to specify the generic characteristics of the IDCSA agents. The agent model proposed is oriented toward services. Agents are capable of carrying out jobs that are offering to other agents (called services). An agent will be selected, the Observation Agent, to discuss the templates of the Agent Model.

- **General information:**
  (i) Name: Observation Agent,
  (ii) Type: Software Agent and Physical Agent,
  (iii) Paper: Environmental,
  (iv) Position: Lowest level agent of the Multiagent System,
  (v) Description: The Observation Agent makes up the data acquiring system: the sensor, the conditioning system, and the transmitter. Through them changes in the process can be detected.
- **Agent objectives:**
  (i) Name: Process Variable Measurement,
  (ii) Type: Persistent, direct objective,
  (iii) Entry parameters: Variables that need to be measured, repository where data is stored,
  (iv) Exit parameters: Measured variable,
  (v) Activation condition: Order of measurement (non-periodic case) or period of time available for measurement,
  (vi) End condition: Measured variable,
  (vii) Representation language: Natural language,
  (vii) Ontology: Control ontology,
  (viii) Description: is associated with the capture of process variable values through carrying out direct or indirect measurements on given variables. The obtained signals should be filtered to eliminate measurement errors, and transported and stored in systems.
- **Agent services:** Process measurement variables, Conversion of variables, and Detection of Quick Changes.
- **Agent capabilities:** contact with the process, make available to the rest of the MAS changes occurring in the environment.

8.6.5 Job model

The elements of this model are ingredients, capacity, environment, and method. A job has associated entry ingredients that permit the job to be executed and at the same time can produce exit ingredients as a product of the execution job. Jobs can be made using a specific technique (classic, intelligent) or hybrid techniques. Below a list of IDCSA jobs is presented.

- **Measurement jobs:**
  - (i) Measure: Sense and condition,
  - (ii) Process the variable.
- **Control jobs:**
  - (i) Decision making,
  - (ii) Obtain control action,
  - (iii) Process control action,
  - (iv) Execute control action.
- **Specialized jobs:**
  - (i) Process specialized agent.
- **Information management job:**
  - (i) Information storage,
  - (ii) Information up-date,
  - (iii) Look for information,
  - (iv) Management of the BD.
- **Location jobs:**
  - (i) Locate resources, agents, or applications,
  - (ii) Assign agent/resource/application.

The observation agent will continue to be utilized to describe the templates that specify the jobs.

Sensing:

- **Objective:** Process variables are registered in a device sensor.
- **Precondition:** The sensor is an element of the lowest level; the only thing that is required is to have an associated process variable.
- **Control structure:** Sensing does not have sub-jobs but rather belongs to a group of jobs that correspond to measuring agents.
- **Execution time:** Continue.
- **Frequency:** Absolute frequency (constant).
- **Content:** The variable is a signal coming from the process, which can be measured and is associated with the state of the process.
- **Surrounding:** Process level.
- **Regulations:** The sensor can be calibrated.
- **Method:** Classic techniques for capturing data are used.

8.6.6 Intelligence model

This model is implemented in those intelligent agents that have reasoning capabilities to decide about the jobs they are going to carry out to solve a situation. For that, the agent can use previous experience or accumulated knowledge through the learning process. According to agent jobs and services defined by IDCSA, their intelligent agents are: Coordinating Agents, Specialized Agents, and Controller Agents. The intelligent model for these agents is generically defined around three elements:

1. **Experience:**
   - (i) Representation: Rules,
   - (ii) Type: Based on cases.

2. **Learning mechanism:**
   (i) Type: Adaptive,
   (ii) Representation Technique: Rules,
   (iii) Learning Source: Process, exit and or failure conditions occurring in the proposal of a control action, proposal of new control, and decision-making parameter values in coordination jobs,
   (iv) Updating: Feedback (using previous experiences for updating knowledge).
3. **Reasoning mechanisms:**
   (i) Information source: Previous results obtained for system control agents,
   (ii) Activation source: control and coordination jobs,
   (iii) Type of inference: Based on rules,
   (iv) Strategic reasoning: Coordinate strategies for the proposed selection of appropriate control algorithms and for decision making in the supervisory and planning level. Confront unknown situations to enrich the experience.

8.6.7 Coordination model

In general, this model describes the coordinating mechanisms between agents. Basically, this model describes the way activities are organized in the IDCSA to reach the objectives. It deals with the planning process and conflict resolution mechanisms that occur between agents. The model is complemented with a description of protocols, ontologies, and communication mechanisms (direct and indirect). In the IDCSA, only predefined planning mechanisms are used. The proposed coordinating model is oriented to services. A service can have associated specific properties (cost, duration, etc.). For the IDCSA case, conversations generated from the predefined planning process are: (i) Obtaining control action, (ii) Manage level objectives, (iii) Sound alarm, (iv) Coordinate humans. Below, the first conservation alone is described:

- When communication exists with the coordinator, the control agent is only in charge of generating the control action (the associated interactions are shown in Table 8.1).

Table 8.1 Interactions of the conversation 'Obtaining Control Action'

Interactions	Services
Request: Control–database	Search for information
Transmit: Control–database	Store decisions
Transmit: Control–coordinator–inferior levels	Coordinate goals
Transmit: current control (level of local control)	Conversion of control signals to physical values

- When no communication exists with the coordinator, the control agent has the possibility of auto-regulating itself. In this case, the control agent, making use of the intelligent model, can identify parameters of the current situation and attempt to define the parameters that remedy said situation.

8.6.8 Communication model

This model describes each interaction between agents (speech acts). In the case of IDCSA, a predefined planning process was used, which determines direct communication carried on between agents. Each interaction between two agents is carried out through sending a message, and has a speech act associated with it, see Table 8.1.

8.7 Networked control system evaluation

The DCSs are being increasingly used in many industrial applications. These systems are often referred as networked control systems. The goal of this paper is to show the network influence on feedback control systems. Two networks are considered: Switched Ethernet network and CAN FB. The first one represents the nondeterministic network and second one represents the deterministic one. Several scenarios are studied to analyze the stability of system according to different network parameters (packet losses, congestion, and frame priority). The *TrueTime* simulator is used in this work.

8.7.1 Introduction

The distributed systems are more and more used in many industrial applications in the context of Flexible Manufacturing Systems (FMS), and Reconfigurable Manufacturing Systems (RMS). In the literature, the DCSs are referenced under two acronyms ICCS: Integrated Communication and Control Systems [467,468] and NCS: Networked Control Systems [8,13,469–471]. Now, the second acronym is commonly used.

The Networked Control Systems are control systems where controllers, sensors, actuators, and other system components communicate over a network (see Figure 8.17). The NCS represents a complex research field due to its pluridisciplinary aspect. NCS studies have to consider simultaneously knowledge on network, computer, and control [472,473].

The insertion of the network communication in the feedback control loop enables to improve, efficiency, flexibility, dependability, and modularity of the systems, provides the features to easy installation and reconfiguration and can also reduce the setup and maintenance costs.

However, this has many impacts. The network implementation in the control system makes the analysis and design of an NCS complex.

Conventional control theories with many ideal assumptions, such as synchronized control and nondelayed sensing and actuation, must be re-evaluated before they can be applied to the NCSs. The network-induced delay makes the traditional study of time-delay systems different. Usually, in these systems the delay is considered

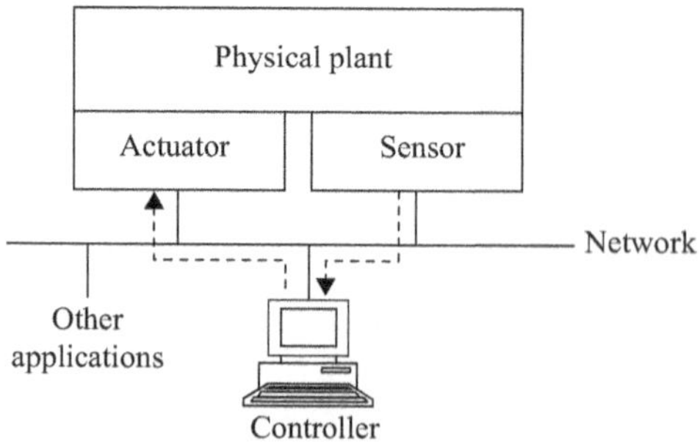

Figure 8.17 Networked control systems framework with other applications on the network

constant or varying in time interval. Another issue caused by the network is the packet loss and the frame scheduling that should be considered during the control system designing. These artefacts can degrade the performance of control systems and can even destabilize the system. These problems are shown in [8,473,474].

Both [468,475] discussed several timing issues such as communication and computation delays, processors jitter, and transient errors existing in NCSs. However, only the case where the total maximum network delay is less than one sampling period was considered.

The network-induced delay is compensated in [476] by using a controller-based Smith predictor. Nevertheless, in their model they make many ideal assumptions to calculate the network delay such as: the communication between sender and receiver is instantaneous when the communication is available, and the transmission and reception instants are known. These assumptions are not always true (e.g., Switched Ethernet).

The delay is compensated in [477] by using the Smith predictor. In this study, the network is a Switched Ethernet network. The measurement of the delay is obtained after the device clock synchronization from the IEEE 1588 protocol.

But the controller-based Smith Predictor model is simple due to lack of robustness. Nowadays this controller is no longer used [478].

Robust control has been proposed to compensate the delay effects. The delay is compensated by using the robust control [477] and the network-induced delay is estimated by using the network calculus theory [479].

In [470], analytic proofs of stability for networked control systems are given for common statistical protocols as well as for a new protocol: Try-Once-Discard (TOD).

In [473] and [478], a relationship is established between the QoS of distributed systems and the performances of the closed-loop control systems. These works consider the design of distributed systems mechanisms and the control systems in a separate way. Only the fieldbuses are considered.

In [480,483,484], a new approach is defined in NCS that is the NCS co-design: the design of controller and scheduling is studied in integrated way.

The networks used in the NCSs are often the fielbuses such as: CAN, FIP, ARINC, ControlNet, etc. Constructors and academic organizations developed many fieldbuses using deterministic protocols.

Their purpose is to satisfy the real-time requirements. However, the application of fieldbuses has been limited due to the high cost of hardware and the difficulty in interfacing with multivendor products. To solve these problems, the computer network technology, especially Ethernet, is being adopted even if the Ethernet protocol is not determinist.

In [471], the network delays are characterized for different industrial networks (two fieldbuses: ControlNet and DeviceNet and Ethernet) and studied the inherent trade-offs between network bandwidth and control sampling rates. They conclude that Ethernet can be used for periodic/nontime-critical and large data size communication, such as communication between or machines. For control systems with short and/or prioritized messages, DeviceNet demonstrates better performance. Nevertheless, the authors were not interested to the Ethernet Switch (Switched Ethernet networks).

In [485], both the Switched Ethernet performances and the controlled systems are evaluated. They observed on the networked control system using the Switched Ethernet that the control performances were affected very little by the network delay. Therefore, even if an NCS uses a controller designed for the centralized control system without taking into account the network parameters, the NCS with the Switched Ethernet network can still satisfy the desired design specifications. Another study [66] has shown the feasibility and effectiveness of real-time control system with Ethernet switch.

Recently, the development of Switched Ethernet shows a very promising prospect for industrial applications due to the elimination of uncertainties (collisions) in the network operation that leads to improved performance.

The adoption of the Switched Ethernet for industrial networking is expected to overcome existing barriers such as high cost, interoperability, and difficulty in application development of the existing FB protocols.

The objective of this paper is to analyze the behavior of control system regarding two kinds of network: a deterministic network: CAN and a nondeterministic network: Ethernet. The interest of this study is to show the importance of the network parameters used in these different networks to control remote applications.

The remaining part of this section is organized as follows: First, the protocols of Switched Ethernet and CAN are presented. The second section describes an example of networked control system. The third section studies the network influence on feedback control systems in two previous networks. The results are based on simulations carried out by using *TrueTime* simulator [486,487]. Finally, conclusions and perspectives are given.

8.8 CAN and Switched Ethernet networks

8.8.1 CAN network

The CAN serial bus system is used in a broad range of embedded as well as automation control systems. The main CAN application fields include cars, trucks, trains, aircraft, factory automation, industrial machine control, building, etc.

CAN network [488] [CAN, 2002] is constituted on serial bus shared by stations by mean of a CSMA/CA scheme with a deterministic collision resolution. The collision resolution is based on priorities associated to identifiers (addresses) of the frame that carry the data (n bytes with $n \leq 8$) of application tasks.

The priority (at which message is transmitted compared to another less urgent message) is specified by the identifier of each message. The priorities are laid down during system design in the form of corresponding binary values and cannot be changed dynamically. The identifier with the lowest binary number has the highest priority. Bus access conflicts are resolved by bit-wise arbitration of the identifiers involved by each station observing the bus level bit for bit. This happens in accordance with the wired mechanism, by which the dominant state overwrites the recessive state. All those stations (nodes) with the recessive transmission and dominant observation lose the competition for bus access. All those losers automatically become receivers of the message with the highest priority and do not reattempt transmission until the bus is available again.

Transmission requests are handled in order of their importance for the system as a whole. This proves especially advantageous in overload situations. Since bus access is prioritized on the basis of the message, it is possible to guarantee low individual latency times in real-time systems.

8.8.2 Switched Ethernet network

Ethernet was developed in the 1970s and emerged in products in the early 1980s. It is now the dominant local area networking solution in the home and office environment. It is fast, easy to install and the interface ICs are cheap. Despite early attempts to use Ethernet as a real-time communication medium in the factories, practitioners were reluctant to adopt this technology because of its intrinsic nondeterminism [489].

Originally, Ethernet uses a shared medium by using, for example, hub technology. In this case, simultaneous access to the medium generate collisions and the transmission is delayed till no collision occurs. It means, in the worse case, when the medium is overloaded, a message could never be transmitted.

Since 1997, new Ethernet versions have been developed and proposed to replace the hub by switches, to connect all the devices in point to point to the switches, to generalize the use of full-duplex mode and to increase the bandwidth. The interest of these technology evolutions is to avoid collisions. But the collision problem is shifted to a congestion problem in switches. The second issue is that the switches generate latencies that have to be taken into account in control systems.

The use of switches to offer real-time guarantees on factory communications has been suggested and analyzed by many authors [479,485,490]. The use of Switched Ethernet architecture in real-time systems led to develop the IEEE 802.1 D and IEEE 802.1Q. They offer the possibility of compensation and reduction delay using prioritization packet procedure.

8.8.3 Network impact on feedback control systems

Many works show the effects of network inside the control loop of the system. Most research in NCS studied the relationships between the network-induced delay and the stability of control systems [481]. And other works [473] showed the impacts of QoS on the stability of feedback control systems. The QoS mechanisms studied are the frame scheduling, the task scheduling, the drop packets, and the protocols. But, only the fieldbuses are considered.

In this section, the goal is to show the network influence on the closed-loop control system. The modeling and simulation are done by using *TrueTime* simulator [486,487]. The considered networks are: Switched Ethernet and CAN networks.

Firstly, a servo problem is chosen. The purpose is to follow the command signal. Consider the PD control of a DC servo described by the following continuous-time transfer function:

$$G(s) = \frac{1000}{s(1+s)} \tag{8.1}$$

Following the specified requirements to have a percentage overshoot less than 5%, $\zeta = 0.7$ the PD parameters are tuned at the following values: $K_p = 1,5, K_d = 0.054$.

Once the PD controller has been designed in the continuous-time domain and with the appropriate sampling period ($h = 10$ ms), we obtained its PD discrete approximation:

$$\begin{aligned}
e(t) &= r(t) - y(t)\\
p(t) &= k_p e(t)\\
i(t) &= i(t-h) + \frac{K_i h}{2}(e(t) - e(t-h))\\
d(t) &= \frac{K_d}{h}(e(t) - e(t-h))\\
u(t) &= p(t) + i(t) + d(t)
\end{aligned} \tag{8.2}$$

The closed-loop control system is distributed on a network (CAN, or Switched Ethernet) and modeled with *TrueTime* simulator (Figure 8.18).

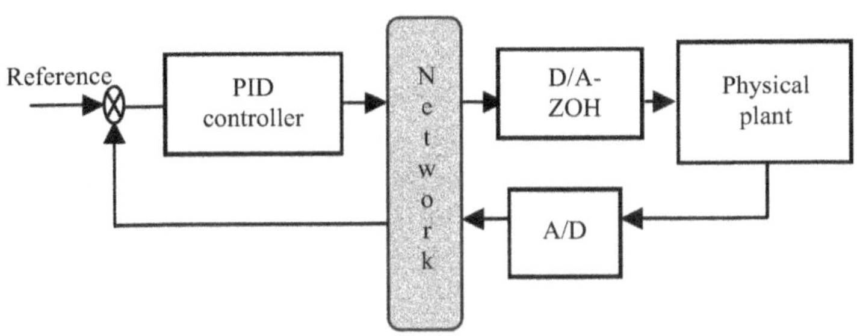

Figure 8.18 Feedback control system model.

8.8.4 Simulation results

In feedback control systems, there are two kinds of networks: Switched Ethernet and CAN are considered to show the influence of network on feedback control loop. The chosen parameters are:

- For the Switched Ethernet: bit rate is 100 Mbit/s, the frame size is 64 Bytes.
- For the CAN networks: bit rate is 1 Mbit/s, the frame size is 8 Bytes.

Three cases are studied:

- Network is used with ideal assumptions
- Network losses packets (information)
- Network is shared by other applications (interference disturbances with drop information case)

In all cases, we assume that the congestion don't generate the drop packets. It means that the buffers in network devices are well dimensioning.

8.8.5 Network used with ideal assumptions

First, the system is simulated in the ideal case. It means, the network introduces no packet losses and the delay depends only due to traffic generated by the real-time system. The network is then not shared with other applications. Figure 8.3 shows the behavior of the system on CAN network and Figure 8.4 on the Switched Ethernet network. In the two cases, the outputs follow the references and the systems are stable. On the CAN experimentations, all the priority configurations of the messages between controller and sensor have been tested. These priority configurations have no impacts on the stability of system (Figures 8.19 and 8.20).

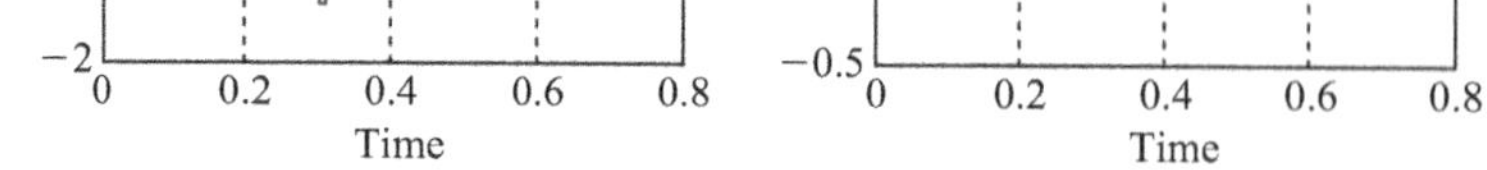

Figure 8.19 Output (y) and controller (u) results of the regulation application on the CAN network

8.8.6 Information lost

In this section, three percentages of packet losses are analyzed, respectively 5%, 10%, and 15%. In the first case, Figure 8.21 shows a small overshot but the system is stable. And in the second case (Figure 8.22), the system becomes instable, but the system over Switched Ethernet is less sensitive to the packet lost than the one over CAN network. In the last case (Figure 8.23), the system is fully unstable whatever the network used.

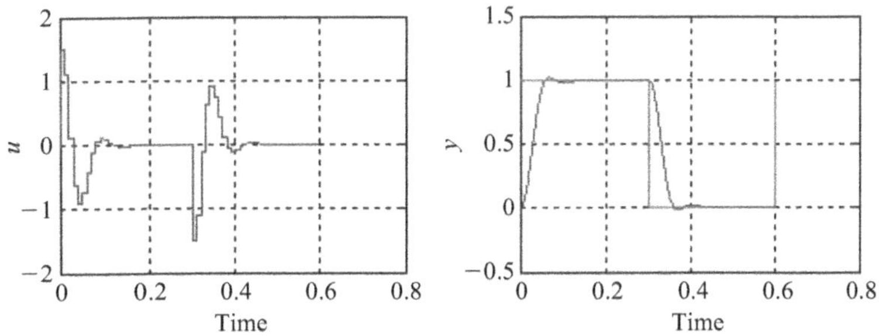

Figure 8.20 Output (y) and controller (u) results of the regulation application on the Switched Ethernet network

8.8.7 Shared network

In this section, the network supports the traffic of other applications. This traffic overloads the medium and this leads to increase in the delay of the packets exchanged between the controller and the actuator and sensor of Servo system. The bandwidth occupation of traffic overload is constant, in all evaluated scenario.

The simulations are analyzed with 1%, 10% bandwidth occupation of interference. The period of these perturbations is 7 ms. For CAN network, two cases are evaluated. The first configuration applies priority 1 (high priority) to the controller and priority 2 to the remote process, and priority 3 to the overload traffic. In the second configuration, the controller has priority 2, the process has priority 3, and the overload traffic has priority 1. For the Switched Ethernet, the scheduling mechanism is FIFO based on store and forward. *TrueTime* simulator does not implement the classification of service in switch and then it is not possible to associate a priority on an Ethernet frame.

Finally, the overload traffic is modeled according to two frame sizes: 10 and 1,500 Bytes.

On the CAN experimentations, when the bandwidth occupation of overload is equal to 1% the control system is always stable (both with short and long packet size). When overload is equal to 10%, the results depend on the priority configurations:

- In case of the overload traffic is tagged with the highest priority, we observe important overshot (>15%).
- In case of overload traffic is tagged with lowest priority, the system is stable.
- In case of overload traffic is tagged with lowest priority and the packets size is equal to 1,500, the system is dramatically instable.

The whole experimentations with Switched Ethernet networks show that the system is always stable.

When size of the interference is long, 1,500 Bytes (in our case), the Switched Ethernet is more efficient (see Figures 8.24 and 8.25).

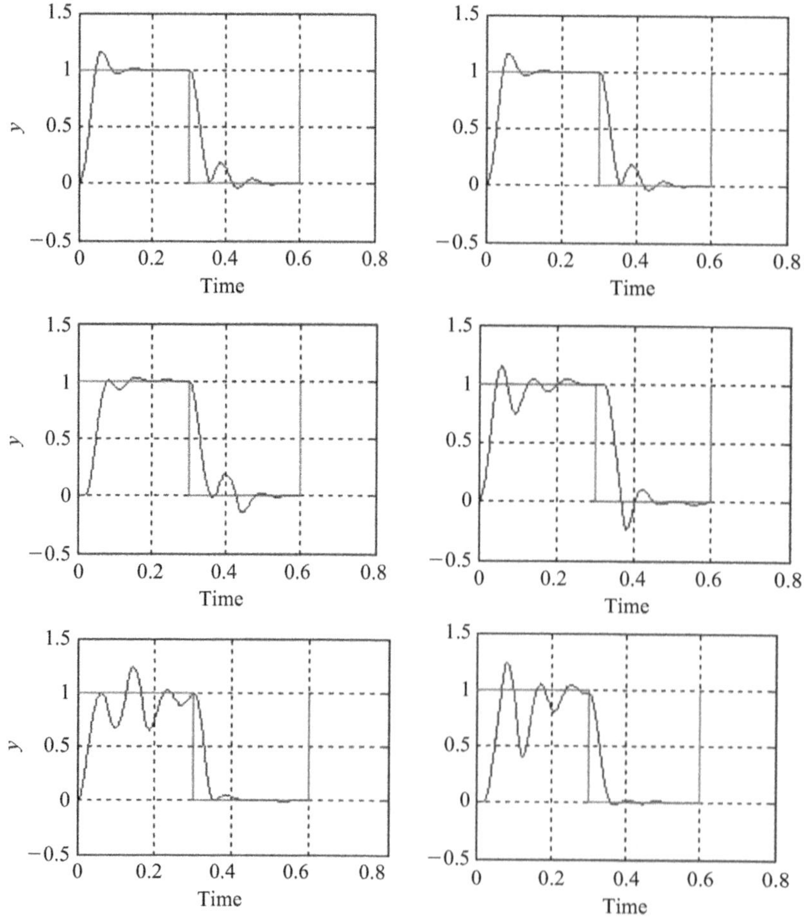

Figure 8.21 Simulation case of networked control system: 1-Switched Ethernet: (in top at left) 5% packets losses, (in middle at left) 10% packets losses (in bottom at left) 15% packets losses. 2-CAN (in top at right) 5% packets losses, (in middle at right) 10% packets losses, (in bottom at right) 15% packets losses

8.9 Design consideration

Increasingly faced with the requirements of safety, environmental sustainability, and profitability, chemical process operation is relying extensively on highly automated control systems. This realization has motivated extensive research, over the last 40 years, on the development of advanced operation and control strategies to achieve economically optimal plant operation by regulating process variables at appropriate values. With respect to process control, control systems traditionally utilize dedicated,

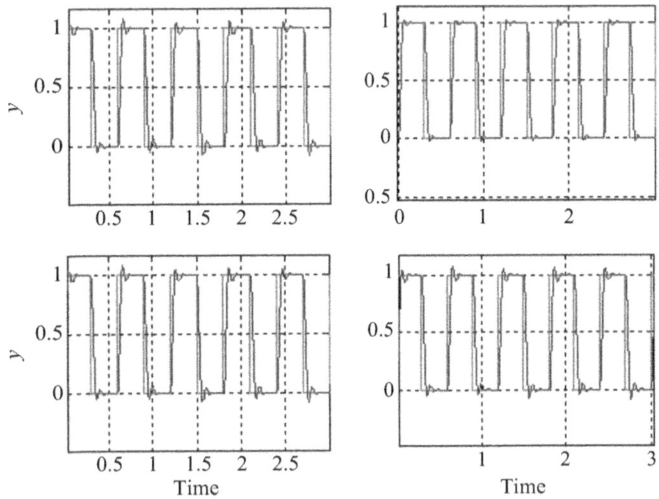

Figure 8.22 Simulation case of networked control system with 10 bytes packet size: 1-Switched Ethernet (at left) 1% bandwidth occupation of interference, 2-CAN (in top at right) 1% of occupation and first priority configuration is applied, (in bottom at right) 1% of occupation and second priority configuration is applied

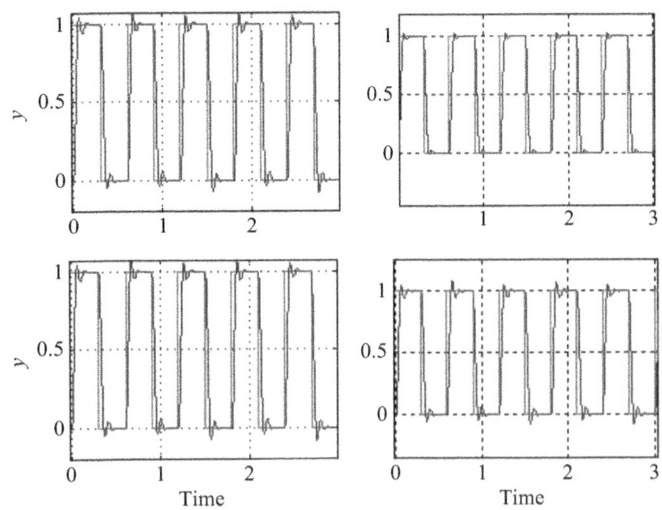

Figure 8.23 Simulation case of networked control system with 10 bytes packet size: 1-Switched Ethernet (at left) 10% bandwidth occupation of interference, 2-CAN: (at the top at right) 10% of occupation and first priority configuration is applied, (in bottom at right) 1% of occupation and second priority configuration is applied

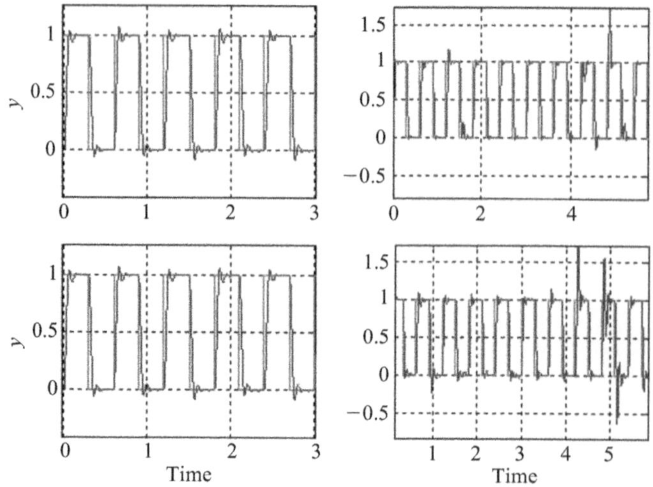

Figure 8.24 Simulation case of networked control system with 1,500 bytes packet size: 1-Switched Ethernet (at left) 1% bandwidth occupation of interference, 2-CAN (in top at right) 1% of occupation and first priority configuration is applied, (in bottom at right) 1% of occupation and second priority configuration is applied

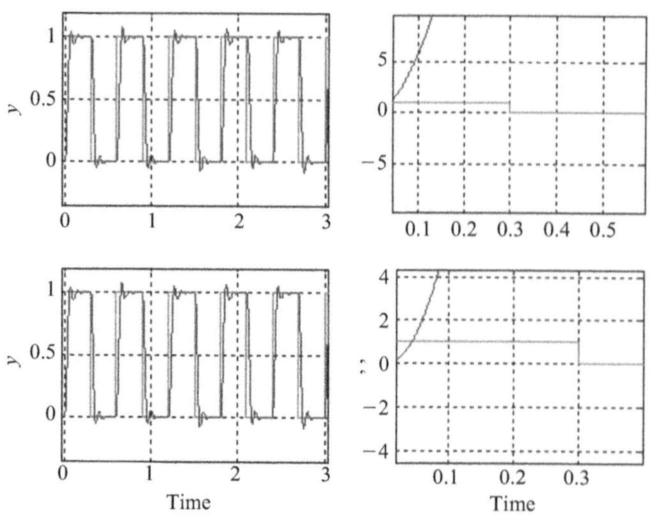

Figure 8.25 Simulation case of networked control system with 1,500 bytes packet size: 1-Switched Ethernet (at left) 10% bandwidth occupation of interference, 2-CAN (in top at right) 10% of occupation and first priority configuration is applied, (in bottom at right) 10% of occupation and second priority configuration is applied

point-to-point wired communication links using a small number of sensors and actuators to regulate appropriate process variables at desired values. While this paradigm to process control has been successful, chemical plant operation could substantially benefit from an efficient integration of the existing, point-to-point control networks (wired connections from each actuator or sensor to the control system using dedicated local area networks) with additional networked (wired or wireless) actuator or sensor devices that have become cheap and easy to install. Such an augmentation in sensor information, actuation capability and network-based availability of wired and wireless data is now well underway in the process industry and clearly has the potential to dramatically improve the ability of the single-process and plant-wide model-based control systems to optimize process and plant performance. Network-based communication allows for easy modification of the control strategy by rerouting signals, having redundant systems that can be activated automatically when component failure occurs, and in general, it allows having a high-level supervisory control over the entire plant. However, augmenting existing control networks with real-time wired or wireless sensor and actuator networks challenges many of the assumptions made in the development of traditional process control methods dealing with dynamical systems linked through ideal channels with flawless, continuous communication. In the context of networked control systems, key issues that need to be carefully handled at the control system design level include data losses due to field interference and time delays due to network traffic as well as due to the potentially heterogeneous nature of the additional measurements. In the context of control system architectures, augmenting dedicated, local control systems with control systems that utilize real-time sensor and actuator networks gives rise to the need to coordinate separate control systems that operate on a process. However, the rigorous design of cooperative, distributed control architectures for non-linear processes is a challenging task that cannot be addressed with traditional process control methods dealing with the design of centralized control systems. To design cooperative, DCSs, key fundamental issues that need to be addressed include the design of the individual control systems and of their communication strategy so that they efficiently cooperate in achieving the closed-loop plant objectives. Motivated by the above, this book presents general methods for the design of networked and distributed predictive control systems, accompanied by their application to non-linear process networks.

8.10 Networked and distributed control architectures

To provide concrete motivation for the control problems addressed in this book, we discuss below the general concept of networked and distributed process control using block diagrams and a chemical process example.

8.10.1 Networked control architectures

Traditionally, the different components (i.e., sensor, controller, and actuator) in a control system are connected via wired, point-to-point links, and the control laws are designed and operate based on local continuously sampled process output measurements. For a system with multiple control loops, the controllers, in general, are designed to work in a decentralized fashion. Figure 8.26 shows a traditional control system with two control loops. In Figure 8.26, two local control systems (i.e., LCS 1 and LCS 2) are designed based on two different continuously sampled outputs, y_1 and y_2, of the system. The two controllers do not exchange information and operate in a decentralized fashion.

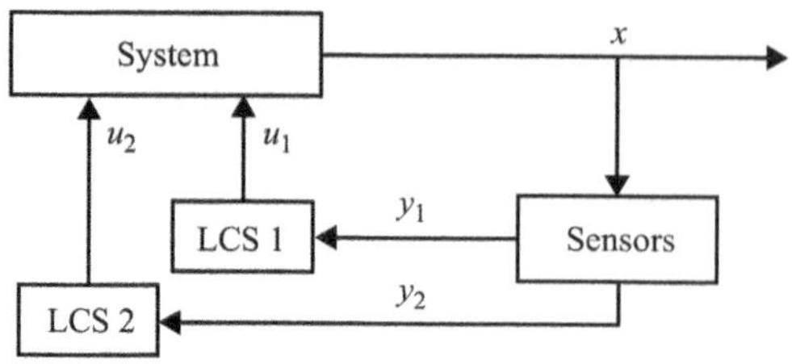

Figure 8.26 A traditional control system with two control loops.

Traditionally, the different components (i.e., sensor, controller, and actuator) in a control system are connected via wired, point-to-point links, and the control laws are designed and operate based on local continuously sampled process output measurements. For a system with multiple control loops, the controllers, in general, are designed to work in a decentralized fashion. Figure 8.26 shows a traditional control system with two control loops. In Figure 8.26, two local control systems (i.e., LCS 1 and LCS 2) are designed based on two different continuously sampled outputs, y_1 and y_2, of the system. The two controllers do not exchange information and operate in a decentralized fashion a system that previously was difficult or impossible to access because of physical or economical reasons may be now available via networked devices like, for example, networked sensors deployed over chemical plants. The additional information may be used to improve the closed-loop performance and the fault tolerance of a control system. However, because of the nature of the additional sensing (e.g., concentration vs. temperature measurements) and the fact that this information is collected and transmitted through real-time wired or wireless networks, a control system should also be able to handle heterogeneous (e.g., continuous, asynchronous, and delayed) measurements. To take advantages of the use of networks in the transmission of information and to use the additional information provided by networked devices, one approach is to design an NCS that takes data losses, delays, and heterogeneous measurements explicitly into account to replace the local control loops. Figure 8.27 shows this kind of NCS design for the system shown in Figure 8.26. In Figure 8.27, an NCS is designed to replace the two local controllers in Figure 8.26 taking into account all the available measurements (i.e., originally available measurements y_1, y_2, and additional measurement y_3). The key issues in the design of such an NCS include the handling of data losses, time-varying delays, and the utilization of heterogeneous measurements.

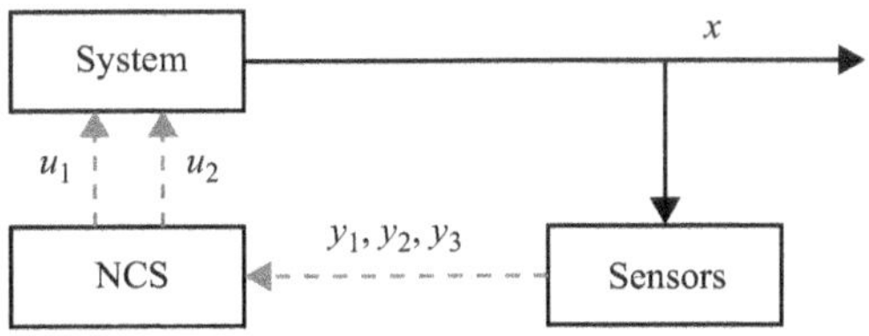

Figure 8.27 A networked configuration for the system shown in Figure 8.26 (dashed lines denote measurements and control actions transmitted via real-time communication networks)

Instead of replacing the local control loops, an alternative to the above networked control configuration is to design an NCS to augment the local control loops to take advantage of the additional measurements to manipulate additional control inputs or adjust the control actions of the existing local controllers to improve the closed-loop performance. The networked control configuration resulting in this case is shown in Figure 8.28. The main question is how to design the NCS to maintain the closed-loop stability achieved by the local controllers while improving the closed-loop performance.

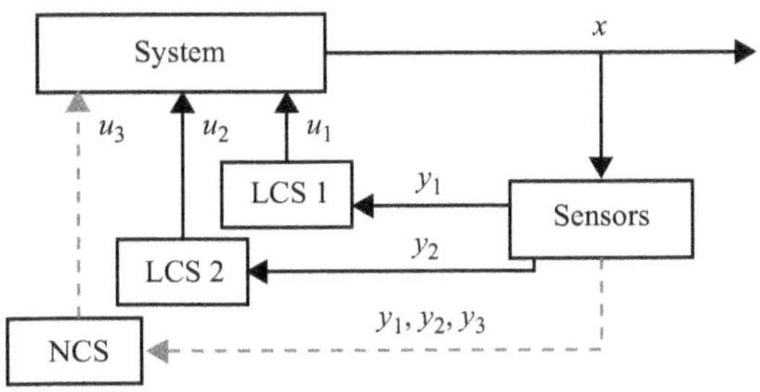

Figure 8.28 An alternative networked configuration for the system shown in Figure 8.26 (dashed lines denote measurements and control actions transmitted via real-time communication networks)

8.10.2 Cooperative, distributed control architectures

Consider the second networked control configuration shown in Figure 8.28. In this configuration, there is no communication between the networked controller and the two local controllers. In this sense, the three controllers work in a decentralized fashion. When the local controllers are designed via classical (e.g., proportional-integral-derivative (PID) control), geometric or Lyapunov-based control methods for which an explicit formula for the calculation of the control action is available, and the networked controller is designed via model-based control methods, like model predictive control (MPC), the coupling between the networked controller and the local controllers may be taken into account if the networked controller is carefully designed. However, when the local controllers are designed via MPC for which there is no explicit controller formula to calculate the future control actions, it is necessary to establish some, preferably small, communication between the different controllers so that they can coordinate their actions, which leads to the design of DCSs.

Figure 8.29 shows such a control configuration for the system shown in Figure 8.26. In this DCS, an LCS is designed to determine u_1 and u_2 and an NCS is designed to calculate u_3 based on all the information available via networks. To coordinate the control actions, the two controllers communicate to exchange information that could be future input trajectories the two controllers will apply or/and system measurements. In this case, we need to consider how the distributed controllers should communicate, what information they need to exchange, and how to coordinate their actions to achieve stability of the entire closed-loop system.

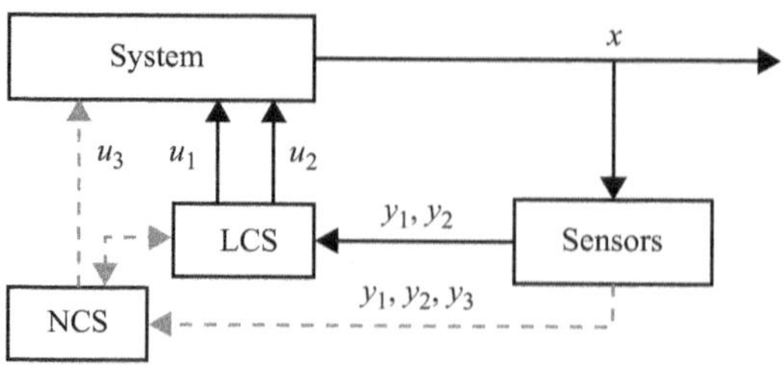

Figure 8.29 A distributed control configuration for the system shown in Figure 8.26 (dashed lines denote information transmitted via real-time communication networks)

In the distributed control configuration shown in Figure 8.29, the control inputs are distributed into the two controllers by their functionalities; that is, the LCS determines u_1 and u_2 to ensure the closed-loop stability, and the NCS determines u_3 to improve the closed-loop performance. An alternative to this kind of decomposition of the control inputs is to decompose the inputs spatially; that is, a distributed controller is designed for each control input (or each subsystem) as shown in Figure 8.30. In the distributed control configuration of Figure 8.30, three NCSs are designed to

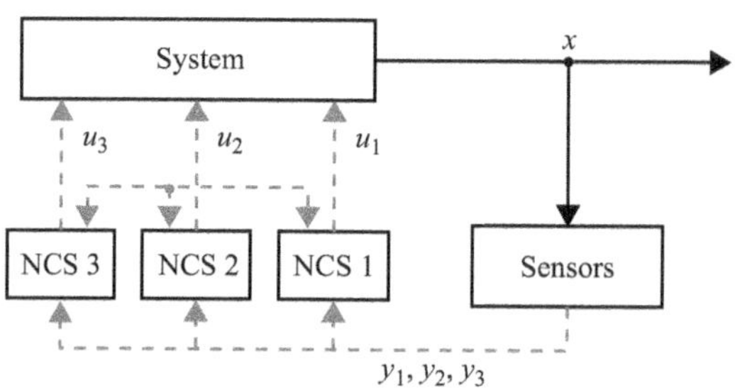

Figure 8.30 An alternative distributed control configuration for the system shown in Figure 8.26 (dashed lines denote information transmitted via real-time communication networks)

manipulate the three control inputs, respectively, based on all the available measurements. The three controllers communicate to coordinate their actions. This type of distributed control configuration is more flexible in the control loop selection compared with the one shown in Figure 8.29.

8.10.3 A reactor–separator process example

Consider a three-vessel, reactor–separator process consisting of two continuously stirred tank reactors (CSTRs) and a flash tank separator shown in Figure 8.31. A feed stream to the first CSTR F_{10} contains the reactant A that is converted into the desired product B. The desired product B can then further react into an undesired side product C. The effluent of the first CSTR along with additional fresh feed F_{20} makes up the inlet to the second CSTR. The reactions $A \rightarrow B$ and $B \rightarrow C$ take place in the two CSTRs in series before the effluent from CSTR 2 is fed to the flash tank. The overhead vapor from the flash tank is condensed and recycled to the first CSTR, and the bottom product stream is removed. A small portion of the overhead is purged before being recycled to the first CSTR.

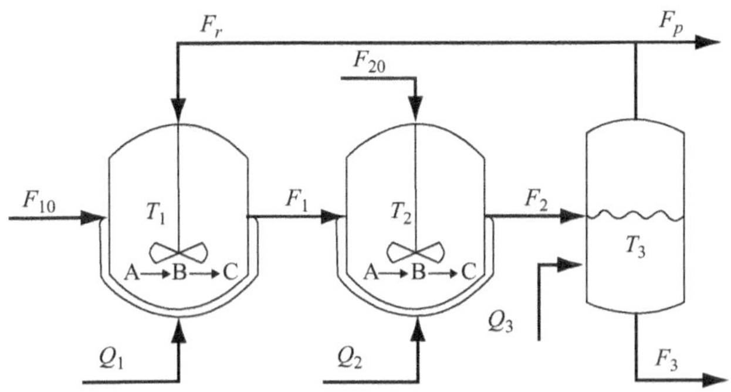

Figure 8.31 Reactor–separator process with recycle.

The control objective is to stabilize the process at a desired operating steady state and achieve an optimal level of closed-loop performance. To accomplish the control objective, we may design three local single loop controllers to manipulate the three heat inputs, Q_1, Q_2, Q_3, based on continuous temperature measurements of the three vessels. The three local controllers may be designed via PID control. This control configuration is shown in Figure 8.32, which is the common traditional local control system configuration for a process shown in Figure 8.31. This local control configuration corresponds to the control architecture shown in Figure 8.26.

In the reactor–separator process, the additional information that we have access to because of additionally deployed networked sensors could be the species concentration measurements of each component in the three vessels. These measurements are subject to sampling delays and network transmission data package dropouts and they may not be available at every sampling time. To use the additional information, we may design an NCS to replace the three local control loops. This networked

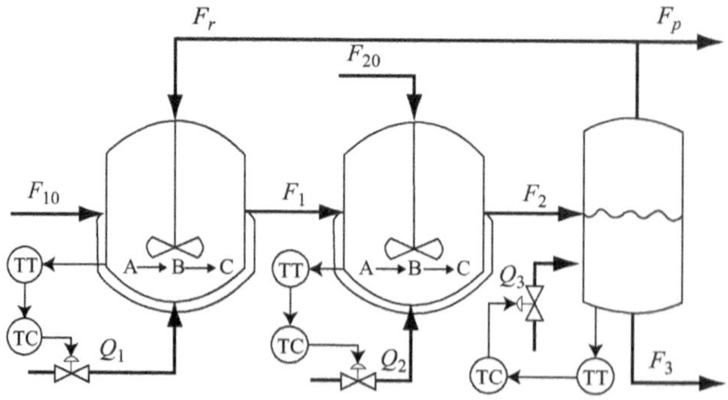

Figure 8.32 Local control configuration for the reactor–separator process.

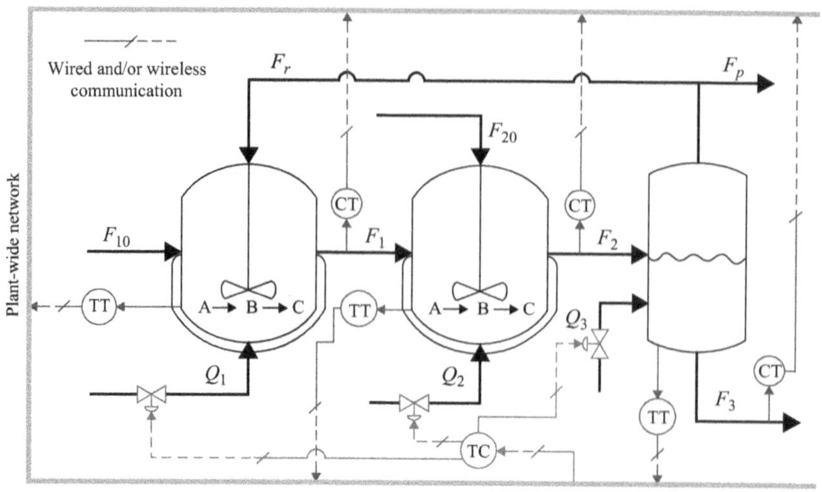

Figure 8.33 A networked control configuration for the reactor–separator process. In this configuration, a networked control system is designed to replace the three local control loops in the local control configuration

control configuration of the reactor–separator process is shown in Figure 8.33 that corresponds to the control architecture shown in Figure 8.27.

Instead of replacing the local control loops, an alternative to the above networked control configuration is to design an NCS to augment the local control loops to take advantage of the additional species concentration measurements as well as of the temperature measurements to adjust additional manipulated inputs, for instance, the feed flow rate to the second vessel, F_{20}. This networked control configuration of the reactor–separator process is shown in Figure 8.34 that corresponds to the control architecture shown in Figure 8.28.

Figure 8.35 shows a distributed control configuration for the reactor–separator process. In this design, two networked controllers are designed to manipulate the three heat inputs and the feed flow rate to vessel 2, respectively, and communicate through

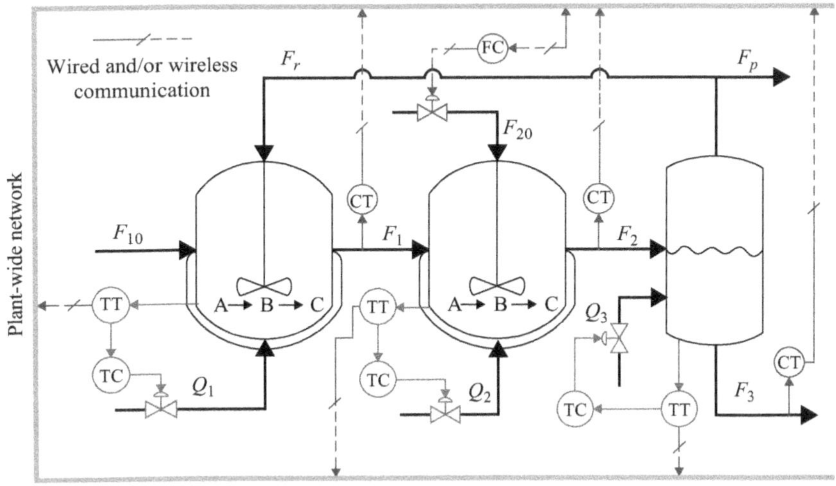

Figure 8.34 A networked control configuration for the reactor–separator process. In this configuration, a networked control system in addition to the three local controllers is designed to improve the closed-loop performance

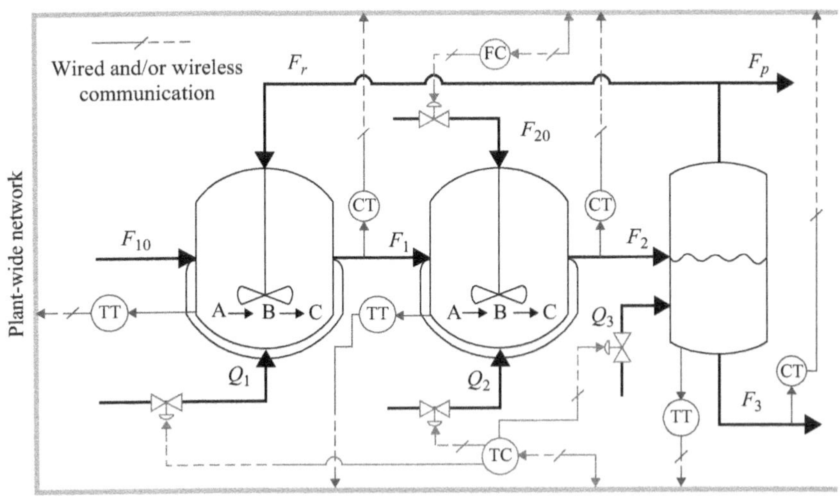

Figure 8.35 A distributed control configuration for the reactor–separator process. In this configuration, the two networked controllers communicate via the plant-wide network to coordinate their actions

the plant-wide network to exchange information and coordinate their actions. This control configuration corresponds to the one shown in Figure 8.29.

Figure 8.36 shows the alternative distributed control configuration corresponding to Figure 8.30 for the reactor–separator process. In this design, four networked controllers are designed to manipulate the four control inputs and communicate through the plant-wide network to exchange information and coordinate their actions.

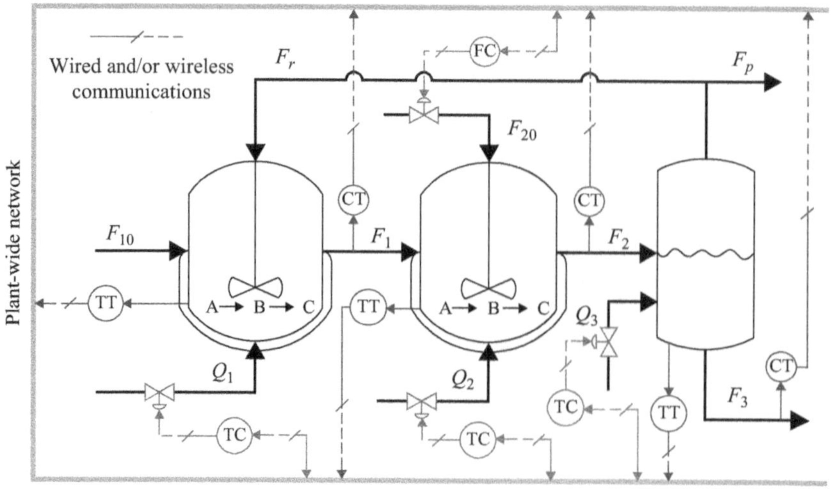

Figure 8.36 A distributed control configuration for the reactor–separator process. In this configuration, four networked controllers are designed to manipulate the four control inputs and communicate via the plant-wide network to coordinate their actions

8.11 Notes

This work proposes a reference model for DCSs based on MAS whose architecture is inspired in a hierarchical scheme. The use of MAS incorporates collaborative, organizational, and intelligent aspects that permit the control system to have emerging behavior. The use of MASINA methodology permitted the specification of the IDCSA model. The IDCSA model allows model DCS in a way that we can include intelligent components in our system. Two main aspects can be considered in our model: we can distribute the different control tasks among the control agents of IDCSA, and the intelligence is part of the control agents.

In this paper, we give, at first, an overview on the NCS research field.

And in order to show the impacts of network on the controlled systems, many simulation carried out with *TrueTime* are realized.

The obtained results show the behavior of NCS depending on the kind of network. Then NCS study has to integrate the specificities of networks such: MAC protocol, frame scheduling, and bandwidth packets size.

The limitation of CAN bandwidth doesn't enable to use it when the real-time process shares the same medium with other applications. In this case the use of Switched Ethernet architectures seems more suitable (due to high bandwidth).

In the future works we will use the prioritization packet procedures to compensate the delay due to the insertion of network in closed-loop control systems in the framework of Switched Ethernet network.

Chapter 9
Appendix

In appendix, we provide some mathematical notations, collect useful algebraic inequalities and lemmas, and present an introduction to some relevant topics that have been extensively used throughout the book.

9.1 Notations

Throughout the different chapters, we denote by $\Re$ the field of real numbers and $\Re^{m\times n}$, the set of $m \times n$, is the set of real matrices. Let $\Re_+$ stand for the nonnegative real numbers. Let $\mathbb{C}$ be the field of complex numbers and denote $\hat{\mathbb{C}} = \{s \in \mathbb{C} : \Re(s) < 0\}$, $\mathbb{C} = \{s \in \mathbb{C} : \mathbf{Re}(s) \leq 0\}$.

The transpose of a vector x and a matrix M is denoted by x^T and M^T, respectively. A matrix $M \in \Re^{n\times n}$ is symmetric if $M = M^T$.

- Let $M \in \Re^{m\times n}$, then $M[i:j, k:l]$ denotes a matrix of dimension $(j-i+1) \times (l-k+1)$ obtained by extracting rows i to j and columns k to l from the matrix M, with $m \geq j \geq i \geq 1, n \geq k \geq l \geq 1$.
- I_m denotes the identity matrix of dimension m, $I_m \in \Re^{m\times m}$.
- Let $\lambda_i(M)$ denote the ith eigenvalue of $M \in \Re^{n\times n}$, $i = 1, \ldots, n$. The spectrum of M will be denoted by $\mathcal{S}(M) = \{\lambda_1(M), \ldots, \lambda_n(M)\}$.

Definition 9.1: Let matrices $A, B \in \Re^{n\times n}$ be symmetric.

- A is positive definite if $x^TAx > 0$ for all nonzero $x \in \Re^n$, and A is positive semi-definite if $x^TAx \geq 0$ for all nonzero $x \in \Re^n$. We denote this by $A > 0$ and $A \geq 0$, respectively.
- A is negative (semi)definite if $-A$ is positive (semi)definite.
- $A < B$ and $A \leq B$ mean $A - B < 0$ and $A - B \leq 0$, respectively.

Definition 9.2: A matrix $M \in \Re^{n\times n}$ is called Hurwitz (or stable) if all its eigenvalues have negative real part, that is, $\lambda_i(M) \in \hat{\mathbb{C}}, i = 1, \ldots, n$.

9.1.1 Kronecker products

Let $A \in \Re^{m\times n}$ and $B \in \Re^{p\times q}$. Then $A \otimes B$ denotes the *Kronecker product* of A and B and given by

$$A \otimes B = \begin{bmatrix} a_{11}B & \cdots & a_{1n}B \\ \vdots & \ddots & \vdots \\ a_{m1}B & \cdots & a_{mn}B \end{bmatrix} \in \Re^{mp\times nq} \tag{9.1}$$

Proposition 9.1: *Consider two matrices $A = \alpha I_n$ and $B \in \Re^{n\times n}$. Then $\lambda_i(A+B) = \alpha + \lambda_i(B), i = 1, \ldots, n$.*

Proof. Take any eigenvalue $\lambda_i(B)$ and the corresponding eigenvector $v_i \in \mathbb{C}^n$. Then $(A+B)v_i = Av_i + Bv_i = \alpha v_i + \lambda_i v_i = (\alpha + \lambda_i)v_i$. □

Proposition 9.2: *Given $A, C \in \Re^{m\times m}$ and $B \in \Re^{n\times n}$, consider two matrices $\bar{A} = I_n \otimes A$ and $\bar{C} = B \otimes C$, where $\bar{A}, \bar{C} \in \Re^{nm\times nm}$. Then $\mathcal{S}(A + \lambda_i(B)C)$, where $\lambda_i(B)$ is the ith eigenvalue of B.*

Proof. Let $v \in \mathbb{C}^n$ be an eigenvector of B corresponding to $\lambda(B)$, and $u \in \mathbb{C}^m$ be an eigenvector of $M = (A + \lambda(B)C)$ with $\lambda(M)$ as the associated eigenvalue. Consider the vector $v \otimes u \in \mathbb{C}^{nm}$. Then

$$\begin{aligned} (\bar{A} + \bar{C})(v \otimes u) &= v \otimes Au + Bv \otimes Cu \\ &= v \otimes Au + \lambda(B)v \otimes Cu \\ &= v \otimes (Au + \lambda(B)Cu) \end{aligned}$$

Since $(A + \lambda(B)C)u = \lambda(M)u$, we get $(\bar{A} + \bar{C})(v \otimes u) = \lambda M(v \otimes u)$. □

In the sequel, Ker(**M**) will be used to denote the orthogonal complement of **M**. A block-diagonal matrix with submatrices $\mathbf{X}_1, \mathbf{X}_2, \ldots, \mathbf{X}_p$ on its diagonal is denoted by $\text{diag}\{\mathbf{X}_1, \mathbf{X}_2, \ldots, \mathbf{X}_p\}$. We use $\mathbf{S}^{n\times n}$ to denote the real, symmetric $n \times n$ matrices, and $\mathbf{S}_+^{n\times n}$ to denote positive definite matrices. If $\mathbf{M} \in \mathbf{S}^{n\times n}$, then $\mathbf{M} > 0$ $(M \geq 0)$ indicates that **M** is positive definite (positive semidefinite), and $\mathbf{M} < 0$, $M \leq 0$ denotes a negative definite (negative semidefinite) matrix. Given a symmetric matrix **A**, its inertia is defined as $In(\mathbf{A}) = (\pi(\mathbf{A}), \nu(\mathbf{A}), \delta(\mathbf{A}))$ are the number of positive, negative, and zero eigenvalues of **A**. For a temporal and spatial variable $x(t,s)$, **T** and **S** are the temporal and spatial forward shift operators and defined as

$$\mathbf{Tx}(t,s) = \mathbf{x}(t+1,s), \quad \mathbf{Sx}(t,s) = \mathbf{x}(t,s+1)$$

With slight abuse of notation, the symbols **T** and **S** are also used to denote the time and spatial differential operators for continuous interconnected systems. The space of square summable functions is denoted by $\mathcal{L}_2$, that is, for any $u \in \mathcal{L}_2$, $\|u\|_2 := [\sum_{t=0}^{\infty}\sum_{s=-\infty}^{\infty} \mathbf{u}^T(t,s)\mathbf{u}(t,s)]^{1/2}$ is finite.

9.1.2 Matrix lemmas

The first lemma relates the inertia of a nonsingular symmetric matrix $\mathbf{N}$ on a subspace with the inertia of its inverse $\mathbf{N}^{-1}$ on the complementary subspace. It can be utilized to derive the dual formulation from its original test. For its proof, readers are referred to [491].

Lemma 9.1: *Suppose that N is symmetric and nonsingular, and $\mathcal{S}$ is a negative subspace of $\mathbf{N}$ with maximal dimension. Then*

$$N^{-1} > 0 \text{ on } \mathcal{S}_{\perp}$$

The second result is basically an extension of the well-known elimination lemma [492,493] to quadratic matrix inequality. It is convenient for eliminating controller parameters from the synthesis conditions. Its proof follows closely to the one in [493,494,505], and is omitted here for lack of space.

Lemma 9.2: *Assume a nonsingular symmetric matrix*

$$\begin{bmatrix} \mathbf{Q} & \mathbf{S} \\ \bullet & \mathbf{R} \end{bmatrix}$$

has its inverse

$$\begin{bmatrix} \tilde{\mathbf{Q}} & \tilde{\mathbf{S}} \\ \bullet & \tilde{\mathbf{R}} \end{bmatrix}$$

with $In(\tilde{\mathbf{Q}}) = In(-\mathbf{R})$ and $\delta(\mathbf{R}) = 0$. The quadratic inequality

$$[\mathbf{I} \quad \mathbf{B}^T\mathbf{X}^T\mathbf{A} + \mathbf{C}^T] \begin{bmatrix} \mathbf{Q} & \mathbf{S} \\ \bullet & \mathbf{R} \end{bmatrix} \begin{bmatrix} \mathbf{I} \\ \mathbf{A}^T\mathbf{X}\mathbf{B} + \mathbf{C} \end{bmatrix} < 0 \tag{9.2}$$

has a solution $\mathbf{X}$ if and only if

$$[\mathbf{B}_{\perp}^T \quad \mathbf{B}_{\perp}^T\mathbf{C}^T] \begin{bmatrix} \mathbf{Q} & \mathbf{S} \\ \bullet & \mathbf{R} \end{bmatrix} \begin{bmatrix} \mathbf{B}_{\perp} \\ \mathbf{C}\mathbf{B}_{\perp} \end{bmatrix} < 0 \tag{9.3}$$

and

$$[-\mathbf{A}_{\perp}^T\mathbf{C} \quad \mathbf{A}_{\perp}^T] \begin{bmatrix} \tilde{\mathbf{Q}} & \tilde{\mathbf{S}} \\ \tilde{\mathbf{S}}^T & \tilde{\mathbf{R}} \end{bmatrix} \begin{bmatrix} -\mathbf{C}^T\mathbf{A}_{\perp} \\ \mathbf{A}_{\perp} \end{bmatrix} > 0 \tag{9.4}$$

9.2 Elements of graph theory

In this section, some preliminary knowledge of graph theory [316,501] is introduced for the analysis throughout the book.

Let $G(V,E,A)$ be a directed graph of order n, where $V = \{s_1, \ldots, s_n\}$ is the set of nodes, $E \subseteq V \times V$ is the set of edges, and $A = [a_{ij}] \in \Re^{n\times n}$ is a weighted adjacency matrix. The node indexes belong to a finite index set $I = \{1, 2, \ldots, n\}$. An edge of G is denoted by $e_{ij} = (s_i, s_j)$, where the first element s_i of the e_{ij} is said to be the tail of the edge and the other s_j to be the head. The adjacency elements associated with the edges are positive, that is $e_{ij} \in E \Leftrightarrow a_{ij} > 0$. If a directed graph has the property that $a_{ij} = a_{ji}$ for any $i, j \in I$, the directed graph is called undirected. The Laplacian with the directed graph is defined as $L = \Delta - A \in \Re^{n\times n}$, where $\Delta = [\Delta_{ij}]$ is a diagonal matrix with $\Delta_{ii} = \sum_{j=1}^{n} a_{ij}$. An important fact of L is that all the row sums of L are zero and thus 1 is an eigenvector of L associated with the zero eigenvalue. The set of neighbors of node s_i is denoted by $N_i = \{s_j \in V : (s_i, s_j) \in E\}$. A directed path is a sequence of ordered edges of the form $(s_{i1}, s_{i2}), (s_{i2}, s_{i3}), \ldots,$ where $s_{ij} \in V$ in a directed graph. A directed graph is said to be strongly connected, if there is a directed path from every node to every other node. Moreover, a directed graph is said to have spanning trees, if there exists a node such that there is a directed path from every other node to this node.

9.2.1 Basic results

Lemma 9.3 [248]: *If the graph G has a spanning tree, then its Laplacian L has the following properties:*

- *Zero is a simple eigenvalue of L, and* 1_n *is the corresponding eigenvector, that is* $L1_n = 0$.
- *The rest* $n-1$ *eigenvalues all have positive real parts. In particular, if the graph G is undirected, then all these eigenvalues are positive and real.*

Lemma 9.4 [330]: *Consider a directed graph G. Let* $D \in \Re^{n\times|E|}$ *be the 01-matrix with rows and columns indexed by the nodes and edges of G, and* $E \in \Re^{|E|\times n}$ *be the 01-matrix with rows and columns indexed by the edges and nodes of G, such that*

$$D_{uf} = \begin{cases} 1 & \text{if the node } u \text{ is the tail of the edge } f \\ 0 & \text{otherwise} \end{cases}$$

$$E_{uf} = \begin{cases} 1 & \text{if the node } u \text{ is the head of the edge } f \\ 0 & \text{otherwise} \end{cases}$$

where $|E|$ *is the number of the edges. Let* $Q = diag\{q_1, q_2, \ldots, q_{|E|}\}$, *where* $q_p (p = 1, \ldots, |E|)$ *is the weight of the pth edge of G (i.e., the value of the adjacency matrix on the pth edge). Then the Laplacian of G can be transformed into* $L = DQ(D^T - E)$.

9.2.2 Laplacian spectrum of graphs

This section is a concise review of the relationship between the eigenvalues of a Laplacian matrix and the topology of the associated graph. We refer the reader to [11,63,499,500] for a comprehensive treatment of the topic. We list a collection of properties associated with undirected graph Laplacians and adjacency matrices, which will be used in subsequent sections of the paper.

A graph $\mathcal{G}$ is defined as

$$\mathcal{G} = (\mathcal{V}, \mathcal{A}) \tag{9.5}$$

where $\mathcal{V}$ is the set of nodes (or vertices) $\mathcal{V} = \{1, \ldots, N\}$ and $\mathcal{A} \subseteq \mathcal{V} \times \mathcal{V}$ the set of edges (i,j) with $i \in \mathcal{V}, j \in \mathcal{V}$. The degree d_j of a graph vertex j is the number of edges that start from j. Let $d_{\max}(\mathcal{G})$ denote the maximum vertex degree of the graph $\mathcal{G}$.

9.2.3 Properties of adjacency matrix

We denote $\mathbf{A}(\mathcal{G})$ by the (0, 1) adjacency matrix of the graph $\mathcal{G}$. Let $\mathbf{A}_{ij} \in \mathbb{R}$ be its i,j element, then $\mathbf{A}_{i,i} = 0$, $\forall i = 1, \ldots, N$, $\mathbf{A}_{i,j} = 0$ if $(i,j) \notin \mathcal{A}$ and $\mathbf{A}_{i,j} = 1$ if $(i,j) \in \mathcal{A}$, $\forall i,j = 1, \ldots, N$, $i \neq j$. We will focus on *undirected* graphs, for which the adjacency matrix is symmetric.

Let $\mathcal{S}(\mathbf{A}(\mathcal{G})) = \{\lambda_1(\mathbf{A}(\mathcal{G})), \ldots, \lambda_N(\mathbf{A}(\mathcal{G}))\}$ be the spectrum of the adjacency matrix associated with an undirected graph $\mathcal{G}$ arranged in nondecreasing semi-order.

- *Property 1:* $\lambda_N(\mathbf{A}(\mathcal{G})) \leq d_{\max}(\mathcal{G})$.

This property together with Proposition 3.1 implies

- *Property 2:* $\gamma_i \geq 0, \forall \gamma_i \in \mathcal{S}(d_{\max} I_N - \mathbf{A})$.

We define the Laplacian matrix of a graph $\mathcal{G}$ in the following way:

$$L(\mathcal{G}) = \mathbf{D}(\mathcal{G}) - \mathbf{A}(\mathcal{G}) \tag{9.6}$$

where $\mathbf{D}(\mathcal{G})$ is the diagonal matrix of vertex degrees d_i (also called the valence matrix). Eigenvalues of Laplacian matrices have been widely studied by graph theorists. Their properties are strongly related to the structural properties of their associated graphs. Every Laplacian matrix is a singular matrix. By Gershgorin's theorem [298], the real part of each nonzero eigenvalue of $L(\mathcal{G})$ is strictly positive.

For undirected graphs, $L(\mathcal{G})$ is a symmetric, positive semidefinite matrix, which has only real eigenvalues. Let $\mathcal{S}(L(\mathcal{G})) = \{\lambda_1(L(\mathcal{G})), \ldots, \lambda_N(L(\mathcal{G}))\}$ be the spectrum of the Laplacian matrix L associated with an undirected graph $\mathcal{G}$ arranged in nondecreasing semi-order. Then

- *Property 3:*
  - $\lambda_1(L(\mathcal{G})) = 0$ with corresponding eigenvector of all ones, and $\lambda_2(L(\mathcal{G}))$ iff $\mathcal{G}$ is connected. In fact, the multiplicity of 0 as an eigenvalue of $L(\mathcal{G})$ is equal to the number of connected components of $\mathcal{G}$.
  - The modulus of $\lambda_i(L(\mathcal{G}))$, $i = 1, \ldots, N$ is less then N.

The second smallest Laplacian eigenvalue $\lambda_2(L(\mathcal{G}))$ of graphs is probably the most important information contained in the spectrum of a graph. This eigenvalue, called the algebraic connectivity of the graph, is related to several important graph invariants, and it has been extensively investigated.

Let $L(\mathcal{G})$ be the Laplacian of a graph $\mathcal{G}$ with N vertices and with maximal vertex degree $d_{\max}(\mathcal{G})$. Then properties of $\lambda_2(L(\mathcal{G}))$ include

- *Property 4:*
  - $\lambda_2(L(\mathcal{G})) \leq (N/(N-1)) \min\{d(v), v \in \mathcal{V}\}$
  - $\lambda_2(L(\mathcal{G})) \leq v(\mathcal{G}) \leq \eta(\mathcal{G})$
  - $\lambda_2(L(\mathcal{G})) \geq 2\eta(\mathcal{G})(1 - \cos(\pi/N))$
  - $\lambda_2(L(\mathcal{G})) \geq 2(\cos\frac{\pi}{N} - \cos 2\frac{\pi}{N})\eta(\mathcal{G}) - 2\cos\frac{\pi}{N}(1 - \cos\frac{\pi}{N})d_{\max}(\mathcal{G})$

where $v(\mathcal{G})$ is the vertex connectivity of the graph $\mathcal{G}$ (the size of a smallest set of vertices whose removal renders $\mathcal{G}$ disconnected) and $\eta(\mathcal{G})$ is the edge connectivity of the graph $\mathcal{G}$ (the size of a smallest set of edges whose removal renders $\mathcal{G}$ disconnected) [299].

Further relationships between the graph topology and Laplacian eigenvalue locations are discussed in [295] for undirected graphs. Spectral characterization of Laplacian matrices for directed graphs can be found in [298].

A lemma about Laplacian L associated with a balanced digraph G is given hereafter.

Lemma 9.5: *If G is balanced, then there exists a unitary matrix:*

$$V = \begin{pmatrix} \frac{1}{\sqrt{n}} & * & \cdots & * \\ \frac{1}{\sqrt{n}} & * & \cdots & * \\ \vdots & \vdots & & \vdots \\ \frac{1}{\sqrt{n}} & * & \cdots & * \end{pmatrix} \in C^{m\times n} \tag{9.7}$$

such that

$$V^*LV = \begin{pmatrix} 0 & \\ & H \end{pmatrix} = \Lambda \in C^{n\times n}$$
$$H \in C^{(n-1)\times(n-1)} \tag{9.8}$$

Moreover, if G has a globally reachable node, $H + H^*$ is positive definite.

Proof. Let $V = [\zeta_1, \zeta_2, \ldots, \zeta_n]$ be a unitary matrix where $\zeta_i \in C^n (i = 1, \ldots, n)$ are the column vectors of V and

$$\zeta_1 = (1/\sqrt{n})1 = (1/\sqrt{n}, 1/\sqrt{n}, \ldots, 1/\sqrt{n})^T$$

Notice that if G is balanced, it implies that $\zeta_1^* L = 0$. Then we have

$$\begin{aligned} V^*LV &= V^*L[\zeta_1, \zeta_2, \ldots, \zeta_n] \\ &= \begin{pmatrix} \zeta_1^* \\ \zeta_2^* \\ \vdots \\ \zeta_n^* \end{pmatrix} [0_n, L\zeta_2, \ldots, L\zeta_n] \\ &= \begin{pmatrix} 0 & 0_{n-2}^T \\ \bullet & H \end{pmatrix} \end{aligned}$$

Furthermore, if G has a globally reachable node, then $L + L^T$ is positive semidefinite, see Theorem 7 in [247]. Hence, $V^*(L + L^T)V$ is also positive semidefinite.

From Lemma 5.4, zero is a simple eigenvalue of L and, therefore, $H + H^*$ is positive definite. □

9.3 Minimum mean square estimate

Given a random variable Y that depends on another random variable X. Of interest is the *minimum mean square error estimate* (MMSEE) which, simply stated, is $\hat{X}$ the estimate of X such that the mean square error given by

$$\mathbb{E}[X - \hat{X}]^2$$

is minimized where the expectation is taken over the random variables X and Y. One of the standard results is given below:

Proposition 9.3: *The minimum mean square error estimate is given by the conditional expectation* $[X|Y = y]$.

Proof. Consider the functional form of the estimator as $g(Y)$. Let $f_{X,Y}(x,y)$ denote the joint probability density function of X and Y. Then the cost function is given by

$$\begin{aligned} C := \mathbb{E}[X - \hat{X}]^2 &= \int_x \int_y (x - g(y))^2 f_{X,Y}(x,y)dx\, dy \\ &= \int_y dy f_Y(y) \int_x (x - g(y))^2 f_{X|Y}(x \mid y)dx \end{aligned}$$

Taking the derivative of the cost function with respect to the function $g(y)$

$$\begin{aligned} \frac{\partial C}{\partial g(y)} &= \int_y dy\, f_Y(y) \int_x 2(x - g(y)) f_{X|Y}(x \mid y)dx \\ &= 2\int_y dy\, f_Y(y) \left(g(y) - \int_x x f_{X|Y}(x \mid y)dx \right) \\ &= 2\int_y dy\, f_Y(y)(g(y) - \mathbb{E}[X \mid Y = y]) \end{aligned}$$

Therefore, the only stationary point is $g(y) = [X \mid Y = y]$ and can be easily verified that it is a minimum. □

Remark 9.1: *It is noted that the result established in Proposition 9.3 holds for vector random variables as well. Observe that MMSE estimates are important because for Gaussian variables, they coincide with the Maximum likelihood (ML) estimates. It is a standard result that for Gaussian variables, the MMSE estimate is linear in the state value.*

In what follows, we will assume zero mean values for all the random variables with R_X being the covariance of X and R_{XY} being the cross-covariance between X and Y.

Proposition 9.4: *The best linear MMSE estimate of X given $Y = y$ is*

$$\hat{x} = R_{XY}R_Y^{-1}y$$

with the error covariance

$$P = R_X - R_{XY}R_Y^{-1}R_{YX}$$

Proof. Let the estimate be $\hat{x} = Ky$. Then the error covariance is

$$\begin{aligned} P &:= \mathbb{E}[(x - Ky)(x - Ky)^T \\ &= R_X - KR_{YX} - R_{XY}K^t + KR_YK^T \end{aligned}$$

Differentiating P with respect to K and setting it equal to zero yields

$$-2R_{XY} + 2KR_Y^{-1}$$

The result follows immediately. □

Extending Proposition 9.3 to the case of linear measurements $y = Hx + v$, we have the following standard result.

Proposition 9.5: *Let $y = Hx + v$, where H is a constant matrix and v is a zero mean Gaussian noise with covariance R_V independent of X. Then the MMSE estimate of X given $Y = y$ is*

$$\hat{x} = R_XH^T(HR_XH^T + R_V)^{-1}y$$

with the corresponding error covariance

$$P = R_X - R_XH^T(HR_XH^T + R_V)^{-1}HR_X$$

9.4 Stability notions

In this section, we present some definitions and results pertaining to stability of dynamical systems. A detailed account of this topic can be found in [495].

Definition 9.3: A function of x and t is a *carathedory function* if, for all $t \in \Re$, it is continuous in x and for all $x \in \Re^n$, it is Lebesgue measurable in t.

9.4.1 Practical stabilizability

Given the uncertain dynamical system

$$\begin{aligned} \dot{x}(t) &= [A + \Delta A(r) + M]x(t) + [B + \Delta B(s)]u(t) \\ &\qquad + Cv(t) + H(t, x, r), x(0) = x_o \end{aligned} \tag{9.9}$$

$$y(t) = x(t) + w(t) \tag{9.10}$$

where $x \in \Re^n, u \in \Re^m, y = \in \Re^n, v \in \Re^s, w \in \Re^n$ are the state, control, measured state, disturbance, and measurement error of the system, respectively, and $r \in \Re^p$,

$s \in \Re^q$ are the uncertainty vectors. System (9.9) and (9.10) is said to be *practically stabilizable* if, given $\mathbf{d} > 0$, there is a control law $g(\cdot,\cdot) : \Re^m \times \Re \to \Re^m$, for which, given any admissible uncertainties r, s, disturbances $w \in \Re^n, v \in \Re^s$, any initial time $t_o \in \Re$, and any initial state $x_o \in \Re^n$, the following five conditions hold:

1. *The closed-loop system*

$$\begin{aligned}\dot{x}(t) &= [A + \Delta A(r) + M]x(t) + [B + \Delta B(s)]g(y,t) \\ &\quad + Cv(t) + H(t,x,r) \end{aligned} \tag{9.11}$$

   possesses a solution $x(\cdot) : [t_o, t_1] \to \Re^n, x(t_o) = x_o$
2. *Given any* $v > 0$ *and any solution* $x(\cdot) : [t_o, t_1] \to \Re^n, x(t_o) = x_o$ *of system (9.11) with* $||x_o|| \leq v$, *there is a constant* $d(v) > 0$ *such that* $||x(t)|| \leq d(v), \forall t \in [t_o, t_1]$.
3. *Every solution* $x(\cdot) : [t_o, t_1] \to \Re^n$ *can be continued over* $[t_o, \infty)$.
4. *Given any* $\bar{d} \geq d$, *any* $v > 0$ *and solution* $x(\cdot) : [t_o, t_1] \to \Re^n, x(t_o) = x_o$ *of system (9.11) with* $||x_o|| \leq v$, *there exists a finite time* $T(\bar{d}, v) < \infty$, *possibly dependent on* v *but not on* t_o, *such that* $\|x(t)\| \leq \bar{d}, \forall t \geq t_o + T(\bar{d}, v)$.
5. *Given any* $\bar{d} \geq d$ *and any solution* $x(\cdot) : [t_o, t_1] \to \Re^n, x(t_o) = x_o$ *of system (9.11) there is a constant* $\delta(\bar{d}) > 0$ *such that* $\|x(t_o)\| \leq \delta\bar{d}$ *implies* $\|x(t)\| \leq \bar{d}, \forall t \geq t_o$.

9.4.2 Razumikhin stability

A continuous function $\alpha : [0, a) \mapsto [0, \infty)$ is said to belong to class K if it is strictly increasing and $\alpha(0) = 0$. Further, it is said to belong to class K_∞ if $a = \infty$ and $\lim_{r\to\infty} \alpha(r) = \infty$.

Consider a time-delay system

$$\dot{x}(t) = f(t, x(t - d(t)) \tag{9.12}$$

with an initial condition

$$x(t) = (t), \quad t \in [-\bar{d}, 0]$$

where the function vector $f : \Re^+ \times \mathcal{C}_{[-\bar{d},0]} \mapsto \Re^n$ takes $\mathcal{R}\times$ (bounded sets of $\mathcal{C}_{[-\bar{d},0]}$) into bounded sets in $\Re^n$; $d(t)$ is the time-varying delay and $\bar{d} := \sup_{t\in\Re^+}\{d(t)\} < \infty$. The symbol $\mathcal{C}_{[a,b]}$ represents the set of $\Re^n$-valued continuous function on $[a, b]$.

Lemma 9.6: *If there exist class* $\mathcal{K}_\infty$ *functions* $\zeta_1(\cdot)$ *and* $\zeta_2(\cdot)$, *a class* $\mathcal{K}$ *function* $\zeta_3(\cdot)$ *and a function* $V_1(\cdot) : [-\bar{d}, \infty] \times \Re^n \mapsto \Re^+$ *satisfying*

$$\zeta_1(||x||) \leq V_1(t,x) \leq \zeta_2(||x||), \quad t \in \Re^+, \ x \in \Re^n$$

such that the time derivative of V_1 *along the solution of the system (9.12) satisfies*

$$\dot{V}_1(t,x) \leq -\zeta_3(||x||) \quad \text{if} \quad V_1(t+d, x(t+d)) \leq V_1(t, x(t)) \tag{9.13}$$

for any $d \in [-\bar{d}, 0]$, *then system (9.12) is uniformly stable. If in addition*

$$\zeta_3(\tau) > 0, \quad \tau > 0$$

and there exists a continuous nondecreasing function $\xi(\tau) > 0, \tau > 0$ *such that (9.13) is strengthened to*

$$\dot{V}_1(t,x) \leq -\zeta_3(||x||) \quad \text{if} \quad V_1(t+d, x(t+d)) \leq \xi(V_1(t, x(t))) \tag{9.14}$$

for any $d \in [-\bar{d}, 0]$, *then system (9.12) is uniformly asymptotically stable. Further, if in addition,* $\lim_{\tau\to\infty} \zeta_1(\tau) = \infty$, *then system (9.12) is globally uniformly asymptotically stable.*

The proof of this lemma can be found in [30].

Lemma 9.7: *Consider system (9.12). If there exists a function*

$$V_o(x) = x^t P x, \quad P > 0$$

such that for $d \in [-\bar{d}, 0]$ *the time derivative of* V_o *along the solution of system (9.12) satisfies*

$$\dot{V}_o(t,x) \leq -q_1 \|x\|^2 \quad \text{if} \quad V_o(x(t+d)) \leq q_2 V_o(x(t)) \tag{9.15}$$

for some constants $q_1 > 0$ *and* $q_2 > 1$, *then the system (9.12) is globally uniformly asymptotically stable.*

Proof. Since $P > 0$, it is clear that

$$\lambda_{\min}(P)\|x\|^2 \leq V_o(x) \leq \lambda_{\max}(P)\|x\|^2$$

Let $\zeta_1(\tau) = \lambda_{\min}(P)\tau^2$ and $\zeta_2(\tau) = \lambda_{\max}(P)\tau^2$. It is easy to see that both $\zeta_1(\cdot)$ and $\zeta_2(\cdot)$ are class $\mathcal{K}_\infty$ functions and

$$\zeta_1(\|x\|) \leq V_0(x) \leq \zeta_2(\|x\|), \quad x\Re^n$$

Further, let $\zeta_3(\cdot) = -q_1\tau^2$ and $\xi(\tau) = q_2\tau$. It is evident from $q_1 > 0$ and $q_2 > 1$ that for $\tau > 0$.

$$\xi(\tau) > \text{ and } \zeta_3(\tau) > 0$$

Hence, the conclusion follows from (9.15). □

9.5 Basic inequalities

All mathematical inequalities are proved for completeness. They are termed facts due to their high frequency of usage in the analytical developments.

9.5.1 Schur complements

Given a matrix Ω composed of constant matrices Ω_1, Ω_2, Ω_3, where $\Omega_1 = \Omega_1^t$ and $0 < \Omega_2 = \Omega_2^t$ as follows

$$\Omega = \begin{bmatrix} \Omega_1 & \Omega_3 \\ \Omega_3^t & \Omega_2 \end{bmatrix}$$

We have the following results:

(A) $\Omega \geq 0$ if and only if either

$$\begin{cases} \Omega_2 \geq 0 \\ \Pi = \Upsilon \Omega_2 \\ \Omega_1 - \Upsilon \Omega_2 \Upsilon^t \geq 0 \end{cases} \tag{9.16}$$

or

$$\begin{cases} \Omega_1 \geq 0 \\ \Pi = \Omega_1 \Lambda \\ \Omega_2 - \Lambda^t \Omega_1 \Lambda \geq 0 \end{cases} \tag{9.17}$$

hold where Λ, Υ are some matrices of compatible dimensions.

(B) $\Omega > 0$ if and only if either

$$\begin{cases} \Omega_2 > 0 \\ \Omega_1 - \Omega_3 \Omega_2^{-1} \Omega_3^t > 0 \end{cases}$$

or

$$\begin{cases} \Omega_1 \geq 0 \\ \Omega_2 - \Omega_3^t \Omega_1^{-1} \Omega_3 > 0 \end{cases}$$

hold where Λ, Υ are some matrices of compatible dimensions.

In this regard, matrix $\Omega_3 \, \Omega_2^{-1} \Omega_3^t$ is often called the Schur complement $\Omega_1(\Omega_2)$ in Ω.

Proof. **(A)** To prove (9.16), we first note that $\Omega_2 \geq 0$ is necessary. Let $z^t = [z_1^t z_2^t]$ be a vector partitioned in accordance with Ω. Thus we have

$$z^t \Omega z = z_1^t \Omega_1 z_1 + 2 z_1^t \Omega_3 z_2 + z_2^t \Omega_2 z_2 \tag{9.18}$$

Select z_2 such that $\Omega_2 z_2 = 0$. If $\Omega_3 z_2 \neq 0$, let $z_1 = -\pi \Omega_3 z_2, \pi > 0$. Then it follows that

$$z^t \Omega z = \pi^2 z_2^t \Omega_3^t \Omega_1 \Omega_3 z_2 - 2\pi z_2^t \Omega_3^t \Omega_3 z_2$$

which is negative for a sufficiently small $\pi > 0$. We thus conclude $\Omega_1 z_2 = 0$ that then leads to $\Omega_3 z_2 = 0, \forall z_2$ and consequently

$$\Omega_3 = \Upsilon \Omega_2 \tag{9.19}$$

for some Υ.

Since $\Omega \geq 0$, the quadratic term $z^t \Omega z$ possesses a minimum over z_2 for any z_1. By differentiating $z^t \Omega z$ from (9.18) with respect to z_2^t, we get

$$\frac{\partial (z^t \Omega z)}{\partial z_2^t} = 2\Omega_3^t z_1 + 2\Omega_2 z_2 = 2\Omega_2 \Upsilon^t z_1 + 2\Omega_2 z_2$$

Setting the derivative to zero yields

$$\Omega_2 \Upsilon z_1 = -\Omega_2 z_2 \tag{9.20}$$

Using (9.19) and (9.20) in (9.18), it follows that the minimum of $z^t\Omega z$ over z_2 for any z_1 is given by

$$\min_{z_2} z^t\Omega z = z_1^t[\Omega_1 - \Upsilon\Omega_2\Upsilon^t]z_1$$

which prove the necessity of $\Omega_1 - \Upsilon\Omega_2\Upsilon^t \geq 0$.

On the other hand, we note that the conditions (9.16) are necessary for $\Omega \geq 0$ and since together they imply that the minimum of $z^t\Omega z$ over z_2 for any z_1 is nonnegative, they are also sufficient.

Using similar argument, conditions (9.17) can be derived as those of (9.16) by starting with Ω_1.

The proof of **(B)** follows as direct corollary of **(A)**. □

9.5.2 Bounding inequalities

Given matrices $0 < \mathcal{Q}^t = \mathcal{Q}, \mathcal{P} = \mathcal{P}^t$, then it follows that

$$-\mathcal{P}\mathcal{Q}^{-1}\mathcal{P} \leq -2\mathcal{P} + \mathcal{Q} \tag{9.21}$$

This can be easily established by considering the algebraic inequality

$$(\mathcal{P} - \mathcal{Q})^t\mathcal{Q}^{-1}(\mathcal{P} - \mathcal{Q}) \geq 0$$

and expanding to get

$$\mathcal{P}\mathcal{Q}^{-1}\mathcal{P} - 2\mathcal{P} + \mathcal{Q} \geq 0 \tag{9.22}$$

which when manipulating, yields (9.21). An important special case is obtained when $\mathcal{P} \equiv I$, that is

$$-\mathcal{Q}^{-1} \leq -2I + \mathcal{Q} \tag{9.23}$$

This inequality proves useful when using Schur complements to eliminate the quantity $\mathcal{Q}^{-1}$ from the diagonal of an LMI without alleviating additional math operations.

Lemma 9.8: *The matrix inequality*

$$-\Lambda + S\Omega^{-1}S^t < 0 \tag{9.24}$$

holds for some $0 < \Omega = \Omega^t \in \Re^{n\times n}$, *if and only if*

$$\begin{bmatrix} -\Lambda & S\mathcal{X} \\ \bullet & -\mathcal{X} - \mathcal{X}^t + \mathcal{Z} \end{bmatrix} < 0 \tag{9.25}$$

holds for some matrices $\mathcal{X} \in \Re^{n\times n}$ *and* $\mathcal{Z} \in \Re^{n\times n}$.

Proof. (⇒) By Schur complements, inequality (9.24) is equivalent to

$$\begin{bmatrix} -\Lambda & S\Omega^{-1} \\ \bullet & -\Omega^{-1} \end{bmatrix} < 0 \tag{9.26}$$

Setting $\mathcal{X} = \mathcal{X}^t = \mathcal{X} = \Omega^{-1}$, we readily obtain inequality (9.25).

($\Leftarrow$) Since the matrix $[I \quad S]$ is of full rank, we obtain

$$\begin{bmatrix} I \\ S^t \end{bmatrix}^t \begin{bmatrix} -\Lambda & S\mathcal{X} \\ \bullet & -\mathcal{X} - \mathcal{X}^t + \mathcal{Z} \end{bmatrix} \begin{bmatrix} I \\ S^t \end{bmatrix} < 0$$

$$\Leftrightarrow \ -\Lambda + S\mathcal{Z}S^t < 0 \Leftrightarrow -\Lambda + S\Omega^{-1}S^t < 0, \mathcal{Z} = \Omega^{-1} \tag{9.27}$$

which completes the proof. □

Lemma 9.9: *The matrix inequality*

$$A\mathcal{P} + \mathcal{P}A^t + D^t\mathcal{R}^{-1}D + \mathcal{M} < 0 \tag{9.28}$$

holds for some $0 < \mathcal{P} = \mathcal{P}^t \in \Re^{n\times n}$,, *if and only if*

$$\begin{bmatrix} A\mathcal{V} + \mathcal{V}^tA^t + \mathcal{M} & \mathcal{P} + A\mathcal{W} - \mathcal{V} & D^t\mathcal{R} \\ \bullet & -\mathcal{W} - \mathcal{W}^t & 0 \\ \bullet & \bullet & -\mathcal{R} \end{bmatrix} < 0 \tag{9.29}$$

holds for some $\mathcal{V} \in \Re^{n\times n}$ *and* $\mathcal{W} \in \Re^{n\times n}$.

Proof. ($\Rightarrow$) By Schur complements, inequality (9.28) is equivalent to

$$\begin{bmatrix} A\mathcal{P} + \mathcal{P}A^t + \mathcal{M} & D^t\mathcal{R} \\ \bullet & -\mathcal{R} \end{bmatrix} < 0 \tag{9.30}$$

Setting $\mathcal{V} = \mathcal{V}^t = P, \mathcal{W} = \mathcal{W}^t = \mathcal{R}$, it follows from Lemma 9.8 with Schur complements that there exists $\mathcal{P} > 0$, $\mathcal{V}$, $\mathcal{W}$, such that inequality (9.29) holds.

($\Leftarrow$) In a similar way, Schur complements to inequality (9.29) imply that

$$\begin{bmatrix} A\mathcal{V} + \mathcal{V}^tA^t + \mathcal{M} & \mathcal{P} + A\mathcal{W} - \mathcal{V} & D^t\mathcal{R} \\ \bullet & -\mathcal{W} - \mathcal{W}^t & 0 \\ \bullet & \bullet & -\mathcal{R} \end{bmatrix} < 0$$

$$\Leftrightarrow \begin{bmatrix} I \\ A \end{bmatrix} \begin{bmatrix} A\mathcal{V} + \mathcal{V}^tA^t + \mathcal{M} + D^t\mathcal{P}^{-1}D & \mathcal{P} + A\mathcal{W} - \mathcal{V} \\ \bullet & -\mathcal{W} - \mathcal{W}^t \end{bmatrix} \begin{bmatrix} I \\ A \end{bmatrix}^t < 0$$

$$\Leftrightarrow A\mathcal{P} + \mathcal{P}A^t + D^t\mathcal{P}^{-1}D + \mathcal{M} < 0, \mathcal{V} = \mathcal{V}^t \tag{9.31}$$

which completes the proof. □

Lemma 9.10: *Let* $0 < L = L^t$ *and* X, Y *be given matrices with appropriate dimensions. Then it follows that the inequality*

$$L(z) + X(z)PY(z) + Y^t(z)P^tX^t(z) > 0 \tag{9.32}$$

holds for some P *and* $z = z_o$ *if and only if the following inequalities*

$$X_*^t(z)L(z)X_*(z) > 0, \quad Y_*^t(z)L(z)Y_*(z) > 0 \tag{9.33}$$

hold with $z = z_o$.

It is significant to observe that feasibility of matrix inequality (9.32) with variables P and z is equivalent to the feasibility of (9.33) with variable z and thus the matrix variable P has been eliminated from (9.32) to form (9.33). Using Finsler's lemma, we can express (9.33) in the form

$$L(z) - \beta X(z)X^t(z) > 0, \quad L(z) - \beta Y(z)Y^t(z) > 0 \tag{9.34}$$

for some $\beta \in \Re$.

The following is a statement of the reciprocal projection lemma.

Lemma 9.11: *Let $P > 0$ be a given matrix. The following statements are equivalent:*

- $\mathcal{M} + Z + Z^t < 0$
- *The LMI problem*

$$\begin{bmatrix} \mathcal{M} + P - (V + V^t) & V^t + Z^t \\ V + Z & -P \end{bmatrix} < 0$$

is feasible with respect to the general matrix V.

The following lemmas show how to produce equivalent LMIs by elimination procedure.

Lemma 9.12: *There exists $\mathcal{X}$ such that*

$$\begin{bmatrix} \mathcal{P} & \mathcal{Q} & \mathcal{X} \\ \bullet & \mathcal{R} & \mathcal{Z} \\ \bullet & \bullet & \mathcal{S} \end{bmatrix} > 0 \tag{9.35}$$

if and only if

$$\begin{bmatrix} \mathcal{P} & \mathcal{Q} \\ \bullet & \mathcal{R} \end{bmatrix} > 0, \quad \begin{bmatrix} \mathcal{R} & \mathcal{Z} \\ \bullet & \mathcal{S} \end{bmatrix} > 0 \tag{9.36}$$

Proof. Since LMIs (9.36) form sub-blocks on the principal diagonal of LMI (9.35), necessity is established. To show sufficiency, apply the congruence transformation

$$\begin{bmatrix} I & 0 & 0 \\ \bullet & I & 0 \\ 0 & -V^t R^{-1} & I \end{bmatrix}$$

to LMI (9.35), it is evident that (9.35) is equivalent to

$$\begin{bmatrix} \mathcal{P} & \mathcal{Q} & \mathcal{X} - \mathcal{Q}\mathcal{R}^{-1}\mathcal{Z} \\ \bullet & \mathcal{R} & 0 \\ \bullet & \bullet & \mathcal{S} - \mathcal{Z}^t\mathcal{R}^{-1}\mathcal{Z} \end{bmatrix} > 0 \tag{9.37}$$

Clearly (9.36) is satisfied for $\mathcal{X} = \mathcal{Q}\mathcal{R}^{-1}\mathcal{Z}$ if (9.36) is satisfied in view of Schur complements. □

Lemma 9.13: *There exists* $\mathcal{X}$ *such that*

$$\begin{bmatrix} \mathcal{P} & \mathcal{Q} + \mathcal{X}\mathcal{G} & \mathcal{X} \\ \bullet & \mathcal{R} & \mathcal{Z} \\ \bullet & \bullet & \mathcal{S} \end{bmatrix} > 0 \tag{9.38}$$

if and only if

$$\begin{bmatrix} \mathcal{P} & \mathcal{Q} \\ \bullet & \mathcal{R} - \mathcal{V}\mathcal{G} - \mathcal{G}^t\mathcal{V}^t + \mathcal{G}^t\mathcal{Z}\mathcal{G} \end{bmatrix} > 0$$
$$\begin{bmatrix} \mathcal{R} - \mathcal{V}\mathcal{G} - \mathcal{G}^t\mathcal{V}^t + \mathcal{G}^t\mathcal{Z}\mathcal{G} & \mathcal{V} - \mathcal{G}^t\mathcal{Z} \\ \bullet & \mathcal{Z} \end{bmatrix} > 0 \tag{9.39}$$

Proof. Applying the congruence transformation

$$\begin{bmatrix} I & 0 & 0 \\ 0 & I & 0 \\ 0 & -\mathcal{G} & I \end{bmatrix}$$

to LMI (9.38) and using Lemma 9.12, we readily obtain the results. □

Lemma 9.14: *There exists* $0 < \mathcal{X}^t = \mathcal{X}$ *such that*

$$\begin{bmatrix} \mathcal{P}_a + \mathcal{X} & \mathcal{Q}_a \\ \bullet & \mathcal{R}_a \end{bmatrix} > 0$$
$$\begin{bmatrix} \mathcal{P}_c - \mathcal{X} & \mathcal{Q}_c \\ \bullet & \mathcal{R}_c \end{bmatrix} > 0 \tag{9.40}$$

if and only if

$$\begin{bmatrix} \mathcal{P}_a + \mathcal{P}_c & \mathcal{Q}_a & \mathcal{Q}_c \\ \bullet & \mathcal{R}_a & 0 \\ \bullet & \bullet & \mathcal{R}_c \end{bmatrix} > 0 \tag{9.41}$$

Proof. It is obvious from Schur complements that LMI (9.40) is equivalent to

$$\mathcal{R}_a > 0, \mathcal{R}_c > 0$$
$$\Xi = \mathcal{P}_a + \mathcal{P}_c - \mathcal{Q}_a\mathcal{R}_a^{-1}\mathcal{Q}_a^t - \mathcal{Q}_c\mathcal{R}_c^{-1}\mathcal{Q}_c^t > 0 \tag{9.42}$$

On the other hand, LMI (9.40) is equivalent to

$$\mathcal{R}_a > 0, \quad \mathcal{R}_c > 0$$
$$\Xi_a = \mathcal{P}_a + \mathcal{X} - \mathcal{Q}_a\mathcal{R}_a^{-1}\mathcal{Q}_a^t > 0 \tag{9.43}$$
$$\Xi_c = \mathcal{P}_c - \mathcal{X} - \mathcal{Q}_c\mathcal{R}_c^{-1}\mathcal{Q}_c^t > 0$$

It is readily evident from (9.42) and (9.43) that $\Xi = \Xi_a + \Xi_c$ and hence the existence of $\mathcal{X}$ satisfying (9.43) implies (9.42). By the same token, if (9.42) is satisfied, $\mathcal{X} = \mathcal{Q}_a\mathcal{R}_a^{-1}\mathcal{Q}_a^t - \mathcal{P}_a - \frac{1}{2}\Xi$ yields $\Xi_a = \Xi_c = \Xi_a = \frac{1}{2}\Xi$ and (9.43) is satisfied. □

9.6 Linear matrix inequalities

It has been shown that a wide variety of problem arising in system and control theory can be conveniently reduced to a few standard convex or quasi-convex optimization problems involving linear matrix inequalities (LMIs). The resulting optimization problems can then be solved numerically very efficiently using commercially available interior-point methods.

9.6.1 Basics

One of the earliest LMIs arises in Lyapunov theory. It is well known that the differential equation

$$\dot{x}(t) = Ax(t) \tag{9.44}$$

has all of its trajectories converge to zero (stable) id and only if there exists a matrix $P > 0$ such that

$$A^tP + AP < 0 \tag{9.45}$$

This leads to the LMI formulation of stability, that is, *a linear time-invariant system is asymptotically stable if and only if there exists a matrix* $0 < P = P^t$ *satisfying the LMIs*

$$A^tP + AP < 0, \quad P > 0$$

Given a vector variable $x \in \Re^n$ and a set of matrices $0 < G_j = G_j^t \in \Re^{n\times n}$, $j = 0, \ldots, p$, then a basic compact formulation of a linear matrix inequality is

$$G(x) := G_0 + \sum_{j=1}^{p} x_j G_j > 0 \tag{9.46}$$

Note that (9.46) implies that $v^tG(x)v > 0 \; \forall 0 \neq v \in \Re^n$. More importantly, the set $\{x \mid G(x) > 0$ is convex. Non-linear (convex) inequalities are converted to LMI form using Schur complements in the sense that

$$\begin{bmatrix} Q(x) & S(x) \\ \bullet & R(x) \end{bmatrix} > 0 \tag{9.47}$$

where $Q(x) = Q^t(x), R(x) = R^t(x), S(x)$ depend affinely on x, is equivalent to

$$R(x) > 0, \quad Q(x) - S(x)R^{-1}(x)S^t(x) > 0 \tag{9.48}$$

More generally, the constraint

$$Tr[S^t(x)P^{\;1}(x)S(x)] < 1, \quad P(x) > 0$$

where $P(x) = P^t(x) \in \Re^{n\times n}, S(x) \in \Re^{n\times p}$ depend affinely on x, is handled by introducing a new (slack) matrix variable $Y(x) = Y^t(x) \in\in \Re^{p\times p}$ and the LMI (in x and Y):

$$TrY < 1, \quad \begin{bmatrix} Y & S(x) \\ \bullet & P(x) \end{bmatrix} > 0 \tag{9.49}$$

Most of the time, our LMI variables are matrices. It should be clear from the foregoing discussions that a quadratic matrix inequality (QMI) in the variable P can be readily expressed as linear matrix inequality (LMI) in the same variable.

9.6.2 Some standard problems

Here we provide some common convex problems that we encountered throughout the book. Given an LMI $G(x) > 0$, the corresponding LMI problem (LMIP) is to

> *find a feasible* $x \equiv x^f$ *such that* $G(x^f) > 0$, *or determine that the LMI is infeasible.*

It is obvious that this is a convex feasibility problem.

The generalized eigenvalue problem (GEVP) is to minimize the maximum generalized eigenvalue of a pair of matrices that depend affinely on a variable, subject to an LMI constraint. GEVP has the general form

$$\begin{aligned} &\textit{minimize } \lambda \\ &\textit{subject to } \lambda B(x) - A(x) > 0, \quad B(x) > 0 \\ &C(x) > 0 \end{aligned} \tag{9.50}$$

where A, B, C are symmetric matrices that are affine functions of x. Equivalently stated

$$\begin{aligned} &\textit{minimize } \lambda_M[A(x), B(x)] \\ &\textit{subject to } B(x) > 0, \quad C(x) > 0 \end{aligned} \tag{9.51}$$

where $\lambda_M[X, Y]$ denotes the largest generalized eigenvalue of the pencil $\lambda Y - X$ with $Y > 0$. This problem is quasi-convex optimization problem since the constraint is convex and the objective, $\lambda_M[A(x), B(x)]$ is quasi-convex.

The eigenvalue problem (EVP) is to minimize the maximum eigenvalue of a matrix that depends affinely on a variable, subject to an LMI constraint. EVP has the general form

$$\begin{aligned} &\textit{minimize } \lambda \\ &\textit{subject to } \lambda I - A(x) > 0, \quad B(x) > 0 \end{aligned} \tag{9.52}$$

where A, B are symmetric matrices that are affine functions of the optimization variable x. This is problem is convex optimization problem.

EVPs can appear in the equivalent form of minimizing a linear function subject to an LMI, that is

$$\begin{aligned} &\textit{minimize } c^t x \\ &\textit{subject to } G(x) > 0 \end{aligned} \tag{9.53}$$

where $G(x)$ is an affine function of x. Examples of $G(x)$ include

$$PA + A^tP + C^tC + \gamma^{-1}PBB^tP < 0, \quad P > 0$$

It should be stressed that the standard problems (LMIPs, GEVPs, EVPs) are tractable, from both theoretical and practical viewpoints:

- They can be solved in polynomial time.
- They can be solved in practice very efficiently using commercial softwares.

9.6.3 The S-procedure

In some design applications, we faced the constraint that some quadratic function be semi-negative whenever some other quadratic function is negative. In such cases, this constraint can be expressed as an LMI in the data variables defining the quadratic functions.

Let $G_o, \ldots, G_p$ be quadratic functions of the variable $\xi \in \Re^n$:

$$G_j(\xi) := \xi^t R_j \xi + 2u_j^t \xi + v_j, \quad j = 0, \ldots, p, R_j = R_j^t$$

We consider the following condition on $G_o, \ldots, G_p$:

$$G_o(\xi) \leq 0 \quad \forall \xi \quad \text{such that} \quad G_j(\xi) \geq 0, \quad j = 0, \ldots, p \tag{9.54}$$

It is readily evident that if there exist scalars $\omega_1 \geq 0, \ldots, \omega_p \geq 0$ such that

$$\forall \xi, \quad G_o(\xi) - \sum_{j=1}^{p} \omega_j G_j(\xi) \geq 0 \tag{9.55}$$

then inequality (9.54) holds. Observe that if the functions $G_o, \ldots, G_p$ are affine, then Farkas lemma states that (9.54) and (9.55) are equivalent. Interestingly enough, inequality (9.55) can written as

$$\begin{bmatrix} R_o & u_o \\ \bullet & v_o \end{bmatrix} - \sum_{j=1}^{p} \omega_j \begin{bmatrix} R_j & u_j \\ \bullet & v_j \end{bmatrix} \geq 0 \tag{9.56}$$

The foregoing discussions were stated for nonstrict inequalities. In case of strict inequality, we let $R_o, \ldots, R_p \in \Re^{n \times n}$ be symmetric matrices with the following qualifications:

$$\xi^t R_o \xi > 0 \quad \forall \xi \quad \text{such that} \quad \xi^t G_j \xi \geq 0, \quad j = 0, \ldots, p \tag{9.57}$$

Once again, it is obvious that there exist scalars $\omega_1 \geq 0, \ldots, \omega_p \geq 0$ such that

$$\forall \xi, \quad G_o(\xi) - \sum_{j=1}^{p} \omega_j G_j(\xi) > 0 \tag{9.58}$$

then inequality (9.57) holds. Observe that (9.58) is an LMI in the variables $R_o, \omega_1, \ldots, \omega_p$.

It should be remarked that the **S**-procedure deals with nonstrict inequalities allows the inclusion of constant and linear terms. In the strict version, only quadratic functions can be used.

9.7 Some formulas on matrix inverses

This concerns some useful formulas for inverting matrix expressions in terms of the inverses of its constituents.

9.7.1 Inverse of block matrices

Let A be a square matrix of appropriate dimension and partitioned in the form

$$A = \begin{bmatrix} A_1 & A_2 \\ A_3 & A_4 \end{bmatrix} \tag{9.59}$$

where both A_1 and A_4 are square matrices. If A_1 is invertible, then

$$\Delta_1 = A_4 - A_3 A_1^{-1} A_2$$

is called the Schur complement of A_1. Alternatively, if A_4 is invertible, then

$$\Delta_4 = A_1 - A_2 A_4^{-1} A_3$$

is called the Schur complement of A_4.

It is well known that matrix A is invertible if and only if either

$$A_1 \quad \textit{and} \quad \Delta_1 \quad \textit{are invertible}$$

or

$$A_4 \quad \textit{and} \quad \Delta_4 \quad \textit{are invertible}$$

Specifically, we have the following equivalent expressions

$$\begin{bmatrix} A_1 & A_2 \\ A_3 & A_4 \end{bmatrix}^{-1} = \begin{bmatrix} \Upsilon_1 & -A_1^{-1} A_2 \Delta_1^{-1} \\ -\Delta_1^{-1} A_3 A_1^{-1} & \Delta_1^{-1} \end{bmatrix} \tag{9.60}$$

or

$$\begin{bmatrix} A_1 & A_2 \\ A_3 & A_4 \end{bmatrix}^{-1} = \begin{bmatrix} \Delta_4^{-1} & -\Delta_4^{-1} A_2 A_4^{-1} \\ -A_4^{-1} A_3 \Delta_4^{-1} & \Upsilon_4 \end{bmatrix} \tag{9.61}$$

where

$$\begin{aligned} \Upsilon_1 &= A_1^{-1} + A_1^{-1} A_2 \Delta_1^{-1} A_3 A_1^{-1} \\ \Upsilon_4 &= A_4^{-1} + A_4^{-1} A_3 \Delta_4^{-1} A_2 A_4^{-1} \end{aligned} \tag{9.62}$$

Important special cases are

$$\begin{bmatrix} A_1 & 0 \\ A_3 & A_4 \end{bmatrix}^{-1} = \begin{bmatrix} A_1^{-1} & 0 \\ -A_4^{-1} A_3 A_1^{-1} & A_4^{-1} \end{bmatrix} \tag{9.63}$$

and

$$\begin{bmatrix} A_1 & A_2 \\ 0 & A_4 \end{bmatrix}^{-1} = \begin{bmatrix} A_1^{-1} & -A_1^{-1} A_2 A_4^{-1} \\ 0 & A_4^{-1} \end{bmatrix} \tag{9.64}$$

9.7.2 Matrix inversion lemma

Let $A \in \Re^{n\times n}$ and $C \in \Re^{m\times m}$ be nonsingular matrices. Then we have

$$[A + BCD]^{-1} = A^{-1} - A^{-1}B[DA^{-1}B + C^{-1}]^{-1}DA^{-1} \tag{9.65}$$

Proof. Consider the block matrix

$$S = \begin{bmatrix} A & B \\ C & D \end{bmatrix}$$

By performing the LDU and UDL decomposition of S and equating them, we get

$$\begin{bmatrix} I & 0 \\ CA^{-1} & 0 \end{bmatrix}\begin{bmatrix} A & 0 \\ 0 & D - CA^{-1}B \end{bmatrix}\begin{bmatrix} I & A^{-1}B \\ 0 & I \end{bmatrix}$$
$$= \begin{bmatrix} I & BD^{-1} \\ 0 & I \end{bmatrix}\begin{bmatrix} A - BD^{-1}C & 0 \\ 0 & D \end{bmatrix}\begin{bmatrix} I & 0 \\ D^{-1}C & I \end{bmatrix}$$

Inverting both sides yields

$$\begin{bmatrix} I & -A^{-1}B \\ 0 & I \end{bmatrix}\begin{bmatrix} A^{-1} & 0 \\ 0 & (D - CA^{-1}B)^{-1} \end{bmatrix}\begin{bmatrix} I & 0 \\ -CA^{-1} & 0 \end{bmatrix}$$
$$= \begin{bmatrix} I & 0 \\ -D^{-1}C & I \end{bmatrix}\begin{bmatrix} (A - BD^{-1}C)^{-1} & 0 \\ 0 & D^{-1} \end{bmatrix}\begin{bmatrix} I & -BD^{-1} \\ 0 & I \end{bmatrix}$$

Equating the (1,1) block shows

$$[A - BCD]^{-1} = A^{-1} + A^{-1}B[D - CA^{-1}B]^{-1}CA^{-1}$$

Substituting $C \to -D$ and $D \to C^{-1}$, we readily obtain (9.65). □

References

[1] Y. Sun and N. H. El-Farra, 'Quasi-decentralized Model-based Networked Control of Process Systems', *Computers and Chemical Engineering*, vol. 32, pp. 2016–2029, 2008.

[2] M. S. Mahmoud, *Decentralized Control and Filtering in Interconnected Dynamical Systems*, CRC Press, Taylor and Francis Group, New York, 2010.

[3] M. S. Mahmoud, 'Decentralized Stabilization of Interconnected Nonlinear Systems with Time-Varying Delays', *IEEE Trans. Automatic Control*, vol. 54, no. 11, pp. 2663–2668, 2009.

[4] H. Cuiand and E. Jacobsen, 'Performance Limitations in Decentralized Control', *J. Process Control*, vol. 12, pp. 485–494, 2002.

[5] M. Baldea, P. Daoutidis, and A. Kumar, 'Dynamics and Control of Integrated Networks with Purge Streams', *The American Institute of Chemical Engineers Journal*, vol. 52, pp. 1460–1472, 2006.

[6] Y. Sun and N. H. El-Farra, 'Resource-Aware Quasi-Decentralized Control of Nonlinear Plants over Communication Networks', *Proc. American Control Conference*, St. Louis, MO, vol. 1, pp. 154–159, 2009.

[7] A. T. Al-Hammouri1, M. S. Branicky, V. Liberatore, and S. M. Phillips, 'Decentralized and Dynamic Bandwidth Allocation in Networked Control Systems', *IEEE Int. Parallel & Distributed Processing Symposium*, Boston, Massachusetts, pp. 8–18, 2006.

[8] W. Zhang, M. S. Branicky, and S. M. Phillips, 'Stability of Networked Control Systems', *IEEE Control Systems Magazine*, pp. 88–99, 2001.

[9] H. Chan and U. Ozguner, 'Closed-Loop Control of Systems over a Communication Network with Queues', *Proc. American Control Conference*, vol. 1, pp. 811–815, 1995.

[10] R. W. Brennan and O. William, 'A Simulation Test-Bed to Evaluate Multi-Agent Control Of Manufacturing Systems', *Proc. 32nd Winter Simulation Conference*, Orlando, FL, pp. 1747–1756, 2000.

[11] B. Marieke, G. Cloosterman, N. Wouw, W. H. Heemels, and H. Nijmeijer, 'Stability of Networked Control Systems with Uncertain Time-Varying Delays', *IEEE Trans. Automatic Control*, vol. 54, pp. 1575–1580, 2009.

[12] Y. Halevi and A. Ray, 'Integrated Communication and Control Systems, Part I: Analysis', *J. Dynamic Systems, Measurements and Control*, vol. 110, pp. 367–373, 1988.

[13] J. Nilsson, *Real-Time Control Systems with Delays*, Ph.D. Dissertation, Dept. Automatic Control, Lund Inst. Technology, Sweden, 1998.

[14] P. Ogren, E. Fiorelli, and N. E. Leonard, 'Cooperative Control of Mobile Sensor Networks: Adaptive Gradient Climbing in a Distributed Environment', *IEEE Trans. Automatic Control*, vol. 49, pp. 1292–1302, 2004.

[15] C. Meng, T. Wang, W. Chou, S. Luan, Y. Zhang, and Z. Tian, 'Remote Surgery Case: Robot-assisted Tele Neuro-Surgery', *IEEE Int. Conference on Robotics and Automation*, vol. 1, pp. 819–823, 2004.

[16] J. P. Hespanha, M. L. McLaughlin, and G. Sukhatme, 'Haptic Collaboration over the Internet', *Proc. 5th Phantom Users Group Workshop*, Aurora, Colorado, 2000.

[17] K. Hikichi, H. Morino, I. Arimoto, K. Sezaki, and Y. Yasuda, 'The Evaluation of Delay Jitters for Haptics Collaboration over the Internet', *Proc. IEEE Global Telecommunication Conference*, vol. 2, pp. 1492–1496, 2002.

[18] S. Shirmohammadi and N. H. Woo, 'Evaluating Decorators for Haptic Collaboration over the Internet', *Proc. 3rd IEEE Int. Workshop Haptic, Audio and Visual Env. Application*, Ottawa, Ont., Canada, pp. 105–109, 2004.

[19] P. Seiler and R. Sengupta, 'Analysis of Communication Losses in Vehicle Control Problems', *Proc. American Control Conference*, vol. 2, pp. 1491–1496, 2001.

[20] P. Seiler and R. Sengupta, 'An $\mathcal{H}_2$ Approach to Networked Control', *IEEE Trans. Automatic Control*, vol. 50, pp. 356–364, 2005.

[21] A. Albert, 'Comparison of Event-Triggered and Time-Triggered Concepts With Regard To Distributed Control Systems', *Proc. Embedded World*, Nurnberg, Germany, pp. 235–252, 2004.

[22] J. Nilsson, B. Bernhardsson, and B. Wittenmark, 'Some Topics in Real-Time Control', *Proc. American Control Conference*, vol. 4, pp. 2386–2390, 1998.

[23] M. Torngren, 'Fundamentals of Implementing Real-Time Control Applications in Distributed Computer Systems', *Real-Time Systems*, vol. 14, pp. 219–250, 1998.

[24] J. Nilsson, B. Bernhardsson, and B. Wittenmark , 'Stochastic Analysis and Control of Real-Time Systems with Random Time Delays', *Automatica*, vol. 34, pp. 57–64, 1998.

[25] A. Mihael, C. Boldisor, and V. Comnac, 'Influence of Variable Time Delay on the Performances of Networked Control Systems', *Interdisciplinary in Engineering Scientific Bulletin*, PetruMaior University ofTirguMures, vol. 5, pp. 56–61, 2009.

[26] F. Lian, J. Moyne, and D. Tilbury, 'Network Design Consideration for Distributed Control Systems', *IEEE Trans. Control Systems Technology*, vol. 10, pp. 297–307, 2002.

[27] M. H. Behnam, *Flexible Scheduling for Real Time Control Systems based on Jitter Margin*, Master Thesis, Mälardalens Högskola, Västerås, 2005.

[28] B. Rahmani, A. Markazi, and N. Mozayani, 'Real Time Prediction of Time Delays in a Networked Control System', *Proc. Int. Symposium on*

Communications, Control and Signal Processing, St. Julian's, Malta, pp. 1242–1245, 2008.

[29] L. Schenato, 'Optimal Estimation in Networked Control Systems Subject to Random Delay and Packet Drop', *IEEE Trans. Automatic Control*, vol. 5, pp. 1311–1317, 2008.

[30] M. S. Mahmoud, *Decentralized Systems with Design Constraints*, Springer-Verlag, UK, 2011.

[31] Y. Guo, D. J. Hill, and Y. Wang, 'Nonlinear Decentralized Control of Large-Scale Power Systems', *Automatica*, vol. 36, pp. 1275–1289, 2000.

[32] D. D. Siljak and D. M. Stipanovic, 'Robust Stabilization of Nonlinear Systems: The LMI Approach', *Math Problems in Engineering*, vol. 6, pp. 461–493, 2000.

[33] X. G. Yan, C. Edwards, and S. K. Spurgeon, 'Decentralized Robust Sliding Mode Control for a Class of Nonlinear Interconnected Systems by Static Output Feedback', *Automatica*, vol. 40, pp. 113–120, 2004.

[34] S. Elloumi and N. B. Braiek, 'Robust Decentralized Control for Multi-Machine Power System: The LMI Approach', *Proc. Int. Conference on Systems, Man and Cybernetics*, Yasmine Hammamet, Tunisia, vol. 6, pp. 2013–2017, 2002.

[35] K. S. Narenda and N. O. Oleng, 'Exact Output Tracking in Decentralized Adaptive Control Systems', *IEEE Trans. Automatic Control*, vol. 47, pp. 390–395, 2002.

[36] R. Scattolini, 'Architectures for Distributed and Hierarchical Model Predictive Control—A Review', *J. Process Control*, vol. 19, pp. 723–731, 2009.

[37] A. Rantzer, 'Using Game Theory for Distributed Control Engineering', *Proc. 3rd World Congress of the Game Theory Society*, Evanston, Illinois, July 2008.

[38] A. S. Tlili and N. B. Braiek, 'Decentralized Observer-based Guaranteed Cost Control for Nonlinear Interconnected Systems', *Int. J. Control and Automation*, vol. 2, pp. 29–46, 2009.

[39] Y. Zhu and P. R. Pagilla, 'Decentralized Output Feedback Control of a Class of Large-Scale Interconnected Systems', *IMA J. Math Control and Information*, vol. 24, 57–69, 2007.

[40] Y. Sun and N. H. El-Farra, 'Quasi-Decentralized Networked Control of Process Systems', *Proc. American Control Conference*, Seattle, Washington, pp. 1507–1512, 2008.

[41] Y. Sun and N. H. El-Farra, 'Quasi-Decentralized Control of Process Systems Using Wireless Sensor Networks with Scheduled Sensor Transmissions', *Proc. American Control Conference*, pp. 3390–3396, 2009.

[42] Y. Sun and N. H. El-Farra, 'Quasi-decentralized Scheduled Output Feedback Control of Process Systems Using Wireless Sensor Networks', *IFAC Symposium on Advanced Control of Chemical Processes*, Koç University Campus, Turkey, 2009

[43] N. Bauer, M. Donkers, W. Heemels, and N. Wouw, 'An Approach to Observer-Based Decentralized Control under Periodic Protocols', *Proc. 29th Benelux*

Meeting on Systems and Control, Heeze, The Netherlands, pp. 2125–2131, 2010.

[44] J. Wei, 'Stability Analysis of Decentralized Networked Control Systems', *World Congress on Intelligent Control and Automation*, vol. 7, pp. 5477–5482, 2008.

[45] S. Yüksel and T. Basar, 'Communication Constraints for Decentralized Stabilizability with Time-Invariant Policies', *IEEE Trans. Automatic Control*, vol. 52, pp. 1060–1066, 2007.

[46] S. Yüksel and T. Basar, 'Optimal Signaling Policies for Decentralized Multi-controller Stabilizability Over Communication Channels', *IEEE Trans. Automatic Control*, vol. 52, pp. 1969–1974, 2007.

[47] A. S. Matveev and A. V. Savkin, 'Decentralized Stabilization of Linear Systems via Limited Capacity Communication Networks', *Proc. 44th IEEE Conference on Decision and Control, and the European Control Conference*, pp. 1155–1161, 2005.

[48] C. Zhang and G. E. Dullerud, 'Decentralized Control with Communication Bandwidth Constraints', *Proc. American Control Conference*, New York City, USA, pp. 1711–1716, 2007.

[49] A. S. Matveev and V. Savkin, 'Decentralized Stabilization of Networked Systems Under Data-Rate Constraints', *Proc. 17th World Congress*, COEX, Korea, pp. 8034–8038, 2008.

[50] S. Yüksel and T. Basar, 'Decentralized Multi-Controller Stabilizability over Communication Channels', *Proc. European Control Conference*, Kos, Greece, 2007.

[51] S. Yüksel, O. Imer, and T. Basar, 'Constrained State Estimation and Control over Communication Networks', *Proc. Conf. Information Sciences and Systems*, Princeton, NJ, pp. 409–414, 2004.

[52] J. B. Rawlings and B. T. Stewart, 'Coordinating Multiple Optimization-based Controllers: New Opportunities and Challenges', *J. Process Control*, vol. 18, pp. 839–845, 2008.

[53] S. Ohand and S. S. Sastry, 'Approximate Estimation of Distributed Networked Control Systems', *Proc. American Control Conference*, pp. 997–1002, 2007.

[54] Q. Zong, W. Liu, and X. He, 'Fault Diagnosis of Distributed Networked Control Systems with Stochastic Communication Logic', *Intelligent Control and Automation*, pp. 6877–6881, 2008.

[55] S. Oh and S. Sastry, 'Distributed Networked Control System with Lossy Links: State Estimation and Stabilizing Communication Control', *Proc. 45th IEEE Conference on Decision and Control*, pp. 1942–1947, 2006.

[56] X. W. Lemmon, 'Asymptotic Stability in Distributed Event-Triggered Networked Control Systems with Delays', *Proc. American Control Conference*, Baltimore, MD, pp. 1362–1367, 2010.

[57] X. W. Lemmon, 'Event-Triggered Broadcasting across Distributed Networked Control Systems', *Proc. American Control Conference*, Seatle, Washington, pp. 3139– 3144, 2008.

[58] T. Alpcan, L. Pavel, and N. Stefanovic, 'A Control Theoretic Approach to Non-cooperative Game Design', *Proc. 48th IEEE Conference on Decision and Control*, Shanghai, China, pp. 8575–8580, 2009.

[59] A. Gattami and B. Bernhardsson, 'Minimax Team Decision Problems', *Proc. American Control Conference*, New York City, USA, pp. 766–771, 2007.

[60] Y. Tipsuwan and M. Chow, 'Control Methodologies in Networked Control Systems', *Control Engineering Practice*, vol. 11, pp. 1099–1111, 2003.

[61] S. Mittal and A. S. Siddiqui, 'Networked Control System: Survey and Directions', *J. Engineering Research and Studies*, vol. 1, pp. 35–50, 2010.

[62] C. C. Chen, S. Hirche, and M. Buss, 'Sampled-Data Networked Control Systems with Random Time Delay', *Automatica*, vol. 40, pp. 1441–1446, 2004.

[63] Q. Ling and M. D. Lemon, 'Optimal Dropout Compensation in Networked Control Systems', *Proc. 42nd IEEE Conference on Decision and Control*, Maui, HI, vol. 1, pp. 670–675, 2003.

[64] C. Robinson and P. Kumar, 'Networked Control Systems with Packet Delays and Losses', *Proc. 47th IEEE Conference on Decision and Control*, Cancun, Mexico, vol. 5, pp. 4602–4607, 2008.

[65] P. A. Kawka and A. G. Alleyn, 'Stability and Feedback Control of Wireless Networked Systems', *Proc. American Control Conference*, Illinois Univ., Champaign, IL, vol. 4, pp. 2953–2959, 2005.

[66] K. Ji and W. Kim, 'Real-Time Control of Networked Control Systems via Ethernet', *Int. J. Control, Automation, and Systems*, vol. 3, pp. 591–600, 2005.

[67] N. Vatanskib, J. Georges, Ca. Aubrun, E. Rondeaua, and L. Jounelab, 'Control Compensation Based on Upper Bound Delay in Networked Control Systems', *Proc. 17th Int. Symposium on Mathematical Theory of Networks and Systems*, pp. 364–369, 2006.

[68] E. Witrant, P. Park, M. Johansson, C. Fischione, and K. Johansson, 'Predictive Control over Wireless Multi-Hop Networks', *IEEE Int. Conference Control Applications*, pp. 1037–1042, 2007.

[69] N. J. Ploplys, P. A. Kawka, and A. G. Alleyne, 'Closed Loop Control over Wireless Networks', *IEEE Control Systems Magazine*, vol. 24, pp. 58–71, 2004.

[70] C. Peng and D. Yue, 'State Feedback Controller Design of Networked Control Systems with Parameter Uncertainty and State-delay', *Asian J. Control*, vol. 8, pp. 385–392, 2006.

[71] A. Willig, K. Matheus, and A. Wolisz, 'Wireless Technology in Industrial Networks', *Proc IEEE*, vol. 93, pp. 1130–1151, 2005.

[72] M. Drew, X. Liu, A. Goldsmith, and K. Hedrick, 'Networked Control System Design over a Wireless LAN', *Proc. 44th IEEE Conference on Decision and Control, and the European Control Conference*, pp. 6704–6709, 2005.

[73] F. D. Pellegrini, D. Miorandi, S. Vitturi, and A. Zanella, 'On the Use of Wireless Networks at Low Level of Factory Automation Systems', *IEEE Trans. Industrial Informatics*, vol. 2, pp. 129–143, 2006.

[74] I. F. Akyildiz and I. H. Kasimoglu, 'Wireless Sensor and Actor Networks: Research Challenges', *Ad Hoc Networks Journal*, vol. 2, pp. 351–367, 2004.

[75] M. S. Hasan, H. Yu, and A. Carrington, 'Overview of Wireless Networked Control Systems over Mobile Ad-hoc Network', *Proc. 14th Int. Conference on Automation and Computing*, 2008.

[76] C. Perkins and E. M. Royer, 'Ad-Hoc On Demand Distance Vector Routing in WMCSA', *Proc. 2nd IEEE Workshop on Mobile Computing Systems and Applications*, pp. 415–425, 1999.

[77] D. Johnson and D. A. Maltz, 'Dynamic Source Routing in Ad Hoc Wireless Networks', *Mobile Computing*, pp. 153–181, 1996.

[78] B. Frankovic and I. Budinsk, 'Creation of Intelligent Distributed Control System Based on Multi-Agent Technology', *J. Electrical Engineering*, vol. 60, pp. 29–33, 2009.

[79] L. Sheng and Y. Pan, 'Distributed Control for Consensus of Networked Multi-Agent Robotic Systems with Time Delays', *Int. J. Information and Systems Science*, vol. 5, pp. 161–178, 2009.

[80] P. Bohner, 'A Multi-Agent Approach to Distributed Control for Task-level Programs for Cooperating Manipulators', *Proc. Int. Conference on Intelligent Robots and Systems*, vol. 1, pp. 1088–1094, 1994.

[81] W. B. Dunbar and R. M. Murray, 'Model Predictive Control of Coordinated Multi-Vehicle Formations', *Proc. Conference on Decision and Control*, pp. 4631–4636, 2002.

[82] W. B. Dunbar and R. M. Murray, 'Distributed Receding Horizon Control for Multi-Vehicle Formation Stabilization', *Automatica*, vol. 42, pp. 549–558, 2006.

[83] H. Yamaguchi, 'A Cooperative Hunting Behavior by Mobile Robot Troops', *Int. J. Robotics Research*, vol. 18, pp. 931–940, 2005.

[84] T. B. Curtin, J. G. Bellingham, J. Catipovic, and D. Webb, 'Autonomous Oceanographic Sampling Networks', *Oceanography*, pp. 86–94, 1993.

[85] T. R. Smith, H. Hmann, and N. E. Leonard, 'Orientation Control of Multiple Underwater Vehicles', *Proc. 40th Conference on Decision and Control*, pp. 4598–4603, 2001.

[86] L. Buzogany, M. Pachter, and J. Azzo, 'Automated Control of Aircraft in Formation Height', *Proc. AIAA Conference Guidance, Navigation and Control*, pp. 1349–1370, 1993.

[87] J. Wolfe, D. Chichka, and J. Speyer, 'Decentralized Controllers for Unmanned Aerial Vehicle Formation Height', *Proc. AIAA Conference Guidance, Navigation and Control*, pp. 3833–3851, 1996.

[88] K. Rohloff, J. Ye, J. Loyall, and R. Schantz, 'A Hierarchical Control System for Dynamic Resource Management', *Proc. IEEE Real-Time and Embedded Technology and Applications Symposium*, RTAS, California, Paper no. 15, April 4–7, 2006.

[89] B. T. Jevtovi'c and M. R. Matausek, 'A Predictive-Adaptive Hierarchical Control System of Bucket-Wheel Excavator: Theory and Experimental Results', Facta Universitatis (NIS), Ser: Elect. Eng., vol. 18, pp. 493–503, 2005.

[90] F. N. Bailey, M. Brdys, K. Malinowski, and P. Tatjewski, *Control and Coordination in Hierarchical Systems*, Wiley and Sons, London, 1980.

[91] F. Amigoni and N. Gatti, 'A Formal Framework for Connective Stability of Highly Decentralized Cooperative Negotiations', *Autonomous Agents and Multi-agent Systems*, vol. 15, pp. 253–279, 2007.

[92] M. A. Brdys, M. Grochowski, T. Gminski, K. Konarczak, and M. Drewa, 'Hierarchical Predictive Control of Integrated Wastewater Treatment Systems', *Control Engineering Practice*, vol. 16, pp. 751–767, 2008.

[93] E. J. Van Henten and J. Bontsema, 'Time-Scale Decomposition of an Optimal Control Problem in Greenhouse Climate Management', *Control Engineering Practice*, vol. 17, pp. 88–96, 2009.

[94] M. S. Mahmoud, 'Decentralized Reliable Control of Interconnected Systems with Time-Varying Delays', *J. Optimization Theory and Applications*, vol. 143, pp. 497–518, 2009.

[95] K. Rohloff, R. Schantzand, and Y. Gabay, 'High-Level Dynamic Resource Management for Distributed, Real-Time Embedded Systems, *Proc. Summer Computer Simulation Conference*, pp. 49–756, 2007.

[96] T. Knudsen, K. Trangbk, and C. S. Kallese, 'Plug and Play Process Control Applied to a District Heating System', *Proc. European Control Conference*, pp. 336–341, 2009.

[97] 'Control IT, AC800M/C Control Functions - User's Guide', *ABB Automation Technology Products*, 3BSE-021-351-R201, 2001.

[98] L. Pernebo and B. Hansson, 'Plug and Play in Control Loop Design', *ABB Automation Technology Products*, 3BSE-028-758, 2001.

[99] R. Vadigepalli and F. J. Doyle III, 'Structural Analysis of Large-scale Systems for Distributed State Estimation and Control Applications', *Control Engineering Practice*, vol. 11, pp. 895–915, 2003.

[100] R. Vadigepalli and F. J. Doyle III, 'A Distributed State Estimation and Control Algorithm for Plantwide Processes', *IEEE Trans. Control Systems Technology*, vol. 11, pp. 119–128, 2003.

[101] J. H. Braslavsky, R. H. Middleton, and J. S. Freudenberg, 'Feedback Stabilization over Signal-to-Noise Ratio Constrained Channels', *Proc. American Control Conference*, pp. 4903–4908, 2004.

[102] N. Elia and S. K. Mitter, 'Stabilization of Linear Systems with Limited Information', *IEEE Trans. Automatic Control*, vol. 46, pp. 1384–1400, 2001.

[103] J. P. Hespanha, A. Ortega, and L. Vasudevan, 'Towards the Control of Linear Systems with Minimum Bit-Rate', *Proc. Int. Symposium on Mathematical Theory of Networks and Systems*, 2002.

[104] G. N. Nair and R. J. Evans, 'Exponential Stabilizability of Finite-Dimensional Linear Systems with Limited Data Rates', *Automatica*, vol. 39, pp. 585–593, 2003.

[105] S. Tatikonda and S. Mitter, 'Control under Communication Constraints', *IEEE Trans. Automatic Control*, vol. 49, pp. 1056–1068, 2004.

[106] W. S. Wong and R. W. Brockett, 'Systems with Finite Communication Bandwidth Constraints-II: Stabilization with Limited Information Feedback', *IEEE Trans. Automatic Control*, vol. 44, pp. 150–159, 1999.

[107] D. Liberzon and J. P. Hespanha, 'Stabilization of Nonlinear Systems with Limited Information Feedback', *IEEE Trans. Automatic Control*, vol. 50, pp. 910–915, 2005.

[108] G. N. Nair, R. J. Evans, I. M. Y. Mareels, and W. Moran, 'Topological Feedback Entropy and Nonlinear Stabilization', *IEEE Trans. Automatic Control*, vol. 49, pp. 1585–1597, 2004.

[109] C. M. Krishna and Y. H. Lee, Eds., 'Special Issue on Real-Time Systems', *Proc. IEEE*, vol. 91, 2003.

[110] K. Menighed, C. Aubrun, and J. J. Yamé, 'Distributed State Estimation and Model Predictive Control of Linear Interconnected System: Application to Fault Tolerant Control', *Proc. IEEE Conference on Control and Automation*, Nancy Univ., Nancy, France, pp. 936–941, 2009.

[111] A. Abur, 'Distributed State Estimation for Mega Grids', *Proc. 15th Power Systems Computation Conference*, Liege, Belgium, pp. 1–5, 2005.

[112] M. Cloosterman, N. Wouw, W. Heemels, and H. Nijmeijer, 'Stabilization of Networked Control Systems with Large Delays and Packet Dropouts', *Proc. American Control Conference*, Seatle, Washington, pp. 4991–4995, 2008.

[113] C. E. Shannon, 'A Mathematical Theory of Communication', *The Bell System Technical Journal*, vol. 27, pp. 379–423, 1948.

[114] H. S. Witsenhausen, 'Separation of Estimation and Control for Discrete-Time Systems', *Proc. IEEE*, vol. 59, no. 11, pp. 1557–1566, 1971.

[115] A. Bemporad and M. Morari, 'Robust Model Predictive Control: A Survey', in *Robustness in Identification and Control* (vol. 245), Springer-Verlag, New York, pp. 207–226, 1999.

[116] E. F. Camacho and C. Bordons, *Model Predictive Control in the Process Industry* (2nd edn.), Springer, London, 2004.

[117] D. Q. Mayne, J. B. Rawlings, C. V. Rao, and P. O. M. Scokaert, 'Constrained Model Predictive Control: Stability and Optimality', *Automatica*, vol. 36, no. 6, pp. 789–814, 2000.

[118] M. Morari and J. H. Lee, 'Model Predictive Control: Past, Present and Future', *Comput. Chem. Eng.*, vol. 23, no. 4–5, pp. 667–682, 1999.

[119] N. R. Sandell, P. Varaiya, M. Athans, and M. G. Safonov, 'Survey of Decentralized Control Methods for Large Scale Systems', *IEEE Trans. Automatic Control*, vol. AC-23, pp. 108–128, 1978.

[120] D. D. Siljak, 'Decentralized Control and Computations: Status and Prospects', *Annu. Rev. Contr.*, vol. 20, pp. 131–141, 1996.

[121] L. Acar, 'Some Examples for the Decentralized Receding Horizon Control', *Proc. 31st Conf. Decision and Control*, Tucson, AZ, pp. 1356–1359, 1992.

[122] M. Aicardi, G. Casalino, R. Minciardi, and R. Zoppoli, 'On the Existence of Stationary Optimal Receding-Horizon Strategies for Dynamic Teams with Common Past Information Structures', *IEEE Trans. Automatic Control*, vol. 37, pp. 1767–1771, 1992.

[123] M. Baglietto, T. Parisini, and R. Zoppoli, 'Neural Approximators and Team Theory for Dynamic Routing: A Receding-Horizon Approach', *Proc. 38th Conf. Decision and Control*, Phoenix, AZ, pp. 3283–3288, 1999.

[124] H. El Fawal, D. Georges, and G. Bornard, 'Optimal Control of Complex Irrigation Systems via Decomposition-Coordination and the use of Augmented Lagrangian', *Proc. 1998 IEEE Int. Conf. Systems, Man and Cybernetics* (vol. 4), San Diego, CA, pp. 3874–3879, 1998.

[125] G. Georges, 'Decentralized Adaptive Control for a Water Distribution System', *Proc. 3rd IEEE Conf. Control Applications*, Glasgow, UK, pp. 1411–1416, 1994.

[126] M. Gomez, J. Rodellar, F. Vea, J. Mantecon, and J. Cardona, 'Decentralized Predictive Control of Multireach Canals', *Proc. 1998 IEEE Int. Conf. Systems, Man, and Cybernetics*, San Diego, CA, pp. 3885–3890, 1998.

[127] S. Oschs, S. Engell, and A. Draeger, 'Decentralized vs. Model Predictive Control of an Industrial Glass Tube Manufacturing Process', *Proc. 1998 IEEE Int. Conf. Control Applications*, Trieste, Italy, pp. 16–20, 1998.

[128] S. Sawadogo, R. M. Faye, P. O. Malaterre, and F. Mora-Camino, 'Decentralized Predictive Controller for Delivery Canals', *Proc. 1998 IEEE Int. Conf. Systems, Man, and Cybernetics*, San Diego, CA, pp. 3880–3884, 1998.

[129] A. L. Barrabasi and R. Albert, 'Emerging of Scaling in Random Networks', *Science*, vol. 286, pp. 509–512, 1999.

[130] C. Li and G. Chen, 'Synchronization in General Complex Dynamical Networks with Coupling Delays', *Physica-A: Stat. Mech. and its Apps.*, vol. 343, no. 15, pp. 263–278, 2004.

[131] W. B. Dunbar and R. M. Murray, 'Distributed Receding Horizon Control for Multi-Vehicle Formation Stabilization', *Automatica*, vol. 42, no. 4, pp. 549–558, 2006.

[132] L. B. de Oliveira and E. Camponogara, 'Multi-Agent Model Predictive Control of Signaling Split in Urban Traffic Networks', *Transportation Research Part C: Emerging Technologies*, vol. 18, no. 1, pp. 120–139, 2010.

[133] R. R. Negenborn, G. Hug-Glanzmann, B. De Schutter, and G. Andersson, 'A Novel Coordination Strategy for Multi-Agent Control using Overlapping Subnetworks with Application to Power Systems', *Efficient Modeling and Control of Large-Scale Systems* (J. Mohammadpour, K.M. Grigoriadis, eds.), Springer, Norwell, Massachusetts, pp. 251–278, 2000.

[134] J. Rawlings and B. T. Stewart, 'Coordinating Multiple Optimization-Based Controllers: New Opportunities and Challenges', *J. Process Control*, vol. 18, no. 9, pp. 839–845, 2008.

[135] B. Brogliato, R. Lozano, B. Maschke, and O. Egeland, *Dissipative Systems Analysis and Control: Theory and Applications*, Springer, Berlin, 2006.

[136] G. Scorletti and G. Duc, 'An LMI approach to Dencentralized $\mathcal{H}_\infty$ control', *Int. J. Control*, vol. 74, no. 3, pp. 211–224, 2001.

[137] T. Tran, 'Overriding Control for Stability with Manifest Variables', *Proceedings of 2010 IEEE Int. Conference on Control, Automation, Robotics and Vision*, Singapore, pp. 236–241, 2010.

[138] T. Tran and Q. P. Ha, 'Distributed Model Predictive Control for Complex Interconnected Systems', *Proc. IEEE Int. Conference on Computing and Communication Technologies, RIVF'10*, Vietnam, pp. 284–289, 2010.

[139] T. Tran and Q. P. Ha, 'Networked Control Systems with Accumulative Quadratic Constraint', *Electronics Letters*, vol. 47, no. 2, pp. 108–110, 2011.

[140] S. I. Niculescu, *Delay Effects on Stability—A Robust Control Approach*, Springer, Berlin, 2001.

[141] J. H. Ryua, J. Artigasb, and C. Preusche, 'A Passive Bilateral Control Scheme for a Teleoperator with Time-Varying Communication Delay', *Mechatronics*, vol. 20, no. 7, pp. 812–823, 2010.

[142] D. D. Sourlas and V. Manousiouthakis, 'Best Achievable Decentralized Performance', *IEEE Trans. Automatic Control*, vol. 40, no. 11, pp. 1858–1871, 1995.

[143] M. Rotkowitz and S. Lall, 'A Characterization of Convex Problems in Decentralized Control', *IEEE Trans. on Automatic Control*, vol. 50, no. 12, pp. 1984–1996, 2005.

[144] S. Dashkovskiy, B. S. Ruffer, and F. Wirth, 'An ISS Small Gain Theorem for General Networks', *Mathematics of Control, Signals, and Systems*, vol. 19, no. 2, pp. 93–122, 2007.

[145] T. Keviczky, F. Borrelli, and G. J. Balas, 'Decentralized Receding Horizon Control for Large Scale Dynamically Decoupled Systems', *Automatica*, vol. 42, no. 12, pp. 2105–2115, 2006.

[146] L. Magni and R. Scattolini, 'Stabilizing Decentralized Model Predictive Control of Nonlinear Systems', *Automatica*, vol. 42, no. 7, pp. 1231–1236, 2006.

[147] D. M. Raimondo, L. Magni, and B. Scattolini, 'Decentralized Model Predictive Control of Nonlinear Systems: an Input-to-State Stability approach', *Int. J. Robust and Nonlinear Control*, vol. 17, no. 17, pp. 1651–1667, 2007.

[148] A. Richards and J. How, 'Distributed Robust Receding Horizon Control for Multivehicle Guidance', *Int. J. Control*, vol. 80, no. 9, pp. 1517–1531, 2007.

[149] J. S. Qin and T. A. Badgwell, 'A Survey of Industrial Model Predictive Control Technology', *Control Engineering Practice*, vol. 11, no. 7, pp. 733–764, 2003.

[150] L. Magni, D. M. Raimondo, and F. Allgower, *Nonlinear Model Predictive Control: Towards New Challenging Applications*, LNCIS-384, Springer, Berlin, 2009.

[151] J. C. Willems, 'Dissipative Dynamical Systems—Part I and II', *Archive for Rational Mechanics and Analysis*, vol. 45, pp. 321–393, 1972.

[152] J. C. Willems, 'Dissipative Dynamical Systems', *European J. Control*, vol. 13, nos. 2–3, pp. 134–151, 2007.

[153] J. A. Primbs, V. Nevistic, and J. C. Doyle, 'A Receding Horizon Generalization of Pointwise Min-Norm Controllers', *IEEE Trans. Automatic Control*, vol. 45, no. 5, pp. 898–909, 2000.

[154] S. J. Qin and T. A. Badgwell, 'A Survey of Industrial Model Predictive Control Technology', *Control Eng. Prac.*, vol. 11, no. 7, pp. 733–764, 2003.

[155] M. Morari and J. H. Lee, 'Model Predictive Control: Past, Present and Future', *Proc. Joint 6th Int. Symposium on Process Systems Engineering*

(PSE '97) and 30th European Symposium on Computer-Aided Process Systems Engineering (ESCAPE 7), Trondheim, Norway, 1997.

[156] J. Lu, 'Multi-Zone Control under Enterprise Optimization: Needs, Challenges and Requirements', in *Nonlinear Model Predictive Control* (F. Allgower and A. Zheng, Eds.), Birkhauser, Basel, vol. 26, pp. 393–402, 2000.

[157] J. Lu, 'Challenging Control Problems and Emerging Technologies in Enterprise Optimization', *Control Eng. Prac.*, vol. 11, no. 8, pp. 847–858, August 2003.

[158] E. Camponogara, D. Jia, B. H. Krogh, and S. Talukdar, 'Distributed Model Predictive Control', *IEEE Control System Magazine*, pp. 44–52, February 2002.

[159] A. N. Venkat, J. B. Rawlings, and S. J. Wright, 'Stability and Optimality of Distributed Model Predictive Control', *Proc. 44th IEEE Conference on Decision and Control, and the European Control Conference*, Seville, Spain, pp. 6680–6685, 2005.

[160] E. F. Camacho and C. Bordons, *Model Predictive Control in the Process Industry*, Springer, Berlin, 1995.

[161] X. Cheng and B. H. Krogh, 'Stability Constrained Model Predictive Control with State Estimation', *Proc. American Control Conference*, Pittsburgh, PA, vol. 4, pp. 2493–2498, 1997.

[162] R. R. Negenborn, B. De Schutter, and H. Hellendoorn, 'Multiagent Model Predictive Control of Transportation Networks', *Proc. IEEE Int. Conference on Networking, Sensing and Control (ICNSC 2006)*, Ft. Lauderdale, FL; pp. 296–301, 2006.

[163] T. C. Yang, 'Networked Control Systems: a Brief Survey' *IEE Proc. Control Theory and Applications*, vol. 152, pp. 403–412, 2006.

[164] P. Neumann, 'Communication in Industrial Automation—What is going on?' *Control Engineering Practice*, vol. 15, pp. 1332–1347, 2007.

[165] D. M. Raimondo, L. Magni, and R. Scattolini, 'Decentralized MPC of Nonlinear System: An Input-to-State Stability Approach', *Int. J. Robust and Nonlinear Control*, vol. 17, pp. 1651–1667, 2007.

[166] S. Li, Y. Zhang, and Q. Zhu, 'Nash-Optimization enhanced Distributed Model Predictive Control Applied to the Shell Benchmark Problem', *Information Sciences, Informatics and Computer Science*, vol. 170, pp. 329–349, 2005.

[167] W. B. Dunbar, 'Distributed Receding Horizon Control of Dynamically Coupled Nonlinear Systems', *IEEE Trans. Automatic Control*, vol. 52, pp. 1249–1263, 2007.

[168] A Richards and J. P. How, 'Robust Distributed Model Predictive Control', *Int. J. Control*, vol. 80, pp. 1517–1531, 2007.

[169] D. Jia and B. Krogh, 'Min-Max Feedback Model Predictive Control for Distributed Control with Communication', *Proc. American Control Conference*, Anchorage, pp. 4507–4512, 2002.

[170] J. Liu J, Muñoz de la Pe, B. Ohran, P. D. Christofides, and J. F. Davis, 'A Two-Tier Architecture for Networked Process Control', *Chemical Engineering Science*, vol. 63, pp. 5349–5409, 2008.

[171] J. Liu J, Muñoz de la Pe, B. Ohran, and P. D. Christofides, 'Distributed Model Predictive Control of Nonlinear Process Systems', *AIChE Journal*, vol. 55, pp. 1171–1184, 2009.

[172] T. Basar and G. J. Olsder, *Dynamic Noncooperative Game Theory*, SIAM, Philadelphia, 1999.

[173] R. B. Myerson, *Game Theory, Analysis of Conflict*, Harvard University Press, Cambridge, MA, 1997.

[174] J. Colandairaj, G. W. Irwin, and W. G. Scanlon, 'Analysis and Co-Simulation of an IEEE 802.11b Wireless Networked Control System', *Proc. Irish Signals and Systems Conference*, pp. 35–36, 2006.

[175] L. Magni, G. De Nicolao, R. Scattolini, and F. Allgower, 'Robust MPC for Nonlinear Discrete-Time Systems', *Int. J. Robust and Nonlinear Control*, vol. 13, pp. 229–246, 2003.

[176] T. Alamo, D. Muñoz de la Peña, D. Limon, and E. F. Camacho, 'Constrained Minimax Predictive Control: Modifications of the Objective Function Leading to Polynomial Complexity', *IEEE Trans. Automatic Control*, vol. 50, no. 5, pp. 710–714, 2005.

[177] M. Lazar, W. P. M. H. D. Heemels, Muñoz de la Peña, and T. Alamo, 'Further Results on Robust MPC using Linear Matrix Inequalities', *Proc. Int. Workshop on Assessment and Future Directions of NMPC*, Pavia, Italy, 2008.

[178] I. Kolmanovsky and E. G. Gilbert, 'Theory and Computation of Disturbance Invariant Sets for Discrete-Time Linear Systems', *Mathematical Problems in Engineering, Theory, Methods and Applications*, vol. 4, pp. 317–367, 1998.

[179] S. Rakovic, E. Kerrigan, K. Kouramas, and D. Q. Mayne, 'Robust MPC for Nonlinear Discrete-Time Systems', *IEEE Trans. Automatic Control*, vol. 50, pp. 406–410, 2005.

[180] J. D. Sterman, *Business Dynamics—Systems Thinking and Modelling in a Complex World*. McGraw-Hill, New York, 2000.

[181] W. B. Dunbar and S. Desa, 'Distributed MPC for Dynamic Supply Chain Management', *Int. Workshop on Assessment and Future Directions of NMPC*, Freudenstadt-Lauterbad, Germany, 26–30 August, 2005.

[182] D. P. Bertsekas, *Nonlinear Programming* (2nd ed.), Athena Scientific, Belmont, MA, 1999.

[183] A. N. Venkat, *Distributed model predictive control: Theory and applications*, Ph.D. Dissertation, Dept. Chem. Biol. Eng., Univ. Wisconsin-Madison, Madison, 2006 [Online]. Available: *http://jbrwww.che.wisc.edu/theses/venkat.pdf*

[184] A. J. Wood and B. F. Wollenberg, *Power Generation Operation and Control*, Wiley, New York, 1996.

[185] R. Gupta, A. Agarwal, M. Chow, and W. Wang, 'Characterization of Data Sensitive Wireless Distributed Networked Control Systems', *Proc. IEEE Conference on Advanced Intelligent Mechatronics*, ASME, pp. 1–6, 2007.

[186] M. Rotkowitz and S. Lall, 'A Characterization of Convex Problems in Decentralized Control', *IEEE Trans. Automatic Control*, vol. 51, no. 2, pp. 274–286, 2006.

[187] H. S. Witsenhausen, 'A Counterexample in Stochastic Optimum Control', *SIAM Journal of Control*, vol. 6, no. 1, pp. 131–147, 1968.

[188] A. Sahai and S. Mitter, 'Information and Control: Witsenhausen Revisited', *Lecture Notes in Control and Information Sciences*, vol. 241, pp. 639–654, 1999.

[189] C. H. Papadimitriou and J. N. Tsitsiklis, 'Intractable Problems in Control Theory', *SIAM Journal of Control and Optimization*, vol. 24, pp. 639–654, 1986.

[190] V. Blondel and J. Tsitsiklis, 'A Survey of Computational Complexity Results in Systems and Control', *Automatica*, vol. 36, no. 9, pp. 1249–1274, 2000.

[191] W. S. Levine, T. L. Johnson, and M. Athans, 'Optimal Limited State Variable Feedback Controllers for Linear Systems', *IEEE Trans. on Automatic Control*, vol. AC-16, pp. 785–793, 1971.

[192] C. J. Wenk and C. H. Knapp, 'Parameter Optimization in Linear Systems with Arbitrarily Constrained Controller Structure', *IEEE Trans. on Automatic Control*, vol. AC-25, no. 3, pp. 496–500, 1980.

[193] T. Soderstrom, 'On Some Algorithms for Design of Optimal Constrained Regulators', *IEEE Trans. on Automatic Control*, vol. AC-23, pp. 1100–1101, 1978.

[194] V. Gupta, B. Hassibi, and R. M. Murray, 'On the Synthesis of Control Laws for a Network of Autonomous Agents', *Proceedings of the American Control Conference*, vol. 6, pp. 4927–4932, 2004.

[195] V. Gupta, B. Hassibi, and R. M. Murray, 'A Sub-optimal Algorithm to Synthesize Control Laws for a Network of Dynamic Agents', *Int. J. Control*, vol. 78, no. 16, pp. 1302–1313, November 10, 2005.

[196] T. C. Xinogalas, M. S. Mahmoud, and M. G. Singh, 'Hierarchical Computation of Decentralized Gains for Interconnected Systems', *Automatica*, vol. 18, no. 4, pp. 474–478, July 1982.

[197] M. S. Mahmoud, Y. Chen, and M. G. Singh, 'An Iterative Block Diagonalization Procedure for Decentralized Optimal Control', *Int. J. Systems Science*, vol. 15, no. 5, pp. 563–573, 1984.

[198] B. D. O. Anderson and J. B. Moore, *Linear Optimal Control*, Prentice Hall, New York, 1971.

[199] D. L. Kleinman, T. Fortmann, and M. Athans, 'On the Design of Linear Systems with Piecewise-Constant Feedback Gains', *IEEE Tran. Automatic Control*, vol. AC-13, no. 4, pp. 354–361, 1968.

[200] G. A. de Castro, *Convex Methods for the Design of Structured Controllers*, Ph.D. Thesis, University of California, Los Angeles, 2001.

[201] C. Langbort, R. S. Chandra, and R. D'Andrea, 'Distributed Control Design for Systems Interconnected over an Arbitrary Graph', *IEEE Trans. on Automatic Control*, vol. 49, no. 9, pp. 1502–1519, September 2004.

[202] C. Langbort, V. Gupta, and R. M. Murray, 'Distributed Control over Failing Channels', in P. Tabuada and P. Antsaklis (Editors), Networked Embedded Sensing and Control, Lecture Notes in Control and Information Sciences Series, vol. 331, Springer Verlag, 2006.

[203] Y. C. Ho and K. C. Chu, 'Team Decision Theory and Information Structures in Optimal Control Problems – Part I', *IEEE Trans. on Automatic Control*, vol. AC-17, no. 1, pp. 15–22, 1972.

[204] N. Sandell and M. Athans, 'Solution of some Nonclassical LQG Stochastic Decision Problems', *IEEE Trans. Automatic Control*, vol. AC-19, pp. 108–116, 1974.

[205] C. Fan, J. L. Speyer, and C. R. Jaensch, 'Centralized and Decentralized Solutions of the Linear-Exponential-Gaussian Problem', *IEEE Trans. Automatic Control*, vol. 39, no. 10, pp. 1986–2003, 1994.

[206] P. G. Voulgaris, 'Optimal Control of Systems with Delayed Observation Sharing Patterns via Input-Output Methods', *Systems and Control Letters*, vol. 50, pp. 51–64, 2003.

[207] B. Bamieh and P. G. Voulgaris, 'Optimal Distributed Control with Distributed Delayed Measurement', in *Proc. IFAC World Congress*, Barcelona, Spain, vol. 1, pp. 116–121, 2002.

[208] X. Qi, M. Salapaka, P. Voulgaris, and M. Khammash, 'Structured Optimal and Robust Control with Multiple Criteria: A Convex Solution', *IEEE Trans. Automatic Control*, vol. 49, no. 10, pp. 1623–1640, 2004.

[209] R. D'Andrea and G. E. Dullerud, 'Distributed Control Design for Spatially Interconnected Systems', *IEEE Trans. Automat. Control*, vol. 48, no. 9, pp. 1478–1495, 2003.

[210] B. M. Mirkin and P.-O. Gutman, 'Output-Feedback Coordinated Decentralized Adaptive Tracking: The Case of MIMO Subsystems with Delayed Interconnections', *Int. J. Adapt. Control and Signal Process.*, vol. 19, no. 8, pp. 639–660, 2005.

[211] J. A. Fax and R. M. Murray, 'Information Flow and Cooperative Control of Vehicle Formations', *IEEE Trans. Automatic Control*, vol. 49, no. 9, pp. 1465–1476, Sep. 2004.

[212] B. Bamieh and P. G. Voulgaris, 'A Convex Characterization of Distributed Control Problems in Spatially Invariant Systems with Communication Constraints', *Syst. and Control Lett.*, vol. 54, pp. 575–583, 2005.

[213] M. R. Jovanovic, 'On the Optimality of Localized Distributed Controllers', in *Proc. Amer. Control Conf.*, Portland, OR, vol. 4, pp. 4583–4588, 2005.

[214] G. E. Dullerud and R. D'Andrea, 'Distributed Control of Heterogeneous Systems', *IEEE Trans. Automat. Control*, vol. 49, no. 12, pp. 2113–2128, 2004.

[215] T. Keviczky, F. Borrelli, and G. J. Balas, 'A Study on Decentralized Receding Horizon Control for Decoupled Systems', in *Proc. Am. Control Conf.*, Boston, MA, vol. 5, pp. 4921–4926, 2004.

[216] T. Keviczky, F. Borrelli, and G. J. Balas, 'Stability Analysis of Decentralized Receding Horizon Control for Decoupled Systems', in *Proc. 44th IEEE Conf. Decision Control, Eur. Control Conf.*, Seville, Spain, pp. 1689–1694, 2005.

[217] W. B. Dunbar and R. M. Murray, 'Receding Horizon Control of Multivehicle Formations: A Distributed Implementation', in *Proc. 43rd IEEE Conf. Decision Control*, pp. 1995–2002, 2004.

[218] A. Richards and J. How, 'A Decentralized Algorithm for Robust Constraied Model Predictive Control', in *Proc. Am. Control Conf.*, Boston, MA, pp. 4261–4266, 2004.

[219] F. Borrelli, T. Keviczky, G. J. Balas, G. Stewart, K. Fregene, and D. Godbole, 'Hybrid Decentralized Control of Large Scale Systems', in *Hybrid Systems: Computation and Control* (vol. 3414), Springer Verlag, New York, pp. 168–183, 2005.

[220] D. Stipanovic, G. Inalhan, R. Teo, and C. J. Tomlin, 'Decentralized Overlapping Control of a Formation of Unmanned Aerial Vehicles', *Automatica*, vol. 40, no. 8, pp. 1285–1296, 2004.

[221] B. Bamieh, F. Paganini, and M. Dahleh, 'Distributed Control of Spatially Invariant Systems', *IEEE Trans. Autom. Control*, vol. 47, no. 7, pp. 1091–1107, 2002.

[222] T. Keviczky, F. Borrelli, G. J. Balas, K. Fregene, and D. Godbole, 'Decentralized Receding Horizon Control and Coordination of Autonomous Vehicle Formations', *IEEE Trans. Control Syst. Technol.*, vol. 16, no. 1, pp. 19–33, 2008.

[223] B. D. O. Andreson and J. B. Moore, *Optimal Control: Linear Quadratic Methods,* Prentice Hall, Englewood Cliffs, NJ, 1990.

[224] M. G. Safonov and M. Athans, 'Gain and Phase Margin for Multiloop LQG Regulators', *IEEE Trans. Automatic Control*, vol. AC-22, no. 2, pp. 173–179, 1977.

[225] C. Scherer and S. Weiland, *Linear Matrix Inequalities in Control* (3rd edn.), Springer-Verlag, New York, 2000.

[226] M. Ajtai, J. Komlos, and E. Szemerédi, 'Sorting in c log n Parallel Steps', *Combinatorica*, vol. 19, no. 1, pp. 125–138, 1999

[227] R. M. Tanner, 'Explicit Concentrators from Generalized n-Gons', *SIAM J. Algebra Discr. Meth.*, vol. 5, pp. 287–293, 1984.

[228] M. Tompa, 'Time-Space Tradeoffs for Computing Functions, using Connectivity Properties of their Circuits', *J. Comp. Syst. Sci.*, vol. 20, pp. 118–132, 1980.

[229] L. Valiant, 'Graph Theoretic Properties in Computational Complexity', *J. Comp. Syst. Sci.*, vol. 13, pp. 278–285, 1976.

[230] B. Mohar, 'The Laplacian spectrum of graphs', *Graph Theory, Combinatorics, Application*, vol. 2, pp. 871–898, 1991.

[231] Q. Hui, 'Distributed Semistable LQR control for Discrete-Time Dynamically Coupled Systems', *J. Franklin Institute*, vol. 349 , no. 1, pp. 74–92, 2012.

[232] Q. Hui, W. M. Haddad, and S. P. Bhat, 'Finite-Time Semistability and Consensus for Nonlinear Dynamical Networks', *IEEE Trans. Automatic Control*, vol. 53, pp. 1887–1900, 2008.

[233] Q. Hui and W. M. Haddad, 'Distributed Nonlinear Control Algorithms for Network Consensus', *Automatica*, vol. 44, pp. 2375–2381, 2008.

[234] S. P. Bhat and D. S. Bernstein, 'Nontangency-Based Lyapunov Tests for Convergence and Stability in Systems having a Continuum of Equilibria', *SIAM J. Control and Optimization*, vol. 42, pp. 1745–1775, 2003.

[235] A. Jadbabaie, J. Lin, and A. S. Morse, 'Coordination of Groups of Mobile Autonomous Agents using Nearest Neighbor Rules', *IEEE Trans. Automatic Control*, vol. 48, pp. 988–1001, 2003.

[236] J. A. Marshall, M. E. Broueke, and B. A. Francis, 'Formations of Vehicles in Cyclic Pursuit', *IEEE Trans. Automatic Control*, vol. 49, pp. 1963–1974, 2004.

[237] J. Singh, B. Patra, and S. Singh, 'An Algorithm to Reduce the Time Complexity of Earliest Deadline First Scheduling Algorithm in Real-Time System', *Int. J. Advanced Computer Science and Applications*, vol. 2, pp. 31–37, 2011.

[238] R. Olfati-Saber, 'Flocking for Multiagent Dynamic Systems: Algorithms and Theory', *IEEE Trans. Automatic Control*, vol. 51, pp. 401–420, 2006.

[239] Q. Hui, W. M. Haddad, and S. P. Bhat, 'Semistability Theory for Differential Inclusions with Applications to Consensus Problems in Dynamical Networks with Switching Topology', in *Proc. American Control Conference,* Seattle, WA, pp. 3981–3986, 2008.

[240] Q. Hui, W. M. Haddad, and S. P. Bhat, 'Semistability for Time-Varying Discontinuous Dynamical Systems with Application to Agreement Problems in Switching Networks', in *Proc. IEEE Conference on Decision Control,* Cancun, Mexico, pp. 2985–2990, 2008.

[241] Q. Hui and W. M. Haddad, 'Uniform Semistabiltty for Time-Varying Dynamical Systems and Network Consensus with Time-Dependent Communication Ninks', in *Proc. American Control Conference,* St. Louis, MO, pp. 4050–4055, 2009.

[242] S. Boyd, A. Ghosh, B. Prabhakar, and D. Shah, 'Randomized Gossip Algorithms', *IEEE Trans. information Theory*, vol. 52, pp. 2508–2530, 2006.

[243] L. Xiao, S. Boyd and S. J. Kim, 'Distributed Average Consensus with Least-Mean Square Deviation', *J. Parallel and Distributed Computing* vol. 67, pp. 33–46, 2007.

[244] Q. Hui, 'Optimal Distributed Linear Averaging', *Automatica*, vol. 47, no. 12, pp. 2713–2719, 2011.

[245] L. Xiao and S. Boyd, 'Fast Linear Iterations for Distributed Averaging', *Systems and Control Letters*, vol. 53, no. 1, pp. 65–78, 2004.

[246] D. S. Bernstein, *Matrix Mathematics*, Princeton University Press, Princeton, NJ, 2005.

[247] R. Olfati-Saber and R. M. Murray, 'Consensus Problems in Networks of Agents with Switching Topology and Time-Delays', *IEEE Trans. Automatic Control*, vol. 49, no. 4, pp. 1520–1533, 2004.

[248] W. M. Haddad, Q. Hui, and V. Chellaboina, '$\mathcal{H}_2$ Optimal Semistable Control for Linear Dynamical Systems: an LMI Approach', *Journal of the Franklin Institute*, vol. 348, pp. 2898–2910, 2011.

[249] Q. Hui and W. M. Haddad, '$\mathcal{H}_2$ Optimal Semistable Stabilization for Linear Discrete-Time Dynamical Systems with Applications to Network Consensus', *Int. J. Control*, vol. 82, pp. 456–469, 2009.

[250] W. M. Haddad, Q. Hui, S. G. Nersesov, and V. Chellaboina, 'Thermodynamic Modeling, Energy Equipartition, and Nonconservation of Entropy for Discrete-Time Dynamical Systems', *Advances in Difference Equations*, vol. 2005, no. 3, pp. 275–318, 2005.

[251] A. Berman and R. J. Plemmons, *Nonnegative Matrices in the Mathematical Sciences,* Academic Press, New York, 1979.

[252] A. Berman and R. J. Plemmons, 'Nonnegative Matrices in the Mathematical Sciences', *SIAM Classics in Applied Mathematics*, 1994.

[253] P. De Leenheer and D. Aeyels, 'Stabilization of Positive Linear Systems', *Systems & Control Letters*, vol. 44, no. 4, pp, 259–271, 2001.

[254] M. Ait Rami and F. Tadeo, 'Controller Synthesis for Positive Linear Systems with Bounded Controls', *IEEE Trans. Circuits and Systems II: Express Briefs,* vol. 54, no. 2, pp. 151–155, 2007.

[255] D. Angeli and D. E. Sontag, 'Monotone Control Systems', *IEEE Trans. Automatic Control*, vol. 48, no. 10, pp. 1684–1698, 2003.

[256] L. Moreau, 'Stability of Continuous-Time Distributed Consensus Algorithms', *Proc. 43rd IEEE Conference on Decision and Control*, Bahamas, 2004.

[257] D. G. Luenberger, *Introduction to Dynamic Systems: Theory, Models, and Applications*. Wiley, New York, 1979.

[258] T. Tanaka and C. Langbort, 'KYP Lemma for Internally Positive Systems and a Tractable Class of Distributed $\mathcal{H}_\infty$ Control Problems', *Proc. American Control Conference*, Baltimore, USA, vol. 6, pp. 6238–6243, 2010.

[259] K. Mårtensson and A. Rantzer, 'Gradient methods for iterative distributed control synthesis', *Proc. 48th IEEE Conference on Decision and Control jointly with the 2009 28th Chinese Control Conference, CDC/CCC 2009*, pp. 549–554, 15–18 December 2009.

[260] T. Kato and T. Namerikawa, 'Distributed control for load frequency of power networks based on iterative gradient methods', *Proc. SICE Annual Conference*, Waseda (Tokyo), vol. 2, pp. 1384–1389, 13–18 September 2011.

[261] A. Al-Hammouri, M. Branicky, V. Liberatore, and S. Phillips. 'Decentralized and Dynamic Bandwidth Allocation in Networked Control Systems', *IEEE International Parallel & Distributed Processing Symposium*, Boston, Massachusetts, vol. 1, pp. 8–18, 2006.

[262] L. Schenato, M. Franceschetti, K. Poolla, and S. Sastry, 'Foundations of Control and Estimation over Lossy Networks', *Proc. of the IEEE*, vol. 95, no. 1, pp. 163–187, 2007.

[263] D. Yue, Q. Han, and C. Peng, 'State Feedback Controller Design of Networked Control Systems', *IEEE Trans. Circuits and Systems II: Express Briefs*, vol. 51, no. 11, 2004.

[264] F. Lian, J. Moyne, and D. Tilbury, 'Time Delay Modeling and Sample Time Selection for Networked Control Systems', *Int. Mechanical Engineering Congress and Exposition*, New Nork, NY, vol. 2, pp. 765–770, 2001.

[265] P. Voulgarisa, G. Bianchinib, and B. Bamiehc, 'Optimal H_2 Controllers for Spatially Invariant Systems with Delayed Communication Requirements', *Systems & Control Letters*, vol. 50, pp. 347–361, 2003.

[266] M. Long and W. Chu, 'Robust Stabilization of Discrete-Time Systems with Time-Varying Delays', *Proc. American Control Conference*, Portland, OR, vol. 2, pp. 1334–1339, 2005.

[267] G. Chen and F. L. Lewis, 'Distributed Adaptive Tracking Control for Synchronization of Unknown Networked Lagrangian Systems', *IEEE Trans. Systems, Man, and Cybernetics: Part B*, vol. 41, no. 3, pp. 805–816, 2011.

[268] P. D. Wilde and G. Briscoe, 'Stability of Evolving Multiagent Systems', *IEEE Trans. Systems, Man, and Cybernetics: Part B*, vol. 41, no. 4, pp. 1149–1157, 2011.

[269] Y. Zhao, C. Zhang, and H. J. Gao, 'A New Approach to Guaranteed Cost Control of T-S Fuzzy Dynamic Systems With Interval Parameter Uncertainties', *IEEE Trans. Systems, Man, and Cybernetics: Part B*, vol. 39, no. 6, pp. 1516–1527, 2009.

[270] L. Schenato, B. Sinopoli, M. Franceschetti, K. Poolla, and S. S. Sastry, 'Foundations of Control and Estimation over Lossy Networks', *Proc. of the IEEE*, vol. 95, pp. 163–187, 2007.

[271] Z. G. Wu, P. Shi, H. Su, and J. Chu, 'Delay-Dependent Stability Analysis for Switched Neural Networks With Time-Varying Delay', *IEEE Trans. Systems, Man, and Cybernetics: Part B*, vol. 41, no. 6, pp. 1522–1530, 2011.

[272] H. Zhang, C. Dang, and J. Zhang, 'Decentralized Fuzzy H_∞ Filtering for Nonlinear Interconnected Systems With Multiple Time Delays', *IEEE Trans. Systems, Man, and Cybernetics: Part B*, vol. 40, no. 4, pp. 1197–1203, 2010.

[273] Y. Sun and N. H. El-Farra, 'Quasi-Decentralized Model-Based Networked Control of Process Systems', *Computers and Chemical Engineering*, vol. 32, pp. 2016–2019, 2008.

[274] Y. Sun and N. H. El-Farra, 'Quasi-Decentralized State Estimation and Control of Process Systems over Communication Networks', *47th IEEE Conference on Decision and Control, 2008*, Cancun, Mexico, vol. 6, pp. 5468–5475, 9–11 December 2008.

[275] T. Vicsek, A. Cziroo'k, E. Ben-Jacob, I. Cohen, and O. Shochet, 'Novel Type of Phase Transitions in a System of Self-Driven Particles', *Phys. Rev. Lett.*, 1995, vol. 75, no. 6, pp. 1226–1229, 1995.

[276] A. Jadbabale, J. Lin, and A. S. Morse, 'Coordination of Groups of Mobile Agents using Nearest Neighbor Rules', *IEEE Trans. Autom. Control*, vol. 48, no. 6, pp. 988–1001, 2003.

[277] W. Ren and R. W. Beard, 'Consensus Seeking in Multiagent Systems under Dynamically Changing Interaction Topologies', *IEEE Trans. Automatic Control*, vol. 50, no. 5, pp. 655–661, 2005.

[278] Y. Hong, G. R. Chen, and L. Bushnell, 'Distributed Observers Design for Leader-Following Control of Multi-agent Networks', *Automatica*, vol. 44, no. 3, pp. 846–850, 2008.

[279] P. Lin and Y. M. Jia, 'Further Results on Decentralised Coordination in Networks of Agents with Second-Order Dynamics', *IET Control Theory Appl.*, vol. 3, no. 7, pp. 957–970, 2009.

[280] G. Xie and L. Wang, 'Consensus Control for a Class of Networks of Dynamic Agents', *Int. J. Robust and Nonlinear Control*, vol. 17, no. 10–11, pp. 941–959, 2007.

[281] S. J. Corte, 'Distributed Algorithms for Reaching Consensus on General Functions', *Automatica*, vol. 44, no. 3, pp. 726–737, 2008.

[282] R. Carli, F. Fagnani, A. Spernzon, and S. Zamprieri, 'Communication Constraints in the Average Consensus Problem', *Automatica*, vol. 43, no. 3, pp. 671–684, 2007.

[283] G. Uo, 'Integrated Communication and Control Systems with Occasional Information Feedback', *Cyberntics and Systems*, vol. 39, no. 8, pp. 843–873, 2008.

[284] G. Guo and H. Jin, 'A Switching System Approach to Actuator Assignment with Limited Channels', *Int. J. Robust and Nonlinear Control*, vol. 20, no. 12, pp. 1407–1426, 2010.

[285] G. Guo and X. Liu, 'Observability and Controllability of Systems with Limited Data Rate', *Int. J. Systems Science*, vol. 40, no. 4, pp. 327–334, 2009.

[286] A. Rahmani, M. Ji, M. Mesbahi, and M. Egerstedt, 'Controllability of Multiagent Systems from a Graph-Theorectic Perspective', *SIAM J. Control Optimization*, vol. 48, no. 1, pp. 162–186, 2009.

[287] Z. K. Li, Z. S. Duan, G. R. Chen, and L. Huang, 'Consensus of Multi-agent Systems and Synchronization of Complex Networks: A Unified Viewpoint', *IEEE Trans. Circuits and Systems I Reg. Papers*, vol. 57, no. 1, pp. 213–224, 2010.

[288] S. E. Tuna, 'Synchronizing Linear Systems via Partial-State Coupling', *Automatica*, vol. 44, no. 8, pp. 2179–2184, 2008.

[289] S. E. Tuna, 'Conditions for Synchronizability in Arrays of Coupled Linear Systems', *IEEE Trans. Automatic Control*, vol. 54, no. 10, pp. 2416–2420, 2009.

[290] J. H. Wang, D. Z. Cheng, and X. M. Hu, 'Consensus of Multi-agent Linear Dynamic Systems', *Asian J. Control*, vol. 10, no. 2, pp. 144–155, 2008.

[291] Z. S. Duan, G. R. Chen, and L. Huang, 'Synchronization of Weighted Networks and Complex Synchronized Regions', *Phys. Lett. A*, vol. 372, no. 21, pp. 3741–3751, 2008.

[292] W. Lu and T. Chen, 'New Approach to Synchronization Analysis of Linearly Coupled Ordinary Differential Systems', *Physica D*, vol. 213, no. 2, pp. 214–230, 2006.

[293] X. F. Wang and G. R. Chen, 'Synchronization in Scale-Free Dynamical Networks: Robustness and Fragility', *IEEE Trans. Circuits and Systems I: Fundam. Theory Appl.*, vol. 49, no. 1, pp. 54–62, 2002.

[294] L. Scardavi and S. Sepulchre, 'Synchronization in Networks of Identical Linear Systems', *Automatica*, vol. 45, no. 11, pp. 2557–2562, 2009.

[295] R. S. Smith and F. Y. Hadaegh, 'Control of Deep-space Formation-Flying Spacecraft; Relative Sensing and Switched Information', *J. Guidance and Control Dynamics*, vol. 28, no. 1, pp. 106–114, 2005.

[296] K. Ogata, *Modern Control Engineering*, Prentice Hall, Englewood Cliffs, NJ, 1996.

[297] R. Horn and C. Johnson, *Matrix Analysis*, Cambridge University Press, New York, 1985.

[298] D. Angeli and P.-A. Bliman, 'Convergence Speed of Unsteady Distributed Consensus: Decay Estimate Along the Settling Spanning-Trees', *SIAM J. Control Optimization*, vol. 48, no. 1, pp. 1–32, 2009.

[299] J. Zhou and Q. Wang, 'Convergence Speed in Distributed Consensus over Dynamically Switching Random Networks', *Automatica*, vol. 45, no. 6, pp. 1455–1461, 2009.

[300] S. P. Boyd, L. El. Ghaoui, E. Feron, and V. Balakrishnan, *Linear Matrix Inequalities in System and Control Theory*, SIAM, Philadelphia, PA, 1994 [SIAM studies in applied mathematics].

[301] T. Iwasaki and R. E. Skelton, 'All Controllers for the General $\mathcal{H}_\infty$ Control Problem: LMI Existence Conditions and State Space Formulas', *Automatica*, vol. 30, no. 8, pp. 1307–1317, 1994.

[302] G. Lafferriere, A. Williams, J. Caughman, and J. J. P. Veerman 'Decentralized Control of Vehicle formations', *System and Control Letters*, vol. 54, pp. 899–910, 2005.

[303] W. Ren, 'Consensus based Formation Control Strategies for Multi-vehicle Systems', *Proc. American Control Conf.*, Minneapolis, MN, USA, pp. 4237–4242, 2006.

[304] Y. Kuramoto, 'Cooperative Dynamics of Oscillator Community', *Prog. Theor. Phys. Suppl.*, vol. 79, pp. 223–240, 1984.

[305] S. H. Strogatz, 'From Kuramoto to Crawford: Exploring the Onset of Synchronization in Populations of Coupled Oscillators', *Physica D*, vol. 143, pp. 1–20, 2000.

[306] V. Gazi and K. M. Passino, 'Stability Analysis of Social Foraging Swarms', *IEEE Trans. Syst., Man, Cyber., Part B: Cyber.*, vol. 34, pp. 539–557, 2004.

[307] V. Gazi and K. M. Passino, 'Stability Analysis of Swarms', *IEEE Trans. Automatic Control*, vol. 48, no. 4, pp. 692–697, 2003.

[308] P.-A. Belman and G. Ferrari-Trecate, 'Average Consensus Problems in Networks of Agents with Delayed Communications', *Automatica*, vol. 44, pp. 1985–1995, 2008.

[309] F. Xiao and L. Wang, 'Consensus Problems for High Dimensional Multiagent Systems', *IET Contr. Theory Appl.*, vol. 1, no. 3, pp. 830–837, 2007.

[310] S. Low, F. Paganini, and J. Doyle, 'Internet Congestion Control', *IEEE Control Systems Magazine*, vol. 32, no. 1, pp. 28–43, 2002.

[311] L. Wang and F. Xiao, 'A New Approach to Consensus Problems in Discrete-Time Multiagent Systems with Time-Delays', *Sci. China Ser. F*, vol. 50, no. 4, pp. 625–635, 2007.

[312] D. Yanakiev and I. Kanellakopoulos, 'A Simplified Framework for String Stability Analysis in AHS', *Proc. 13th IFAC World Congress*, San Francisco, CA, vol. Q, pp. 177–182, 1996.

[313] R. Bachmayer and N. E. Leonard, 'Vehicle Network for Gradient Descent in a Sampled Environment', *Proc. 41st Decision and Control Conference*, Anchorage, AK, vol. 1, pp. 112–117, 2002.

[314] L. Moreau, 'Stability of Multiagent Systems with Time-Dependent Communication Links', *IEEE Trans. Automatic Control*, vol. 50, no. 2, pp. 169–182, 2005.

[315] R. Bellman and K. Cooke, *Differential Difference Equations*, Academic, New York, 1963.

[316] C. Godsil and G. Royle, *Algebraic Graph Theory*, Springer-Verlag, New York, 2001.

[317] W. Ren, 'On Consensus Algorithms for Double Intregrator Dynamics', *IEEE Trans. Automatic Control*, vol. 53, no. 6, pp. 1503–1509, 2008.

[318] J. Bang-Jensen and G. Gutin, *Digraphs Theory, Algorithms and Applications*, Springer-Verlag, New York, 2002.

[319] M. S. Mahmoud, *Robust Control and Filtering for Time-Delay Systems*, Marcel-Dekker, New York, 2003.

[320] G. Sun and L. Wang, 'Consensus Problems in Networks of Agents with Double-Integrator Dynamics and Time-Varying Delays', *Int. J. Control*, vol. 82, no. 10, pp. 1937–1945, 2009.

[321] Y. P. Tian and C. L. Liu, 'Robust Consensus Next Term of Multiagent Systems with Diverse Input Delays and Asymmetric Interconnection Perturbations', *Automatica*, vol. 45, no. 5, pp. 1347–1353, 2009.

[322] J. F. Sturm, 'Using SeDuMi 1.02, a MATLAB Toolbox for Optimization over Symmetric Cones', *Optimization Methods Software*, vol. 11–12, no. 1–4, pp. 625–653, 1999.

[323] W. Ren and R. W. Beard, 'Consensus Seeking in Multiagent Systems under Dynamically Changing Interaction Topologies', *IEEE Trans. Automtic Control*, vol. 50, no. 5, pp. 655–661, 2005.

[324] V. D. Blondel, J. M. Hendrickx, A. Olshevsky, and J. N. Tsitsiklis, 'Convergence in Multiagent Coordination, Consensus, and Flocking', *Proc. IEEE Conf. on Decision and Control, and the European Control Conf.*, Seville, Spain, vol. 4, pp. 2996–3000, 2005.

[325] H. G. Tanner and D. K. Christofoilakis, 'State Synchronization in Local-Interaction Networks is Robust with Respect to Time Delays', *Proc. IEEE Conf. on Decision and Control, and the European Control Conf.*, Seville, Spain, vol. 6, pp. 4945–4950, 2005.

[326] P. Lin and Y. Jia, 'Average-Consensus in Networks of Multiagents with both Switching Topology and Coupling Time-Delay', *Physica A*, vol. 387, no. 1, pp. 303–313, 2008.

[327] P. Lin, Y. Jia, J. Du, and S. Yuan, 'Distributed Control of Multiagent Systems with Second-Order Agent Dynamics and Delay-Dependent Communications', *Asian J. Control*, vol. 10, no. 2, pp. 254–259, 2008.

[328] W. Ren, 'On Consensus Algorithms for Double-Integrator Dynamics', *Proc. 46th IEEE Conf. on Decision and Control*, New Orleans, LA, vol. 4, pp. 2295–2300, 2007.

[329] Y. Hong, L. Gao, D. Cheng, and J. Hu, 'Tracking Control for Multiagent Consensus with an Active Leader and Variable Topology', *Automatica*, vol. 42, no. 7, pp. 1177–1182, 2006.

[330] P. Lin, Y. Jia, and L. Lin, 'Distributed Robust $\mathcal{H}_\infty$ Consensus Control in Directed Networks of Agents with Time-Delay', *Systems and Control Lett.*, vol. 57, no. 8, pp. 643–653, 2008.

[331] K. C. Goh, L. Turan, M. S. Safonov, G. P. Papavassilopoulos, and J. H. Ly, 'Baffine Matrix Inequality Properties and Computational Methods', *Proc. American Control Conf.*, Baltimore, MD, pp. 850–855, 1994.

[332] S. Ibaraki and M. Tomizuka, 'Rank Minimization Approach for Solving BMI Problems with Random Search', *Proc. American Control Conf.*, Arlington, VA, vol. 3, pp. 1870–1875, 2001.

[333] S. Kanev, C. Scherer, M. Verhagen, and B. De Schutter, 'Robust Output-Feedback Controller Design via Local BMI Optimization', *Automatica*, vol. 40, no. 7, pp. 1115–1127, 2004.

[334] D. Liberson, *Theory of Functional Differential Equations*, Birkhaüser, Boston, 2003.

[335] B. Sinopoli, C. Sharp, L. Schenato, S. Schaffert, and S. Sastry, 'Distributed Control Applications within Sensor Networks', *Proc. IEEE: Special Issue on Distributed Sensor Networks*, vol. 91, no. 8, pp. 1235–1246, 2003.

[336] A. Speranzon, C. Fischione, K. H. Johansson, and A. Sangiovanni, 'A Distributed Minimum Variance Estimator for Sensor Networks', *IEEE J. Selected Areas in Communications*, vol. 26, no. 4, pp. 609–621, 2008.

[337] R. Carli, A. Chiuso, L. Schenato, and S. Zampieri, 'Distributed Kalman Filtering based on Consensus Strategies', *IEEE J. Selected Areas in Communications*, vol. 26, no. 4, pp. 622–632, 2008.

[338] A. Speranzon, C. Fischione, and K. H. Johansson, 'Distributed and Collaborative Estimation over Wireless Sensor Networks', *Proc. IEEE Conference on Decision and Control*, 2006.

[339] E. G. Gilbert and K. T. Tan, 'Linear Systems with State and Control Constraints: The Theory and Application of Maximal Output Admissible Sets', *IEEE Trans. Automatic Control*, vol. 36, pp. 1008–1020, 1991.

[340] E. D. Sontag, 'Smooth Stabilization Implies Coprime Factorization', *IEEE Trans. Automatic Control*, vol. 34, pp. 435–443, 1989.

[341] D. Estrin, R. Govindan, J. Eidemann, and S. Kumar, 'Next Century Challenges: Scalable Coordination in Sensor Networks', *Proc. ACM/IEEE Int. Conference on Mobile Computing and Networking*, Seattle, WA, vol. 2, pp. 872–876, 1999.

[342] Z.-Q. Luo, M. Gatspar, J. Liu, and A. Swami, Eds., *IEEE Signal Processing Magazine: Special Issue on Distributed Signal Processing in Sensor Networks*, vol. 23, no. 4, 2006.

[343] H. Gharavi and P. R. Kumar, Eds., *Proc. IEEE: Special Issue on Sensor Networks and Applications*, vol. 91, 2003.

[344] H. Gharavi and P. R. Kumar, 'Toward a Theory of In-Network Computation in Wireless Sensor Networks', *IEEE Commun. Mag.*, vol. 44, no. 1, pp. 97–107, 2006.

[345] D. P. Bertsekas and J. N. Tsitsiklis, *Parallel and Distributed Computation: Numerical Methods*, Athena Scientific, 1997.

[346] R. Carli, F. Fagnani, A. Speranzon, and S. Zampieri, 'Communication Constraints in the Average Consensus Problem', *Automatica*, vol. 44, no. 4, pp. 671–684, 2008.

[347] Y. Shang, W. Ruml, Y. Zhang, and M. Fromherz, 'Localization from Connectivity in Sensor Networks', *IEEE Trans. Parallel and Distributed Systems*, vol. 15, no. 11, pp. 961–974, 2004.

[348] Y.-W. Hong, L. Cheow, and A. Scaglione, 'A Simple Method to Reach Detection Consensus in Massively Distributed Sensor Networks', *Proc. IEEE Int. Symposium on Information Theory*, Chicago, IL., vol. 3, pp. 3455–3460, 2004.

[349] L. Xiao, S. Boyd, and S. Lall, 'A Scheme for Robust Distributed Sensor Fusion based on Average Consensus', *Proc. IEEE IPSN*, Bois, ID, vol. 2, pp. 674–679, 2005.

[350] D. P. Spanos, R. Olfati-Saber, and R. M. Murray, 'Approximate distributed Kalman filtering in sensor networks with quantifiable performance', *Proc. IEEE Conference on Decision and Control*, Seville, Spain, vol. 1, pp. 112–128, 2005.

[351] R. Olfati-Saber and J. S. Shamma, 'Consensus Filters for Sensor Networks and Distributed Sensor Fusion', *Proc. IEEE Conference on Decision and Control*, Seville, Spain, vol. 2, pp. 816–822, 2005.

[352] L. Xiao, S. Boyd, and S. J. Kim, 'Distributed Average Consensus with Least-Mean-Square Deviation', *J. Parallel and Distributed Computing*, 2006.

[353] L. Xiao and S. Boyd, 'Fast Linear Iterations for Distributed Averaging', *System and Control Letter*, vol. 53, no. 1, pp. 65–78, 2004.

[354] A. Pikovsky, M. Rosenblum, and J. Kurths, *Synchronization: A Universal Concept in Nonlinear Sciences*, Cambridge University Press, 2003.

[355] S. Barbarossa and G. Scutari, 'Decentralized Maximum Likelihood Estimation for Sensor Networks Composed of Nonlinearly Coupled Dynamical System', *IEEE Trans. Signal Processing*, vol. 55, no. 7, pp. 3456–3470, 2007.

[356] F. Zhao, J. Liu, L. Guibas, and J. Reich, 'Collaborative Signal and Information Processing: an Information-Directed Approach', *Proc. IEEE*, vol. 91, no. 8, pp. 1199–1309, 2003.

[357] J. Liu, J. Reich, and F. Zhao, 'Collaborative In-Network Processing for Target Tracking', *European Journal on Applied Signal Processing*, no. 4, pp. 378–391, 2003.

[358] G. Di Stefano, F. Graziosi, and F. Santucci, 'Distributed Positioning Algorithm for Ad-Hoc Networks', *Int. Workshop on Ultra Wideband Systems (IWUWBS)*, University of Oulu, Finland, pp. 326–331, 2003.

[359] A. Giridhar and P. Kumar, 'Distributed clock synchronization over wireless networks: Algorithm and analysis', *Proc. 45th IEEE Conference on Decision and Control*, San Diego, CA, vol. 4, pp. 3356–3362, 2006.

[360] P. Alriksson and A. Rantzer, 'Distributed Kalman Filtering using Weighted Averaging', *Proc. 17th Int. Symposium on Mathematical Theory of Networks and Systems*, Koyoto, Japan, vol. 2, pp. 1209–1214, 2006.

[361] L.-M. S. C. of the IEEE Computer Society, 'Wireless LAN Medium Access Control (MAC) and Physical Layer Specification', *IEEE std 802.11-1997 edition, Tech. Rep.*, IEEE, New York, NY, USA, 1997.

[362] C. Tomlin, *STARMAC, Stanford Testbed of Autonomous Rotor Craft for Multiagent Control*, Research Project, Available: http://hybrid.stanford.edu/starmac/overview.

[363] S. Boyd and L. Vandenberghe, *Convex Optimization*, Cambridge University Press, 2004.

[364] L. Ljung, *System Identification: Theory for the User* (2nd edn.), Prentice Hall, 1998.

[365] M. Cloosterman, N. Wouw, M. Heemels, and H. Nijmeijer, 'Robust Stability of Networked Control Systems with Time-varying Network-induced Delays', *Proc. 45th IEEE Conference on Decision and Control*, San Diego, CA, vol. 6, pp. 4980–4985, 2006.

[366] A. Alouani, J. Gray, and D. McCabe, 'Theory of distributed estimation using multiple asynchronous sensors', *IEEE Trans. Aerospace and Electronic Systems*, vol. 41, no. 2, pp. 717–722, 2005.

[367] Y. Bar-Shalom, X. Li, and T. Kirubarajan, *Estimation with Applications to Tracking and Navigation*, Wiley, 2001.

[368] D. Hall and J. Llinas, *Handbook of Multisensor Data Fusion*, CRC Press, 2001.

[369] Y. Bar-Shalom, 'Update with Out-of-Sequence Measurements in Tracking: Exact Solution', *IEEE Trans. Aerospace and Electronic Systems*, vol. 38, no. 3, pp. 769–777, 2002.

[370] S. Zhang and Y. Bar-Shalom, 'Optimal Removal of Out-of-Sequence Measurements From Tracks Using the IF-Equivalent Measurement', *Proc. 49th IEEE Conference on Decision and Control (CDC)*, Atlanta, GA, pp. 1312–1317, December 15–17, 2010.

[371] S. Sun and Z. Deng, 'Multi-Sensor Optimal Information Fusion Kalman Filter', *Automatica*, vol. 40, no. 6, pp. 1017–1023, 2004.

[372] Y. Xia, Li Li, G. P. Liu, and P. Shi, '$\mathcal{H}_\infty$ Predictive Control of Networked Control Systems', *Int. J. Control*, vol. 84, no. 6, pp. 1080–1097, 2011.

[373] Y. Xia, J. Shang, J. Chen, and G. P. Liu, 'Networked Data Fusion with Packet Losses and Variable Delays', *IEEE Trans. Systems, Man, and Cybernetics: Part B*, vol. 39, no. 5, pp. 1107–1120, 2009.

[374] L. Yan, D. Zhou, M. Fu, and Y. Xia, 'State Estimation for Asynchronous Multirate Multisensor Dynamic Systems with Missing Measurements', *IET Signal Processing*, vol. 4, no. 6, pp. 728–739, 2010.

[375] L. Yan, B. Liu, and D. Zhou, 'Asynchronous Multirate Multisensor Information Fusion Algorithm', *IEEE Trans. Aerospace and Electronic Systems*, vol. 43, no. 3, pp. 1135–1146, 2007.

[376] B. Sinopoli, L. Schenato, M. Franceschetti, K. Poolla, M. Jordan, and S. Sastry, 'Kalman Filtering with Intermittent Observations', *IEEE Trans. Automatic Control*, vol. 49, no. 9, pp. 1453–1464, 2004.

[377] A. Alouani and T. Rice, 'On Asynchronous Data Fusion', *Proc. 26th Southeastern Symposium on System Theory*, Athens, pp. 143–146, 1994.

[378] L. Yan, B. Liu, and D. Zhou, 'The Modeling and Estimation of Asynchronous Multirate Multisensor Dynamic Systems', *Aerospace Science and Technology*, vol. 10, no. 1, pp. 63–71, 2006.

[379] L. Yan, H. Shi, M. Du, and Z. Zhu, 'Asynchronous Multirate Multisensor State Fusion Estimation with Incomplete Measurements', *Proc. 4th Int. Conference on Wireless Communications, Networking and Mobile Computing (WiCOM'08)*, pp. 1–4, 2008.

[380] L. Zhang, X. Wu, Q. Pan, and H. Zhang, 'Multiresolution Modeling and Estimation of Multisensor Data', *IEEE Trans. Signal Processing*, vol. 52, no. 11, pp. 3170–3182, 2004.

[381] M. Mallick, S. Coraluppi, and C. Carthel, 'Advances in Asynchronous and Decentralized Estimation', in *Proc. IEEE Aerospace Conference*, vol. 4, pp. 1873–1879, 2001.

[382] X. B. Chen, W. B. Xu, T. Y. Huang, X. Y. Ouyang, and S. S. Stankovic, 'Pair-Wise Decomposition and Coordinated Control of Complex Systems', *Information Sciences*, vol. 185, pp. 78–99, 2012.

[383] M. Saleem, G. A. Di Caro, and M. Farooq, 'Swarm Intelligence-based Routing Protocol for Wireless Sensor Networks: Survey and Future Directions', *Information Sciences*, vol. 181, pp. 78–99, 2011.

[384] S. T. Chen and Y. N. Chang, 'Robust Anti-Windup Controller Design of Time-Delay Fuzzy Systems with Actuator Saturations', *Information Sciences*, vol. 181, pp. 3225–3245, 2011.

[385] G. W. Zheng, P. Shi, S. Hongy, and J. Chu, '$\ell_2 - \ell_\infty$ Filter Design for Discrete-Time Singular Markovian Jump Systems with Time-Varying Delays', *Information Sciences*, vol. 181, pp. 5537–5545, 2011.

[386] F. S. Cattivelli and A. H. Sayed, 'Diffusion Strategies for Distributed Kalman Filtering and Smoothing', *IEEE Trans. Automatic Control*, vol. 55, pp. 1520–1533, 2010.

[387] L. Chai, B. Hu, and P. Jiang, 'Distributed State Estimation Based on Quantized Observations in a Bandwidth Constrained Sensor Network', *Proc. 7th World Congress on Intelligent Control and Automation*, Chongqing, China, vol. 3, pp. 2411–2415, 2008.

[388] M. Farina, G. Ferrari-Trecate, and R. Scattolini, 'Distributed Moving Horizon Estimation for Sensor Networks', *Proc. 1st IFAC Workshop on Estimation and Control of Networked Systems*, Venice, Italy, pp. 126–131, 2009.

[389] M. Kamgarpour and C. Tomlin, 'Convergence Properties of a Decentralised Kalman Filter', *Proc. 47th IEEE Conference on Decision and Control*, Cancun, Mexico, pp. 3205–3210, 2008.

[390] R. Olfati-Saber, 'Kalman-Consensus Filter Optimality Stability and Performance', *Proc. 48th IEEE Conference Decision and Control*, Shanghai, China, pp. 7036–7042, 2009.

[391] W. Yu, G. Chen, Z. Wang, and W. Yang, 'Distributed Consensus Filtering in Sensor Networks', *IEEE Trans. Systems Man and Cybernetics-B*, vol. 39, pp. 1568–1577, 2009.

[392] P. Shi, M. S. Mahmoud, K. Nguang, and A. Ismail, 'Robust Filtering for Jumping Systems with Mode Dependent Delays', *Signal Processing*, vol. 86, pp. 140–152, 2006.

[393] L. Wu, P. Shi, H. Gao, and C. Wang, '$\mathcal{H}_\infty$ Filtering for 2D Markovian Jump Systems', *Automatica*, vol. 44, pp. 1849–1858, 2008.

[394] H. Zhang, G., Feng, G., Duan, and X. Lu, '$\mathcal{H}_\infty$ Filtering for Multiple-Time-Delay Measurements', *IEEE Trans. Signal Processing*, vol. 54, pp. 1681–1688, 2006.

[395] F. S. Cattivelli and A. H. Sayed, 'Diffusion Strategies for Distributed Kalman Filtering: Formulation and Performance Analysis', *Proc. Cognitive Information Processing*, Santorini, Greece, vol. 1, pp. 36–41, 2008.

[396] R. Olfati-Saber, 'Distributed Kalman Filtering for Sensor Networks', *Proc. 46th IEEE Conference Decision and Control*, New Orleans, LA, pp. 5492–5498, 2007.

[397] R. Olfati-Saber and J. S. Shamma, 'Consensus Filters for Sensor Networks and Distributed Sensor Fusion', *Proc. 44th IEEE Conference Decision and Control, and the European Control Conference*, Seville, Spain, pp. 6698–6703, 2005.

[398] B. Shen, Z. Wang, and Y. S. Hung, 'Distributed $\mathcal{H}_\infty$-Filtering in Sensor Networks with Multiple Missing Measurements: the Finite-horizon Case', *Automatica*, vol. 46, pp. 1682–1688, 2010.

[399] V. Ugrinovskii, 'Distributed Robust Filtering with $\mathcal{H}_\infty$ Consensus of Estimations', *Proc. 2010 American Control Conference*, Baltimore, MD, vol. 2, pp. 1374–1379, 2010.

[400] H. Gao and T. Chen, '$\mathcal{H}_\infty$ Estimation for Uncertain Systems with Limited Communication Capacity', *IEEE Trans. Automatic Control*, vol. 52, pp. 2070–2084, 2007.

[401] X. He, Z. Wang, and D. Zhou, 'Robust Fault Detection for Networked Systems with Communication Delay and Data Missing', *Automatica*, vol. 45, pp. 2634–2639, 2009.

[402] Z. Wang, D. W. C. Ho, Y, Liu, Y., and X. Liu, 'Robust $\mathcal{H}_\infty$ Control for a Class of Nonlinear Discrete Time-delay Stochastic Systems with Missing Measurements', *Automatica*, vol. 45, pp. vol. 684–691, 2009.

[403] Z. Wang, F. Yang, D. W. C. Ho, and X. Liu, 'Robust $\mathcal{H}_\infty$ Filtering for Stochastic Time-delay Systems with Missing Measurements', *IEEE Trans. Signal Processing*, vol. 54, pp. 2579–2587, 2006.

[404] J. Xiong and J. Lam, 'Stabilization of Linear Systems Over Networks with Bounded Packet Loss', *Automatica*, vol. 43, pp. 80–87, 2007.

[405] Y. Zhao, J. Lam, and H. Gao, 'Fault Detection for Fuzzy Systems with Intermittent Measurement', *IEEE Trans. Fuzzy Systems*, vol. 17, pp. 398–410, 2009.

[406] K. You and L. Xie, 'Minimum Data Rate for Mean Square Stabilisability of Linear Systems with Markovian Packet Losses', *IEEE Trans. Automatic Control*, vol. 56, no. 4, pp. 772–785, 2011.

[407] Y. Chu and K. Glover, 'Bounds of the Induced Norm and Model Reduction Errors for Systems with Repeated Scalar Nonlinearities', *IEEE Trans. Automatic Control*, vol. 44, pp. 4215–4226, 1999.

[408] Y. Chu, 'Further Results for Systems with Repeated Scalar Nonlinearities', *IEEE Trans. Automatic Control*, vol. 44, pp. 2031–2035, 2001.

[409] H. Gao, J. Lam, and C. Wang, 'Induced l2 and Generalised H2 Filtering for Systems with Repeated Scalar Nonlinearities', *IEEE Trans. Signal Processing*, vol. 53, pp. 4215–4226, 2005.

[410] M. S. Mahmoud, *Robust Control and Filtering for Time-Delay Systems*, Marcek-Dekker, New York, 2000.

[411] H. Tamura, 'Decentralised Optimisation for Distributed-lag Models of Discrete Systems', *Automatica*, vol. 11, pp. 593–602, 1975.

[412] X. He, Z. Wang, Y. D. Ji, and D. Zhou, 'Fault Detection for Discrete-Time Systems in a Networked Environment', *Int. J. Systems Science*, vol. 41, pp. 937–945, 2010.

[413] D. Hall and J. Llinas, 'An Introduction to Multi-Sensor Data Fusion', *Proc. IEEE*, vol. 85, no. 1, pp. 6–23, 1997.

[414] R. Viswanathan and P. Varshney, 'Distributed Detection with Multiple Sensors Part I: Fundamentals', *Proc. IEEE*, vol. 85, no. 1, pp. 54–63, 1997.

[415] H. Durrant-Whyte, B. Rao, and H. Hu, 'Toward a Fully Decentralized Architecture for Multi-Sensor Data Fusion', *Proc. IEEE Int. Conference on Robotics and Automation*, vol. 2, pp. 1331–1336, 1990.

[416] S. Grime, H. F. Durrant-Whyte, and P. Ho, 'Communication in Decentralized Data-Fusion Systems', *Proc. American Control Conference*, pp. 3299–3303, 1992.

[417] F. Bourgault and H. F. Durrant-Whyte, 'Communication in General Decentralized Filters and the Coordinated Search Strategy', *Proc. 7th Int. Conference on Information Fusion*, Stockholm, Sweden, pp. 279–285, July 2004.

[418] D. P. Spanos, R. Olfati-Saber, and R. M. Murray, 'Distributed Sensor Fusion using Dynamic Consensus', *Proc. 16th IFAC World Congress*, Czech Republic, pp. 114–121, July 2005.

[419] R. Olfati-Saber, 'Distributed Kalman Filter with Embedded Consensus Filters', *Proc. 44th IEEE Conference on Decision and Control and European Control Conference*, Seville, Spain, vol. 3, pp. 2543–2549, December 2005.

[420] K. Chang, T. Zhi, and R. Saha, 'Performance Evaluation of Track Fusion with Information Matrix Filter', *IEEE Trans. Aerospace and Electronic Systems*, vol. 38, no. 2, pp. 455–466, 2002.

[421] S. Sun, 'Multi-Sensor Optimal Information Fusion Kalman Filter for Discrete Multi-Channel Arma Signals', *IEEE Int. Symposium on Intelligent Control*, pp. 377–382, 2003.

[422] X. R. Li, Y. M. Zhu, J. Wang, and C. Z. Han, 'Optimal Linear Estimation Fusion-Part I: Unified Fusion Rules', *IEEE Trans. Information Theory*, vol. 49, no. 9, pp. 2192–2208, 2003.

[423] Y. Bar-Shalom, editor, *Multitarget-Multisensor Tracking: Advanced Applications*, Atech House, Norwood, MA, vol. 1, 1990.

[424] C. Y. Chong, K. C. Chang, and S. Mori, 'Distributed Tracking in Distributed Sensor Networks', *Proc. American Control Conf.*, Seattle, WA, June 1986.

[425] H. R. Hashmipour, S. Roy, and A. J. Laub, 'Decentralized Structures for Parallel Kalman Filtering', *IEEE Trans. Automatic Control*, vol. 33, no. 1, pp. 88–93, 1988.

[426] C. Y. Chong, S. Mori, and K. C. Chang, 'Distributed Multitarget Multisensor Tracking', *In Multitarget-Multisensor Tracking: Advanced Applications* (Bar-Shalom, Ed.) (vol. 1), Atech House, Norwood, MA, 1990.

[427] Y. M. Zhu, Z. You, J. Zhao, K. S. Zhang, and X. K. Li, 'The Optimality for the Distributed Kalman Filtering Fusion with Feedback', *Automatica*, vol. 37, pp. 1489–1493, 2001.

[428] Y. Gao, W. L. Wang, and Z. L. Deng, 'Information Fusion Estimation of Noise Statistics for Multisensor Systems', *Chinese Control and Decision Conference*, Guilin, pp. 1127–1131, 2009.

[429] Claudio, L., A. Souza, and R. M. Palhares, 'Parameter Estimation on Linear Time-Varying Systems', *J. Franklin Institute*, vol. 348, no. 4, pp. 777–789, 2011.

[430] T. Li and Y. Zhang, 'Fault Detection and Diagnosis for Stochastic Systems via Output PDFs', *J. Franklin Institute*, vol. 348, no. 6, pp. 1140–1152, 2011.

[431] N. Orani, A. Pisanoand, and E. Usai, 'Fault Diagnosis for the Vertical Three-Tank System via High-Order Sliding-Mode Observation', *J. Franklin Institute*, vol. 347, no. 6, pp. 923–939, 2010.

[432] H. R. Karimi, M. Zapateiro, and M. Luo, 'A Linear Matrix Inequality Approach to Robust Fault Detection Filter Design of Linear Systems with Mixed Time-Varying Delays and Nonlinear Perturbations', *J. Franklin Institute*, vol. 347, no. 6, pp. 957–973, 2010.

[433] M. Basin, and D. Calderon-Alvarez, 'Optimal Filtering over Linear Observations with Unknown Parameters', *J. Franklin Institute*, vol. 347, no. 6, pp. 988–1000, 2010.

[434] D. Zhang and L. Yu, '$\mathcal{H}_\infty$ Filtering for Linear Neutral Systems with Mixed Time-Varying Delays and Nonlinear Perturbations', *J. Franklin Institute*, vol. 347, no. 7, pp. 1374–1390, 2010.

[435] M. Basin, P. Shi, and P. Soto, 'Central Suboptimal $\mathcal{H}_\infty$ Filtering for Nonlinear Polynomial Systems with Multiplicative Noise', *J. Franklin Institute*, vol. 347, no. 9, pp. 1740–1754, 2010.

[436] G. L. Tao, X. H. Guan, and Z. L. Deng, 'Self-Tuning Distributed Fusion Kalman Filter with Asymptotic Global Optimality', *Proc. 29th Chinese*

Control Conference, Beijing Conference Center, Beijing, China, pp. 1268–1272, July 29–31, 2010.

[437] E. Song, Y. Zhu, and J. Zhou, 'The Optimality of Kalman Filtering Fusion with Cross-Correlated Sensor Noises', *Proc. 43rd IEEE Conference on Decision and Control*, Atlantis, Paradise Island, Bahamas, pp. 4637–4642, December 14–17, 2004.

[438] L. Ljung, *System Identification: Theory for the User* (2nd edn.), Prentice Hall, NY, 1999.

[439] M. Liggins, C. Y. Chong, I. Kadar, M. G. Alford, V. Vannicola, and S. Thomopoulos, 'Distributed Fusion Architectures and Algorithms for Target Tracking', *Proc. IEEE*, vol. 85, no. 1, pp. 95–107, 1997.

[440] D. I. Schizas, G. B., Giannakis, and Z. Q. Luo, 'Distributed Estimation using Reduced-Dimensionality Sensor Observations', *IEEE Trans. Signal Processing*, vol. 55, no. 8, pp. 4285–4299, 2007.

[441] E. B. Song, Y. M. Zhu, and J. Zhou, 'Sensors' Optimal Dimensionality Compression Matrix in Estimation Fusion', *Automatica*, vol. 41, no. 12, pp. 2131–2139, 2005.

[442] J. Xiao, A. Ribeiro, Z. Luo, and G. Giannakis, 'Distributed Compression-Estimation using Wireless Sensor Networks', *IEEE Signal Processing Magazine*, pp. 27–46, July 2006.

[443] H. Durrant-Whyte, 'Data Fusion in Sensor Networks', *Fourth Int. Symposium on Information Processing in Sensor Networks*, Annapolis, MD, pp. 3324–3329, April 2005.

[444] L. Yingting and Z. Yunmin, 'Distributed Kalman Filtering Fusion with Packet Loss or Intermittent Communications from Local Estimators to Fusion Center', *Proc. 29th Chinese Control Conference*, Beijing, China, pp. 4768–4775, July 29–31, 2010.

[445] D. Henriksson, A. Cervin, and K. E. Arzen, 'TrueTime: Simulation of Control Loops Under Shared Computer Resources', *15th IFAC World Congress on Automatic Control*, Barcelona, Spain, July 2002.

[446] M. Andersson, D. Henriksson, A. Cervin, and K. E. Arzen, 'Simulation of Wireless Networked Control Systems', *Proceedings of the 44th IEEE Conference on Decision and Control, and the European Control Conference*, Seville, Spain, December 12–15, 2005.

[447] M. D. Biasi, C. Snickars, K. Landernas, and A. Isaksson, 'Simulation of Process Control with WirelessHART Networks Subject to Clock Drift', *IEEE Int. Computer Software and Applications Conference*, Turku, Finland, pp. 1355–1360, 2008.

[448] A. Cervin, D. Henriksson, B. Lincoln, and J. Eker, 'How Does Control Timing Affect Performance?, Analysis and Simulation of Timing Using Jitterbug and TrueTime', *IEEE Control Systems Magazine*, pp. 16–30, June 2003.

[449] Z. Dezong, L. Chunwen, and R. Jun, 'Fuzzy Speed Control and Stability Analysis of a Networked Induction Motor System with Time Delays and Packet Dropouts', *Nonlinear Analysis: Real World Applications* vol. 12, pp. 273–287, 2011.

[450] A. Fadaei and K. Salahshoor, 'Design and Implementation of a New Fuzzy PID Controller for Networked Control Systems', *ISA Trans.*, vol. 47, pp. 351–361, 2008.

[451] Z. N. Gao, Q. W. Chen, and W. L. Hu, 'A New Experimental Platform for Networked Control Systems based on CAN and Switched-Ethernet', *Information Technology Journal*, vol. 10, no. 1, pp. 219–230, 2011.

[452] S. M. Hasan, H. Yu, and A. Carrington, 'Overview of Wireless Networked Control Systems over Mobile Ad-hoc Network', *Proc. 14th Int. Conference on Automation and Computing*, Brunel University, West London, UK, September 2008.

[453] C. Peng, D. Yue, Z. Gu, and F. Xia, 'Sampling Period Scheduling of Networked Control Systems with Multiple-Control Loops', *Mathematics and Computers in Simulation*, vol. 79, pp. 1520–1511, 2009.

[454] D. Peng, H. Zhang, C. Huang, J. Lin, and H. Li, 'Study of Immune PID Networked Control System Based on TrueTime', *J. Networks*, vol. 6, no. 6, pp. 912–915, 2011.

[455] K. Kubo, K. Miyazawa, T. Shimomura, K. Kamada, and N. Okabe, 'SICE-ICASE', *Int. Joint Conference*, Bexco, Busan, Korea, October 18–21, 2006.

[456] M. M. D. Santos, M. R. Stemmer, and F. Vasques, 'Evaluationa of the Timing Properties of Two Control Networks: CAN and PROFIBUS', *IEEE Int. Symposium on Industrial Electronics,* vol. 2, pp. 874–879, 2003.

[457] C. Snickars, *Design of a WirelessHART Simulator for Studying Delay Compensation in Networked Control Systems*, M.Sc. Project, KTH Electrical Engineering, Stockholm, Sweden, February 2008.

[458] A. B. Soglo and Y, Xianhui, 'Networked Control System Simulation Design and Its Application', *Tshinghua Science and Technology*, vol. 11, no. 3, pp. 287–294, 2006.

[459] J. Wang, W. X. Zheng, and T. Chen, 'Identification of Linear Dynamic Systems Operating in a Networked Environment', *Automatica*, vol. 45, pp. 273–2772, 2009.

[460] A. Wei, Y. Chen, and J. Wu, 'Simulation Study of TCP/IP Communication Based on Networked Control Systems', *Proc. 6th World Congress on Intelligent Control and Automation*, Dalian, China, June 21–23, 2006.

[461] A. Cervin and B. Lincoln, *Jitterbug 1.23 Reference Manual*, Department of Automatic Control, LTH, Lund University, July 1, 2010.

[462] M. Wooldridge and N. Jennings, 'Intelligent Agents: Theory and Practice', *Knowledge Engineering Review*, vol. 2, no. 10, pp. 115–152, 1995.

[463] F. Hidrobo and J. Aguilar, 'Resources Management System for Distributed Platforms Based on Multi-Agent Systems', *Proc. 5th Int. Conference on Artificial Intelligence, Simulation and Planning-AIS' 2000*, Firenze, Italy, pp. 60–69, 2000.

[464] B. Franek and C. Gaspar, 'SMI++ Object Oriented Framework for Designing and Implementing Distributed Control Systems', *IEEE Trans. Nuclear Science*, vol. 45, pp. 1946–1950, 1998.

[465] G. Haring, G. Kotsis, A. Puliafito, and O. Tomarchio, 'A Transparent Architecture for Agent Based Resource Management', *Proc. IEEE Intl. Conf. On Intelligent Engineering*, Nants, France, pp. 338–342, 1998.

[466] J. Aguilar, M. Cerrada, F. Hidrobo, G. Mousalli, F. Rivas, and A. Díaz, 'Application of the Agents Reference Model for Intelligent Distributed Control Systems', in N. Matorakis, L. Pecorelli (eds), *Advances in Systems Sciences Measurement*, Springer-Verlag, Berlin, pp. 204–210, 2001.

[467] A. Ray 1989, 'Introduction to Networking for Integrated Control Systems', *IEEE control Systems Magazine*, pp. 76–79, January 1989.

[468] B. Wittenmark, J. Nilsson, and M. Törngren, 'Timing Problems in Real-Time Control Systems', *Proc. American Control Conference*, Seattle, Washington, 1995.

[469] M. S. Branicky, S. M. Phillips, and W. Zhang, 'Stability of Networked Control Systems: Explicit Analysis of Delay', *Proc. American Control Conference*, Chicage, IL, June 2000.

[470] G.C. Walsh, H. Ye, and L. Bushnell, 'Stability Analysis of Networked Control Systems', *Proc. American Control Conference*, Chicago, pp. 2876–2880, June 1999.

[471] F.-L. Lian, *Analysis, Design, Modelling and Control of Networked Control Systems*, Ph.D. Dissertation, Mechanical Engineering University of Michigan, 2001.

[472] F.-L. Lian, J. Moyne, and D. Tilbury, 'Analysis and Modelling of Networked Control Systems: MIMO Case with Multiple Time Delays', *Proc. American Control Conference* Arlington, Virginia, June 2001.

[473] G. Juanole and G. Mouney, 'QOS in Real Time Distributed Systems and Process Control Applications', *Proc. Workshop on Networked Control Systems (NeCST)*, European Project, Corsica, France, October 2005.

[474] Y. Halevi and A. Ray, 'Performance Analysis of Integrated Communication and Control Systems Networks', *ASME, J. Dynamic Systems, Measurement and Control*, vol. 112, pp. 365–632, 1990.

[475] M. Sanfridson, *Timing Problems in Distributed Real-Time Computer Control Systems*, Technical Report Mechatronics Lab, Stockholm, Sweden, 2000.

[476] P. H. Bauer, M. Sichitiu, C. Lorand, and K. Premaratne, 'Total Delay Compensation in LAN Control Systems and Implications for Scheduling', *LAN Control Systems and Implications for Scheduling of the American Control Conference*, Arlington, VA, vol. 6, pp. 4300–4305, June 2001.

[477] J. P. Georges, N. Vatanski, E. Rondeau, C. Aubrun, and S. L. Jämsä-Jounela 'Control Compensation based on Upper Bound Delay in Networked Control Systems', *Proc. 17th Int. Symposium on Mathematical Theory of Networks Systems*, Kyoto, July 2006.

[478] J. K. Cheong, *Delay Modelling and Controller Design for Networked Control Systems*, Master of Applied Science Thesis, Dept. of Electrical and Computer Engineering in the University of Toronto, September 2003.

[479] J. P. Georges, T. Divoux, and E. Rondeau, 'Conforting the Performances of Switched Ethernet Network with Industrial Constraints by Using the Network Calculus', *Int. J. Communication Systems*, vol. 18, pp. 877–903, 2005.

[480] O. Sename, D. Simon, and D. Robert, 'Feedback Scheduling for Real-Time Control of Systems with Communication Delays', *IEEE Conference ETFA'03*, Lisbonne, Portugal, 2003.

[481] M. S. Branicky, V. Liberatore, and S. M. Phillips, 'Networked Control System Co-Simulation for Co-Design', *Proc. American Control Conference*, Denver, USA, vol. 4, pp. 3341–3346, 2003.

[482] M. S. Branicky, S. M. Phillips, and W. Zhang, 'Scheduling and Feedback Co-Design for Networked Control Systems', *Proc. IEEE Conference on Decision and Control*, Las Vegas, Nevada, vol. 2, pp. 1211–1217, December 2002.

[483] J. Eker and A. Cervin, 'A MATLAB Toolbox for Real-Time and Control Systems Co-design', *Proc. Sixth Int. Conference on Real-Time and Control Systems and Application*, Hong Kong, China, vol. 1, pp. 320–326, 1999.

[484] A. Cervin, *Integrated Control and Real-Time Scheduling*, Ph.D. Dissertation, Dept. of Automatic Control Lund Institute of Technology, Lund, April 2003.

[485] K. C. Lee and S. Lee, 'Performance Evaluation of Switched Ethernet for Real-Time Industrial Communications', *Computer Standards and Interfaces*, vol. 24, pp. 411–423, 2002.

[486] D. Henriksson, A. Cervin, and K.-E. Årzén, 'TrueTime: Simulation of Control Loops under Shared Computer Resources', *Proc. 15th Triennial IFAC World Congress*, Barcelona, Spain, 2002.

[487] D. Henriksson, A. Cervin, and K.-E. Årzén, 'TrueTime: Real-Time Control System Simulation with Matlab/Simulink', *Proc. Nordic MATLAB Conference*, Copenhagen, Denmark, October 2003.

[488] G. Juanole, C. Calmettes, G. Mouney, and M. Peca, 'On the Implementation of a Process Control System on a CAN Network: Linking the process Control Parameters to the Network Parameters', *Proc. IEEE Int. Conference on Emerging Technologies and Factory Automation, ETFA'05*, Catania, Italy, September 2005.

[489] J-D. Decotignie, 'Ethernet Based Real-Time and Industrial Communications', *Proc. IEEE*, vol. 93, no. 6, 2005.

[490] P. Pedreiras, R. Leite, and L. Almaida, 'Characterizing the Real Time Behavior of Prioritized Switched Ethernet', *Proc. 2nd Int. workshop Real-time LAN's in the internet Age*, Vienna, Austria, 2003.

[491] C. W. Scherer, 'LPV Control and Full Block Multiplier', *Automatica*, vol. 37, pp. 361–375, 2001.

[492] A. K. Packard, 'Gain Scheduling via Linear Fractional Transformations', *System and Conrol Letters*, vol. 22, no. 2, pp. 79–92, 1994.

[493] R. D' Andrea, G. E. Dullerud, and S. Lall, 'Convex $\mathcal{L}_2$ Synthesis for Multidimensional Systems', *Proc. 37th IEEE Conf. Decision and Control*, Tampa, FL, pp. 1883–1888, December 1998.

[494] R. D. Andrea, 'Extension of Parrott's Theorem to Nondefinite Scalings', *IEEE Trans. Autom. Control*, vol. 45, no. 5, pp. 937–940, 2001.

[495] A. A. Bahnasawi and M. S. Mahmoud, *Control of Partially-Known Dynamical Systems*, Springer-Verlag, Berlin, 1989.

[496] M. C. Gennaro and A. Jadbabaie, 'Decentralized Control of Connectivity for Multi-Agent Systems', *Proc. 45th IEEE Conference on Decision and Control*, San Diego, CA, vol. 4, pp. 3628–3633, 2006.

[497] R. M. Murray, K. J. Astrorm, S. P. Boyd, R. W. Brockett, and G. Stein, 'Control in an Information Rich World', *IEEE Control System Magazine*, vol. 23, pp. 20–33, 2003.

[498] I. Dumitrache and I. Necoara, 'Distributed Control in Networked Systems', *Control Engineering and Applied Informatics*, vol. 11, pp. 3–5, 2009.

[499] R. Merris, 'Laplacian Matrices of Graphs: A Survey', *Linear Algebra Application*, vol. 197, pp. 143–176, 1994.

[500] R. Agaev and P. Chebotarev, 'On the Spectra of Nonsymmetric Laplacian Matrices', *Linear Algebra Application*, vol. 399, no. 5, pp. 157–168, 2005.

[501] B. Bollobás, *Modern Graph Theory*, Springer, New York, 2002.

[502] P. Gren, E. Fiorelli, and N. E. Leonard, 'Cooperative Control of Mobile Sensor Networks: Adaptive Gradient Climbing in a Distributed Environment', *IEEE Trans. Automatic Control*, vol. 49, no. 8, pp. 1292–1302, 2004.

[503] A. Y. Al-Rayyah, and M. S. Mahmoud, 'Decentralized Reliable Control of Interconnected Time-Delay Systems against Sensor Failures', *J. Optimization Theory and Applications*, vol. 147, pp. 318–336, 2010.

[504] M. Ohlin, D. Henriksson, and A. Cervin, *TRUETIME 1.5 – Reference Manual*, Department of Automatic Control, Lund University, January 2007.

[505] R. A. Horn and C. R. Johnson, *Matrix Analysis*, Cambridge University Press, 1985.

[506] Z. Liu, H. Zhang, P. Smith, and Q. Hui, *Hierarchical Optimization Strategies for Sensor Network Deployment*, 2012 World Automation Congress, Puerto Vallarta, Mexico, June 2012.

[507] J. Schendel, M. Donkers, W. Heemels, and N. Wouw. 'On Dropout Modeling for Stability Analysis of Networked Control Systems', *Proc. American Control Conference*, pp. 555–561, 2010.

[508] L. Li and F. Wang, 'Control and Communication Synthesis in Networked Control Systems', *Int. J. Intelligent Control and Systems*, vol. 13, pp. 81–88, 2008.

[509] D. Nesic and D. Liberzon, 'A Unified Approach to Controller Design for Systems with Quantization and Time Scheduling', *Proc. 46th IEEE Conference on Decision and Control*, New Orleans, LA, vol. 4, pp. 3939–3944, 2007.

[510] D. Dacic and D. Nesic, 'Simultaneous Controller and Protocol Design for Networked Control', *Proc. IEEE Conf. on Decision and Control*, San Diego, CA, vol. 1, pp. 508–514, 2006.

[511] Y. Bar-Shalom and L. Campo, 'The Effect of Common Process Noise on Two-Sensor Fused-Track Covariance', *IEEE Trans. Aerospace and Electronic Systems*, vol. AES 22, no. 6, pp. 803–805, 1986.

[512] A. Smith, H. Balakrishnan, M. Goraczko, and N. B. Priyantha, 'Tracking Moving Devices with the Cricket Location System', *Proc. Int. Conference on Mobile Systems, Applications and Services*, Boston, MA, June 2004.

[513] D. S. Moore and G. P. McCabe, *Introduction to the Practice of Statistics* (5th ed.), W. H. Freeman, 2006.

[514] J. Z. Lu, 'Challenging Control Problems and Emerging Technologies in Enterprise Optimization', *Proc. 6th IFAC Symposium Dyn. Contr. Proc. Syst.*, Jejudo Island, Korea, pp. 29–40, 2001.

[515] N. L. Ricker, 'Decentralized Control of the Tennessee Eastman Challenge Process', *J. Proc. Control*, vol. 6, no. 4, pp. 205–221, 1996.

[516] P. L. Lee, H. Li, and I. T. Cameron, 'Decentralized Control Design for Nonlinear Multi-Unit Plants: A Gap Metric Approach', *Chem. Eng. Sci.*, vol. 55, pp. 3743–3758, 2000.

[517] S. Skogestad, 'Self-Optimizing Control: The Missing Link between Steady-State Optimization and Control', *Comput. Chem. Eng.*, vol. 24, pp. 569–575, 2000.

[518] C. K. Chui and G. Chen, *Kamlan Filtering With Real-Time Applications* (vol. 17), Springer-Verlag, New York, 1987.

[519] A. G. O. Mutambara, *Decentralized Estimation and Control for Multisensor Systems*, CRC Press, Boca Raton, FL, 1998.

[520] A. G. O. Mutambara and H. F. Durrant-Whyte, 'Estimation and Control for a Modular Wheeled Mobile Robot', *IEEE Trans. Control Systems Technology*, vol. 8, no. 1, pp. 35–46, 2000.

[521] P. S. Maybeck, *Stochastic Models, Estimation and Control*, Academic Press, Boston, MA, vol. 1, 1997.

[522] K.-E. Arzen, B. Bernhardsson, J. Eker, A. Cervin, P. Persson, K. Nilsson, and L. Sha, *Integrated Control and Scheduling*, Dept. of Automatic Control, Lund Institute of Technology, August, 1999.

Index